Lecture Notes in Mechanical Engineering

Lecture Notes in Mechanical Engineering (LNME) publishes the latest developments in Mechanical Engineering—quickly, informally and with high quality. Original research reported in proceedings and post-proceedings represents the core of LNME. Volumes published in LNME embrace all aspects, subfields and new challenges of mechanical engineering. Topics in the series include:

- Engineering Design
- Machinery and Machine Elements
- Mechanical Structures and Stress Analysis
- Automotive Engineering
- Engine Technology
- Aerospace Technology and Astronautics
- Nanotechnology and Microengineering
- Control, Robotics, Mechatronics
- MEMS
- Theoretical and Applied Mechanics
- Dynamical Systems, Control
- Fluid Mechanics
- Engineering Thermodynamics, Heat and Mass Transfer
- Manufacturing
- Precision Engineering, Instrumentation, Measurement
- Materials Engineering
- Tribology and Surface Technology

To submit a proposal or request further information, please contact the Springer Editor of your location:

China: Dr. Mengchu Huang at mengchu.huang@springer.com
India: Priya Vyas at priya.vyas@springer.com
Rest of Asia, Australia, New Zealand: Swati Meherishi at swati.meherishi@springer.com
All other countries: Dr. Leontina Di Cecco at Leontina.dicecco@springer.com

To submit a proposal for a monograph, please check our Springer Tracts in Mechanical Engineering at http://www.springer.com/series/11693 or contact Leontina.dicecco@springer.com

Indexed by SCOPUS. The books of the series are submitted for indexing to Web of Science.

More information about this series at http://www.springer.com/series/11236

Parvathy Rajendran · Nurul Musfirah Mazlan ·
Aslina Anjang Ab Rahman ·
Nurulasikin Mohd Suhadis ·
Norizham Abdul Razak ·
Mohd Shukur Zainol Abidin
Editors

Proceedings of International Conference of Aerospace and Mechanical Engineering 2019

AeroMech 2019, 20–21 November 2019,
Universiti Sains Malaysia, Malaysia

Springer

Editors
Parvathy Rajendran
School of Aerospace Engineering
Universiti Sains Malaysia
Nibong Tebal, Pulau Pinang, Malaysia

Aslina Anjang Ab Rahman
School of Aerospace Engineering
Universiti Sains Malaysia
Nibong Tebal, Pulau Pinang, Malaysia

Norizham Abdul Razak
School of Aerospace Engineering
Universiti Sains Malaysia
Nibong Tebal, Pulau Pinang, Malaysia

Nurul Musfirah Mazlan
School of Aerospace Engineering
Universiti Sains Malaysia
Nibong Tebal, Pulau Pinang, Malaysia

Nurulasikin Mohd Suhadis
School of Aerospace Engineering
Universiti Sains Malaysia
Nibong Tebal, Pulau Pinang, Malaysia

Mohd Shukur Zainol Abidin
School of Aerospace Engineering
Universiti Sains Malaysia
Nibong Tebal, Pulau Pinang, Malaysia

ISSN 2195-4356 ISSN 2195-4364 (electronic)
Lecture Notes in Mechanical Engineering
ISBN 978-981-15-4755-3 ISBN 978-981-15-4756-0 (eBook)
https://doi.org/10.1007/978-981-15-4756-0

This Springer imprint is published by the registered company Springer Nature Singapore Pte Ltd.
The registered company address is: 152 Beach Road, #21-01/04 Gateway East, Singapore 189721, Singapore

Preface

On behalf of the organizing and scientific committee, we are delighted to welcome the delegates from all over the world to second International Conference of Aerospace and Mechanical Engineering, AeroMech'19 on 20 till 21 November 2019 held at Light Hotel, Seberang Jaya, Penang, Malaysia.

Organized in a beautiful place located at the north of Peninsular Malaysia, AeroMech'19 is an international podium to present the state of the art research & development in Aerospace and Mechanical Engineering, while exchanging ideas, thoughts and information between academia and industry.

This year, our theme is Recent Advancements and Innovations in Aerospace & Mechanical Engineering. Being the second international conference organized by the School of Aerospace Engineering, we please to have accepted 53 papers and managed to attract almost 145 authors' participation from all over the world.

We are also delighted to have two distinguished keynote speakers, Prof. Dr. Thanigaiarasu Subramanian, Professor and Head of Department of Aerospace Engineering, MIT, Anna University, and Mr. Mohd Faizal Allaudin, Co-Founder and CEO of FAAS Engineering and Consultancy (FAAS) to address our conference aspiration.

Professor Dr. Thanigaiarasu Subramanian will deliver a talk on the Mixing Enhancement Techniques In High Speed Jet Controls, while Mr. Mohd Faizal Allaudin will be presenting on Challenges in Aerospace and Mechanical Industry.

We hope all delegates will participate actively in the conference by sharing their scope of interest within the aerospace and mechanical engineering areas. Our gratitude and best wishes to the Dean of School Aerospace Engineering, the entire conference committee team, and reviewers who have made the conference a success.

Thank you.

Nibong Tebal, Malaysia

Parvathy Rajendran
aeparvathy@usm.my
Nurul Musfirah Mazlan
Nurulasikin Mohd Suhadis
Aslina Anjang Ab Rahman
Mohd Shukur Zainol Abidin
Norizham Abdul Razak

Contents

A Review on Alternative Fuel Spray Characteristics

Qummare Azam, Ahmed M. Alhaj, Ayub A. Janvekar, and Nurul Musfirah Mazlan

Abstract Air transport is already a well-known transportation in assuring the need of worldwide social connection and commercial purposes. Nowadays huge demand on aviation fuel increases severe tense due to depletion of fossil fuel. Furthermore, the increases of greenhouse gas emission from the fossil fuel reduces the environment quality and human life. Hence, an attention towards alternative aviation fuel from renewable sources has become a priority as it could offer greener aviation. However alternative fuel must be compatible in comparison of the conventional fuel in terms of its physical properties such as viscosity, energy density, sustainability, affordability and availability. Those properties will affect spray characteristics and leads to combustion quality. Therefore, this study presents a review on the different alternative fuels and their compatibility in the aircraft engine. The importance of spray characteristic on the engine and different experimental approach to examine the spray characteristics of fuel particularly the penetration length, spray morphology, cone angle and atomization of the fuel also are presented. This work is beneficial in providing understanding on alternative fuels and spray atomization examination.

Keywords Alternative fuel · Spray characteristics · Spray penetration · Fuel blends

1 Introduction

Environmental concern, social pressure, cost concern, and availability of fossil fuel, caused an interest toward alternative fuel. An alternative fuel presents an opportunity to minimize environmental issues and processing cost for aviation fuel [1]. The need of petroleum-based fuel is expected to increase by 1.3% per year till 2030 [2].

Q. Azam · A. M. Alhaj · N. M. Mazlan (✉)
School of Aerospace Engineering, Universiti Sains Malaysia, Engineering Campus, 14300 Nibong Tebal, Malaysia
e-mail: nmusfirah@usm.my

A. A. Janvekar
School of Mechanical Engineering, Universiti Sains Malaysia, Engineering Campus, 14300 Nibong Tebal, Malaysia

P. Rajendran et al. (eds.), *Proceedings of International Conference of Aerospace and Mechanical Engineering 2019*, Lecture Notes in Mechanical Engineering,
https://doi.org/10.1007/978-981-15-4756-0_1

Alternative options have to be found to improve the transportation sector by taking into account reconsideration of vehicle design [3]. A development of fuel processing technology and replacement of conventional fuel with alternative fuel are some of the options. Thus, several governments have started different programs to encourage the use of renewable sources as a drop in fuel in transportation including aircraft engine.

The alternative fuel is developed by different method and particular renewable sources of biomass. Thermal cracking, transesterification, Fischer-Tropsch and hydro-processing are common chemical processes used by industries to produce alternative jet fuels. It was discovered that blends up to 50% of alternative jet fuel with conventional aviation fuel met with the aviation fuel requirement such as fuel viscosity, soot point, energy density and freezing point [4]. The features of the alternative fuel affect the spray geometry for a clean and efficient combustion.

Compatibility of biofuels in reducing gas emissions from gas turbine engine is proven. Allouis et al. [1] observed the impact of pure vegetable oil on a micro gas turbine engine and found a comparable of CO, NO_x, and total particulate matter to the pure diesel oil. Habib et al. [2] observed an improvement (reduction) in static thrust and thrust-specific fuel consumption, increased in thermal efficiency as well as reduction in NO and CO emissions when burning soy, canola and recycled methyl ester and their blends with Jet A in a 30 kW gas turbine engine. Blending butanol with gasoline/Jet A is preferred because of the similarity in its properties with gasoline/Jet-A. Butanol/Jet-A blends provided reduction in operational thrust range but improved NOx and CO emission indices when compared with Jet-A [3]. Gawron and Białecki [4] tested a blending of Jet A-1 with HEFA on a miniature turbojet engine and observed big improvement in fuel consumption and smaller emission indices of CO, CO_2 and NO_x. Generation of gaseous emissions such as nitrogen oxide (NO_x), carbon monoxide (CO), unburnt hydrocarbon (UHC), and particulate matter (PM) were influenced by factors such as fuel properties and combustion performance mainly atomization behavior.

Although the aforementioned studies discussed on the effect of alternative fuel on combustion emissions thus concluded the significant effect of the fuels for environment, however to the authors knowledge, studies on the atomization characteristics of the alternative fuels which leads to the combustion performance is still lacking. Therefore, this study will provide a review comprising of different alternative fuels, atomization of alternative fuels and various experimental approaches to analyze the characteristics. The organization of the present review as follows. First, various alternative fuels are provided in Sect. 2, the atomization of alternative fuels is presented in Sect. 3, and finally various experimental approaches conducted to characterize atomization is presented in Sect. 4.

2 Alternative Fuels

Aviation responsibles for 2–3% of CO_2 emission to the environment [5] which provides great impact on climate change and global warming [6]. Instead of impact on environment, the need for biofuels in the aviation sector is necessary to reduce over-dependency on conventional fossil fuel. Biofuels is made of biomass sources such as plant crops, animal fats, and waste cooking oil.

Milano et al. [7] listed various conversion processes in producing biofuels particularly from algae which are through chemical conversion, thermochemical conversion, and biochemical. In chemical conversion, the process involved transesterification and esterification. The output of these processes is biodiesel. Gasification, thermochemical liquefaction, pyrolysis and direct combustion are processes involved in thermochemical conversion. In these process, bio-oil, hydrogen, fuel gas, and solid residue will be produced. Finally, in biochemical conversion, it involves process of anaerobic digestion which produces methane and hydrogen; alcoholic fermentation which produces alcohol and keton; and interesterification which produces biodiesel.

However, not all biofuels are suitable to be used for transportation particularly in aircraft engine due to properties limitation which will influence combustion performance. As for transportation consuming gasoline and diesel fuel, bioethanol and biodiesel can be replaced the fossil fuel. However, the replacing fuel must comply with two biodiesel standards which are ASTMD6751 and EN14214. Both standards served as a guideline and assessment on biodiesel quality [7]. While for aviation, ASTM D7566-18 was a standard that provides guidelines in producing biojet fuel. ASTM D7566-18 has identified five types of synthesized paraffinic kerosene namely FT-SPK, FT-SPK/A, HEFA-SPK, SIP-SPK and ATK-SPK as blending components with conventional jet fuel [8]. Only 50% volume is permitted to blend with conventional jet fuel.

Bio-synthetic paraffinic kerosene (Bio-SPK) is a types of derived alternative fuel in which the sustainable sources such as bio-derived oils containing triglycerides and fatty acids are converted chemically into Bio-SPK type of fuel which has similar types of molecules to the current jet fuel. In addition, the production of Bio-SPK fuel complies with ASTM D7566 standard for several properties such as density, lubricity, high heating value, freezing point, sulphur content and energy density [5]. The ignition of Bio-SPKs outcomes has lesser NO_x and carbon dioxide emissions in comparison of other conventional fuel. As a results, Bio-SPK fuel has been noticed by IATA due to appropriate applicable option as alternative of conventional fuels for commercialization [6]. Since 2009 several successful flight tests were conducted to test capability of blending 50% alternative fuel produced through Bio-SPK in one of the aircraft engines.

Hydro-processed Esters and Fatty Acid (HEFA) also is a chemical processing technology in which the alternative jet fuel is produced from renewable sources such as vegetable oils, algae etc. ASTM has approved HEFA as the properties are matched with the alternative fuel specification D7566 [5]. It is considered as main processed fuel that is developed from renewable sources. Many reputed airlines such

Table 1 Specification of Jet A, SPK and HEFA fuel [10]

Fuel and properties	Jet A	SPK fuel	HEFA fuel
Density (kg/m^3)	788.0	775	779.9
Freezing point (°C)	−47	−63.5	−63
Viscosity (mm^2/s)	2.992	3.336	5.004
Sulphur content (%)	0.003	<0.0001	<0.0001
Low heating value (MJ/kg)	43.45	42.00	43.70
Aromatics (% v/v)	15.8	25	9.4

as Aeroméxico, Air China and Japan Airlines have tested this fuel on commercial aircraft by using blends of till 90% processed fuel produced using this technology [8].

Properties of the fuel such as cetane number influence ignition and combustion characteristics of the fuel [9] while oleic acid (C18) content depict ignition quality, combustion heat, viscosity and lubricity [8]. Table 1 shows properties of SPK and HEFA in comparison to conventional jet fuel, Jet-A.

The mixing of fuels in different ratios was tested in the numerous previous researches at various conditions to investigate the spray characteristics. Table 2 shows the list of the properties of blended HEFA and Jet-A fuels as used in Starck et al. [11]. Results obtained show that the properties of the fuels changed after blending process. The impact of physical properties changes have significant role on combustion. The fuel was blend according to an experimental setup as shown in Fig. 1 thus produce homogenous blending between different types of fuels.

Table 2 Fuel properties comparison of Jet A-1, HEFA and blends [11]

HEFA derived Jet fuel	100% Jet A-1	90% Jet A-1/10% HEFA	80% Jet A-1/20% HEFA	70% Jet A-1/30% HEFA	100% HEFA
Flash point (°C)	40.5	43	43.5	45	67
Freezing point (°C)	−47	−49	−46.5	−44.5	−49.6
Lubricity (mm)	0.64	–	–	–	0.910
Density (Kg/m^3)	775–840	800	797	794	773.5
Smoke point (mm)	25/19	25	26	28	>50

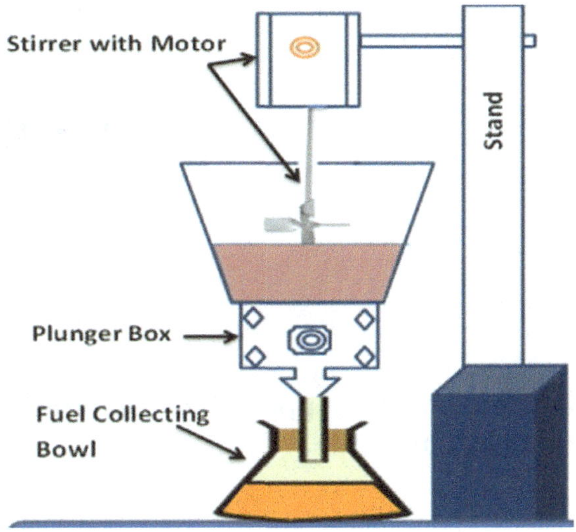

Fig. 1 Schematic diagram of fuel blending apparatus [12]

3 Atomization of a Spray

Atomization is a process where the liquid is converted into small drops which influence by surface tension and the action of internal and external forces [13]. Increasing of average droplet diameter causes increases of nitrogen oxide (NO), carbon monoxide (CO), unburned hydrocarbon (UHC) and particulate matter (PM) although the tendency is not constant depending on the changes of equivalence coefficient, ϕ [14].

Quality of the atomization is characterized through several parameters such as spray penetration, spray cone angle, dispersion, and patternation. Size of the spray angle plays a vital role during the atomization process as it leads to the region between fuel and antioxidant. For small spray cone angle, the region between fuel and antioxidant is small while for wide spray angle, significant overlapping between fuel and antioxidant observed thus caused instability in the gas turbine power output [15]. On the other hand, droplet size and spray penetration length are important parameters to calculate vaporization time [16]. Maximum distance the spray can reach when injected either into stagnant or crossflowing air is defined as spray penetration [13]. Figure 2 describes the penetration length and angle formation of fuel spray. Regardless whether the evaporation is considered, Zhu et al. [17] observed an increment in the penetration length with the spray angle. They also observed reduction in droplet size as spray angle increases. Fuel properties has significant effect on spray behavior particularly droplet geometry and droplet breakup process. Fuel density reduces linearly, and viscosity reduces exponentially due to increasing the fuel temperature. Wang et al. [18] observed influence of increases temperature with decreases of fuel density resulted to better atomisation. Additionally, the fuel density is inversely proportional with the length of the penetration [19]. Significant effect of the fuel properties on spray performance also observed in Zhou et al. [20]

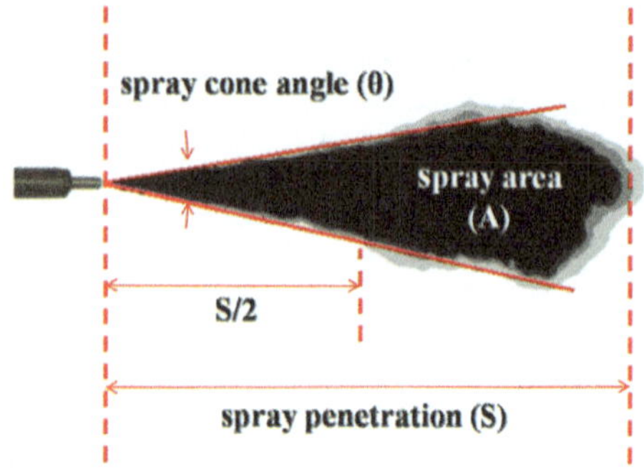

Fig. 2 Spray tip penetration and spray angle [21]

4 Experimental Approach to Analyze Atomization Behavior

Analysis of atomization behavior is crucial as it influence combustion performance. Studies were performed both numerically and experimentally to analyze the atomization behavior. However, in this present paper, only experimental approach will be discussed in detail which includes the experimental set-up and related parameters that necessary in characterizing the atomization behavior. Lee et al. [22] set-up an experimental work which consists of Particle Image Velocimetry (PIV) system, fuel jet injector, fuel flow regulator and fuel filter to investigate the physical properties of fuel, and to analyze fuel atomization when the fuel is injected at high pressure (Fig. 3). In this set up, biodiesel blended fuels were injected at high pressure through inclined level of tubes to the fuel rail section and propagates to generate the flow. The swirling or straight of the ejected turbulent flow consists of high pressure force because of the specific nozzle shape of orifice and outflow resulted by output nozzle orifice into a vacuum region. The fuel droplets initially break into linear accumulated structured then converted into very fine spray. In their experiment, analysis of the mixture of biodiesel-blended fuel on the injection profile was measured based on method suggested by Bosch [23]. The sensor probe for pressure detection was fixed

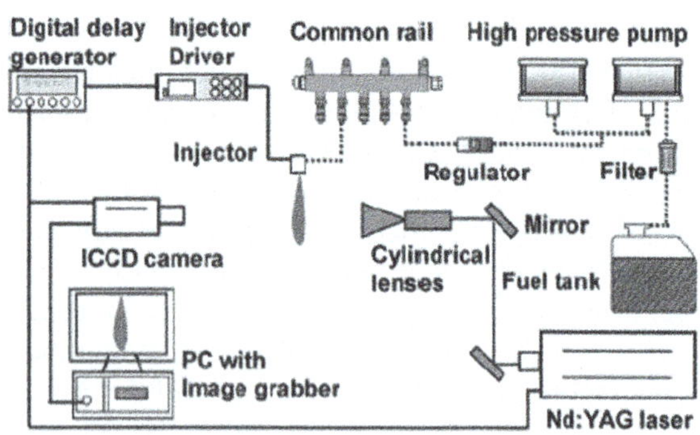

Fig. 3 Spray visualization experimental rig [22]

in a tube filled with the tested fuels. The probe was used to observe the pressure difference as per change of rate of time after initiate propagation of fuel injection. Pressure difference obtained by the sensor was used to develop injection profile of the spray. The pressure within the tube is fixed by some extent of pressure. In this study, measurement on spray penetration, SMD and mean velocity distributions were obtained. Results obtained depicted increases of droplet size with the mixing ratio due to higher viscosity and surface tension compared to diesel fuel. Increases of biodiesel in the mixture decreases HC and CO emissions whilst increases in NO_x which associated with oxygen content in the biodiesel and ignition delay.

Similar setup performed in Ghahremani et al. [19] to evaluate spray characteristic and atomization pattern of Norouzak biodiesel and its blends with conventional diesel at various injection pressure. Fuel was injected from fuel tank to combustion chamber through an injection system which comprises of high pressure pump, regulator, common rail and injector. The injector has an internal diameter of 0.5 mm. However in this setup, the pressure transmitter was used to control the injection pressure in supplying the desired pressure. Meanwhile, Geng et al. [24] used a high speed camera test system and a Malvern laser PSA to measure spray tip penetration and droplet Sauter mean diameter of biodiesel-ethanol fuel blends at injection pressure of 30 MPa. The nozzle has a single a diameter of 30 mm and 1.464 mm length. By using this method, microscopic characteristics of biodiesel were obtained. Main finding obtained from this study was increases of spray penetration and reduction of spray cone angle with injection pressure.

A simpler set up was proposed in Vouros et al. [25] who constructed an experimental set up to measure atomization by circulating liquid mixtures as shown in Fig. 4. Dimension of the chamber is 1000 mm height with a 400 × 400 mm^2 and were made by a glass window for optical access. The liquid flow in the circulation loop was maintained using an explosion-proof Danfoss oil pump. The pressure was in the range of 5–12 bars. This set up used a rotameter, pressure gauges, a sheathed thermocouple and control valves to monitor and control the flow. A generic full spray

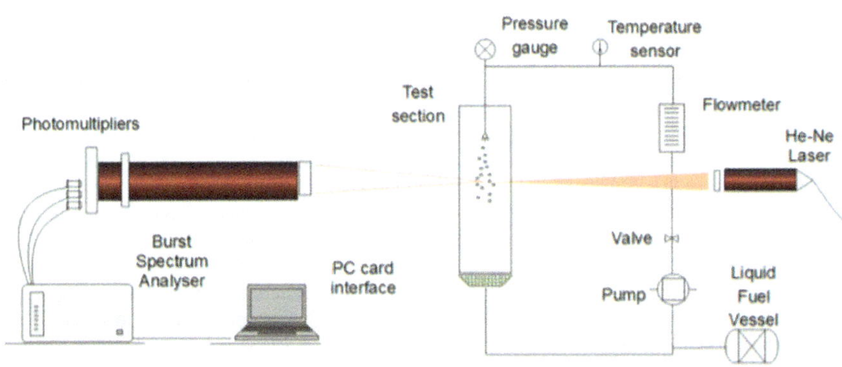

Fig. 4 Schematic diagram for atomization [25]

cone angle was used to develop fuel spray. During the experiment, fuel was accumulated at the bottom of test section and was recirculated back through the pump. Size and velocity of the droplet were measured using phase Doppler anemometry (PDA) technique which determines droplet size and droplet velocity by measuring the phase shift of droplets and through standard "fringe" mode laser respectively. This setup was different with aforementioned setup in which the fuels were heated during passing through the pump to allow comparisons with similar data. Fuels used for the evaluation were mixture of gas-to-liquid (GTL) with aromatics and napthenes.

5 Conclusion

Depletion in petroleum based fuel and increasing amount of toxic emissions in the environment has grown attention on alternative sources. Many factors led dispersion of gaseous emissions. One of the factors are atomization of the spray. Fuel sprays are among the most complicated but essential for the fuel combustion research. As interest on alternative fuel increases due to the aforementioned reasons, this paper provides a review of different type of alternative fuels and presents their compatibility in aircraft engine. This work also presents the essential of atomization characteristics on combustion performance as well as different experimental approaches which are beneficial to analyze combustion performance specifically spray atomization and spray penetration of the alternative fuels particularly for the fuels that have been tested in commercial aircraft.

Acknowledgments The authors would like to thank Universiti Sains Malaysia Research University Grant (Grant No: 1001/PAERO/8014019) and USM Fellowship, for the support to this study.

References

1. Allouis C, Amoresano A, Capasso R, Langella G, Niola V, Quaremba G (2018) The impact of biofuel properties on emissions and performances of a micro gas turbine using combustion vibrations detection. Fuel Process Technol 179:10–16
2. Habib Z, Parthasarathy R, Gollahalli S (2010) Performance and emission characteristics of biofuel in a small-scale gas turbine engine. Appl Energy 87:1701–1709
3. Mendez CJ, Parthasarathy RN, Gollahalli SR (2014) Performance and emission characteristics of butanol/Jet A blends in a gas turbine engine. Appl Energy 118:135–140
4. Gawron B, Białecki T (2018) Impact of a Jet A-1/HEFA blend on the performance and emission characteristics of a miniature turbojet engine. Int J Environ Sci Technol 15:1501–1508
5. Blakey S, Rye L, Wilson CW (2011) Aviation gas turbine alternative fuels: a review. Proc Combust Inst 33:2863–2885
6. Yilmaz N, Atmanli A (2017) Sustainable alternative fuels in aviation. Energy 140:1378–1386
7. Milano J, Ong HC, Masjuki HH, Chong WT, Lam MK, Loh PK, Vellayan V (2016) Microalgae biofuels as an alternative to fossil fuel for power generation. Renew Sustain Energy Rev 58:180–197

8. Yang J, Xin Z, He Q, Corscadden K, Niu H (2019) An overview on performance characteristics of bio-jet fuels. Fuel 237:916–936
9. Johnson MB, Wen Z (2009) Production of biodiesel fuel from the *Microalga schizochytrium* limacinum by direct transesterification of algal biomass. Energy Fuels 23:5179–5183
10. Rahmes TF, Kinder JD, Henry TM, Crenfeldt G, LeDuc GF, Zombanakis GP, Abe Y, Lambert DM, Lewis C, Juenger JA (2009) Sustainable bio-derived synthetic paraffinic kerosene (Bio-SPK) jet fuel flights and engine tests program results. Report no AIAA 7002
11. Starck L, Pidol L, Jeuland N, Chapus T, Bogers P, Bauldreay J (2016) Production of hydroprocessed esters and fatty acids (HEFA)—optimisation of process yield. Oil Gas Sci Technol Rev IFP Energies Nouvelles 71:10
12. Kareddula VK, Puli RK (2018) Influence of plastic oil with ethanol gasoline blending on multi cylinder spark ignition engine. Alexandria Eng J 57:2585–2589
13. Lefebvre AH, McDonell VG (2017) Atomization and sprays. 2nd edn. CRC press
14. Jankowski A, Kowalski M (2015) Influence of the quality of fuel atomization on the emission of exhaust gases toxic components of combustion engines. J KONBiN 36
15. Gong J-S, Fu W-B (2007) The experimental study on the flow characteristics for a swirling gas–liquid spray atomizer. Appl Therm Eng 27:2886–2892
16. Han I-S, Chung C-B, Riggs JB (2000) Modeling of a fluidized catalytic cracking process. Comput Chem Eng 24:1681–1687
17. Zhu L, Luo F, Qi YY, Wei M, Ge JR, Liu WL, Li GL, Jen TC (2018) Effects of spray angle variation on mixing in a cold supersonic combustor with kerosene fuel. Acta Astronaut 144:1–11
18. Wang C, Sahu A, Coratella C, Xu C, Saul J, Xu H (2019) Spray characteristics of a gasoline-diesel blend (ULG75) using high-speed imaging techniques. Fuel 239:677–692
19. Ghahremani AR, Jafari M, Ahari M, Saidi MH, Hajinezhad A, Mozaffari AA (2018) Spray characteristics and atomization behavior of bio-diesel (Norouzak) and diesel fuel blends. Part Sci Technol 36:270–281
20. Zhou G, Zhou J, Fang Y, Yang X (2019) Properties effect of blending fischer-tropsch aviation fuel on spray performances. Energy 179:1082–1093
21. Pan M, Huang R, Liao J, Jia C, Zhou X, Huang H, Huang X (2019) Experimental study of the spray, combustion, and emission performance of a diesel engine with high n-pentanol blending ratios. Energy Convers Manage 194:1–10
22. Lee CS, Park SW, Kwon SI (2005) An experimental study on the atomization and combustion characteristics of biodiesel-blended fuels. Energy Fuels 19:2201–2208
23. Bosch W (1966) The fuel rate indicator: a new measuring instrument for display of the characteristics of individual injection. SAE technical paper, 22
24. Geng L, Xie Y, Wang J, Liu W, Li C, Wang C (2019) Experimental and numerical analysis of the spray characteristics of biodiesel–ethanol fuel blends. Simulation, 1–12
25. Vouros A, Vouros A, Panidis T (2017) Spray characteristics of alternative aviation fuel blends. Aerospace 4:18

Modelling of Crack Propagation for Embedded Crack Structure

H. A. Haziq Aiman, M. R. M. Akramin, M. N. M. Husnain, and M. S. Shaari

Abstract Inner or embedded cracks were created in the structures of nuclear power plant by fatigue after years of operation time. Embedded cracks which generated in nuclear power plant is modelled as circular cracks in a plane normal to tension loading direction in Probabilistic S-version Finite Element Model (Prob-SFEM). Prob-SFEM which can generate the model with fatigue loading and crack growth due to number of fatigue cycle realistically by using uncertainty parameters from Monte Carlo method. Fatigue life and crack growth were generated from ProbS-FEM simulation. A simple model for embedded cracked structure was shown in this paper. Crack propagation of the embedded crack was analyse using deterministic S-version Finite Element Model and ProbS-FEM. Comparison between those two methods were investigated in this paper.

Keywords Embedded crack · S-FEM · ProbS-FEM

1 Introduction

Analysis of embedded or internal crack was presented as it happens in many industries such as in the power plants [1], aging aircrafts [2], civil engineering structures [3], etc. In the long period, embedded crack might generate in the structure of nuclear power plant by fatigue [4] and the analysis of this subsurface crack is significant to ensure the toughness of the structures. Initial defect might create at some area of structure where the local stress concentration take place [1]. Estimation of fatigue fracture is significant for the integrity of nuclear power plant structure. Prediction of crack growth and fatigue life are considered to avoid catastrophic failure. Experimental test to analyse the crack [5] is difficult hence the simulation is an option. Finite element method (FEM) was introduced to overcome this difficulty but failed due to

H. A. Haziq Aiman · M. R. M. Akramin (✉) · M. N. M. Husnain · M. S. Shaari
Faculty of Mechanical and Automotive Engineering Technology, Universiti Malaysia Pahang, 26600 Pekan, Pahang, Malaysia
e-mail: akramin@ump.edu.my

P. Rajendran et al. (eds.), *Proceedings of International Conference of Aerospace and Mechanical Engineering 2019*, Lecture Notes in Mechanical Engineering,
https://doi.org/10.1007/978-981-15-4756-0_2

the absent of remeshing process which is needed for modelling the growing of crack configurations.

Lately, a new method was developed to solve this problem which is known as deterministic S-version Finite Element Method (S-FEM). S-FEM is capable to clarify the remeshing process and capable to estimate the crack paths [5]. However, S-FEM ssdoes not provide solution for the uncertainty approach in the fatigue crack growth analysis [6]. The Probabilistic S-version Finite Element Method (ProbS-FEM) was introduced to overcome the uncertainties in fatigue crack growth by including the uncertainties quantification feature in the S-FEM [7].

In this paper, S-FEM and ProbS-FEM methods are conducted on embedded crack model to predict crack growth and fatigue life. Comparison of both methods results are studied and analyse in this paper.

2 Basic Deterministic S-version Finite Element Model (S-FEM)

S-FEM is defined as the S-version Finite Element Model and was introduced by J. Fish (1992) [8]. The embedded crack in a structure was designed and modelled as a global mesh and a local mesh as shown in Fig. 1. A coarser mesh was generated at the global area, Ω^G while a denser mesh was used for the local area, Ω^L. Course mesh is significant to model the global area, Ω^G where global area does not contain crack. A crack was modelled in the local area, Ω^L by using a denser mesh at the crack tip [1].

Then, local area is overlaid on the global area and it is attached with one another. The displacement function in each area is determined independently. The displacement for the overlaid area is defined as the total of the displacements for the local and global area. The displacement of local area is presumed to be zero to enable the continuity at the boundary between global area and local area, Γ^{GL}. This is shown in below equation:

$$\begin{aligned} u_i = \{u_i^G i \\ \in \Omega^G - \Omega^L u_i^G + u_i^L i \\ \in \Omega^L u_i^L = 0 i \in \Gamma^{GL} \end{aligned} \quad (1)$$

These displacement functions can be related to principle of virtual work and expressed as:

$$\begin{aligned} &\int_{\Omega^G} \delta u_{i,j}^G D_{ijkl} u_{k,l}^G d\Omega + \int_{\Omega^L} \delta u_{i,j}^G D_{ijkl} u_{k,l}^L d\Omega \\ &+ \int_{\Omega^L} \delta u_{i,j}^L D_{ijkl} u_{k,l}^G d\Omega + \int_{\Omega^L} \delta u_{i,j}^L D_{ijkl} u_{k,l}^L d\Omega = \int_{\Gamma^t} \delta u_i^G t_i d\Gamma^{tG} \end{aligned} \quad (2)$$

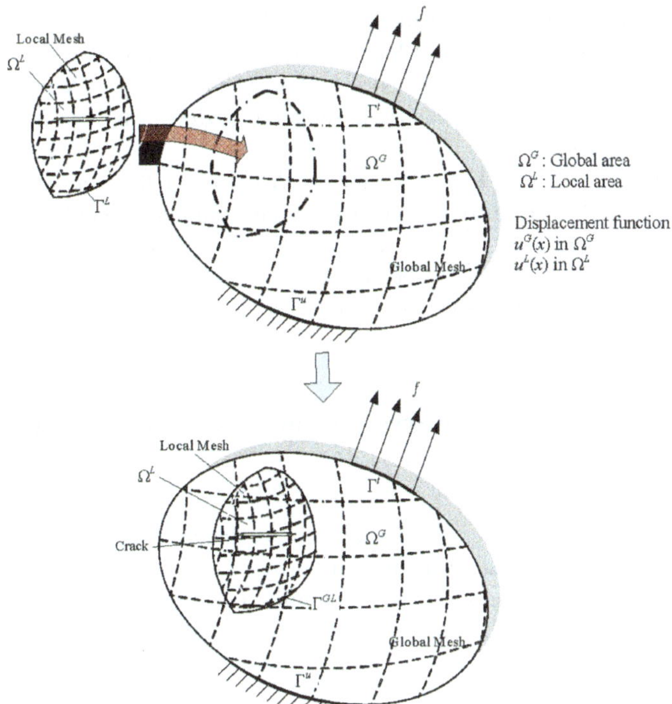

Fig. 1 Relation between the global mesh and the local mesh in S-FEM

Then, the matrix form of S-FEM is received as:

$$[[K^{GG}][K^{GL}][K^{LG}][K^{LL}]]\{\{u^G\}\{u^L\}\} = \{\{F^G\}\{F^L\}\} \quad (3)$$

where

$$\begin{aligned}
\left[K^{GG}\right] &= \int_{\Omega^G} [B_G]^T[D][B_G]d\Omega^G \\
\left[K^{GL}\right] &= \int_{\Omega^L} [B_G]^T[D][B_L]d\Omega^L \\
\left[K^{LG}\right] &= \int_{\Omega^L} [B_L]^T[D][B_G]d\Omega^L \\
\left[K^{LL}\right] &= \int_{\Omega^L} [B_L]^T[D][B_L]d\Omega^L \\
\{F_G F_L\} &= \left\{ \int_{\Omega^G} [N]^T\{f_G\}d\Omega \ \int_{\Omega^L} [N]^T\{f_L\}d\Omega \right\}
\end{aligned} \quad (4)$$

Here, $\left[K^{GG}\right]$, $\left[K^{LL}\right]$, $\left[K^{LG}\right]$ and $[K^{GL}]$ are the stiffness matrix at global, local, local-global and local-global area respectively while u^G and u^L are defined as the displacement for global and local area respectively. $\{F^G\}$ and $\{F^L\}$ are represented the nodal forces at the global and local areas, respectively. Meanwhile, $[B_G]$ and $[B_L]$ are the displacement–strain matrices defined in the global and local areas, respectively. $[D]$ and $[N]$ are defined as the material properties and the shape function matrix, respectively.

The stiffness matrix of $[K^{LG}]^T$ is equal to $[K^{GL}]$ which they are symmetrical to each other. $[K^{GL}]$ is the stiffness matrix between local and global areas. Accurate and good FEM results could be obtained if $[K^{GL}]$ is well calculated. Displacements of both local and global areas can be obtained by resolve Eq. (3). The specification of this theory had been published in [1, 6].

3 Basic Probabilistic S-version Finite Element Method (ProbS-FEM)

The ProbS-FEM is developed for the uncertainty analysis of the complex mathematical models of failure and was introduced by Akramin (2016) [6]. Each of the random parameters is developed by a probability density function (PDF). Then, a joint PDF, $f_X(X)$ is created. Hence, the probability of failure (POF), P_f is expressed as formula below:

$$P_f = \int_{g(X)<0} f_X(X)dX \tag{5}$$

where X is defined as the vector of the independent random parameter and $g(X)$ is defined as the performance function. If $g(X) < 0$, it shows the indication of failure. The $g(X)$ and $f_x(X)$ are non-linear and multidimensional which make it difficult to calculate an Eq. (5) [9]. Hence, Monte Carlo simulation (MCS) was introduced and have been used in the ProbS-FEM to ignore the evaluation in Eq. (5).

MCS functioned to give a distribution on final outputs. This Monte Carlo method was used to distribute the parameter uncertainty in the ProbS-FEM. The steps for accounting the parameter uncertainty are as follows; (1) Sampling of random parameters, (2) Generating data from the selected distribution, (3) Using the generated data as a parameter in the model to produce the output and (4) Extracting the probabilistic information.

In the early stage, a random number was selected and computed in the sampling process. Different random numbers were evaluated for different parameters and it was continuously generated. The inverse transformation technique is capable to generate a random parameter from the selected distribution such as Gaussian, Log-normal and Weibull distributions [7, 10–12] and was used once the random number had been selected.

The generated random parameters were computed to allow ProbS-FEM to generate the outputs. The generated random parameters were account into the finite element model and the related results were evaluated. After the new sample had finished computed, a new sampling process started once again. The whole finite element evaluation operation was repeated by taken the new generated data. The probabilistic data was extracted once all the samples had done generated. Every sample created its own output and were collected for the extraction of the probabilistic data. The specification of this theory had been published [7].

4 Result and Discussion

The model was conducted based on deterministic result in Yoshitaka Wada [1]. An embedded crack and mathematical model were used in the S-FEM and ProbS-FEM simulation to predict the crack growth. The material properties of the model were simulated referring to the specification in Table 1. An embedded crack model with material of A533B steel was used. The load applied on the model was normal to tension loading direction. The maximum load was 50 MPa with a stress ratio of 0.1. The size of crack initiation area was modelled in a circular local mesh. The crack propagation was calculated by Paris' law equation;

$$\frac{da}{dN} = C\left(\Delta K_{eq}\right)^{n} \tag{6}$$

where, da is the crack growth rate and N is the number of cycles. The C and n constants are totally depends on the material properties. The embedded crack simulated by ProbS-FEM had used 50 number of samples but only 45 samples were selected. Figure 2 presented the embedded crack growth to predict it fatigue life.

Figures 3 and 4 present graph of comparison between computational simulation of deterministic S-FEM and mean ProbS-FEM results for embedded crack growth with

Table 1 Input distribution for A533B steel

Variable	Distribution	Deterministic/mean value	Standard deviation
Critical stress intensity factor, K_{IC}	Deterministic	29 MPa $\sqrt{m}$	0
Fatigue power parameter, n	Deterministic	3.23	0
Tensile strength, yield	Deterministic	503 MPa	0
Young's modulus, E	Gaussian	206,000 Pa	900
Paris coefficient, C	Gaussian	1.67×10^{-12}	9.40×10^{-12}
Threshold value, ΔK_{th}	Gaussian	5.66 MPa $\sqrt{m}$	0.268
Maximum crack growth increment, da_{max}	Gaussian	2.0 mm	0
Initial crack depth, a_i	Gaussian	2.0 mm	0.5
Initial crack length, c_i	Gaussian	2.0 mm	0.5

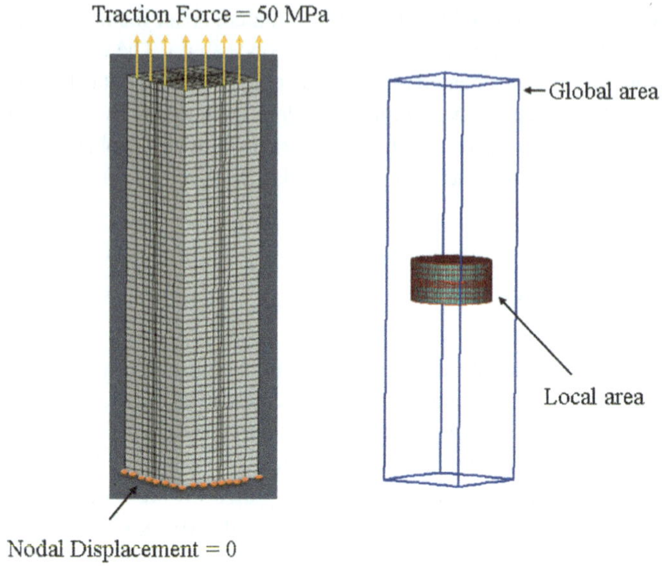

Fig. 2 Global mesh with boundary conditions and overlaid local mesh of embedded crack model

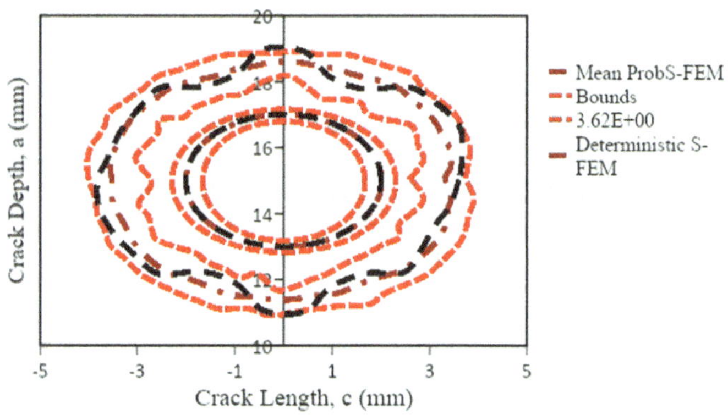

Fig. 3 Comparison between computational simulation, deterministic S-FEM and mean ProbS-FEM results for embedded crack growth with bounds

bounds and 95% confidence bounds respectively. The deterministic results were represented by black colour dash lines while the probabilistic results were represented by red colour dash line. The ProbS-FEM was used to conduct a probabilistic perspective of the results. In the probabilistic approach, 45 number of samples were generated the probabilistic crack growth. The mean of the 45 samples, 95% confidence bounds and bounds were plotted as in Figs. 3 and 4. The 95% confidence bounds were calculated as below:

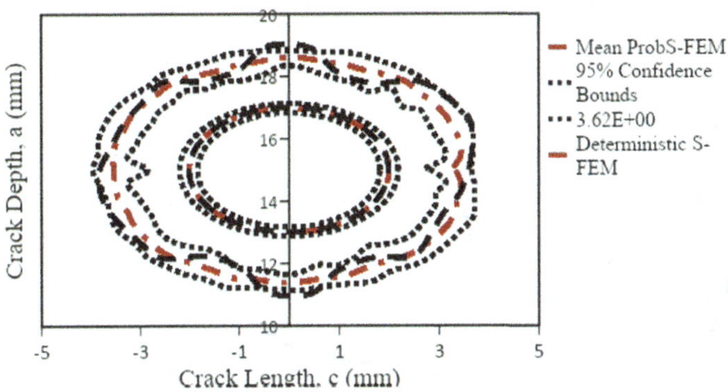

Fig. 4 Comparison between computational simulation, deterministic S-FEM and mean ProbS-FEM results for embedded crack growth with 95% confidence bounds

$$Bounds_{95\%} = \underline{\mu}_X \pm \frac{1.96\underline{\sigma}_X}{\sqrt{N_t}} \tag{7}$$

There are lower and upper confidence bounds and the result for deterministic S-FEM and ProbS-FEM were between them. Then, the bounds represented the maximum and minimum point of embedded crack growth. The result for deterministic S-FEM and ProbS-FEM were between the bounds. The results of deterministic S-FEM are observed to be diverged compared to the ProbS-FEM result but still located between the 95% confidence bounds and the bounds. It is shows that the confidence level of the result is reliable. The calculation of the confidence level can be referred to [13]. Consideration of uncertainty is indicated in this analysis result and created a massive development for more accurate and valid crack growth results.

The ProbS-FEM and deterministic S-FEM were capable to predict the fatigue life of the model [1]. Figure 5 presents the comparison of fatigue life between deterministic S-FEM, ProbS-FEM with 95% confidence bounds, bounds and S-FEM by Yoshitaka Wada [1]. The deterministic S-FEM reached a maximum point at 8.63 mm crack length with 101,686,000 cycles. Meanwhile, mean ProbS-FEM reach a maximum point at 8.04 mm crack length with 21,354,982 cycles. Vary from Yoshitaka Wada's result, which reach a maximum point at 8.62 mm crack length with 201,489,325 cycles. Mean of fatigue life was calculated by referring to the average of forty five samples.

The ProbS-FEM result was located between the 95% confidence bounds where the deterministic and Yoshitaka Wada's results are not in the bounds as shown in Fig. 5. The 95% confidence bounds present the range of possibilities for the fatigue life where deterministic and Yoshitaka Wada's results here is far from the 95% confidence bounds. ProbS-FEM created the minimum and maximum cycles before the unstable crack growth happened. The ProbS-FEM considered the uncertainty of material properties, the initial crack length and the initial crack depth in the mathematical model while the deterministic S-FEM neglected the uncertainty in its

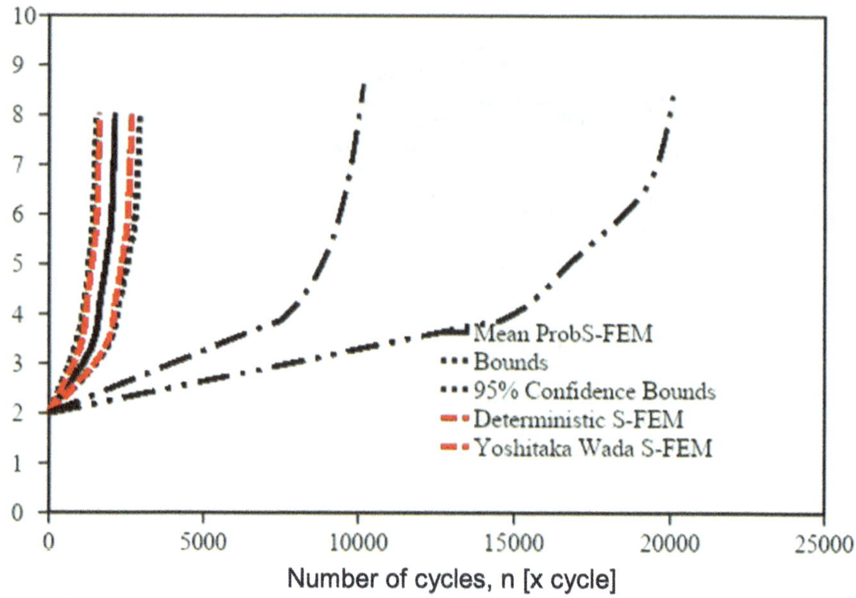

Fig. 5 The comparison of fatigue life between deterministic S-FEM, ProbS-FEM and Yoshitaka Wada S-FEM [1]

calculation. Hence, ProbS-FEM results was closer to the reliability that occurred in the real engineering application.

5 Conclusion

By using deterministic S-FEM and ProbS-FEM simulation in the embedded crack cases, comparison of results between them was achieved. The embedded crack initiation and propagation from deterministic S-FEM and ProbS-FEM results are found valid due to located in between the 95% percentile bounds. The beach marks produced from deterministic S-FEM and ProbS-FEM were vary as beach mark for deterministic S-FEM had an irregular shape while in beach mark ProbS-FEM had circular shape. Meanwhile, fatigue life of embedded crack in deterministic S-FEM had a significant difference when compared to fatigue life in ProbS-FEM. The problem for this fact might be closely related in the numerical method or the meshing process of the model. When comparing both results with Yoshitaka Wada, it is found that the results have a significance different. It is concluded that the embedded crack model of deterministic S-FEM and ProbS-FEM have errors in the meshing process. Further research, study and investigation for the embedded crack are needed to extend this present work.

Acknowledgements This study was funded by RDU170383 from Universiti Malaysia Pahang (UMP) and Fundamental Research Grant Scheme (FRGS/1/2017/TK03/UMP/02/24) from Kementerian Pendidikan Malaysia (KPM) with number RDU170124.

References

1. Wada Y, Kikuchi M, Yamada S, Serizawa R, Li Y (2014) Fatigue growth of internal flaw: simulation of subsurface crack penetration to the surface of the structure. Eng Fract Mech 123:100–115
2. Molent L, Barter SA (2010) The lead fatigue crack concept for aircraft structural integrity. Proc Eng 2(1):363–377
3. Kong X, Li J (2019) Non-contact fatigue crack detection in civil infrastructure through image overlapping and crack breathing sensing. Autom Constr 99:125–139
4. Wada Y et al (2014) Fatigue growth of internal flaw: simulation of subsurface crack penetration to the surface of the structure. Eng Fract Mech 123:100–115
5. Kamaya M, Miyokawa E, Kikuchi M (2010) Growth prediction of two interacting surface cracks of dissimilar sizes. Eng Fract Mech 77(6):3120–3131
6. Möller B, Graf W, Beer M (2003) Safety assessment of structures in view of fuzzy randomness. Comput Struct 81(15):1567–1582
7. Mohd Akramin MR (2016) Analysis of fatigue surface crack using the probabilistic s-version finite element model. Universiti Kebangsaan Malaysia
8. Kikuchi M, Wada Y, Shimizu Y, Li Y (2012) Crack growth analysis in a weld-heat-affected zone using s-version FEM. Int J Press Vessel Pip 90–91:2–8
9. Chowdhury MS, Song C, Gao W (2014) Probabilistic fracture mechanics with uncertainty in crack size and orientation using the scaled boundary finite element method. Comput Struct 137:93–103
10. Lan C et al (2018) Weibull modeling of the fatigue life for steel rebar considering corrosion effects. Int J Fatigue 111:134–143
11. Jiang H et al (2017) Fatigue life assessment of electromagnetic riveted carbon fiber reinforce plastic/aluminum alloy lap joints using Weibull distribution. Int J Fatigue 105:180–189
12. Toasa Caiza PD, Ummenhofer T (2018) Consideration of the runouts and their subsequent retests into S-N curves modelling based on a three-parameter Weibull distribution. Int J Fatigue 106:70–80
13. Petty MD (2012) Calculating and using confidence intervals for model validation. Proc. fall 2012 Simul. interoperability work, pp 10–14

Mitigation of Carbon Footprint of an Airport in the Kingdom of Saudi Arabia

Vineet Tirth, Saad AL-Mashhour, Mustafa Al-Ani, and Mohannad Alqahtani

Abstract Aviation is a major source of carbon emissions and this industry is growing at a fast pace, drawing the attention of environmentalists. In addition to being a large source of carbon emissions, aviation is also responsible for the deterioration of air quality, noise pollution and consumption of water resources. With the Kingdom of Saudi Arabia (KSA) being the largest country in the middle east, aviation is a major means of mass transport. The number of air travelers is increasing exponentially, resulting in a serious threat to sustainability, with both climate change and the environment in perspective. The per capita carbon footprint of KSA is among the top 20 countries in the world. In this study, an international airport in KSA has been considered with its carbon footprint estimated for 2016. Airports are purposefully located outside the city and require large construction free land for the purpose of security and airport activities. Thus, an airport provides ample space and opportunity to be used for energy harvesting and carbon mitigation. In this work, a three-tier process to neutralize the carbon emissions of the airport has been suggested to transform it into a carbon-neutral airport. The reduction of carbon emissions suggested in this study includes a reduction in emissions, promoting renewable energy, and carbon capture and storage. The proposed methods complement each other and enable a 360° approach for carbon mitigation.

Keywords Aviation · Carbon footprint of airport · Energy harvesting · Environment impact assessment · Global warming · Green and renewable energy

---

V. Tirth (✉)
Mechanical Engineering Department, College of Engineering, King Khalid University, Abha 61411, Kingdom of Saudi Arabia
e-mail: vtirth@kku.edu.sa

S. AL-Mashhour
First Lieutenant Engineer, Special Emergency Forces, Riyadh, Kingdom of Saudi Arabia

M. Al-Ani
Department of Mechanical and Manufacturing Engineering, University of Calgary, Alberta, Canada

M. Alqahtani
First Lieutenant Engineer, Royal Saudi Air Force, Riyadh, Kingdom of Saudi Arabia

P. Rajendran et al. (eds.), *Proceedings of International Conference of Aerospace and Mechanical Engineering 2019*, Lecture Notes in Mechanical Engineering,
https://doi.org/10.1007/978-981-15-4756-0_3

1 Introduction

Major greenhouse gases (GHGs) in the earth's atmosphere are CO_2, N_2O, water vapor, and CH_4. They occur naturally in the air [1]. Some GHGs are man-made which include CFCs, HFCs, per-fluoro carbons, and sulfur-hexafluoride [1]. The primary sources of GHG emissions in the world are shown in Fig. 1.

The GHGs are expressed in equivalent tons of carbon dioxide ($etCO_2$), defined as carbon footprint (CFP). One unit of carbon is equivalent to 3.664 units of CO_2 [3]. There are several units in which one can represent carbon emissions (or CO_2 equivalent); from electricity generation in mass/kWh, from transport in mass/km or mass/liter fuel [4]. In the estimation of CFP of a country, units are primarily tons, million tons (Mt) or giga tons (Gt) per unit time (year) or per person or a group of persons (in millions), etc. Another method to analyze CO_2 is ppm [5]. When air quality is determined, CO_2 is represented in ppm and when emissions from different energy sources and human activities are estimated, CO_2 is represented in mass.

Aviation is a major means of mass transport. There are many reasons for the increase in CFP, one major single source is aviation contributing about 4% of global GHG emissions [6]. CO_2 emissions from aviation are expressed as CO_{2e}

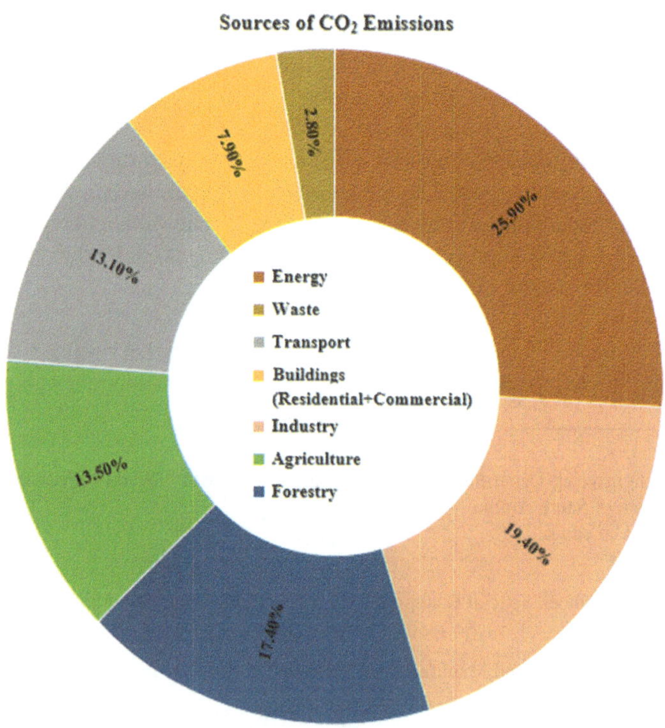

Fig. 1 The primary sources of GHG emissions in the world [2]

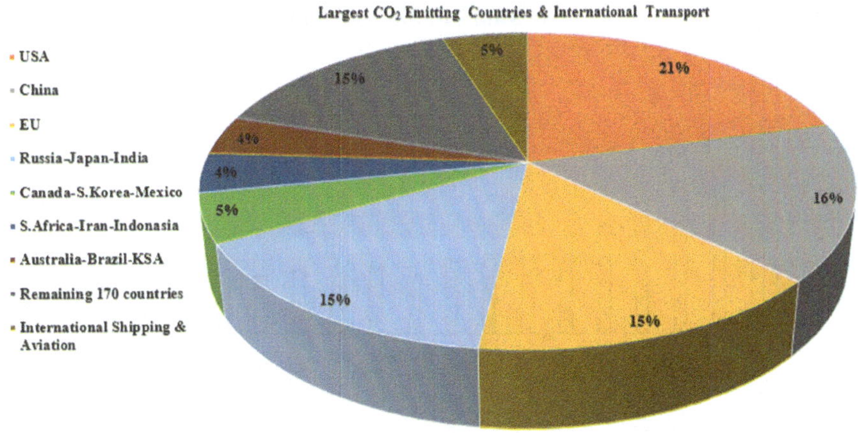

Fig. 2 CO_2 emitting countries and emissions from transport including aviation [8]

per passenger-km. For short distance (<463 km), 259 g/km CO_{2e} per person was estimated for a decade-old aircraft [7]. For long-distance flights, 114 g/km CO_{2e} per passenger was estimated [7]. This data excludes the modern more efficient jets, which have been introduced in recent times, for example, Boeing 787, Airbus A380, etc. If emissions from aviation are not controlled immediately, it is soon going to multiply many folds. A glimpse of CO_2 emissions across the world is represented in Fig. 2 [8].

KSA is extremely vulnerable to climate change due to limited greenery and vegetation. Having 0.4% of the world's population but 1.35% of the world's CO_2 emissions estimated to increase by 68% till 2030 and reach the level 845.51 $MtCO_{2e}$. The per capita, CO_2 emissions of the top 20 countries are plotted in Fig. 3 [9]. KSA is 8th from top but unfortunately, very little efforts are being done to offset its CFP.

Civil aviation in KSA started in 1944 (Hijri 1364) [10]. In the absence of a country-wide rail network for mass transport, air travel in KSA is a popular means of passenger travel as well as cargo transport. The number of air travelers increased by 9.5% in 2015 from 2014. The aircraft operations increased by 9.8% over 2014 and reached 646,693 operations in 2015. The volume of Cargo increased by 14.1% more than 2014 in 2015.

KSA has 27 airports (4 international, 10 regional and 13 domestic) at present serving domestic, international and cargo requirements as shown in Fig. 4 [11]. The total number of passengers traveling by air in KSA from 2006 to 2015 is shown in Fig. 5 [11]. One may observe an exponential increase in air travelers in KSA beyond 2009.

Carbon emissions from aviation may be studied in two categories, from the aircraft, and from airport activities. Airports have free space due to restrictions on human activities and act as service hubs for dealing with aircraft. Thus, providing space and opportunity to be used for neutralizing the high concentration of carbon emissions.

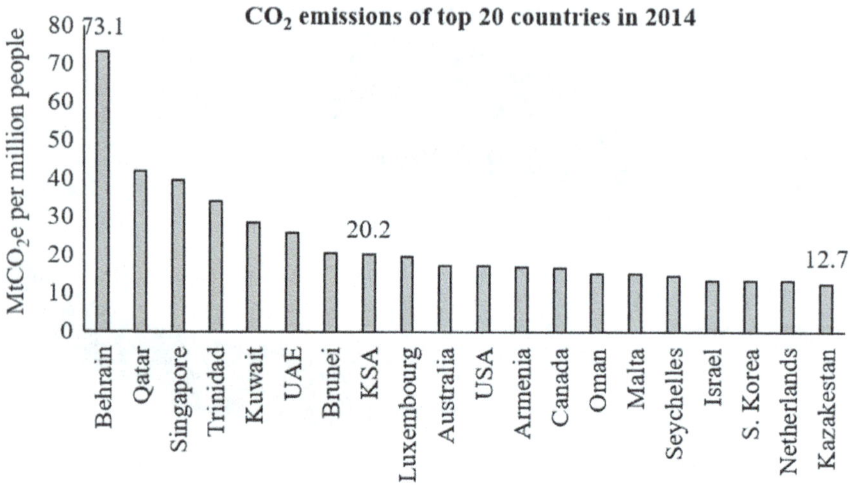

Fig. 3 Per capita CO_2 emissions of the top 20 countries [9]

Fig. 4 27 Airports (4 international, 10 regional and 13 domestic) in KSA [11]

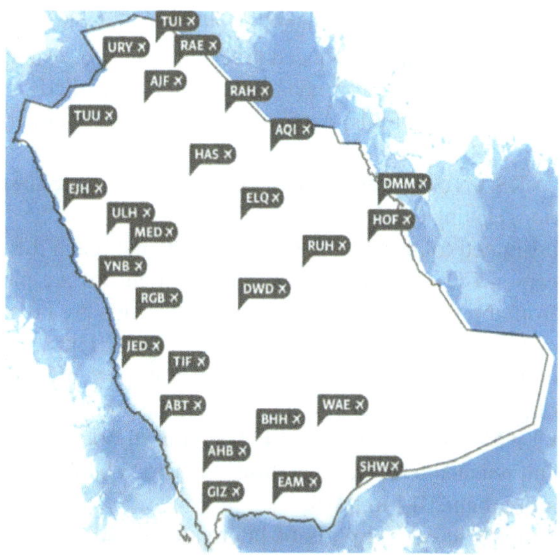

The environmental issues in aviation are mainly carbon emissions in large quantities due to the combustion of aviation fuel, the noise pollution due to the sound of jet engines. The consumption of huge quantity of water during washing and cleaning of the aircraft as well as air pollution at airports due to the combustion of the fuel by the aircraft and the airport activities [12] further add to this tally.

In view of the literature review, it is deduced that aviation is one of the major sources of CO_2 emissions. In a fast-growing economy like KSA, the aviation industry

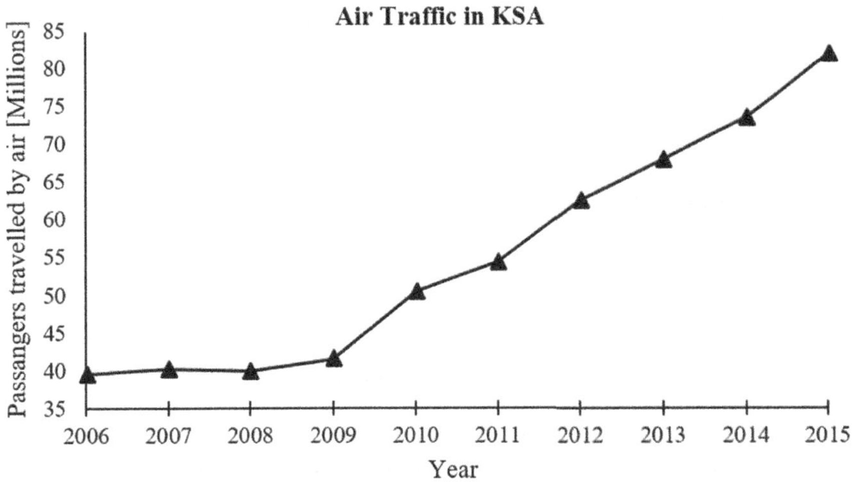

Fig. 5 Passengers traveled by air in KSA from 2006 to 2015

is also growing at a faster rate and hence the CO_2 emissions are increasing. It becomes relevant to concentrate on the estimation of airport CFP and explore the opportunities and methods to offset the aviation CO_2 emission in KSA. In the present work, sources of carbon emissions at Abha international airport in KSA are estimated for the year 2016 and neutralization methods are suggested.

2 Material and Methods

Firstly, sources of energy consumption have been identified then; their percentage share in total carbon emissions of the airport is estimated by conversion. Effective methods to absorb and reduce CO_2 from the air have been studied and feasible methods have been identified in the light of the literature review.

Abha international airport in Abha-Asir, KSA has been considered as a reference airport and methods to reduce carbon emissions at this airport is proposed so that the airport is transformed into 'carbon neutral'. Data is collected from several websites of authorized National and International Organizations like the General Authority of Civil Aviation (GACA) KSA, International Civil Aviation Organization (ICAO), World Bank, U.S. Energy Information Administration (EIA), etc. Calculations reported are made in two stages. Firstly, the carbon emissions of the world in the past, present, and future have been estimated on the country and per capita basis. Secondly, the type of aircraft that operate in KSA, their emission status and the airport capabilities of KSA have been estimated, analyzed, and reported. Finally, solutions to mitigate the CFP have been proposed. The proposed solutions for offsetting the CO_2 emissions at Abha International Airport have been marked in a google map and presented in the latter section of results and discussion.

3 Results and Discussion

The aircraft operating in KSA are shown in Fig. 6 [13]. Abha International Airport has certain limitations such as limited runway length, short apron length, and constrained service activities. Due to these limitations about 95% of the aircraft operated by major airlines namely Saudia, Air Arabia, Egypt Air, Flynas, Flydeal, Flydubai, etc. are Airbus A320s and, Boeing 737s. However, at present, the Abha International airport is undergoing an extension to accommodate larger aircraft. CO_2 emissions from different aircraft operated from Abha Airport with their capacity, average annual occupancy, rate of carbon emission per km at full capacity and at 70% occupancy (estimated annual average) is given in Table 1. As per the Climate Change Performance Index report 2014, published in 2015 [14], aviation is responsible for 4% of CO_2 emissions. KSA is a fast-growing economy and infrastructural development of the country includes expansion of the airports and construction of new airports. The growth in the economy increases the CO_2 emissions linearly [15].

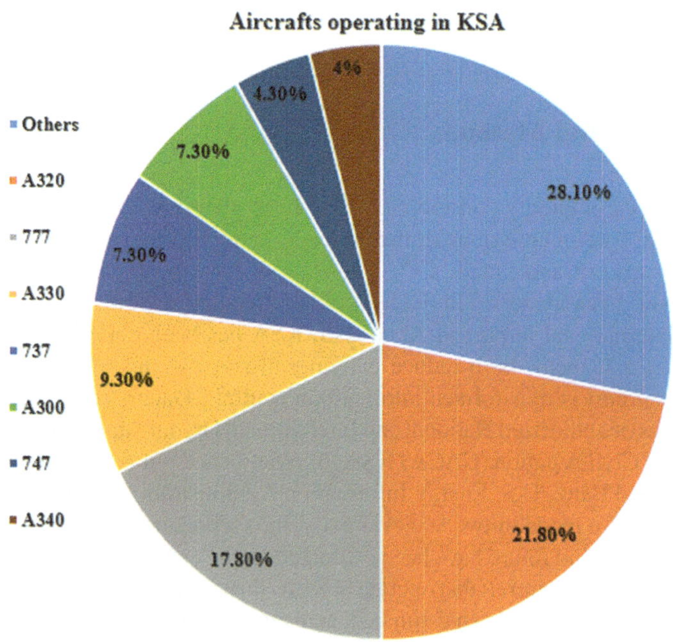

Fig. 6 Aircrafts operating in KSA [13]

Table 1 CO_2 emissions from aircrafts operating in KSA with 100 and 70% occupancy

Aircraft type	Capacity	Cabin factor 70%	CO_2 rate [kg/km]	Emissions per person at 100% occupancy [kg/km]	Emissions per person at 70% occupancy [kg/km]
Bombardier Learjet 45	9	6	1.766	0.196	0.294
Airbus A321	189	132	9.074	0.048	0.069
Airbus A320	150	105	9.149	0.061	0.087
Boeing 737–400	168	118	9.36	0.056	0.079
Boeing 737–600	145	102	7.214	0.050	0.071
Boeing 747–400	409	286	30.638	0.075	0.107
RJ100	99	69	10.159	0.103	0.147
RJ185	84	59	9.425	0.112	0.160
MD90	166	116	11.522	0.069	0.099

Normally, the aircraft with large capacity has fewer emissions per passenger.

Airport carbon emissions are estimated and represented as Carbon Footprint (CFP). The sources of carbon emissions at Abha airport and their percentage contributions are shown in Fig. 7. It includes electricity, aircraft movement, airport activities, and transportation within the airport. The exclusions are human factors and carbon emissions due to transport from the city to the airport. Total carbon emissions at Abha International Airport are 5875 tCO_{2e} in the year 2016. A fraction of 83% of the total emissions is due to the airport activities and aircraft movement, represented by '*A*' in Fig. 7. A fraction of 4% is due to indirect emissions due to purchased electricity, represented by '*I*', and a fraction of 13% of the total emissions is due to direct activities such as energy produced at the airport, cargo and transport within the airport, represented by '*D*'.

Neutralization of the carbon emissions from the atmosphere includes a two-fold approach: (1) absorb CO_2 from the atmosphere (air) and; (2) reduce emissions by higher efficiency and green energy [13]. Both approaches have been considered in this study and solution to make Abha International Airport, a '*carbon neutral*' is proposed. The methods available for CO_2 absorption from the air are shown in Fig. 8. The methods to absorb carbon emissions include CO_2 capture by chemicals such as CaO and NaOH and transform the captured CO_2 as carbonates. There are some plastics, which are reported to sponge up CO_2 from air in a chemical reaction and release it when the plastics are washed with water. Such plastics have been recommended for a candidate material for artificial trees [16]. Algae bioreactors have also been prepared to capture the CO_2 from the air and transform it into hydrocarbons

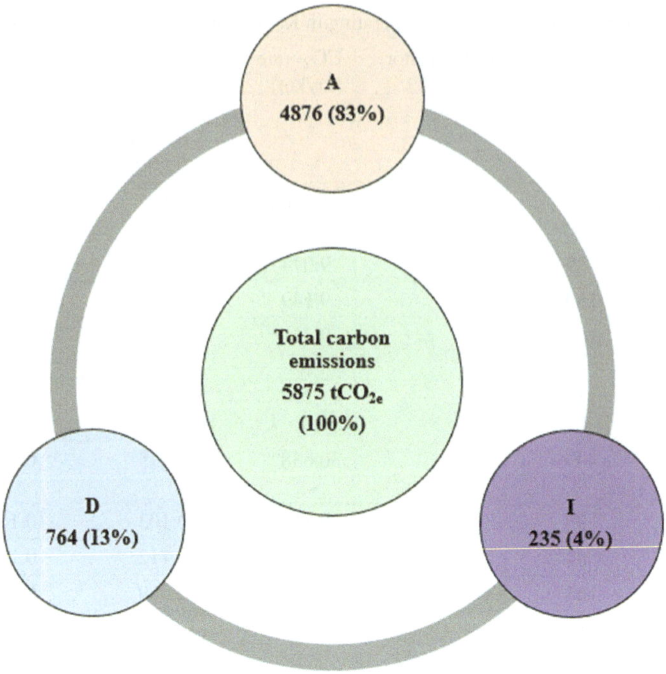

Fig. 7 The carbon footprint of Abha International Airport for 2016

Fig. 8 Methods are available for CO_2 absorption from the air

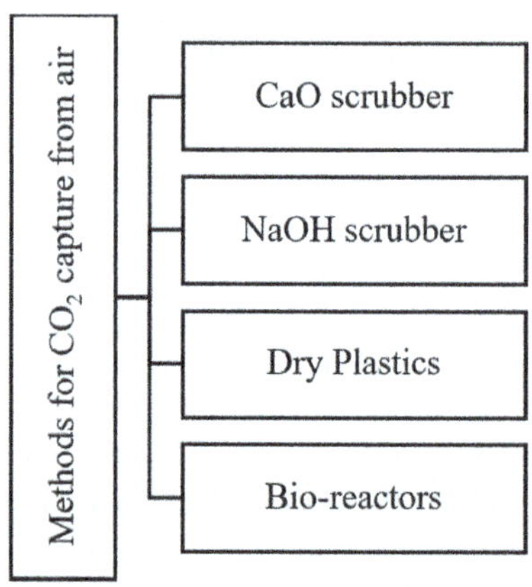

using a photosynthesis reaction. The hydrocarbons can be used as an alternative fuel as a supplement of petroleum [17].

The proposed model of the hybrid chemical scrubber to capture CO_2 from the air is shown in Fig. 9 with four chemical reactions, two exothermic and one endothermic. After energy balance, 179.2 kJ/mol of heat is required to operate the scrubber, which

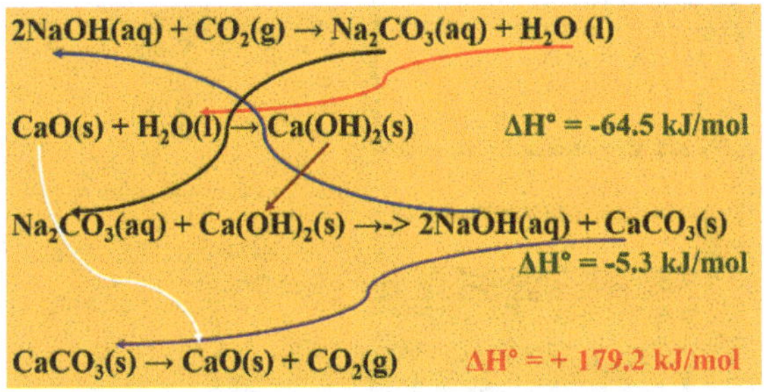

Fig. 9 Proposed model of NaOH–CaO scrubber

may be derived from the solar power system. Further, $CaCO_3$ precipitate obtained from the reaction may be used for construction, ground filling, and excavation work.

Among the above methods to capture CO_2 from the air, simplest, proven and cost-effective are scrubbers based on CaO and NaOH. Other methods are still under development and expensive. The NaOH based scrubber uses aqueous NaOH solution giving H_2O as a by-product and CaO based scrubber needs H_2O to absorb CO_2. The carbon emissions may also be reduced by increasing the energy efficiency and energy conservation. The proposed solution at Abha International Airport is shown in Fig. 10. Map of Abha International airport has been taken from google maps. Employing a hybrid chemical scrubber such that the latter may use the water produced from the former to absorb 20% of the total CO_2 emissions of the airport in a year. The remaining 80% CO_2 is proposed to be neutralized by using a Grid-integrated solar power system proposed at the available land at the airport. The green marks in Fig. 10 show the proposed locations of solar panels and yellow marks show the proposed installation of NaOH–CaO hybrid scrubbers. In a study on renewable energy systems in different cities of KSA [18], it was suggested that renewable energy may be generated at a cost of about 0.609 USD/kWh, which may be further brought down if energy is generated on a large scale.

Both the proposed solutions are proven methods and their effectiveness is quite reliable as the chemical scrubbers have been used for the capture of CO_2 for decades and solar energy is the most abundant green energy. There is little doubt about the effectiveness of the proposed methods. The plenty of unutilized lands available at the airports can accommodate these methods with little constraints. The KSA has an ambitious plan for the installation of solar energy plants up to 41 GW capacity by the year 2032 [19]. The proposal to power the airports by solar energy complements this plan. The proposed reduction in CO_2 emissions is based on the estimation of CO_2 at Abha airport. Based on the feasibility of the methods, it is suggested to capture the CO_2 using chemical scrubbers and to offset the remaining 80% CFP by using solar energy.

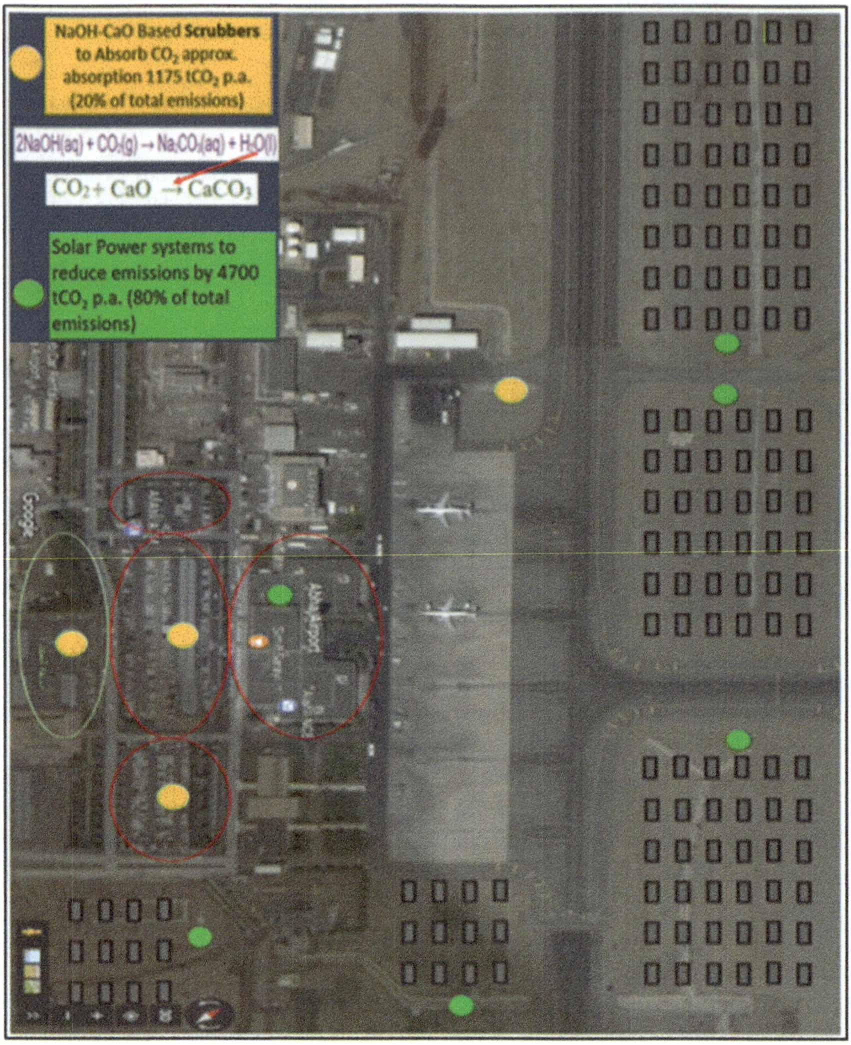

Fig. 10 The proposed solution for carbon-neutral Abha International Airport

4 Conclusions

The following major conclusions are derived from the study:

1. The average equivalent CO_2 emissions from Abha Airport is about 5875 tCO_{2e} in 2016.
2. About 20% (1175 tCO_{2e}) may be reduced by hybrid NaOH–CaO scrubber.
3. Remaining 80% (4700 tCO_{2e}) may be neutralized using a Solar Power System of capacity 8.9 MWh.

4. Modern fuel-efficient aircraft like Boeing 787 and Airbus A380 in full capacity produce 75 g of CO_2 per passenger per km. These are 20–30% more fuel-efficient than the decades-old aircraft. This emission is much less than modern compact cars [13] and so, it is recommended to replace the decades-old aircraft by the efficient ones.

Acknowledgments The authors gratefully acknowledge resources and facilities given by the Mechanical Engineering Department, College of Engineering, King Khalid University, Asir-Abha KSA to conduct the study.

References

1. National Centers for Environmental Information, National Oceanic and Atmospheric Administration (NOAA) USA homepage. https://www.ncdc.noaa.gov/monitoring-references/faq/greenhouse-gases.php. Last accessed 20 Dec 2018
2. Google Earth & Environmental Problems and Solutions. https://sites.duke.edu/tlge_sss29/carbon-dioxide-emissions/carbon-dioxide/. Last accessed 03 Jan 2019
3. Carbon Emissions by Source. http://shrinkthatfootprint.com/explain-carbon-budget. Last accessed 15 Mar 2018
4. US Energy Information Administration. https://www.eia.gov/tools/faqs/faq.php?id=74&t=11. Last accessed 29 Jun 2018
5. Wikipedia. https://en.wikipedia.org/wiki/Carbon_footprint#Measuring_carbon_footprints. Last accessed 18 Feb 2019
6. Allianz. 15 Sources of Greenhouse Gases. https://www.allianz.com/en/about_us/open-knowledge/topics/environment/articles/140912-fifteen-sources-of-greenhouse-gases.html/. Last accessed 30 Sept 2018
7. Pejovic T, Noland RB, Williams V, Toumi R (2008) Estimates of UK CO_2 emissions from aviation using air traffic data. Clim Change 88:367–384
8. International Maritime Emission Reduction Scheme Homepage. http://www.imers.org/case. Last accessed 24 May 2019
9. The Shift Project Data Project, Browse Energy And Climate Data Homepage. http://www.tsp-data-portal.org/TOP-20-CO2-emitters-per-capita#tspQvChart. Last accessed 11 Mar 2018
10. Statistical Yearbook (2014) General Authority of Civil Aviation (GACA). Kingdom of Saudi Arabia, pp 1–29
11. Statistical Yearbook (2015) General Authority of Civil Aviation (GACA). Kingdom of Saudi Arabia, pp 1–25
12. Dempsey P (2010) Environmental law and sustainability in international aviation. Institute of Air & Space Law McGill University
13. CAPA Centre for Aviation (2013) World aviation yearbook 2013 middle east, pp 1–95
14. CCPI (2015) Climate change performance index components overall results, pp 1–12
15. Samargandi N (2017) Sector value addition, technology and CO_2 emissions in Saudi Arabia. Renew Sustain Energy Rev 78:868–877
16. Artificial Trees Could Offset Carbon Dioxide Emissions. http://climatechange.medill.northwestern.edu/2016/11/29/artificial-trees-might-be-needed-to-offset-carbon-dioxide-emissions/. Last accessed 08 Oct 2019
17. This New Bioreactor Uses Algae to Capture as Much Carbon Dioxide as an Acre of Trees. https://www.goodnewsnetwork.org/algae-bioreactor-captures-as-much-carbon-as-acre-of-trees/. Last accessed 08 Oct 2019

18. Al-Sharafi A, Sahin AZ, Ayar T, Yilbasa BS (2016) Techno-economic analysis and optimization of solar and wind energy systems for power generation and hydrogen production in Saudi Arabia. Renew Sustain Energy Rev 55:697–702
19. Sulaiman A, Irfan MA (2017) The techno-economic potential of Saudi Arabia's solar industry. Renew Sustain Energy Rev 69:33–49

Error Analysis on Galerkin Scheme for the Diffusion Problem

Wei Shyang Chang, Haswira Hassan, Hazim Fadli Aminnuddin, Vishal Singh, and Farzad Ismail

Abstract For advection-diffusion problem, the non-unified approach of classic residual distribution (RD) schemes for advection discretization and pure Galerkin scheme for diffusion discretization has seen accuracy degradation and previous work has dealt with the accuracy analysis of the advection part on various RD schemes. The current work focuses on the accuracy analysis on diffusion discretization scheme of pure Galerkin method with the effect of grid skewness variation. Analytically, the accuracy of the pure Galerkin method is analyzed rigorously with Fourier expansion technique for pure diffusion problem in right-running grid with the introduction of grid skewness parameter as a variable. The truncation error study shows that pure Galerkin scheme is second order accurate in space with minimal or no effect from the variation of skewness. Numerical tests were performed on both steady and unsteady diffusion problems and the outcome matches the analytical study. Additional numerical test on randomized grid revealed that the analytical claims on uniform grid can be extended to poor quality grid. This work suggests that the diffusion discretization using pure Galerkin scheme is not the main factor of the degradation of the accuracy of the non-unified RD-Galerkin scheme.

Keywords Diffusion · Pure galerkin method · Skewness variation

---

W. S. Chang (✉) · F. Ismail
School of Aerospace Engineering, Engineering Campus, Universiti Sains Malaysia, 14300 Nibong Tebal, Penang, Malaysia
e-mail: changweishyang@usm.my

H. Hassan · H. F. Aminnuddin
Universiti Tun Hussein Onn Malaysia, 86400 Parit Raja, Johor, Malaysia

V. Singh
Seagate Global Business Services (M) Sdn Bhd, 1-16-7, Suntech@Cybercity Lintang Mayang Pasir 3, Penang, Bayan Baru, 11950 Bayan Lepas, Penang, Malaysia

P. Rajendran et al. (eds.), *Proceedings of International Conference of Aerospace and Mechanical Engineering 2019*, Lecture Notes in Mechanical Engineering,
https://doi.org/10.1007/978-981-15-4756-0_4

1 Introduction

1.1 Background and Motivation

The ultimate goal of any numerical method in computational fluid dynamics (CFD) is to produce sufficiently accurate predictions to various types of fluid flows. Governed by the Navier-Stokes equations, adequate aspects of physics have to be considered and properly modeled in the calculations for more accurate and comprehensive fluid simulations. Without neglecting the viscous effect, the Navier-Stokes equations can be written as a system of quasi-linear equations, which can be represented by a set of advection-diffusion equations in the scalar level. Besides the viscous fluid flow, advection-diffusion equations are also employed in the studies of chemical reactant transport [1], plasma physics [2], contaminant dispersion in shallow lakes [3] and various other sciences, engineering and even finance disciplines [4].

The discretization of the equation can be done in a non-unified (separate) manner for each term as reported in [5, 6]. However, there was an underlying problem regarding the accuracy loses where the order of accuracy drops to first order even though second order discretizations are used for both terms [7]. Recent work documented in [6] attempted this approach with flux-difference Residual Distribution (RD) for advection discretization and the classic Galerkin for diffusion with success in retaining second order accuracy numerically. Recent trend of researches on distorted grids [8, 9] has motivated the authors to invest on further investigation on the variation of skewness of this scheme which revealed a drop of half an order in the accuracy at higher skewness. However, the reason behind the accuracy drop remains undiscovered.

We attempt to use the divide-and-conquer method to investigate the source of the problem due to the non-unified nature of the scheme. Looking at the advection discretization, previous works of [10–12] has shown the detailed analytical analysis on the order of accuracy of some classic RD as well as flux-difference RD schemes with variation of grid skewness but very little has been done on the diffusion part. Hence, it is imperative to uncover the standalone performance of the pure Galerkin scheme which serves as the underlying motivation of the current work. Stemming from the finite element viewpoint, the Galerkin method is well-recognized as a second order accurate method with linear variation of solution [13–15] although there is yet any comprehensive analytical study performed on the truncation error analysis via Taylor's series expansion with the effect of skewness variation of the grid.

This paper is organized as the following. The steps to solve pure diffusion problem numerically with Galerkin method is recalled in Sect. 2. The analytical analysis on the order of accuracy of the scheme via Taylor series expansion, together with the effect of grid skewness variation, are presented in Sect. 3. In Sect. 4, the numerical results are demonstrated and discussed. Lastly, the conclusive findings are presented in Sect. 5.

2 Galerkin Method for Pure Diffusion Problem

The pure diffusion equation is parabolic (elliptic for steady cases) and can be represented as the following.

$$u_t = \nu(\nabla \cdot \nabla u) \tag{1}$$

ν is the positive diffusion coefficient. Since the present work is a supplementary work for the analysis for the flux-difference RD scheme in [6], we retain the main RD framework on the discretization of the integral form of Eq. (1).

Consider a 2-dimensional domain triangulated into sets of triangle $\{T\}$ with the solution of u stored as nodal values in the set of nodes $\{J\}$, The definition of inward scaled normals, $\boldsymbol{n}$ of the cells and median-dual area, S_i in the neighboring cells set $\{V_i\}$ are shown in Fig. 1.

The residual for the element T for a pure diffusion problem is the same to the viscous residual for the advection-diffusion problem and is recalled as follows.

$$\phi^T = \iint \nu(\nabla \cdot \nabla u)dxdy = \oint \nu \nabla u \cdot d\boldsymbol{n} \tag{2}$$

Again we retain the implementation of compact Galerkin scheme for the spatial discretization which solves the weak form of Eq. (2), and the signal to be distributed to each node i is represented by

$$\phi_i^T = \iint t_i \cdot [\nu(\nabla \cdot \nabla u)]dxdy, \tag{3}$$

where t_i is Galerkin test function. After performing integration by parts and using a linear basis function for the test function, Eq. (3) yields the following.

$$\phi_i^T = -\frac{\nu}{4A_T}\sum_{j=1}^{3} u_j n_j \cdot n_i \tag{4}$$

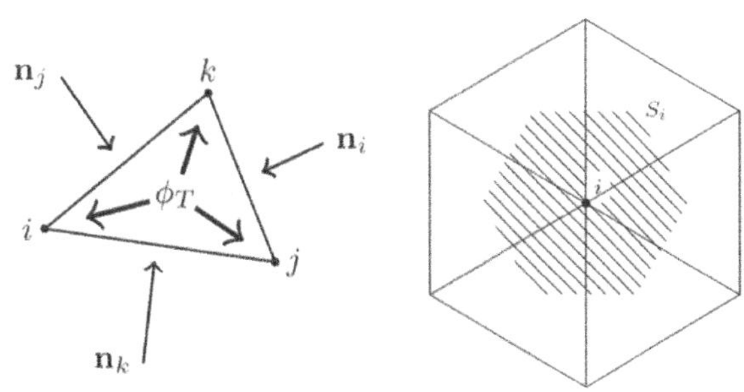

Fig. 1 Triangular cells with inward scaled normals and median-dual cells

A_T is the area of element T and j denotes the each vertex of element T. The semi-discrete form of the conserved governing equation for node i can be written as the following.

$$S_i \frac{du_i}{dt} = -\sum_{T \in V_i} \phi_i^T \tag{5}$$

We discretize the time derivative with pure explicit Forward Euler method for its simplicity and low computational cost which yields full discrete form of Eq. (1) as the following.

$$u_i^{n+1} = u_i^n - \frac{\Delta t}{S_i} \sum_{T \in V_i} \phi_i^T \tag{6}$$

3 Analytical Study on Galerkin Method

3.1 Truncation Error Analysis

In the effort of finding the truncation error for this method, we could write the diffusion equation by combining Eqs. (4) and (6) in the form of

$$u_i^{n+1} = u_i^n - \frac{\Delta t}{S_i} \sum_{T \in U\Delta i} \sum_{j \in T} \frac{\boldsymbol{n}_i^T \cdot \boldsymbol{n}_j^T}{4A_T} u_j^n \tag{7}$$

In steady state, $\mathrm{u}_i^{n+1} = \mathrm{u}_i^n \mathrm{u}_i^{n+1} = \mathrm{u}_i^n$. Therefore, the difference between the second term on the right-hand side and the exact solution is the truncation error (*TE*) for the study of order-of-accuracy of the method which has been explained in the work of [10]. The Taylor series expansion of the neighboring (*j*) points from the main point (*i*) of interest were performed in order to determine the *TE*.

$$TE^{steady} = -\frac{1}{S_i} \sum_{T \in U\Delta i} \sum_{j \in T} \frac{\boldsymbol{n}_i^T \cdot \boldsymbol{n}_j^T}{4A_T} u_j^n - \left(-\nu\left(\frac{\partial^2 u}{\partial x^2} + \frac{\partial^2 u}{\partial y^2}\right)\right) \tag{8}$$

Based on Fig. 2, for right-running (RR) triangular grid, the Taylor series expansion about node 0 is as follows.

$$u_j = \sum_{a=0}^{n} \sum_{b=0}^{n-a} \frac{d^{(a+b)} f}{\partial x^a \partial y^b} \frac{(h)^a (k)^b}{a!b!}, n \to \infty \tag{9}$$

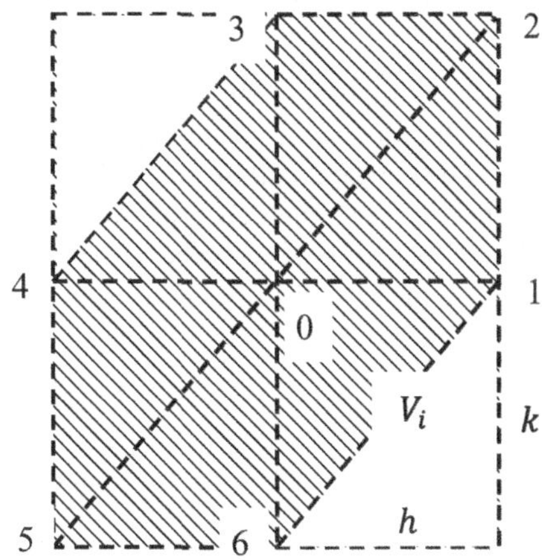

Fig. 2 Right-running grid topology

where we introduce a stretching parameter S, defined as the following.

$$k = hs \rightarrow s = \frac{k}{h} \tag{10}$$

After summing up each signal distribution from every neighboring cells to the center point of interest over the median-dual area (S_i), the *TE* yields

$$TE^{steady} = h^2\left(\frac{\nu u_{xxxx} + \nu s^2 \nu u_{yyyy}}{12}\right) + O\left(h^3\right) \tag{11}$$

For unsteady solution, due to the condition of $u_i^{n+1} \neq u_i^n$, the time derivative term has to be included in the *TE* analysis which yields

$$TE^{unsteady} = -\frac{u_i^{n+1} - u_i^n}{\Delta t} - \frac{1}{S_i}\sum_{T \in U\Delta i}\sum_{j \in T}\frac{n_i^T \cdot n_j^T}{4A_T}u_j^n - \left(-\frac{\partial u}{\partial t} - \nu\left(\frac{\partial^2 u}{\partial x^2} + \frac{\partial^2 u}{\partial y^2}\right)\right) \tag{12}$$

Next, the solution at the next time-step is also expanded using Taylor's series expansion in time

$$TE^{unsteady} = O(\Delta \mathrm{t}) + h^2\left(\frac{\nu u_{xxxx} + \nu s^2 \nu u_{yyyy}}{12}\right) + O\left(h^3\right) \tag{13}$$

The unsteady TE obtained is first order in time and second order in space.

3.2 Order-of-Accuracy on Variation of Skewness

Considering only right-running grid as discussed in the previous subsection, we can test both the analytical and numerical order-of-accuracy of Galerkin scheme for linear case. The skewness can be controlled by systematically varying the length (h) and height (k) of the triangular grid. We consider the skewness (Q) range from $0.3 \leq Q \leq 0.9$ and the skewness can be associated with the stretching parameter as follows, where the in-depth details can be found in the work of [10, 12].

$$s = 2\tan\left(\frac{\pi}{2}Q\right) \tag{14}$$

To obtain the order-of-accuracy analytically, L_2 error is determined as

$$L_2 = \sqrt{\sum_{i=0}^{N_x}\sum_{j=0}^{N_y}\left(Er_{(i,j)}^2 A_{(i,j)}\right)} \tag{15}$$

where $A(i, j)$ is the median-dual area for node (i, j) whilst $N_x = 1/h$ and $N_y = 1/(h\ s)$ is the number of nodes along x and y direction respectively. $Er(i, j)$ represents the local error at point (i, j), which is also the local TE in analytical sense. L_2 is casted as a function of (h, s). Since the analytical order-of-accuracy is associated with logarithmic spacing, h is replaced with 10^{P} to express the logarithmic error function of L_2 as a function of grid size and stretching ratio, $f\ (P, s) = \log L_2$. Subsequently, the order-of-accuracy, OoA(s) can be calculated as the slope of the function f with respect of P, i.e.

$$\text{OoA}(s) = \frac{\partial f(P, s)}{\partial P} \tag{16}$$

4 Numerical Test Cases of Diffusion Problems

For numerical verification, we used the same right-running grids two test cases; steady diffusion and unsteady diffusion with average grid spacing from 0.03 to 0.0125. Figure 3 shows the right-running grid used in the present study; regular grid with $Q = 0.3$, systematically increased to $Q = 0.9$ as highly skewed grid, as well as randomized grid to further test the analysis on very poor quality grid. As for the boundary conditions, we applied Dirichlet boundary condition on all sides of the square domain $[0, 1] \times [0, 1]$ with the exact solutions.

The exact solution for steady diffusion is shown as Eq. (17)

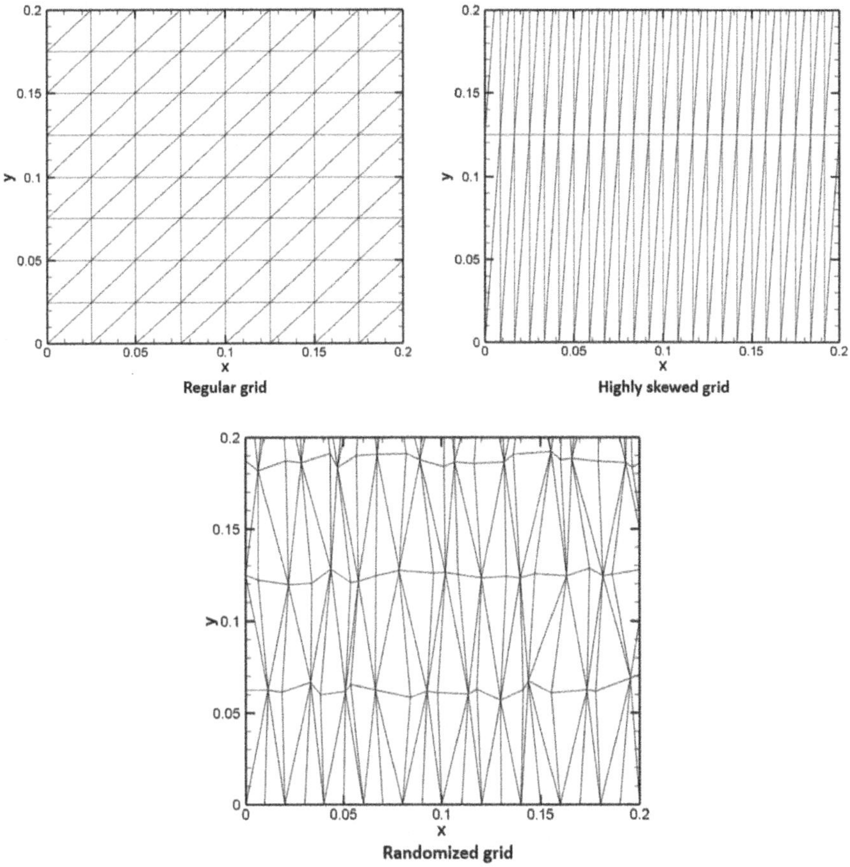

Fig. 3 Computational domain and right-running grid used for both cases

$$u(x, y) = \frac{\sinh(\pi x)\sin(\pi y) + \sinh(\pi y)\sin(\pi x)}{\sinh(\pi)} \tag{17}$$

and the same for unsteady diffusion is

$$u(x, y, t) = Ce^{-\upsilon\pi^2(a^2+b^2)t}\sin(a\pi x)\sin(b\pi y) \tag{18}$$

where $C = 1$, $a = 1$, $b = 2$ and t = [0, 0.1]. Also, both cases employed a constant time step of 1e^{-06} and the viscosity parameter, v = 0.1.

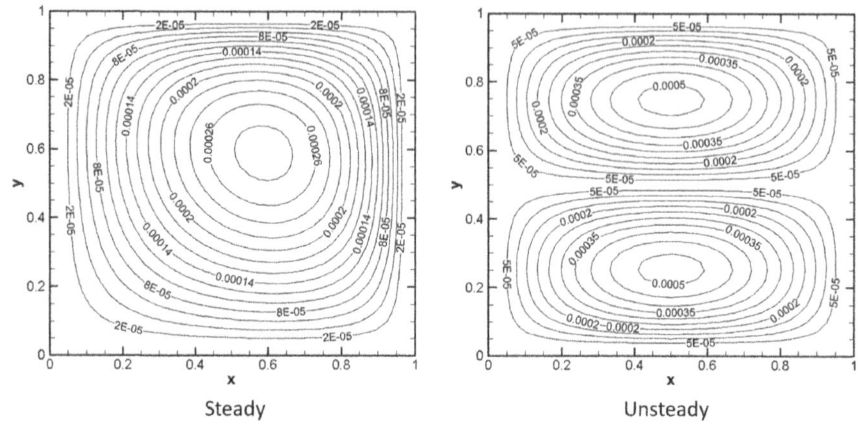

Fig. 4 Error contour plot for steady and unsteady diffusion cases

4.1 Steady and Unsteady Diffusion

Figure 4 shows the error contour plot of the steady and unsteady solution using Galerkin method. Both cases show small error magnitude ranging from the order of 10^{-4} to 10^{-5}, revealing that the Galerkin method is considerably good for pure diffusion on steady and unsteady cases. Aside from the error magnitude, the numerical performance in terms of order of accuracy is assessed by the graph of L_2 error versus the grid spacing as demonstrated in Fig. 5. In both cases, reducing grid sizes yields solutions which practically preserve second order accuracy. Meanwhile, Fig. 6 shows the convergence history for steady diffusion case where the solution starts to converge at pseudo time, t = 0.3.

In addition, the scheme was tested on various grid skewness from 0.3 to 0.9 for both steady and unsteady problems with the same boundary conditions and the results are plotted as in Fig. 7. The numerical outcomes are compared to their analytical results obtained from Sect. 3 with good agreement between them. In overall, the order of accuracy is preserved across the variation of skewness although there is a slight decay in the accuracy towards the highly skewed grids for $Q > 0.7$. Similar patterns are observed for both numerical solutions.

4.2 The Effect of Grid Randomization

In this subsection, all three grids (regular, highly skewed and randomized) are compared on steady diffusion case. Figure 8 shows the graph of the L_2 error versus the grid spacing with the gradient as the order of accuracy. As depicted, all three grids

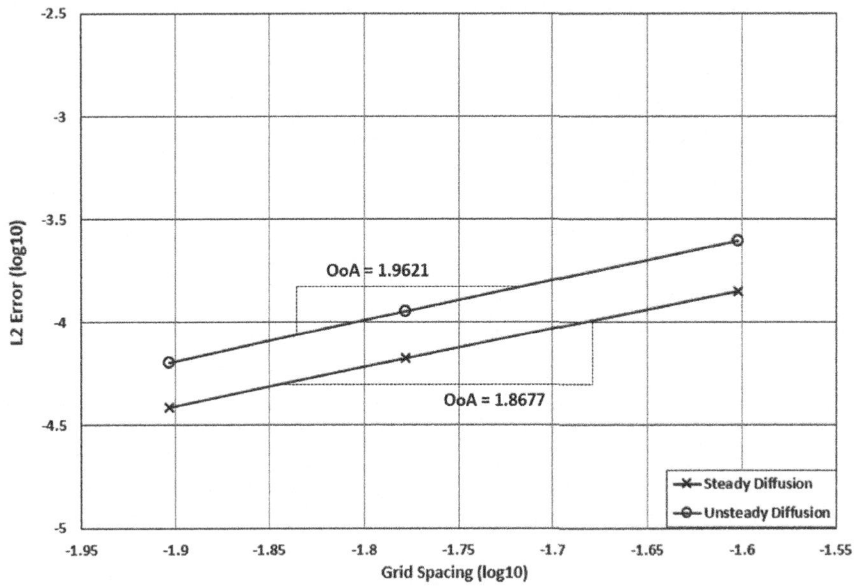

Fig. 5 Numerical L_2 error versus the grid distance in logarithmic scale for steady case

Fig. 6 Convergence plot for steady diffusion case

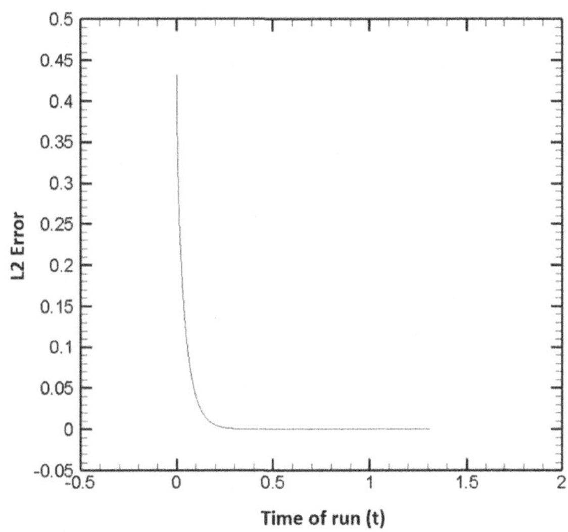

are able to produce results close to second order accurate with randomized grid having the highest error magnitude. These results proved that Galerkin method is not affected from the grid quality degradation even on the randomized grid.

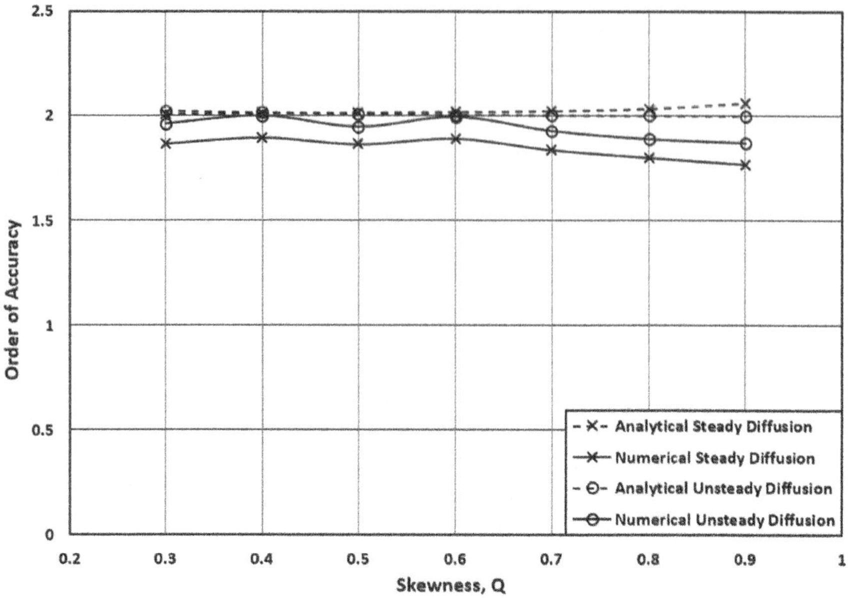

Fig. 7 Order of accuracy with various skewness, Q for both diffusion cases for analytical and numerical results

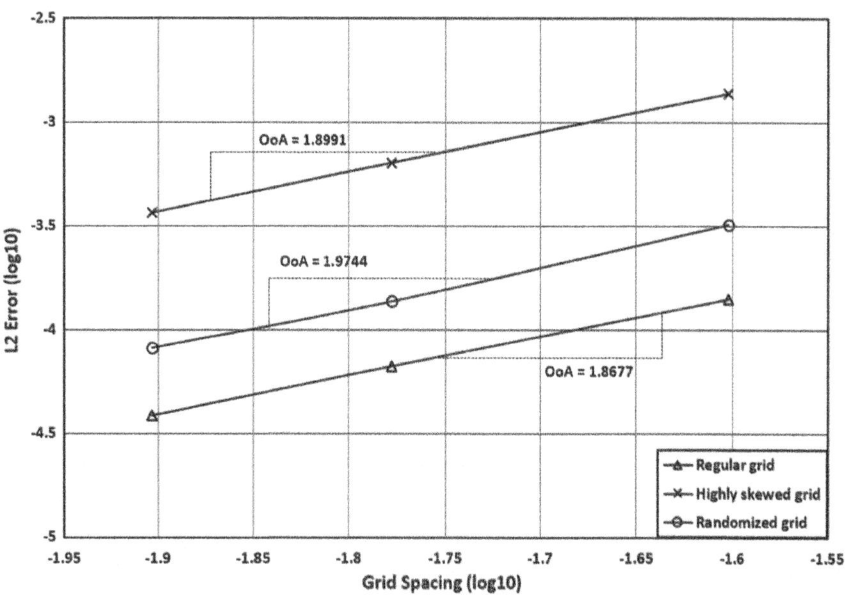

Fig. 8 Numerical L_2 error versus the grid distance in logarithmic scale for various grid on steady diffusion case

5 Conclusion

The pure Galerkin scheme is proven to be an effective numerical method for both steady and unsteady diffusion problems in term of its robustness in retaining its accuracy even at high grid skewness. Using Taylor series expansion and grid skewness parameter, the truncation error analysis demonstrated that the spatial accuracy is of second order throughout the skewness range considered whilst using forward Euler time discretization for unsteady problems can only attain first order temporal accuracy. The numerical results adhered to the analytical claims with negligible differences. Moreover, the results from the randomized grid illustrated that the analysis can be extended to poor quality grid, albeit the analysis was established on structured right-running grid. The findings in this work suggest that the diffusion discretization using Galerkin method via non-unified approach for advection-diffusion problem has the ability to retain second order accuracy even in highly skewed and randomized grids.

References

1. Perez LJ, Hidalgo JJ, Dentz M (2018) Reactive random walk particle tracking and its equivalence with the advection-diffusion-reaction equation. Water Resour Res Banner 55(1):847–855
2. Heene M, Hinojosa AP, Obersteiner M, Bungartz HJ, Pflüger D (2018) EXAHD: an exa-scalable two-level sparse grid approach for higher-dimensional problems in plasma physics and beyond. In: Nagel W, Kröner D, Resch M (eds) High performance computing in science and engineering '17. Springer, Cham, pp 513–529
3. Karahan H (2001) An iterative method for the solution of dispersion equation in shallow water, In: Brebbia CA (ed) Water pollution VI: modelling, measuring and prediction, Wessex Institute of Technology, Southampton, pp 445–453
4. Morton KW (1996) Numerical solution of convection-diffusion problems. Chapman and Hall, London
5. Mazaheri A, Nishikawa H (2015) Improved second-order hyperbolic residual-distribution scheme and its extension to third-order on arbitrary triangular grids. J Comput Phys 300(C):455–491
6. Singh V, Chizari H, Ismail F, Abgrall R (2018) Non-unified compact residual-distribution methods for scalar advection-Diffusion problems. J Sci Comput 76(3):1521–1546
7. Nishikawa H, Roe PL (2004) On high-order fluctuation-splitting schemes for Navier–Stokes equations. In: 3rd ICCFD Conference, Toronto
8. Nishikawa H (2019) Efficient gradient stencils for robust implicit finite-volume solver convergence on distorted grids. J Comput Phys 386:486–501
9. Cao F, Sheng Z, Yuan G (2018) Monotone finite volume schemes for diffusion equation with imperfect interface on distorted meshes. J Sci Comput 76(3):1–23
10. Chizari H, Ismail F (2016) Accuracy variations in residual distribution and finite volume methods on triangular grids. Bull Malays Math Sci Soc 2(5):99–110
11. Ismail F, Chizari H (2017) Developments of entropy-stable residual distribution methods for conservation laws I: scalar problems. J Comput Phys 330:1093–1115
12. Ismail F, Chang WS, Chizari H (2018) On flux-difference residual distribution methods. Bull Malays Math Sci Soc 41(3):1629–1655

13. Nishikawa H (2005) Higher-order discretization of diffusion terms in residual distribution methods. In: 34th VKI CFD lecture series, very high order discretization methods, VKI Lecture Series
14. Tomaich GT (1995) A genuinely multi-dimensional upwinding algorithm for the Navier-Stokes equations on unstructured grids using a compact, highly-parallelizable spatial discretization, PhD Thesis. University of Michigan, Ann Arbor, Michigan
15. Thomee V (2006) Galerkin finite element methods for parabolic problems, 2nd edn. Springer, Berlin Heidelberg

Design and Modeling of Actuation System of Unmanned Tricopter with Thrust-Vectoring Front Tilt Rotors for Sustainable Flying

Riady Siswoyo Jo, Aaron EuGene Tan, Mark Tee Kit Tsun, and Hudyjaya Siswoyo Jo

Abstract Tilt rotor aircrafts combine the advantages of the long mileage of fixed-wing airplanes and the vertical take-off and landing (VTOL) capabilities of rotorcrafts by using tiltable rotors. However, tilt rotor aircrafts suffer from poor autorotation performance and landing stabilitiy. This work proposes an alternative actuation configuration of a tricopter tilt rotor aircraft that aims to optimize the flight duration, VTOL capabilities and stability control. The proposed configuration employs two front tilt rotors that are able to provide thrust-vectoring control with the addition of a fixed tail rotor that produces upward thrust. Dynamic modeling of the proposed actuation configuration is developed to provide insights on how different motions (hovering, ascending, descending, rotating and airplane-like flying) are achieved by manipulating the amounts of thrusts and rotor tilt angles. Preliminary prototype as well as the electronic and control systems are developed and discussed. Proposed configuration has the potential to achieve more sustainable flying.

Keywords Tricopter · UAV · VTOL · Tilt rotor · Actuation · Dynamic model · Sustainable flying · Thrust-vectoring

1 Introduction

Since the first successful pioneering works on aircrafts by the Wright brothers through their Wright Flyer experiments in the early twentieth century [1], engineers and inventors never cease to develop and improve aircraft propulsion methods, structures as well as its dynamic performance. From then on, developments of aircraft propulsion have taken two separate paths to reach what we know today as conventional fixed-wing aircrafts (airplanes) and rotorcrafts.

Forward propulsion of airplanes has the advantage of covering long mileage and high-speed flight, which makes them irreplaceable for commercial flights. Nonetheless, airplanes are inferior in terms of maneuverability as the orientation and altitude

R. Siswoyo Jo (✉) · A. E. Tan · M. Tee Kit Tsun · H. Siswoyo Jo
Swinburne University of Technology Sarawak Campus, 93350 Kuching, Sarawak, Malaysia
e-mail: rsiswoyo@swinburne.edu.my

P. Rajendran et al. (eds.), *Proceedings of International Conference of Aerospace and Mechanical Engineering 2019*, Lecture Notes in Mechanical Engineering,
https://doi.org/10.1007/978-981-15-4756-0_5

are not able to be instantaneously controlled. Rotorcrafts, on the other hand, mainly use powered rotary blades to create thrusts. Helicopter is the common form of a rotorcraft. Most rotorcrafts are able to perform VTOL (vertical take-off and landing), a feature essential to eliminate the need for runways.

Recent developments in UAV technologies and their rapidly rising usages mainly use multiple rotors to perform flights. Multirotor aircrafts generally provide higher controllability, maneuverability and require relatively simple operation to perform VTOL. Rotorcrafts, including multirotor aircrafts, however, have low cruise speed and flight endurance [2]. For similar payload and distance, rotorcrafts consume much more energy than their fixed-wing counterparts, therefore making them less sustainable.

Tilt rotor aircraft concept was then introduced in the 1930s [3] to combine the advantages of the long mileage of airplanes and the VTOL capabilities of rotorcrafts by using tiltable rotors. Tiltable rotors allow for the manipulation of the rotor thrust orientations and thus allowing thrust-vectoring control. The basic operation of a tilt rotor aircraft involves tilting the rotors to create purely vertical thrusts during take-off and landing. Once the aircraft has attained a certain altitude after take-off, the rotors are then tilted to create horizontal thrust, thus converting it to an airplane (Airplane mode flight) [2]. Tilt rotor aircrafts may not need all the rotors to be tiltable. Park et al. [4] described the three modes of tilt rotor flight modes: Helicopter mode (VTOL), Conversion mode (when the tilt angle of the rotors are in between upright and Airplane mode) and the Airplane mode.

It was not until 1989 that a tilt rotor aircraft (Bell/Boeing V-22 Osprey) successfully performed a flight. Bell/Boeing V-22 Osprey employed the configuration that uses only two front tilt rotors without additional fixed rotors [5]. Although both VTOL and Airplane mode flights are possible, the control of aircraft orientation angles during vertical hovering, take-off and landing are difficult. Tilt rotor aircrafts also suffer from poor autorotation and landing stability [6].

In the most recent years, tricopter configuration [7, 8] for tilt rotor aircrafts have become more common as it gives more control during vertical hovering, take-off and landing. These works use two fixed front rotors with tiltable tail rotor, for which the tiltable tail rotor is mainly used for yaw and pitch corrections. However, this configuration does not allow flying in Airplane mode.

There were several works dedicated to the modeling and control of tilt rotor aircrafts of different configurations. The work by Song and Wang [9] proposed a small tilt rotor aircraft with a fixed wing and two front (left and right) tiltable rotors. Another configuration [10] uses one set of tiltable coaxial rotor with a portion of the wing tilting together with the rotor. This allows for the VTOL as well the transition to a true fixed wing aircraft at the Airplane mode. A large-size quad rotor variant of tilt rotor aircraft was studied by Wang and Cai [11], whereby the tilting mechanisms of all four rotors are mechanically coupled and tilting with the same tilt angles. Recent novel tilt rotor/ tilt wing propulsion was proposed by Wang and Zhou [12] by distributing the propulsion to front wing and rear wing. The front wing was equipped with 6 propeller rotors fixed on and distributed along the front wingspan and the rear wing was equipped with 2 propeller rotors fixed on and distributed along the rear wingspan. Both front and rear wings are also tilting at the same tilt angles.

Other aspects of tilt rotor aircrafts have also been subject to research. The work by Saharudin [13] proposed the material selection and structural design and analysis for wings used in tilt rotor UAVs. Recent article [14] discusses how sustainable flying can be achieved by optimizing the flight profile from VTOL to Conversion and Airplane mode.

To this end, we propose an alternative actuation configuration of a tricopter tilt rotor aircraft that aims to allow further optimization of the flight duration, VTOL capabilities and stability control. Section 2 discusses the detailed modeling of the proposed actuation configuration. Sections 3 and 4 cover the electronic control and developed prototype arising from the proposed configuration. Section 5 concludes the work as well as discussion on further works.

2 Modeling of the Actuation System

2.1 Mechanical Configuration

The mechanical configuration of the tricopter considered in this study uses a fixed tail rotor that produces upward thrust and two thrust-vectoring front tilt rotors that produce thrusts depending on the tilt angles of the rotors. An overall illustration of this configuration is shown in Fig. 1. The three rotors are installed on a frame that sits on top of the fuselage where the electronic and control components are placed in.

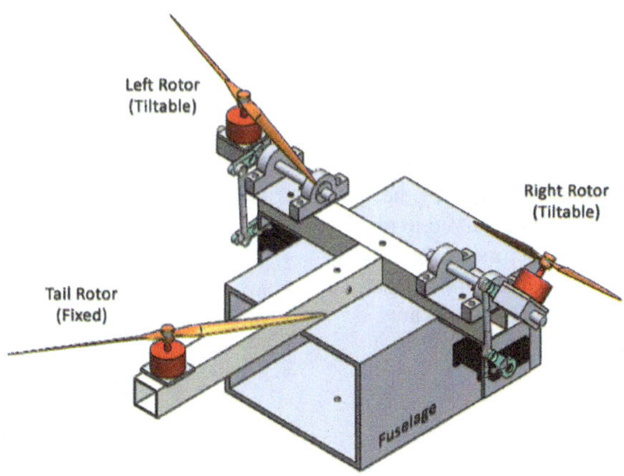

Fig. 1 Isometric view of the tricopter configuration with front tilt rotors

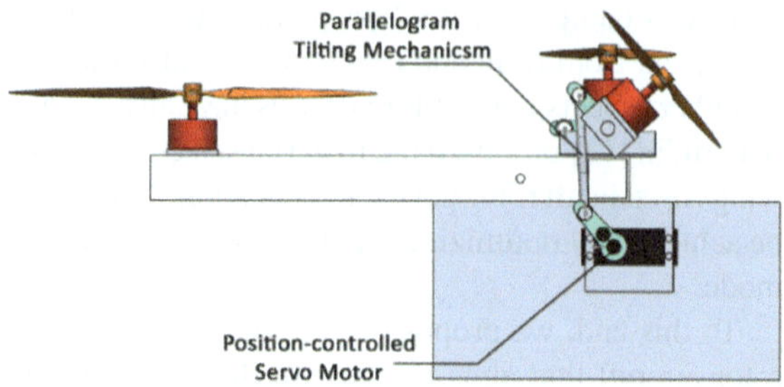

Fig. 2 Side view of the tricopter configuration with front tilt rotors

Each front rotor is installed on a tilting mechanism (see Fig. 2). The rotor carriage acts as a rocker link in the parallelogram mechanism that is driven by a position-controlled servo motor. In this case, the crank angle of the servo motor directly controls the tilt angle of the rotor.

The proposed configuration offers an alternative that combine the advantages of the two front rotors only configuration (as in V-22 Osprey) and the tilt tail rotor tricopters [7, 8] by allowing a more direct control in both VTOL and Airplane modes. The fixed tail rotor are able to generate thrust that stabilizes the aircraft especially during changes of tilt rotor angles.

2.2 *Dynamic Model of the Mechanical System*

In order to investigate the motions resulted from the thrusts produced by the rotors, a dynamic model is developed. A free-body diagram of the proposed tricopter configuration is shown in Fig. 3. In here, the inertias of the system are simplified as lumped masses/weights.

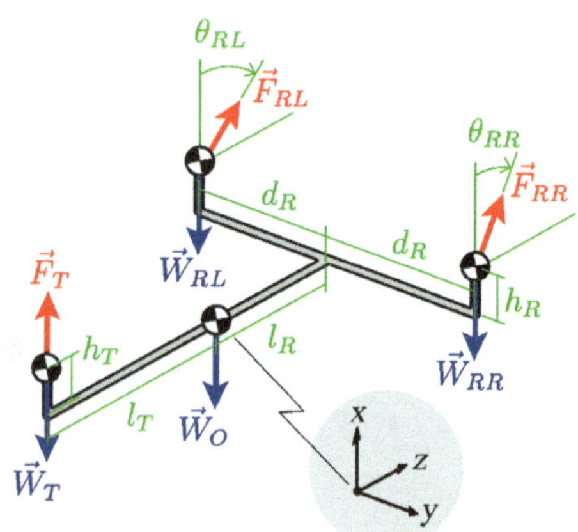

Fig. 3 Free-body diagram of tricopter with front tilt rotors. The system is simplified as lumped masses. Thrusts generated by the rotors are demarked in red. Weight vectors are demarked in blue while dimensions (lengths and angles) are demarked in green

Four weights are used in this model. $\overrightarrow{\mathbf{W}}_O$ is the total weight of the frame and fuselage excluding the rotors. $\overrightarrow{\mathbf{W}}_{RL}$ and $\overrightarrow{\mathbf{W}}_{RR}$ are the total weights of the left and right rotors including the tilting mechanisms, respectively. $\overrightarrow{\mathbf{W}}_T$ is the total weight of the tail rotor. Weight vectors are written in the matrix forms as:

$$\overrightarrow{\mathbf{W}}_i = \begin{bmatrix} m_i g \\ 0 \\ 0 \end{bmatrix}, \quad \text{for } i = O, RL, RR, T \tag{1}$$

The thrusts produced by the left rotor, right rotor and tail rotor (denoted in red in Fig. 3) are written in Eq. (2) at the home position, i.e., position of zero orientation angles and zero displacement. The tilt angles of the left and right rotors are θ_{RL} and θ_{RR}, respectively, with trigonometric functions written in short forms, i.e. $S_\alpha = \sin\alpha$ and $C_a = \cos\alpha$. Altering the tilt angles allow for thrust-vectoring.

$$\overrightarrow{\mathbf{F}}_T = \begin{bmatrix} F_T \\ 0 \\ 0 \end{bmatrix}, \quad \overrightarrow{\mathbf{F}}_{RL} = \begin{bmatrix} F_{RL} C_{\theta_{RL}} \\ 0 \\ F_{RL} S_{\theta_{RL}} \end{bmatrix}, \quad \overrightarrow{\mathbf{F}}_{RR} = \begin{bmatrix} F_{RR} C_{\theta_{RR}} \\ 0 \\ F_{RR} S_{\theta_{RR}} \end{bmatrix} \tag{2}$$

At any given aircraft orientation in terms of Euler angles (roll, pitch and yaw), the thrust orientations change while the weight vectors remain pointing downward (negative x axis). Following the coordinate frame in Fig. 1, the general orientation matrix is given by [15] as:

$$\begin{aligned} \mathbf{RPY}(\phi, \theta, \psi) &= \mathbf{Rot}(z, \phi)\mathbf{Rot}(y, \theta)\mathbf{Rot}(x, \psi) \\ &= \begin{bmatrix} C_\phi C_\theta & C_\phi S_\theta S_\psi - S_\phi C_\psi & C_\phi S_\theta C_\psi + S_\phi S_\psi \\ S_\phi C_\theta & S_\phi S_\theta S_\psi + C_\phi C_\psi & S_\phi S_\theta C_\psi - C_\phi S_\psi \\ -S_\theta & C_\theta S_\psi & C_\theta C_\psi \end{bmatrix} \end{aligned} \tag{3}$$

which results in each thrust vector to be transformed to Eq. (4) for any arbitrary orientation.

$$\overrightarrow{\mathbf{F}}_{\mathrm{RPY},k} = \mathbf{RPY}(\phi, \theta, \psi)\overrightarrow{\mathbf{F}}_k, \quad \text{for } k = RL, RR, T \tag{4}$$

By applying the principles of rigid-body dynamics, two equations of motion can be developed: linear and angular. Taking the fuselage's center of mass (COM) as the reference, the other weight and thrust vectors act as external forces acting on the fuselage. From here, the linear equation of motion is given by Newton's second law as:

$$\sum \overrightarrow{\mathbf{F}}_{\text{acting on } O} = \sum_i \overrightarrow{\mathbf{W}}_i + \sum_k \overrightarrow{\mathbf{F}}_{\mathrm{RPY},k} = m_O \begin{bmatrix} \ddot{x} \\ \ddot{y} \\ \ddot{z} \end{bmatrix} \tag{5}$$

Similarly, the angular equation of motion of the fuselage can be developed by considering the total moments produced by the external forces. The moment arm vectors are written as Eq. (6) at the home position.

$$\vec{\mathbf{r}}_O = \begin{bmatrix} 0 \\ 0 \\ 0 \end{bmatrix}, \quad \vec{\mathbf{r}}_{RL} = \begin{bmatrix} h_R \\ -d_R \\ l_R \end{bmatrix}, \quad \vec{\mathbf{r}}_{RR} = \begin{bmatrix} h_R \\ d_R \\ l_R \end{bmatrix}, \quad \vec{\mathbf{r}}_T = \begin{bmatrix} h_T \\ 0 \\ -l_T \end{bmatrix} \tag{6}$$

At any arbitrary orientation, the moment arm vectors are written as:

$$\vec{\mathbf{r}}_{\text{RPY},i} = \mathbf{RPY}(\phi, \theta, \psi)\vec{\mathbf{r}}_i \tag{7}$$

The angular equation of motion can then be expressed as Eq. (7) by considering taking sum of all moments as the time rate of change of the fuselage's angular momentum. he moment of inertias $I_{O,xx}, I_{O,yy}$ and $I_{O,zz}$ are taken at the three principal axes.

$$\begin{aligned} \sum \vec{\mathbf{M}}_{\text{acting on } O} &= \sum_i \text{cross}\left(\vec{\mathbf{r}}_{\text{RPY},i}, \vec{\mathbf{W}}_i\right) + \sum_k \text{cross}\left(\vec{\mathbf{r}}_{\text{RPY},k}, \vec{\mathbf{F}}_{\text{RPY},k}\right) \\ &= \begin{bmatrix} I_{O,xx}\ddot{\psi} + \left(I_{O,zz} - I_{O,yy}\right)\dot{\theta}\dot{\phi} \\ I_{O,yy}\ddot{\theta} + \left(I_{O,xx} - I_{O,zz}\right)\dot{\phi}\dot{\psi} \\ I_{O,zz}\ddot{\phi} + \left(I_{O,yy} - I_{O,xx}\right)\dot{\psi}\dot{\theta} \end{bmatrix} \end{aligned} \tag{8}$$

The cross() operator is a vector cross product operation in the matrix form, which is given by Liu and Trenkler [16] as:

$$\mathbf{A} \times \mathbf{B} = \text{cross}(\mathbf{A}, \mathbf{B}) = \begin{bmatrix} 0 & -A_z & A_y \\ A_z & 0 & -A_x \\ -A_y & A_x & 0 \end{bmatrix} \begin{bmatrix} B_x \\ B_y \\ B_z \end{bmatrix} \tag{9}$$

2.3 Common Actuating Actions

The dynamic model given in terms of the equations of motion Eqs. 5 and 8 has provided insights on how different motions can be produced by the proposed actuation configuration.

Table 1 summarizes the common actuating actions that can be achieved by varying the thrusts and rotor tilt angles. Here, hovering thrusts refer to the thrusts required for all three rotors to maintain the tricopter's hovering altitude without changing its orientation. Increased thrust and decreased thrust, therefore, refer to thrusts greater and lesser than the hovering thrust, respectively.

Table 1 Common actuating actions

Action	Left rotor	Right rotor	Tail rotor	Rotational axis
Hover	Upright tilt Hovering thrust	Upright tilt Hovering thrust	Hovering thrust	N/A
Ascent	Upright tilt Increased thrust	Upright tilt Increased thrust	Increased thrust	N/A
Descent	Upright tilt Decreased thrust	Upright tilt Decreased thrust	Decreased thrust	N/A
CCW yaw	Increase backward tilt Increased thrust	Increase forward tilt Increased thrust	Hovering thrust	About X+
CW yaw	Increase forward tilt Increased thrust	Increase backward tilt Increased thrust	Hovering thrust	About X−
Up pitch	Upright tilt Increased thrust	Upright tilt Increased thrust	Hovering thrust	About Y+
Down pitch	Upright tilt Decreased thrust	Upright tilt Decreased thrust	Hovering thrust	About Y−
CW roll	Upright tilt Increased thrust	Upright tilt Decreased thrust	Hovering thrust	About Z+
CCW roll	Upright tilt Decreased thrust	Upright tilt Increased thrust	Hovering thrust	About Z−
Airplane mode	Forward tilt thrust for flight speed	Forward tilt thrust for flight speed	Thrust for orientation	N/A

Combination of the three thrust magnitudes and two rotor tilt angles allows motion control of the tricopter's orientation, altitude and speed. Pure linear ascending and descending motions can be achieved by keeping the total moment to be zero while keeping the front rotors to be in upright positions.

The Airplane mode flight is performed by keeping the front rotors to tilt forward from its initially upright position. In the Airplane mode, the tail rotor acts to compensate for the orientation, while the thrusts from the front rotors provide the traverse speed.

3 Electronic and Control System

The three rotors of the tricopter are brushless DC motors attached to propeller blades. The electronic and control system of the tricopter is shown in Fig. 4. The main interfacing controller of the system is HKPilot32 Flight Controller running on 32-bit STM32F427 Cortex M4 ARM-based processor. HKPilot 32 interfaces the actuators, telemetry and user input device.

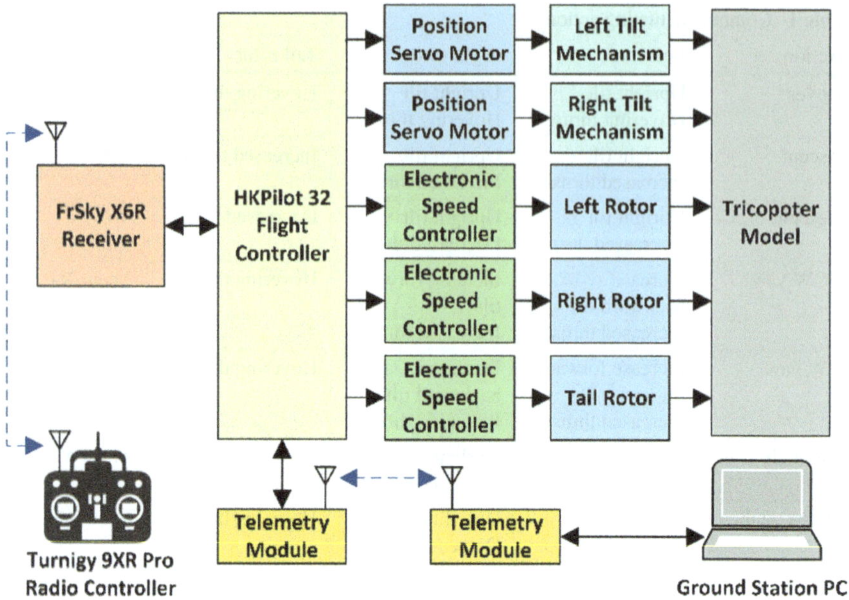

Fig. 4 Block diagram of the electronic and control system of the tricopter with front tilt rotors

Turnigy 9XR Pro Radio Controller is a user interface device where its joystick inputs are mapped to the amount of thrusts to each rotors as well as the rotor tilt angles. The radio controller information is transmitted to FrSky X6R receiver wirelessly. The received information is sent to HKPilot32 using SBUS interface.

The information at the HKPilot32 controller are monitored at the Ground Station PC, which are transmitted wirelessly through Telemetry modules. The HKPilot32 input and output device configurations and calibrations can also be performed at the Ground Station PC.

Each brushless DC motor is driven by an Electronic Speed Controller, where the control signals are provided by HKPilot32. The tilting mechanisms are driven by position-controlled servo motors that receive their control signals from HKPilot32 as well.

4 Developed Prototype and Results

4.1 Prototype and Design Parameters

Figure 5 shows the developed tricopter prototype based on the proposed front-tilt rotor configuration. The mechanical, electronic and control systems are integrated and the prototype has an overall mass of 1.5 kg. At zero tilt angles, the three rotor

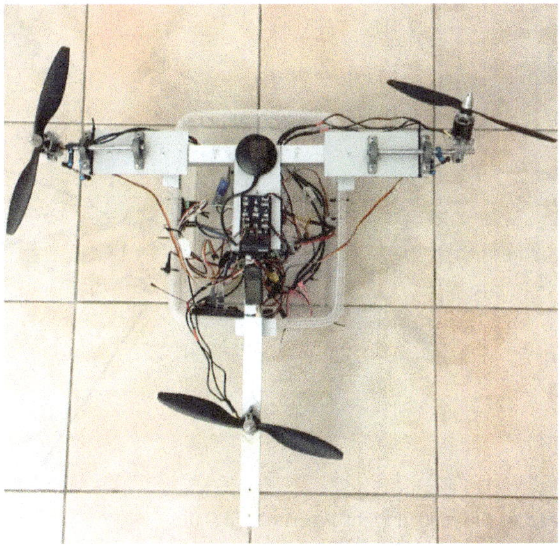

Fig. 5 Developed prototype

shafts make an equilateral triangle with the side length of 600 mm. The left, right and tilt rotors are NTM Prop Drive 28-30S brushless motors. The propeller blades used are APC 9X6E propellers. This combination of rotor and propeller gives a thrust of 1.05 kg per rotor.

4.2 Suspended Flight Test

In order to test the functionality and actuation system of the tricopter, a suspended flight experiment (Fig. 6) is carried out. The tricopter fuselage was suspended at its four corners with polyester ropes. Different amounts of rotor thrusts and tilt angles were applied and the ability of the developed tricopter prototype to tend to produce the motions as tabulated in Table 1 were confirmed.

5 Conclusion and Future Work

In this paper, the mechanical configuration, dynamic modeling, actuation, electronic and control systems of a front tilt rotor tricopter have been discussed. The formulated rigid-body dynamic model has provided insights on how different motions of the tricopter can be produced.

The mechanical configuration proposed in this work provides the advantage of VTOL aircraft and also rather simple transition to Airplane mode flying by varying rotor tilt angles.

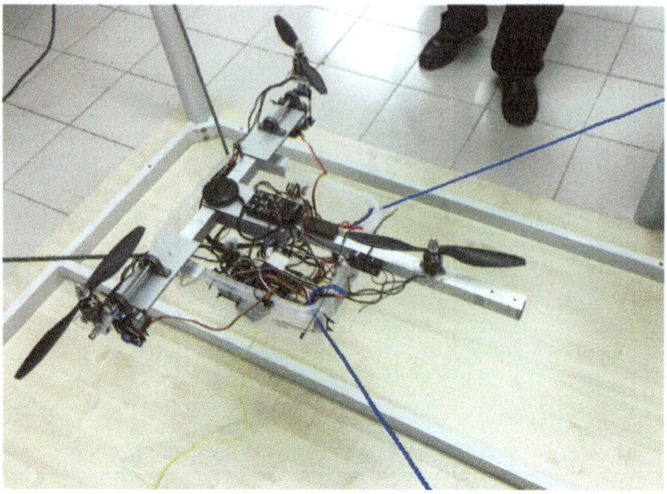

Fig. 6 Suspended flight experimental setup

Current work considers how common actuation (linear and rotational) motions can produced by varying the thrusts and rotor tilt angles, which is a forward dynamic problem. An inverse dynamic model that is able to compute the necessary rotor thrust and tilt angle outputs from a desired flight trajectory, however, still remains a challenging task. Furthermore, the combination of the inverse dynamic model and appropriate control system requires further studies and therefore an essential future work.

Successful implementation of the front tilt rotor tricopter has the prospects of potential applications in miniature level, enterprise drone solution or even commercial aircrafts by striking the right balance to provide maneuverability and flight duration, and hence promote more sustainable flying practices.

References

1. Culick FC (2003) The Wright brothers: first aeronautical engineers and test pilots. AIAA J 41:985–1006
2. Liu Z, He Y, Yang L, Han J (2017) Control techniques of tilt rotor unmanned aerial vehicle systems: a review. Chin J Aeronaut 30(1):135–148
3. George L (1930) Flying machine. US Patent 1,775,861
4. Park S, Bae J, Kim Y, Kim S (2013) Fault tolerant flight control system for tilt-rotor UAV. J Franklin Inst 350:2535–2559
5. Markman S, Holder B (2000) Bell/boeing V-22 Osprey tilt-engine VTOL transport (USA). Straight up: a history of vertical flight
6. Tan W, Li Y, Miao Y, Lv H (2014) The Analysis of airworthiness issues influence to the development of tiltrotor aircraft. Proc Eng 80:602–608
7. Ansari A, Prach A, Bernstein DS (2017) Adaptive trim and trajectory following for a tilt-rotor tricopter. In: 2017 American control conference (ACC), pp 1109–1114

8. Mehndiratta M, Kayacan E, Kayacan E (2018) A simple learning strategy for feedback linearization control of aerial package delivery robot. In: 2018 IEEE conference on control technology and applications (CCTA), pp 361–367
9. Song Y, Wang H (2009) Design of flight control system for a small unmanned tilt rotor aircraft. Chin J Aeronaut 22:250–256
10. Rawal NA (2015) Modelling and analysis of a coaxial tiltrotor. In: 4th Asian/Australian rotorcraft forum, pp, 143–163
11. Wang X, Cai L (2015) Mathematical modeling and control of a tilt-rotor aircraft. Aerosp Sci Technol 47:473–492
12. Wang Y, Zhou Y, Lin C (2019) Modeling and control for the mode transition of a novel tilt-wing UAV. Aerosp Sci Technol 91:593–606
13. Saharudin MF (2016) Development of tilt-rotor unmanned aerial vehicle (UAV): material selection and structural analysis on wing design. IOP Conf Ser Mater Sci Eng 152(1):012017
14. Kasliwal A, Furbush NJ, Gawron JH, McBride JA, Wallington TJ, De Kleine RD, Kim HC, Keoleian DA (2019) Role of flying cars in sustainable mobility. Nat Commun 10(1555):1–9
15. Spong MW, Vidyasagar M (2008) Robot dynamics and control. Wiley
16. Liu S, Trenkler G (2008) Hadamard, Khatri-Rao, Kronecker and other matrix products. Int J Inf Syst Sci 4:160–177

Design of an Air-Duct Pipe Inspection and Cleaning Robot

Adzly Anuar, Juniza Md Saad, and M. N. A. Arimi

Abstract Air-ducts are one of the important elements in a building for making the living environment safe and healthy by ensuring smooth flow of fresh air, often cool air-conditioned, into rooms and spaces inside buildings. Often, over time, the air-ducts and vents can be unclean with dusts, dirt and contaminants. This can affect the performance and efficiency of the heat, ventilation and air conditioning (HVAC) system, and more importantly the people's health. Manual cleaning of the air duct usually very challenging, tedious and time consuming. Considering these conditions, an in-pipe robot that can move along air-duct piping to carry out inspection and cleaning work is designed. This paper describes the design process that was carried out, started out with developing the design requirements and intended functions of the robot. A 3D model was developed using Creo software and stress analysis were conducted. The final design of the robot was produced, utilizing Arduino Uno as the main microcontroller, eight servomotors for the actuators and multiple pressure sensors are used to assist in the navigation and motion.

Keywords Air-duct · Inspection robot · Cleaning robot · 3-D printing · In-pipe robot

1 Introduction

Air-ducts and vents can be very unclean over time with dusts, dirt and contaminants. It can affect the performance and efficiency of the heat, ventilation and air conditioning (HVAC) system and the people's health. Less efficient means more time, energy and money wasted. This is why cleaning the duct is important. Air-ducts come in different shapes and sizes. Either the shape is a circle or rectangle, which create the challenge in traversing through and carry of the cleaning process. The entrance of the air duct

A. Anuar · J. Md Saad (✉)
Institute of Sustainable Energy, Universiti Tenaga Nasional, 43000 Kajang, Malaysia
e-mail: juniza@uniten.edu.my

M. N. A. Arimi
Department of Mechanical Engineering, Universiti Tenaga Nasional, 43000 Kajang, Malaysia

P. Rajendran et al. (eds.), *Proceedings of International Conference of Aerospace and Mechanical Engineering 2019*, Lecture Notes in Mechanical Engineering,
https://doi.org/10.1007/978-981-15-4756-0_6

is also a challenge to accomplish. The robot is required to fit into the opening of the air duct and also remain lightweight.

The aim of this research is to design a small robot that could be used to inspect and clean air duct piping system. The mechanical design of the robot is challenging due to the air duct pipe may vary in size, orientation and arrangement. The entry point for the air duct can also be challenging in terms of geometry and size. Based on the review conducted on air duct cleaning robots, it was found that there are several different ways of cleaning [1–7], which are grouped as manually cleaning by hand, use of a controlled robot with spinning brush or air compressor and use of a controlled robot with dry ice.

2 Literature Review

2.1 Locomotion and Motion Strategy

The most commonly used locomotion in in-pipe inspection robots are wall-pressed, wheel or track-based and snake or inch-worm type. The combination of different type of locomotion are also introduced, known as hybrid locomotion. There are three types of hybrid locomotion systems, which are caterpillar wall-pressed type, wheeled wall-pressed type and wheeled wall-pressing screw type. Each system is design for a specific task. Based on the research, wall pressed type is the most popular locomotion system in in-pipe robot related. Most of the prototypes are able to travel into branches with the same diameter as the pipe. The advantage of caterpillar wheel is preventing motion singularity problem while surpassing branches, while wheeled wall-pressed type is fast in pipe and wheeled wall pressing screw type can navigate the best in curved pipe [8].

A recent movement strategy for an in-pipe inspection robot utilising two mechanisms called Adaptable Quad Arm Mechanism (AQAM) and the Swivel Hand Mechanism (SHM). This would allow the robot to travel through many pipe configurations including branch, elbow and vertical pipes with tight curvature. The SHM allows the robot to change its positioning, and to be exact, allows it to bypass bumps [9].

An in-pipe crawling robot containing a vehicle body movable inside piping along the axis, a couple devices in front and rear of the vehicle body. Each of the devices has a pair of wheels and are steerable about a vertical axis of the vehicle body. When the robot travels in a circular direction inside the piping, the vehicle body is ready and set to a posture to travel in a circular direction. This results in stable travel in a circular direction of the piping [10].

Jetty robot [11] is one of the commercially available system used for inspecting and cleaning air ducts and vents or any place needed to clean that is difficult to access. The robot basically uses wall-press method as the main locomotion system. It consists of multiple tracked arm which can expand and contract in order to fit in the duct size snugly and to pass through. This design allows it to pass through circular,

Table 1 Summary of comparison

Feature	US5392715 [10]	Jetty robot [11]	Danduct [12]
Min and Max size of duct (cm)	Not specified	D36–D130	D40–D120
Weight (kg)	Not specified	35	13
Types of cleaning method	Not specified	Dry ice	Brush and compressed air
Camera	No	Yes	Yes
Wired or wireless	Not specified	Wired	Wired
Suspended inside the duct	No	Yes	No
Controller	Yes	Yes	Yes
Changeable size of duct	Not specified	Yes	Yes

rectangular or square-shaped ducts and it is claimed that it can be used in vertical, horizontal, sloping or S-shaped ducting. It has camera to scan and monitor the air duct while carrying out the cleaning process.

Danduct Multi-Purpose Robot [12] is another example of existing system that is used for cleaning air duct. It has two cameras that can change the view from front to rear. The robot uses wheel locomotion and it is claimed that it can go into curves smoothly due to its 4-wheel drive feature. Rotating brush located at the front of the robot is used to sweep or brush dust inside the air duct. The robot comes with dust extraction fan and filters.

Table 1 shows the comparison of the findings. The findings were compared by the size of air-duct it is able to fit, weight, cleaning methods, camera usage, wired, suspended in the air-duct, controller usage, and changeable size of the duct.

It is concluded that due to difficulties in maneuvering inside the air duct, an automated cleaning mechanism is needed. There are still room of improvement on the existing cleaning robot especially when involving different size of duct and minimizing the size and weight of the designed mechanism.

2.2 Component Review

Component selection is played a major role in ensuring the functionality of the robot fulfilling the selected design requirements. As such, all the feasible options need to be reviewed to ensure the good component selection is carried out. Table 2 shows the major components for the robot and the discussion on the selection.

Table 2 Major components of the robot

Component	Description
Motor	Geared DC motor will be used to drive the robot due to its high torque. Small DC motor will be used to spin the cleaning brush to clean the duct. Micro servomotor will be used for arm movement due to position constraint
Micro-controller	Arduino UNO will be used to control the robot because it is easy to program, an open source product with a large community to help one another and easy to get
Controller for the robot	A wireless PlayStation 2 controller will be used to control the robot because it is cheap and easy to program due to the existing library in the Arduino
Motor driver	Motor driver is needed to control the motors from the Arduino. L298N motor driver will be used as it is cheap and easy to use
Camera	Camera is needed for inspection purposes. Camera that is compatible with Arduino should be chosen to make it easy to program
Sensor	Force sensitive sensor will be used. It is relatively cheap and easy to use

3 Design

As there is a wide variety of the air duct size, shape, orientation and entry geometry, it was decided to narrow down the scope as below:

i. Geometry: circular
ii. Diameter: 25–35 cm
iii. Entrance: direct horizontal or perpendicular upward
iv. Entrance geometry: circular, diameter 25 cm or bigger
v. Cleaning strategy: dry method (spinning brush).

3.1 Functional and Design Requirements

The main function of the robot is to inspect and clean air-duct pipe. It is expected that the robot will be able to perform the following functions:

i. Able to enter and exit the air-duct piping
ii. Able to traverse or move along the pipe
iii. Able to carry out cleaning process inside the pipe.

Based on the required functions and literature review conducted, the following design requirements as shown in Table 3 were developed.

Table 3 Main design requirements

Design requirement	Description
Robot weight	The weight of the robot that will be in the air duct should be between 1 and 3 kg
Adjustment to different sizes of air duct	The robot must be able to adapt to a different sizes of air duct
Turning motion	The robot must be able to turn inside the air duct
Air duct cleanliness	The robot must be able to clean the air duct
Robot control	The robot must be able to control by the user
Process monitoring	The robot must be able to monitor the process of cleaning of the air duct
Required component	The robot must have less component. Reducing the components will reduce the cost, complexity and weight of the robot

3.2 Design Concepts

At the start of the design process, several brainstorming sessions were conducted to generate design concepts based on the required functions and design requirements.

Figure 1 shows the first design concept, uses wall-press approach for locomotion and adapting to changes of direction or size of the air-duct, by having extendable arms

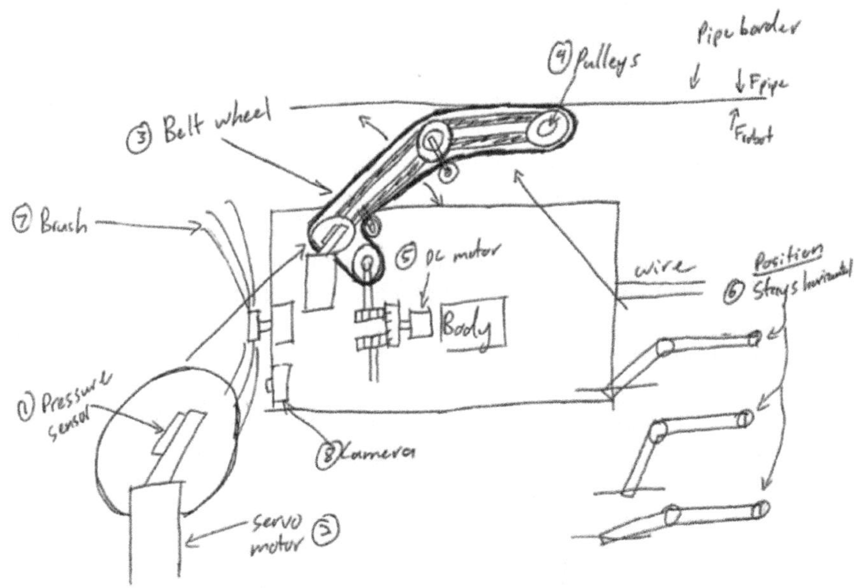

Fig. 1 Sketch of concept design 1

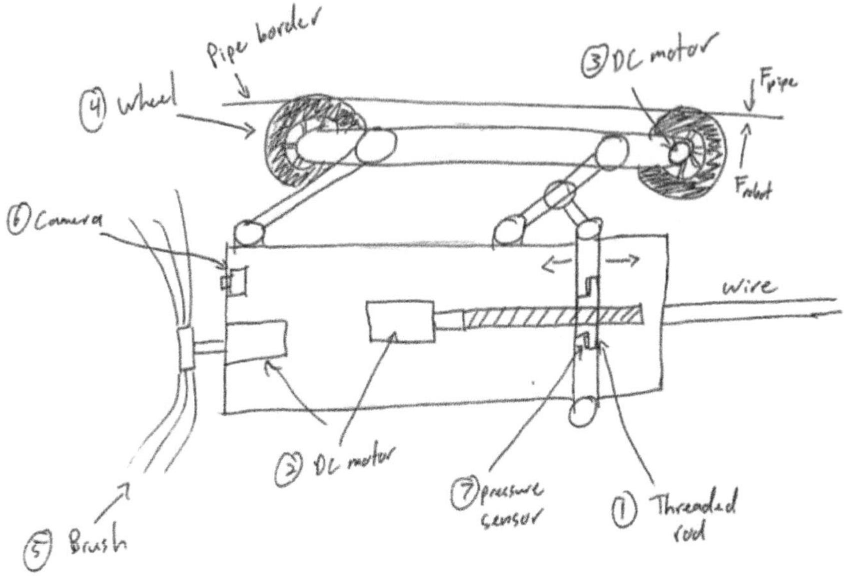

Fig. 2 Sketch of concept design 2

on the side of the robot body. Belt will be attached to each arm providing the traction to move the robot, powered by a single DC motor through small gearing system. The extendibility of each arm is actuated by servo motor, and pressure sensor is used to detect if the arm is sufficiently gripping the inner wall of the pipe.

The second concept as shown in Fig. 2 is using the threaded rod to change the rotation motion of the motor to a linear motion. This is to control the parallelogram mechanism simultaneously. DC motor will be used to control the arm's extension, the cleaning brush, and as a driving source for the four moving arms. Similar as the first concept, the extendable mechanism for the moving arms will be controlled by pressure sensor to ensure they fit in the air duct. The addition of wheels and camera will smoothen the movement in the air duct, also monitoring the cleaning process.

Figure 3 illustrates the final concept sketching in the design process. The locomotion of the final concept, the driving source and cleaning attachments are similar with the previous two concepts. The servo motor and pressure sensor is used to control the movement of the arms in order to make contact with the wall of the air duct. The concept also uses DC motor to control the spinning motion of the brush and servo motor to align the brush at specific angle as required.

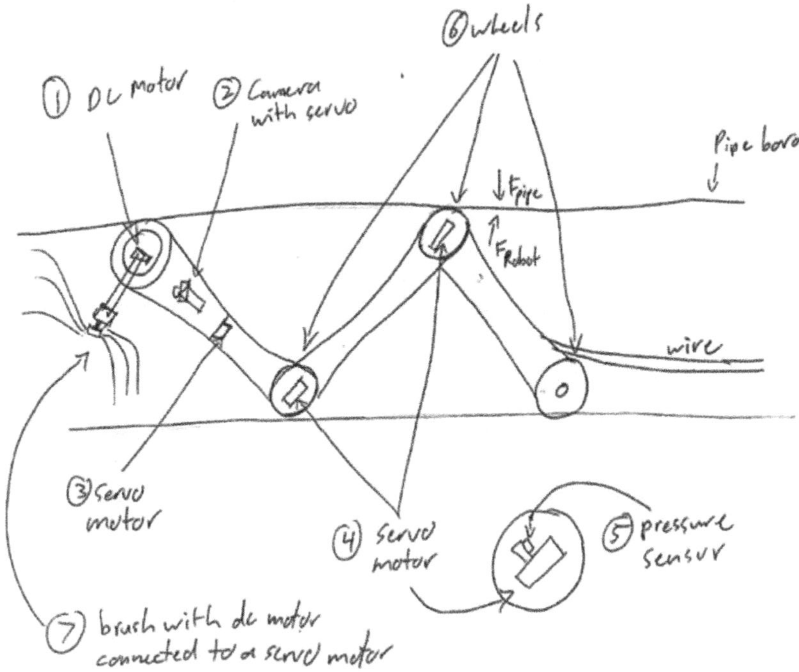

Fig. 3 Sketch of concept design 3

3.3 Concept Selection and Refinement

From the concept designs, the camera, brush and DC motor for the brush was selected because it is the base design for the robot which is to clean the air duct while monitoring it. For the arms, concept design 1 was chosen because it is more stable to have a large surface area for the robot to roll in the air duct. The way to extend the arms are by a servo motors in each of the arms integrated with a pressure sensor to adjust the arm's diameter with the air duct. To drive the robot, four DC motor for each arm are chosen so that the robot can turn easily in the air duct. The sketch of the refined concept design is as shown in Fig. 4.

3.4 Linkage Design

3D design of the first prototype of one arm was made using Creo. The arm was designed for one unit and using rapid prototyping to test the mechanisms and the programming. This will result in less time of error detection than to prototype all four of the arms together. Figure 5 shows the 3D model of one arm. The links are labeled Link 1, Link Top, Link Motor and Link Work. A servo will be attached to

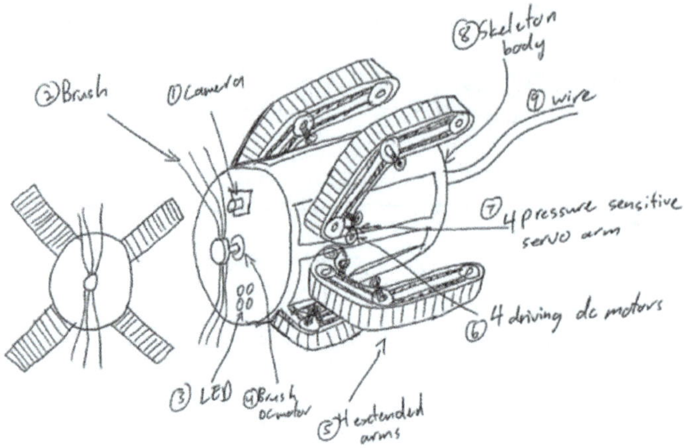

Fig. 4 Sketch of selected and refined concept

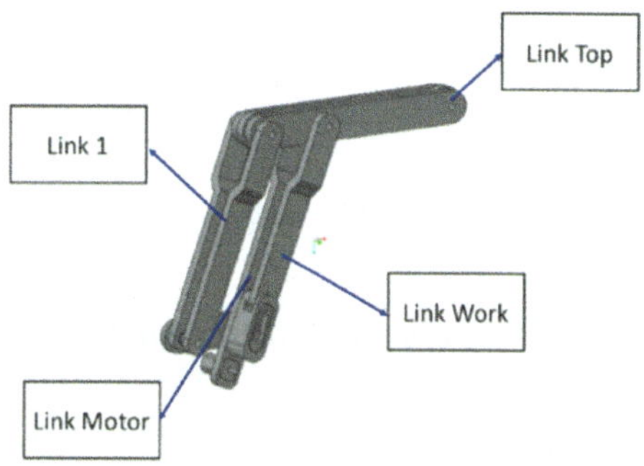

Fig. 5 3D model of the arm

the Link Work to actuate the movement of the link for contacting with the wall of the air duct. A DC motor will be attached to the bottom of Link Motor to drive the belt for moving the robot. A force sensor will be attached in the Link Work to sense the force on the link.

3.5 *Stress Analysis*

Stress analysis is conducted to ensure the link can withstand the expected loading. Figure 6a shows the result of stress analysis on the top link. It can be seen from the figure that the most critical part is in the middle part of the link representing the pivot. The most affected area for the motor link seems similar as the top link as shown in Fig. 6b. Both links showed that the pivot may affect the strength of the links. However

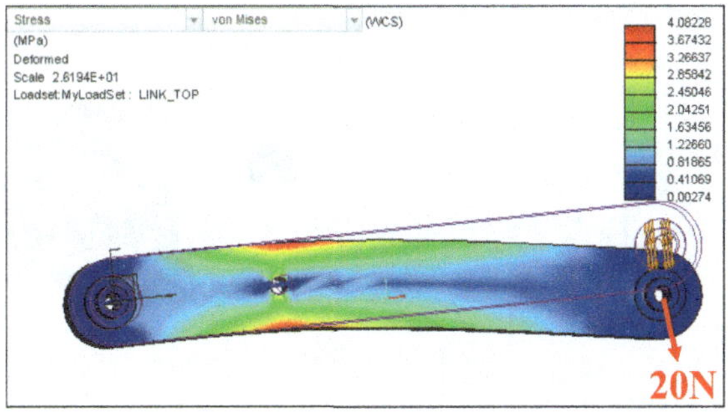

(a)

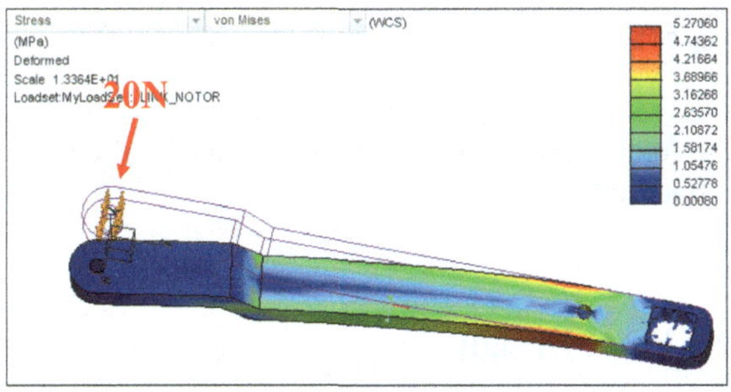

(b)

Fig. 6 **a** Top link and **b** Motor link

based on the data presented, it showed that both links able to withstand the external force.

3.6 Circuit Design

Figure 7a, b shows the block diagram and the circuit design of the arm. Two inputs which is the button and force sensor were connected to a microcontroller which is an Arduino and from there, a micro servo was connected for the output of the arm's actuator.

3.7 The Algorithm

Figure 8 shows the flowchart of the programming. The expected output of the program will be; if the button is pressed, the arm will extend until a force is exerted on the arm. If a force is exerted to the arm which is by placing a hand, the arm will stop extending further maintaining constant force between the arm and the hand. If the hand presses down further (creates more force), the arm will retract back to maintain

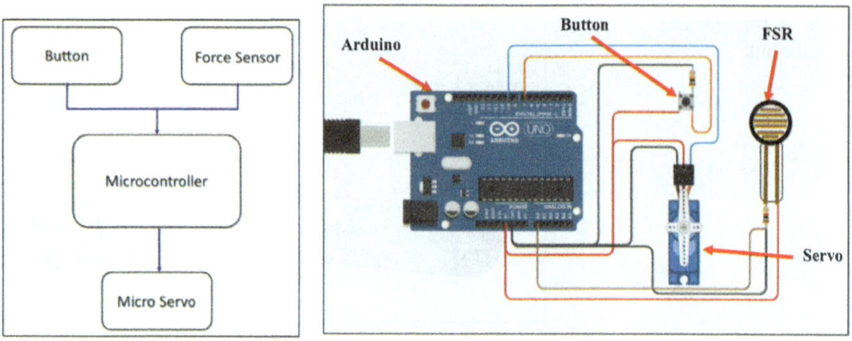

(a) (b)

Fig. 7 **a** Simple block diagram and **b** Circuit design [13]

the force between the arm and the hand. If the button is not pressed, the arm will retract fully.

4 Final Design

Figure 9 shows the final design of the air-duct inspection and cleaning robot. It consisted of a central body which houses the motors and electronic boards. Four arms are attached to the body at 90° to each other. These arms are capable to be extended to snugly fit the diameter of the air-duct. The designed robot is expected to weight about 5 kg including all electronics components, which much lower than current existing robots.

5 Conclusion

The aim of this project is to design and make a prototype of a robot that can clean air ducts and ventilation. This robot was made to help ease the work of cleaning the air ducts. Literature review was made by researching the method of cleaning of the air duct, the robot and the components. It is found that there are many ways to clean the air ducts and also the design of the robot varies from a simple car like design to an extension arm design. Concept designs were made to have a final design concept. Final design concept was obtained from the three conceptual designs. A 3D design of the prototype was successfully designed with Creo. The stress analysis showed that the robot linkage would not fail under expected working condition. The robot is designed to be able to traverse along a circular shaped duct due to the four extended arms. The next step is to produce the prototype of the robot. It is proposed to 3D

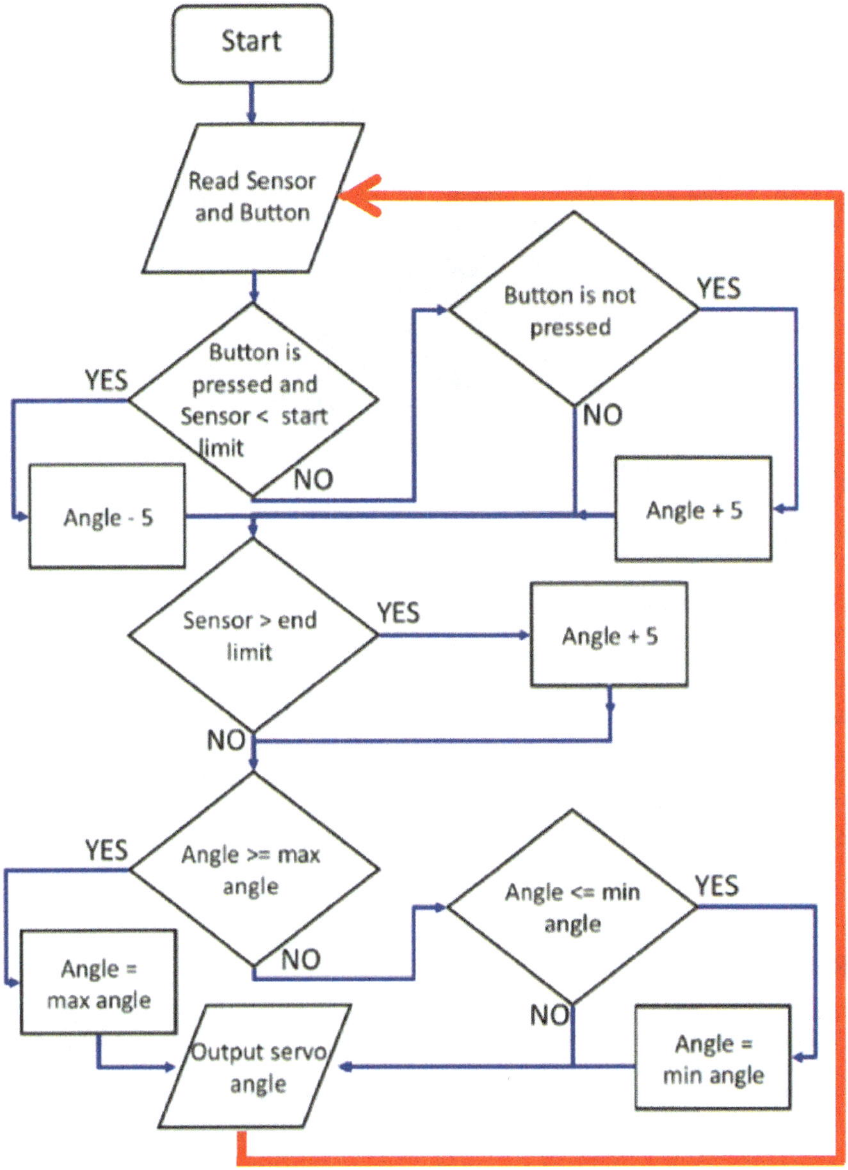

Fig. 8 Flowchart of the algorithm

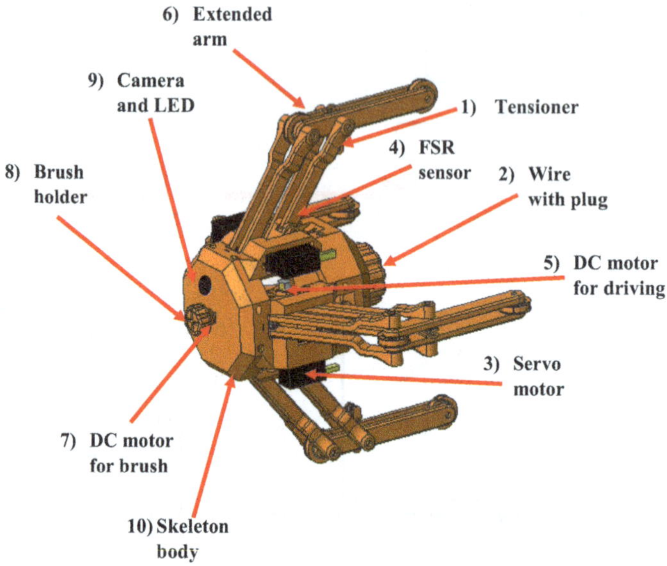

Fig. 9 Fully assembled final design

print the links of each arm and the body frame so that it can be fully tested in an air duct pipe.

Acknowledgements The authors would like to thank all who has providing assistance for this project. The work reported in this paper is partially taken from an undergraduate final year project for Bachelor of Mechanical Engineering at Universiti Tenaga Nasional.

References

1. Wang Y, Zhang J (2006) Autonomous air duct cleaning robot system. In: 49th IEEE international midwest symposium on circuits and systems, San Juan, pp 510–513 (2006)
2. Jeon SW, Jeong W, Park D, Kwon SB (2012) Design of an intelligent duct cleaning robot with force compliant brush. In: 12th international conference on control, automation and systems, JeJu Island, pp 2033–2037
3. Yamamoto M, Enatsu Y, Mohri A (2004) Motion analysis of a cleaner robot for vertical type air conditioning duct. In: IEEE international conference on robotics and automation, vol 5. New Orleans, LA, USA, pp 4442–4447
4. Truong-Thinh N, Ngoc-Phuong N, Phuoc-Tho T (2011) A study of pipe-cleaning and inspection robot. In: International conference on robotics and biomimetics, Phhuket
5. Yoon-Gu K, Dong-Hwan S, Jeon-Il M, Jinung A (2011) Design and implementation of an optimal In-pipe navigation mechanism for a steel pipe cleaning robot. In: 8th international conference on ubiquitous robots and ambient intelligence, Incheon
6. Ye C, Ma S, Li H, Yang J (2010) Development of a pipe cleaning robot for air conditioning system. In: IEEE international conference on robotics and biomimetics, Tianjin, pp 1525–1529

7. Li Z, Zhu J, He C, Wang W (2009) A new pipe cleaning and inspection robot with active pipe-diameter adaptability based on ATmega64. In: 9th international conference on electronic measurement & instruments, Beijing, pp 2-616–2-619
8. Roslin N, Anuar A, Jalal M, Sahari K (2012) A review: hybrid locomotion of In-pipe inspection robot. Proc Eng 41:1456–1458
9. Lee D, Park J, Hyun D, Yook G, Yang H (2012) Novel mechanisms and simple locomotion strategies for an In-pipe robot that can inspect various pipe types. Mech Mach Theory 56:52–63
10. Pelrine R, Park M (1995) In-pipe running robot and method of running the robot. U.S. Patent 5 392 715
11. ROBOT|Jetty Robot. http://www.jettyrobot.com/jetty-system/robot/. Last accessed 7 Sept 2017
12. The most advanced duct cleaning robot for ventilation hygine: Danduct clean. http://www.danduct.com/multi-purpose-robot. Last accessed 7 Sept 2017
13. Tinkercad| Create 3D digital designs with online CAD. https://www.tinkercad.com/. Last accessed 19 Nov 2017

Investigation of Variable Geometry Intake to Mitigate Unwanted Shock-Shock Interactions in a Hypersonic Air-Breathing Propulsion Device

Azam Che Idris, Mohd Rashdan Saad, and Konstantinos Kontis

Abstract Recently there has been a resurgence of interest in achieving practical hypersonic flight especially since China and Russia has announced successful demonstration of their military technology. The key enabler for hypersonic is the scramjet engine as the air breathing propulsion device. Engineering design of a scramjet intake is crucial since unwanted shock-shock interactions inside the intake could impede safe combustion process. In this study, we demonstrate that minimal geometry modification of the intake could significantly increase its performance. A generic scramjet intake that performs sub optimally is shown to benefit greatly by having its cowl deflected inwards at small degree. Pressure-sensitive paint measurement showed that peak pressure due to adverse shock-shock interactions at cowl tip has been reduced significantly.

Keywords Hypersonic · Scramjet · Pressure sensitive paint

1 Introduction

Hypersonic flight for both civil and military application has been the holy grail of aviation since the early days of aviation. Even before the supersonic flight has been demonstrated, researchers have already postulated the concept of hypersonic flight, with free stream flight Mach number of more than 5. At such speed, the hypersonic plane would need to rely on the novel concept of propulsion called the supersonic combustion ramjet (scramjet) which can compress and combust air at high Mach number [1].

A. Che Idris (✉)
Universiti Brunei Darussalam, BE1410 Negara Brunei Darussalam, Brunei
e-mail: azam.idris@ubd.edu.bn

M. R. Saad
Universiti Pertahanan Nasional Malaysia, Kuala Lumpur 57000, Malaysia

K. Kontis
University of Glasgow, Glasgow G128QQ, Scotland, UK

P. Rajendran et al. (eds.), *Proceedings of International Conference of Aerospace and Mechanical Engineering 2019*, Lecture Notes in Mechanical Engineering,
https://doi.org/10.1007/978-981-15-4756-0_7

The scramjet engine relies only on its forebody, the compression ramp/cone intake, for compression, unlike the normal jet engine which has a dedicated compressor for that purpose. This fact is the reason why the performance of the engine depends heavily on the shape of the intake [2]. Many studies have been performed to study the performance of a scramjet intake but most findings were hampered by low-quality flow visualization and limited quantitative data [3–7].

Recently, we have introduced the technique of "final shock" analysis which can provide high-quality flow visualization and performance measurement at the same time [8–10]. Using the method of pressure-sensitive paint, we are able to study the complex shock-boundary layer interactions prevalent in the typical intake compression ramp in order to plan a strategy for flow control in hypersonic regime [11, 12]. Inspired by the drag-reducing riblets structure on shark skin [13], we have improved the performance of a scramjet intake using sub-boundary layer vortex generator placed strategically on the compression ramp [14]. In this latest development of our hypersonic propulsion research campaign, we will show that a small geometrical variation of the intake could significantly affect the shock-shock interaction and thus improve its overall performance.

2 Methodology

2.1 Experimental

The experiments were conducted in the High Supersonic Tunnel in the University of Manchester which is capable to run up to Mach 6 for test duration close to 10 s. In our current setup, the stagnation pressure and total temperature were set to 6.45 bar and 375 K respectively, producing free stream Mach number of around 4.95 with Reynolds number of $13 \times 10^6\ m^{-1}$. A tricolor-wheel consisting of red, green and blue segments was used instead of a slit in a typical Toepler's Z-setup schlieren. The introduction of color would aid in distinguishing the different flow feature that will be observed unlike the more traditional black-white color scheme of a typical schlieren setup.

A generic scramjet intake has been designed with a modular feature allowing for the changing of the cowl segment (see Fig. 1). In this study, we are using the normal cowl (Baseline Case), cowl with 3-degrees inward deflection (Cowl-3 Case) and cowl with 5-degrees inward deflection (Cowl-5 Case) (see Fig. 2).

The setup of pressure-sensitive paint measurement is shown in Fig. 3. This is the exact same setup used in our previous study and details can be found there [8]. The general idea is to paint the internal wall of the scramjet intake with pressure-sensitive paint with the excitation source and camera are peeking through the side window of the wind tunnel working section.

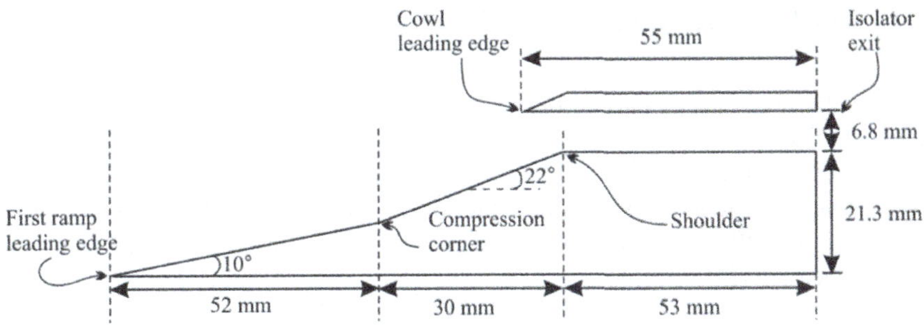

Fig. 1 Sketch of 2D scramjet intake with normal cowl (Baseline Case)

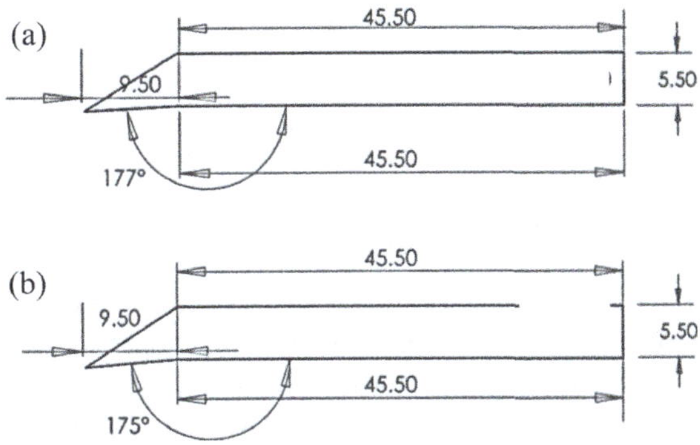

Fig. 2 Sketch (**a**) cowl with 3-deg inward deflection (Cowl-3 Case) and (**b**) cowl with 5-deg inward deflection (Cowl-5 Case)

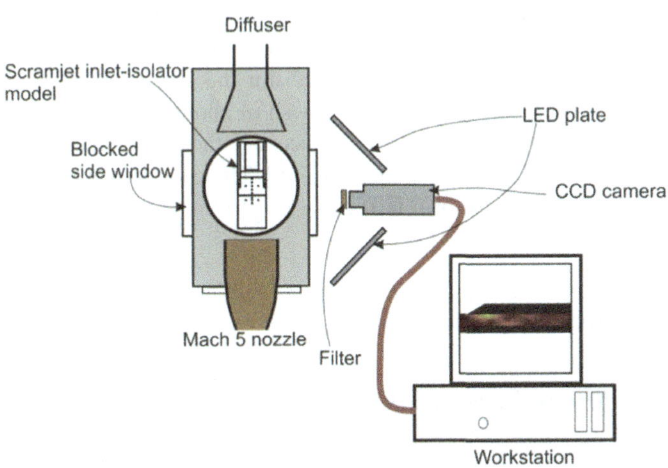

Fig. 3 Setup of PSP system to measure pressure contour on the intake sidewall

2.2 Intake Performance Calculation

The "final shock" analysis introduced in our previous article is briefly presented here [8]. The idea is to measure the rise of pressure across the "final shock" of a shock-train inside a scramjet inlet-isolator. In combination with the shock angle, the stagnation pressure and total temperature at the end of the isolator section can be calculated. Finally, the intake kinetic energy efficiency can be calculated using the formula below [15]:

$$\eta_{KE} = 1 - 5 * \left(\frac{1}{M_o^2}\right)\left[\left(\frac{T_{t3}\left(\frac{P_o}{P_{t3}}\right)^{0.286}}{T_o}\right) - 1\right] \quad (1)$$

2.3 Numerical

Computational Fluid Dynamics (CFD) analysis has been very important for scramjet intake investigation due to the scarcity of experimental apparatus suitable for hypersonic flow [16–19]. However, they are mostly reliant on proprietary codes that are not widely available for other users. In this study, we use Ansys Fluent as the numerical solver because we believe that this code is robust enough for complex thermal-fluid analysis in many situations [8, 20–22]. The solver is a density-based Navier-Stokes solver and the turbulence model utilized in this study is the shear-stress transport k-omega model by Menter. The y^+ value is set as 1 thus limiting the grid size on the surface of the scramjet to about 2 microns. The mass flow rate balance entering and leaving the domain must be less than 1% for the simulation to be considered as converged. Typically, for each case, the mesh of about 52,000 (Medium Density Grid) cells are found to be enough to produce satisfactory results. The structured mesh are shown below for Medium Density Grid. The comparison of numerical and experimental value are also shown in Fig. 4.

3 Results and Discussion

3.1 Shock-Shock Interactions

Color schlieren and numerical schlieren of the Baseline, Cowl-3 and Cowl-5 cases are compared to each other in Fig. 5. In the baseline case, we can observe that the compression shock from the ramp impinged slightly aft the cowl tip. This in turn

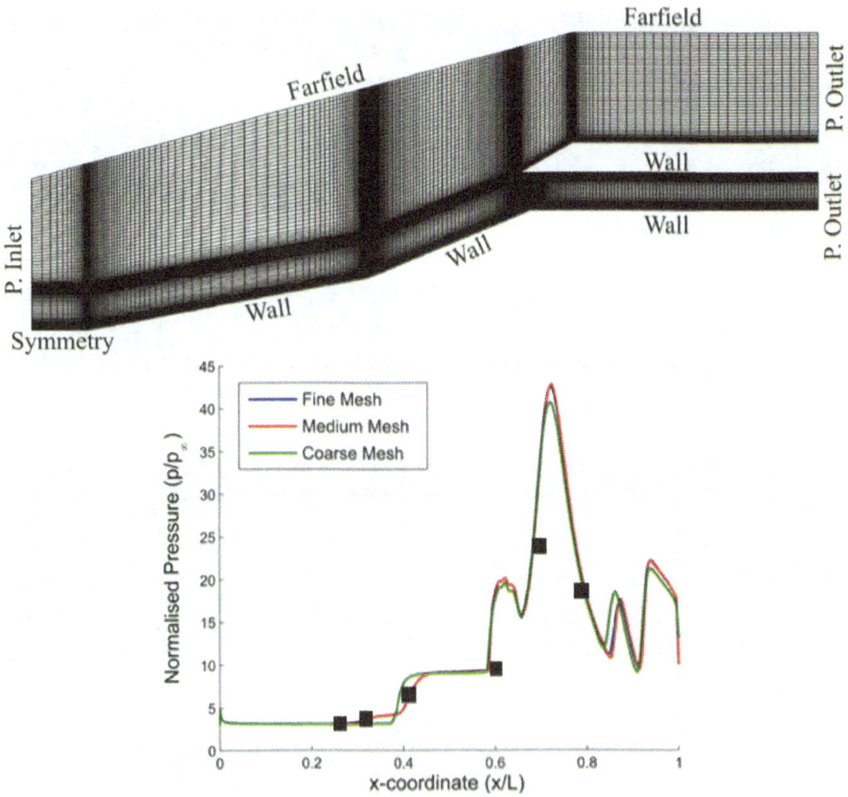

Fig. 4 Structured mesh for medium density grid and grid sensitivity analysis validated with discrete experimental data points

induced the formation of flow separation around the cowl tip area thus decreasing the effective intake cross-section.

The strong shock from cowl-tip caused the development of another shock at the shoulder region. This is a bigger separation than the one at the tip due to relaminarization of the boundary layer at the shoulder which has made it more prone to separation [18]. The separation bubble's height is slightly more than half the cowl height thus the effective intake area has been significantly reduced. The re-attachment shock from this separation has caused another separation further downstream inside the isolator. Overall, the internal flow is very complex with the formation of various shock-shock and shock-boundary layer interactions forming at different locations. This has been captured quite agreeably by the numerical schlieren result shown in Fig. 5b. Slight difference can be noted at the smaller cowl tip separation and larger shoulder separation predicted by the numerical scheme. In the numerical schlieren, it can be observed that the shoulder separation has spread upstream into the compression ramp and this can lead to intake unstart in real-flight condition.

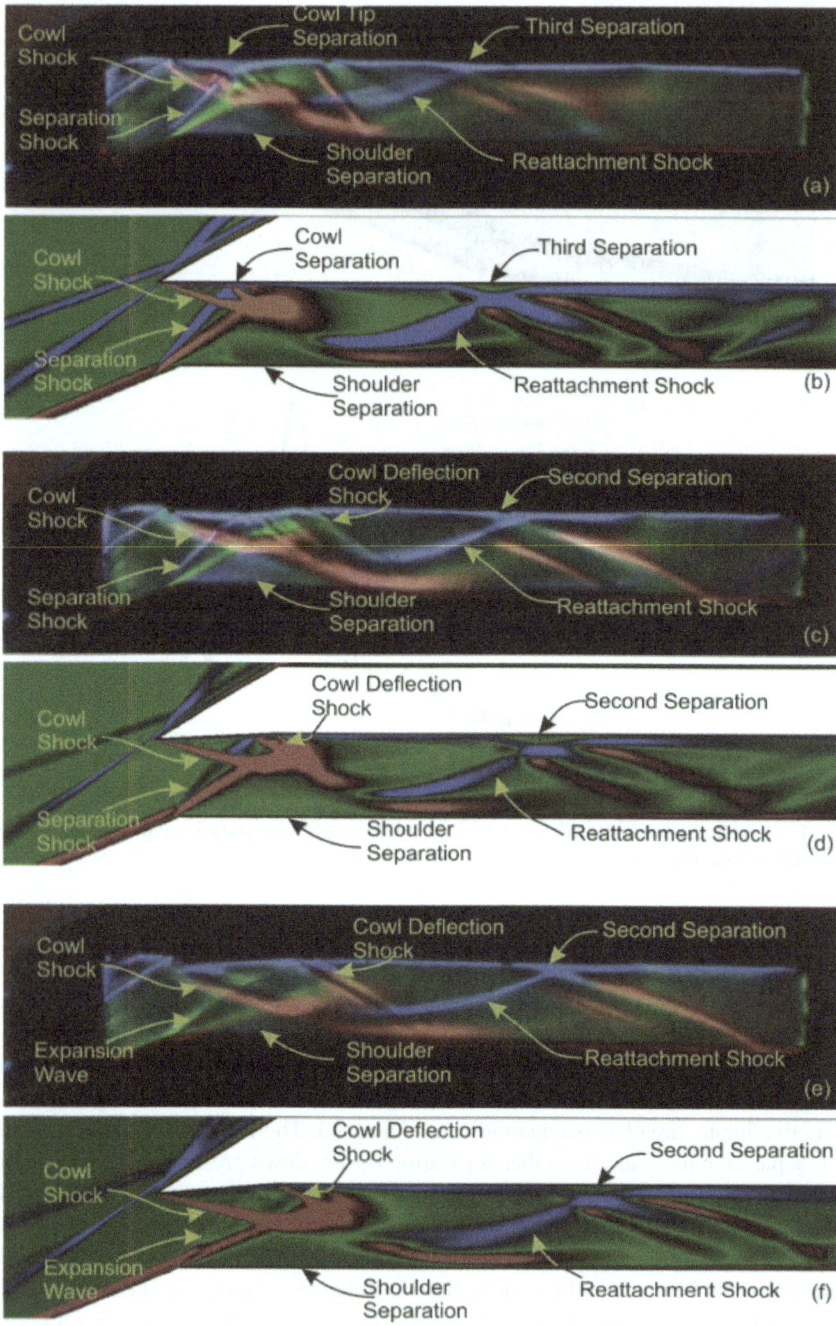

Fig. 5 Experimental and numerical schlieren images of Baseline [(**a**) & (**b**)], Cowl-3 [(**c**) & (**d**)], Cowl-5 [(**e**) & (**f**)]

In the case of Cowl-3, the introduction of inward deflection has eliminated the cowl-tip separation. Additionally, the shoulder separation has been reduced in height and the separation bubble has been confined inside the internal section of the intake. This is due to the reduction of total turning angle the flow need to through to enter the isolator section, thus lowering the strength of the cowl shock. This will reduce the probability of intake unstart. The numerical schlieren result for Cowl-3 agrees very well with the experimental and slight difference is observed only in the size of the separation bubble.

In the case of Cowl-5, the shock from the shoulder (due to separation) has been eliminated and formation of the expansion wave can be observed. The height of the shoulder separation bubble has been reduced substantially thus forcing the flow to turn away from itself in order to navigate around the bubble. The Cowl-5 is expected to have better performance due to the reduction of shock-shock interaction around the shoulder region.

3.2 *Intake Performance*

The pressure contours on the internal sidewall of the intake are compared in Fig. 6. The contours have been produced using pressure-sensitive paint which are calibrated against Kulite pressure readings taken inside the isolator section. On the sidewall surface, for the Baseline case, we can detect two shock waves from the cowl labelled

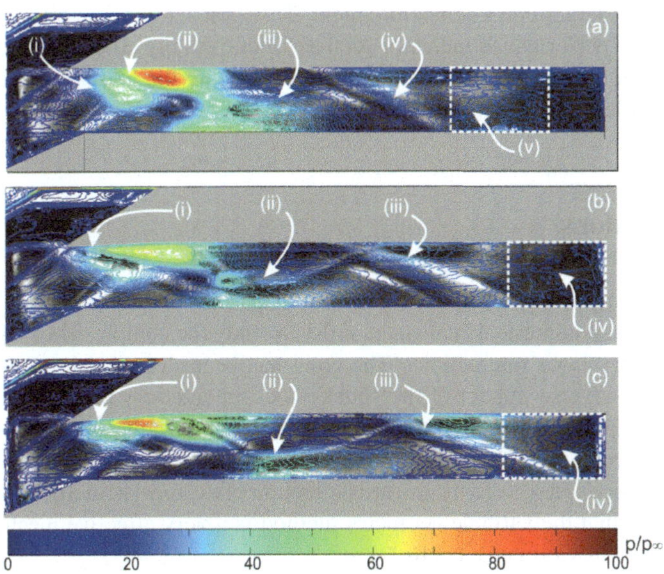

Fig. 6 Pressure contours with the "final shock" clearly marked

Table 1 Comparison of flow intake performance for different case

Case name	M at throat	M_{CFD} at throat	Compression Ratio	Kinetic energy efficiency
Baseline	2.13	2.01	17.7	0.923
Cowl-3	2.23	2.25	19.8	0.944
Cowl-5	2.36	2.39	19.1	0.956

(i) and (ii) in Fig. 5a. This is different from Cowl-3 and Cowl-5 cases which only have one oblique shock emanating from the cowl. This is due to the elimination of cowl-tip separation which in turn lowered the pressure peak on the cowl surface by almost 40%.

The pressure contour has allowed for the calculation of pressure and Mach number ratio using the "final shock analysis" that has been introduced in our previous article [8]. This has greatly reduced the number of measurement techniques to be applied in order to calculate the performance of the hypersonic intake. The calculated Mach number, pressure, pressure recovery, temperature, compression ratio and efficiencies for all cases are presented in the table below. The Mach number predicted using the numerical scheme are also noted in Table 1.

In general, the intake performance improvement was proportional to the cowl deflection angle. The contraction ratio decreased with every cowl deflection but the compression ratio unexpectedly increased, negating our initial expectation. The explanation we could offer is that a series of shock waves is more efficient in compressing the air rather than relying only on a single shock. The pressure recovery improved with larger cowl deflection angle. This translate into an improvement in all readings of performance indicator such as kinetic energy efficiency and entropic efficiency. The Mach number for all cases predicted by the numerical scheme was within 5% or less.

4 Conclusions

A generic scramjet intake designed for operation in Mach 5 flow has been subjected to experimental and numerical analysis in order to study the complex shock-shock and shock-boundary layer interactions. The cowl has been deflected inwards in order to reduce the adverse impact of the shock-shock interaction. Performance indicators showed massive improvement when the deflected cowl diminished some of the shocks at the cowl tip region. The compression ratio increased despite the reduction of wind ingestion area due to the inward deflection of the cowl actually. Thus it could be concluded that the variable geometry method is a very effective mechanism for scramjet intake performance control.

References

1. Roberts KN, Wilson DR (2009) Analysis and design of a hypersonic scramjet engine with a transition mach number of 4.00. In: 47th AIAA aerospace sciences meeting including the new horizons forum and aerospace exposition, pp 5–8
2. Fry RS (2004) A century of ramjet propulsion technology evolution. J Propul Power 20:27–58
3. Gruhn P, Gulhan A (2011) Experimental investigation of a hypersonic inlet with and without sidewall compression. J Propul Power 27:718–729
4. Haberle J, Gulhan A (2008) Investigation of two-dimensional scramjet inlet flowfield at Mach 7. J Propul Power 24:446–459
5. Hohn OM, Gulhan A (2011) Experimental investigation on the influence of sidewall compression on the flow field of a scramjet inlet at Mach 7. In: 17th AIAA international space plances and hypersonic systems and technologies conference, pp 1–17
6. Hohn OM., Gulhan A (2010) Experimental investigation of the influence of yaw angle on the inlet performance at Mach 7. In: 48th AIAA sciences meeting including the new horizons forum and aerospace exposition, pp 1–16
7. Tan H, Sun S, Huang H (2012) Behavior of shock trains in a hypersonic inlet/isolator model with complex background waves. Exp Fluids 53:1647–1661
8. Idris AC, Saad MR, Zare-Behtash H, Kontis K (2014) Luminescent measurement systems for the investigations of a scramjet inlet-isolator. Sensors 14:6606–6632
9. Idris AC, Saad MR, Lo KH, Kontis K (2018) Background-oriented schlieren (BOS) for scramjet inlet-isolator investigation. In: IOP conference series: materials science and engineering, vol 370 (2018)
10. Idris AC, Saad MR, Rahman MRA, Hashim FR, Kontis K (2019) Experimental validation of artificial neural network (ANN) model for scramjet inlet monitoring and control. Int J Recent Technol Eng 7(5S4):558–563 (2019)
11. Saad MR, Zare-Behtash H, Idris AC, Kontis K (2012) Micro-ramps for hypersonic flow control. Micromachines 3(2):364–378
12. Saad MR, Idris AC, Zare-Behtash H, Kontis K (2012) Micro-ramps in Mach 5 hypersonic flow. In: 50th AIAA aerospace sciences meeting including the new horizons forum and aerospace exposition, pp 676–685
13. Sidhu BS, Saad MR, Ahmad KZK, Idris AC (2016) Riblets for airfoil drag reduction in subsonic flow. ARPN J Eng Appl Sci 11(12):7694–7698
14. Idris AC, Saad MR, Kontis K (2019) Microvortex generator for scramjet inlet application. Progress Propuls Phys 11:747–758
15. Heiser WH, Pratt DT (1994) Hypersonic airbreathing propulsion. AIAA, Washington
16. Krause M, Reinartz B, Ballmann J (2007) Numerical investigation and simulation of transition effects in hypersonic intake flows. In: High performance computing in science and engineering '06, pp 391–406
17. Slater JW, Saunders JD (2009) CFD simulation of hypersonic TBCC inlet mode transition. In: AIAA Paper 2009–7349
18. Reinartz BU, Herrmann CD, Ballmann J, Koschel WW (2003) Aerodynamics performance analysis of a hypersonic inlet isolator using computation and experiment. J Propuls Power 19:868–875
19. Throckmorton R, Schetz JA, Jacobsen LS (2010) Experimental and computational investigation of a dynamic starting method for supersonic/hypersonic inlets. In: AIAA 2010–589
20. Rahman MRA, Leong KY, Idris AC, Saad MR, Anwar M (2017) Numerical analysis of the forced convective heat transfer on Al2O3-Cu/water hybrid nanofluid. Heat Mass Transf 53(5):1835–1842
21. Rahman MRA, Leong KY, Idris AC, Saad MR (2017) Thermal fluid dynamics of Al2O3-Cu/water hybrid nanofluid in inclined lid driven cavity. J Nanofluids 6(1):149–154
22. Shuib W, Idris AC, Saad MR, Hasan MF, Rahman MR (2018) Study on the effect of bluff body with slit in the micro-channel combustor. J Adv Res Fluid Mech Thermal Sci 52(2):215–222

Enhancing Productivity of Solar Still with Pyramid Chamber Step-Wise Basin Design

Salah Abdallah and Hanan Saleet

Abstract Improving the performance of the traditional single slope solar still is a pressing need. In this paper, modifications were introduced to the pyramid design, where it has three different faces, and has a Step-wise chamber. Experimental studies were performed in winter where the following interesting results are achieved: the production rate of distilled water of the modified design is up to 85% higher than the traditional design; in addition, temperature gain is 25% higher than traditional design.

Keywords Solar still · Step-wise chamber · Pyramid design

1 Introduction

Safe and fresh drinking water is precious for human kind. But only 3% of total water on the earth is fresh water, and less than 1% of that fresh water is suitable for human and animal consumptions. To mitigate this problem, water purification processes are introduced. Filtration, reverse osmosis, ultraviolet radiation, carbon absorption are used. But the most reliable processes are distillation and boiling [1]. Solar distillation is one of the most efficient methods of harvesting solar energy and then using this energy to provide drinking water to communities. Thus, solar distillation is used to produce pure water and is one of the most economically effective methods. In addition, solar distillation is non-polluting and uses free of cost solar energy to supply pure water to countries with shortage of fresh water [2, 3].

Many research and experiments on the use of distillers were reported in [4–6]. In addition, many researchers presented heat transfer analysis and energy balance analysis of solar stills such as the research sited in [3, 4, 7]. In [3], Saini et al. studied the performance of single slope solar still integrated with SPV module and passive condenser. They built a thermal model of the system and analytically found the

S. Abdallah (✉) · H. Saleet
Applied Science Private University, Amman 1193, Jordan
e-mail: salahabdalah@asu.edu.jo

P. Rajendran et al. (eds.), *Proceedings of International Conference of Aerospace and Mechanical Engineering 2019*, Lecture Notes in Mechanical Engineering,
https://doi.org/10.1007/978-981-15-4756-0_8

maximum overall energy efficiency. The effect of using diverse absorbing materials in solar stills to increase their production rate is cited in [8, 9].

In [10], Abdallah et al. reported the achievement of up to 380% improvements in productivity when installing different modifications on the solar still. A single axis sun-tracking system was connected to a single slope solar still as reported by Abdallah and Badran [11]. They compared its performance to the use of a solar tracking system. The later system increased the production rate of distilled water by up to 22%. Palpandi and Prem Raj [12] created a slope type solar still with 24° tilt angle for the required basin area. Objective of this study is to evaluate the performance improvement of single basin type solar still using different TDS (Total dissolved salts) of water and change in absorbing materials with phase change material. Theoretical analysis was carried out to calculate the incident and useful solar radiation of the location (TIRUNELVELI), tilt angle and the performance improvement of the solar stills. In [13], Abidi, et al. improved the productivity gain by using automatically controlled Heliostats to follow the sun and then collect the solar rays into the solar still. The features of Photovoltaic (PV) system combined with reverse osmosis (RO) desalination technology are analyzed in [14]. The authors compared the water production cost of their system with the system powered by traditional generator such as fossil fuel, they reported a significant reduction in the cost. In addition, many authors reported the achievement of gain in the productivity by experimenting different modifications to the traditional solar still; such as water circulation [15], applying humidification and dehumidification [16]. Abdallah [17] experimented a design where an electrical heater that has a PV power generator, and chamber step-wise design were introduced to the traditional solar still.

This work tackles the main problem of the traditional single slope solar still which is low production rate. To enhance the production rate, modifications are introduced so that the still has a three face chamber with step-wise pyramid design.

2 Solar Still Design

Before presenting the modified solar still design, in the following the traditional solar still functionality is presented

2.1 Traditional Solar Still

When the sun falls through the glass surface, which is located at the upper surface as shown in Fig. 1, the rise in temperature of water and the surrounding metal sheets cause the water to evaporate. The vapor condenses on the inner surface of the glass because of the temperature difference between the glass and the vapor. The condensed water flows by gravity force where it is collected into an external tank.

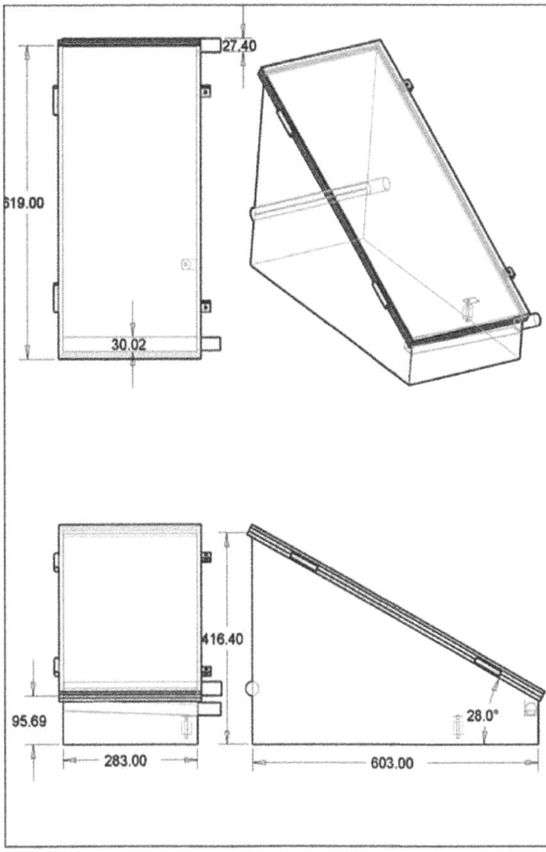

Fig. 1 Traditional design with its dimensions in mm

2.2 *Modified Solar Still Design*

The modified solar still works the same process like the traditional still, with the addition of various system modifications; as illustrated in the following. The modified solar still design has a pyramid shape with three different faces. The inside chamber is divided into five chambers each chamber have a square shape with dimensions staring from (20, 20 cm) to (60, 60 cm). The still has a height 71.6 cm and width 74 cm which make the inclination angle of the glass 44°; as shown in Fig. 2. The thickness of the steel is 1.8 mm for chamber and 3 mm for boundary. The glass thickness is 6 mm.

It is designed to have more than one collecting face to collect the solar radiation. It has three collecting faces to allow the solar radiations to break through the glass in different times: sunrise, noon and sunset. In addition, the new design has stepwise chambers which will increase the surface area exposed to solar radiation as shown in Fig. 3. The still is covered with wood insulation material from backside to decrease heat losses. These design modification increase solar radiation in the basin and assist

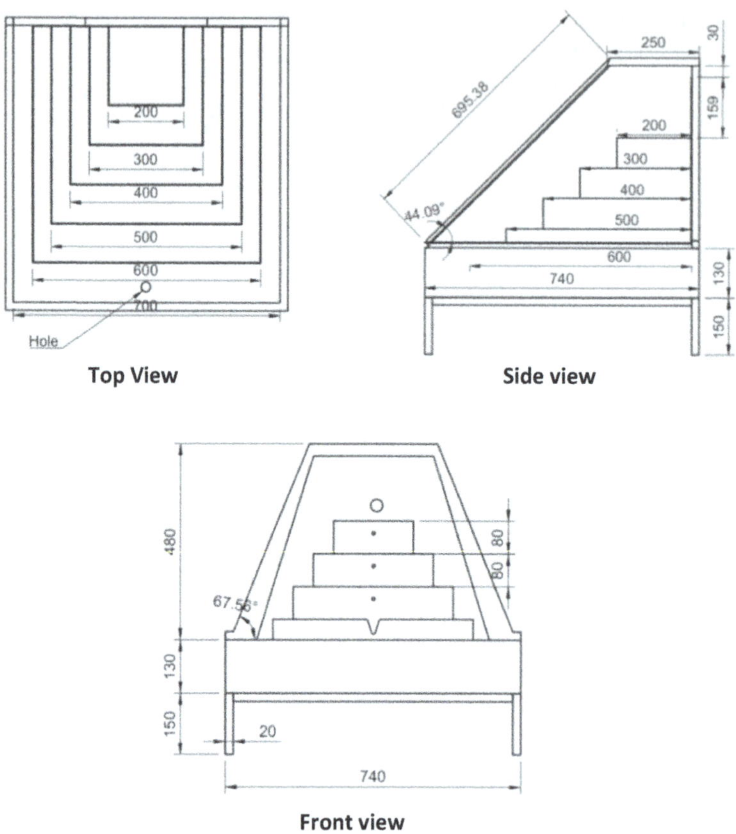

Fig. 2 Different views of the modified design

Fig. 3 Solar still with chamber step-wise basin during manufacturing process

Fig. 4 Real view for both traditional and modified solar stills

the water to evaporate faster. The solar radiation break through the glass cover into the stepwise chambers which is filled with saline water, the kinetic energy of water drops increases and it starts to evaporate and condense on the back surface of glass cover, by gravity the condensed water on the glass will flow down to the collecting tank.

3 Experimental Procedure

Experiments were conducted on the stills, both the traditional and modified stills, at same climate conditions; they were located near each other during the experimentation, as shown in Fig. 4. The experiments were performed in winter in the period from 7AM to 4PM for two days: 25th of December 2018, and 26th of December 2018 at the Applied Science Private University, in the Energy Center. Readings about water temperature, ambient temperature, collected water, and solar radiations were recorded nearly every an hour. Before being used for taking the readings, the measuring devices were tested and calibrated.

4 Results and Discussions

The readings from 26/12/2018 will be used to represent the analysis of the results in the form of curves. Figures 5 and 6 record both solar radiation variation and ambient temperature as a function of time, respectively.

Figure 7 shows the temperature variation inside the chamber of both the traditional still, and the modified still. In addition, Fig. 8 shows the distilled water collected hourly for both the traditional and modified stills. The maximum amount of water was collected nearly around 1:00 pm. During this period, the solar radiation is vertical

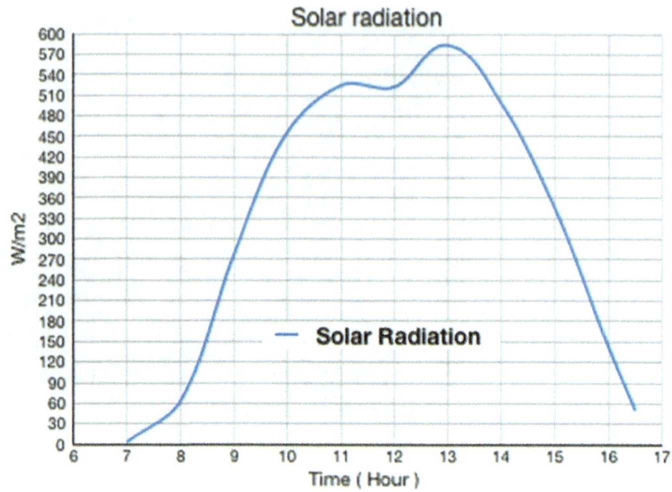

Fig. 5 Solar radiation

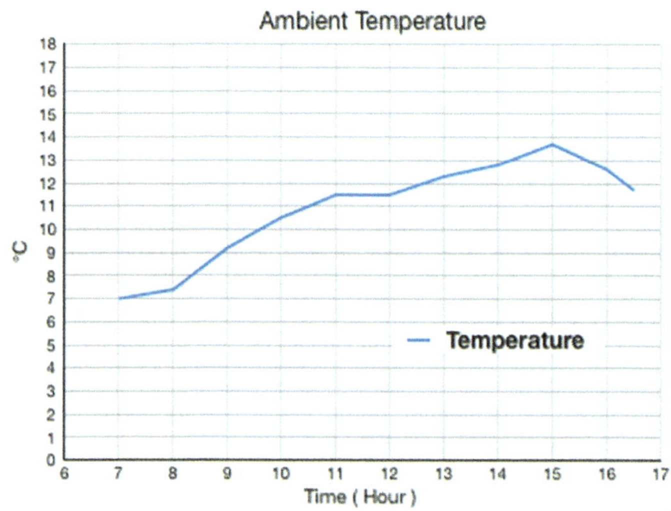

Fig. 6 Ambient temperature

and then the ambient temperature is maximum. The achievements from using the modified design are significant. The production rate gained by the modified still reach up to 88.95%.

The modifications added to the traditional solar still produced a higher thermal performance due to the following: (1) The addition of the chamber step-wise design shortened the distance between the glass cover and the water surface in the chambers; lowering the kinetic energy needed to raise the vapor up and shortening the time

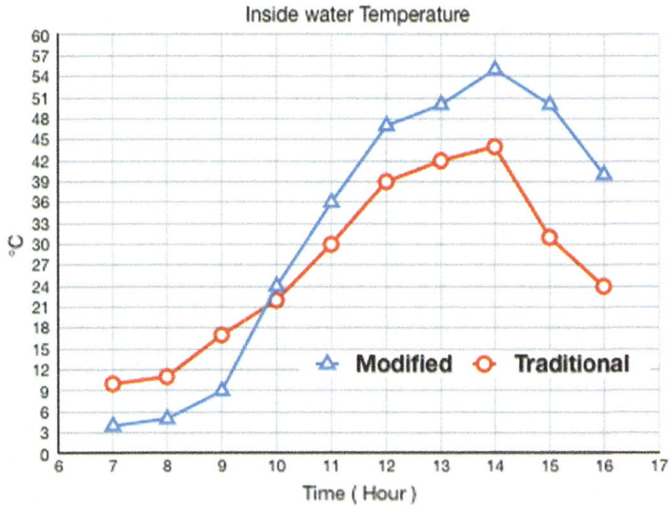

Fig. 7 The variation of temperature of water

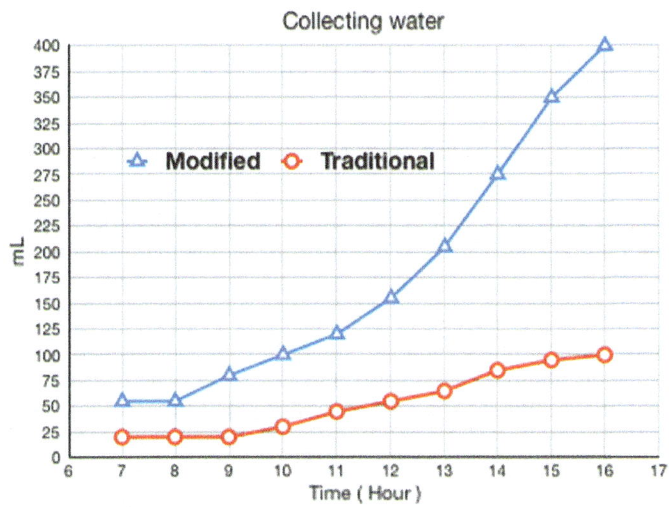

Fig. 8 Pure collecting water

needed to finalize the process. (2) Higher heat and mass transfer surface is provided because of the sheet metal, which has much more surface that is directly facing the sun. (3) The availability of three collecting faces to allow the solar radiations to break through the glass in different times: sunrise, noon and sunset.

The effect of the pyramid shape with three faces and chamber step-wise design introduced to the traditional still was evident through the production gains achieved for different days as shown in Table 1. As calculated by Eq. 1, the production gain for

Table 1 Distilled water collected

Date	Collected pure water, L/m^2/day	Gain (%)
Day 1 (25/12/2018) • Old design • New design	0.717 1.116	55.65
Day 2 (26/12/2018) • Old design • New design	0.588 1.111	88.95

the modified solar still (new design) reached up to 88.95% compared to the traditional still (old design), due to the design modifications; this results from the fact that the modified design can gain temperature up to 25% higher than traditional design. As mentioned earlier the experiments were held during two days in the winter, these achievements will be much better if the measurements were taken in the summer.

$$\text{Gain for day 2} = \frac{1.111 - 0.588}{0.588} \times 100 = 88.95\% \quad (1)$$

5 Conclusions

This work tackles the main problem of the traditional single slope solar still which is low production rate. Modifications are experimented including: the change of the shape to be as a pyramid, with three faces, the addition of chamber step-wise design and the addition of inclined shape. The effects of adding these modifications on the output parameters were investigated using experimental work. These modifications enhance the productivity of distiller by up to 88.95%; this value will be much better if experiments are held in the summer. The results are considerable and significant for solar still applications. Consequently, the arid zones, where insolation levels are high, can use the modified solar still to increase the efficiency of distilled water production.

Acknowledgements The authors are grateful to the Applied Science Private University, Amman, Jordan for the full financial support granted to this research project.

References

1. Hameed K, Khan MM, Ateeq IS, Omair SM, Ahmer M, Wajid A (2013) Solar power water distillation unit. J phys 450(1):1–5
2. Gorjian S, Ghobadian B (2015) Solar desalination: a sustainable solution to water crisis in Iran. Renew Sustain Energy Rev 48:571–584

3. Saini V, Sahota L, Jain VK, Tiwari GN (2019) Performance and cost analysis of a modified built-in-passive condenser and semitransparent photovoltaic module integrated passive solar distillation system. J Energy Storage 24
4. Rani A, Kumar A, Suresh S (2019) Review on thermal modeling of solar desalination systems. Res J Chem Environ 23(4)
5. Manchanda H, Kumar M (2015) A comprehensive decade review and analysis on designs and performance parameters of passive solar still. Renewables 2(17)
6. Kaushal A, Goel V (2010) Solar stills: a review. Renew Sustain Energy Rev 14(1):446–453
7. Tripathi R (2019) Analysis of solar stills by using solar fraction. In: Green energy and technology, In book: Handelsrecht-Schnell erfasst
8. Abdallah S, Abu-Khader M, Badran O (2009) Effect of various absorbing materials on the thermal performance of solar stills. Desalination 242:128–173
9. Nasri B, Benatiallah A, Kalloum S, Benatiallah D (2019) Improvement of glass solar still performance using locally available materials in the southern region of Algeria. Groundwater Sustain Dev 9:100213
10. Abdallah SM, Badran OO, Abu-Khader MM (2008) Performance evaluation of a modified design of a single slope solar still. Desalination 219:222–230
11. Abdallah SM, Badran OO (2008) Sun-tracking system for productivity enhancement of solar still. Desalination 220:669–676
12. Palpandi K, Prem Raj R (2015) Performance test on solar still for various TDS water and phase change materials. IJIREST Int J Innov Res Sci Eng Technol 4(13)
13. Abidi M, Ben Jabrallah S, Corriou JP (2016) Optimal design of a solar desalination unit with heliostats, Hindawi. Int J Chem Eng 2017
14. Shouman1 ER, Sorour MH, Abulnour AG (2015) Economics of renewable energy for water desalination in developing countries. J Eng Sci Technol Rev 8(5):227–231
15. El-Agouz SA (2014) Experimental investigation of stepped solar still with continuous water circulation. Energy Convers Manag 86:186–193
16. El-Nasr AM, Kamal M, Saad H, Elhelaly M (2015) Water desalination using solar energy: humidification and dehumidification principle. Innov Energy Res
17. Abduallah S (2018) Productivity enhancement of solar still with PV powered heating coil and chamber step-wise basin. J Ecol Eng 119(2):8–15

Mechanical Characteristics of Paraffin Wax-HTPB Based Hybrid Rocket Fuel

Sri Nithya Mahottamananda and P. N. Kadiresh

Abstract Paraffin wax has 3–4 times higher regression rate compared to conventional polymeric fuels used in hybrid rocket applications. Hydroxyl-terminated polybutadiene (HTPB) suffers poor mechanical properties and has lower tensile strength, Young's modulus compared to pure paraffin wax. The mechanical properties of HTPB can suitably be improved by mixing paraffin wax with HTPB. In this study, paraffin wax and HTPB are mixed with different weight ratios to form nine different compositions viz, 100P, 90P, 80P, 70P, 60P, 50P, 40P, 25P and 0P (where P denotes the percentage of paraffin wax and the remaining forms the HTPB system). To understand the effect of addition of paraffin wax with HTPB, nine fuel samples were prepared and experimentally tested for their mechanical properties. It is observed that addition of paraffin wax content in solid fuel composition increases maximum stress and Young's modulus in the range of 16–74% and 15–97% respectively. It is also observed that addition of paraffin wax content in solid fuel composition decreases percentage of elongation by 98–57%. The surface morphology and chemical composition were examined using Scanning Electron Microscope (SEM) and Fourier Transform Infrared Spectroscopy (FTIR) to understand the blending of paraffin wax + HTPB and to find out the occurrence of chemical reaction during mixing and curing processes.

Keywords Mechanical properties · Paraffin wax-HTPB based fuel · Hybrid rocket fuel

1 Introduction

A hybrid rocket motor which uses propellants in two different states of matter: one solid and the other either gas or liquid. Hybrid propulsion systems make use of fuel and oxidizer stored in different physical state prior to combustion and thus forms

S. N. Mahottamananda (✉) · P. N. Kadiresh
Department of Aerospace Engineering, B.S. Abdur Rahman Crescent Institute of Science & Technology, Chennai, Tamilnadu, India
e-mail: mahottamananda@gmail.com

P. Rajendran et al. (eds.), *Proceedings of International Conference of Aerospace and Mechanical Engineering 2019*, Lecture Notes in Mechanical Engineering,
https://doi.org/10.1007/978-981-15-4756-0_9

a separate class of propulsion from conventional all-solid or all-liquid propulsion system [1]. Hybrid rocket system is having attractive features like restart capability, throttling ability, improved safety, low impact on environment and low development cost [2–5]. Hybrid rockets also are attractive since they are much less sensitive to fractures in the fuel grain that might occur during storage or transportation, when compared to solid rocket motor. These features of hybrid rocket attract to develop various applications like missile, sounding rocket, space applications, upper stage and human space flight. Hybrid rockets are also having some drawbacks like low regression rate, oxidizer-fuel ratio shift during combustion, diffusion flame controlled combustion and low volumetric loading and poor combustion efficiency [6–9].

HTPB is most widely used fuel in hybrid rockets. HTPB fuel suffers with poor mechanical properties, high heat of ablation leading to low regression rate of fuel grain in hybrid rocket operation. Low regression rate leads to low mass flow generation and results in low thrust which is a major problem in hybrid rocket motors. This limits development and applications of hybrid fuels. The casting of large size paraffin fuel grain that can sustain the rocket inertial loads, hoop stress due to radial combustion pressure and thrust are quite challenging task for developers. Many enhancement strategies have been employed to overcome poor mechanical strength of the paraffin-based fuels. Sinha et al. reported dynamic mechanical properties of HTPB-paraffin wax blend using dynamic mechanical analyzer. Blending the paraffin with HTPB can favour the improvement in thermal stability and mechanical strength of paraffin [10, 11]. Maruyama et al. [12] tested mechanical properties of paraffin wax added with ethylene vinyl acetate (EVA) based formulation. They observed 20 wt% EVA additive gives 1.6 times increased maximum strength compared to other formulation. Recently, Rajiv Kumar et al. [13] observed mechanical properties of 20 wt% EVA give the tensile strength 50% higher than other formulation. Kim et al. [14] performed tensile and compressive test of PE addition to paraffin wax fuel. They observed the significant improvement in tensile and compressive strength when 10 wt% PE was added to paraffin wax. Furthermore, DeSain et al. [15] reported addition of 4 wt% low-density polyethylene (LDPE) in paraffin wax the tensile strength and percentage elongation were improved.

Mechanical properties play major role in selection of fuel formulation in hybrid rocket motor. In this research, an effort is made to improve the mechanical properties of HTPB by adding paraffin wax. Present work involves blending of paraffin wax with HTPB and studying their mechanical properties. Paraffin wax weight percentage is varied from 0 to 100% with HTPB fuel. The tension test was carried out using the Instron Universal testing machine (UTM) of 9000 Series and Effect of increasing weight percentage of paraffin wax in HTPB fuel on mechanical properties is studied.

Table 1 Compositions of paraffin-HTPB based solid fuel

Sample name	Paraffin wax (wt%)	HTPB system (HTPB + DOA + TDI + Glycerol) (wt%)
100P	100	0
90P	90	10(7.2 + 2.2 + 0.56 + 0.04)
80P	80	20(14.4 + 4.4 + 1.12 + 0.08)
70P	70	30(21.6 + 6.6 + 1.68 + 0.12)
60P	60	40(28.8 + 8.8 + 2.24 + 0.16)
50P	50	50(36.0 + 11.0 + 2.80 + 0.20)
40P	40	60(43.2 + 13.2 + 3.36 + 0.24)
25P	25	75(54.0 + 16.5 + 4.20 + 0.30)
0P	0	100(72.0 + 22.0 + 5.60 + 0.40)

2 Experimentation

2.1 *Preparation of Fuel Sample*

Nine different fuel compositions of HTPB and Paraffin wax were prepared. The fuel compositions were 100P, 90P, 80P, 70P, 60P, 50P, 40P, 25P and 0P where P denotes the percentage of paraffin wax and the rest form the HTPB system. Nine solid fuel compositions are given in Table 1. In the propellant mixer, mixing chamber temperature was controlled by hot water bath facility. These facilities would avoid sudden solidification during mixing of paraffin wax with HTPB. The fuel samples were prepared by following procedures. Initially, the paraffin wax was melted then mixed with Hydroxyl Terminated Polybutadiene (HTPB) systems. The mixing chamber temperature was maintained at 60 °C. With the help of stirrer type mixer, paraffin wax and HTPB were mixed uniformly. The prepared fuel sample was poured into dog bone shape aluminium die and kept for one day in open atmospheric curing and then in oven. The curing time for prepared samples was five days at a temperature of 60 °C. After curing it was carefully removed from the die. 100P sample indicates pure paraffin wax, and 0P is pure HTPB, The macrostructure of prepared fuel samples are shown in Fig. 1.

2.2 *Density Analysis and Scanning Electron Microscopy (SEM)*

The prepared solid fuel sample's quality was estimated using simple density measurement. The solid fuel density of prepared fuel was determined by Archimedes principle, a gravimetric method. The theoretical fuel density was estimated by additive rule of mixture. The porosity indicates the presence of voids, air bubbles, cracks

Fig. 1 Macrostructure of 80P fuel samples obtained from dog bone shape die

in prepared solid fuel sample. SEM analysis was carried out using JEOL Model JSM-6390LV microscope under a 5 kV beam, to understand the blending of paraffin wax and HTPB. The moisture in fuel samples was removed before the experiment.

2.3 *Fourier Transform Infrared Spectroscopy (FTIR)*

FTIR studies were carried out using FTIR-JASCO 6390 to find out the occurrence of chemical reaction during mixing and curing processes.

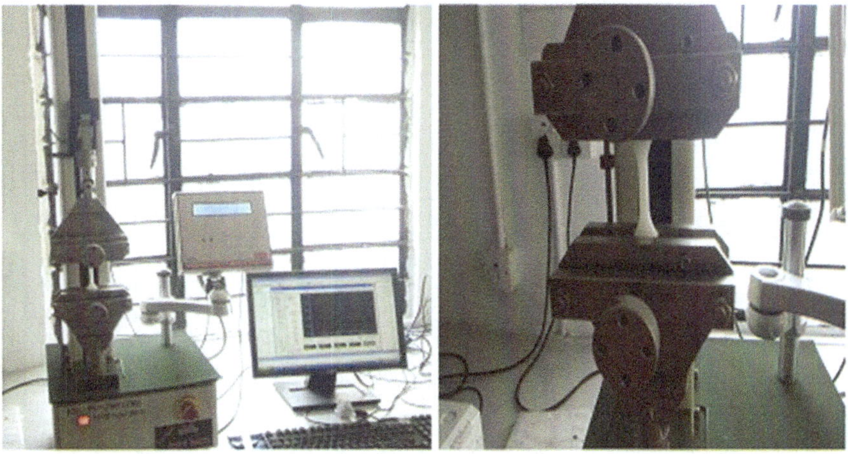

Fig. 2 Universal testing machine-9000 series

2.4 *Mechanical Characteristics*

The tension test was carried out using the Instron UTM machine of 9000 Series, Fig. 2. This test was carried out to determine the tensile strength, maximum stress, strain and Young's modulus of the sample. Three samples were prepared and tested for each fuel compositions. The shape of the test piece is made in accordance with ASTM standard D412. Figure 3 shows the die prepared as per the ASTM standard D412. The specimen was held between two wedge grips, which were attached to the crosshead. The upper crosshead was fixed with the vertical columns and remains stationary. The lower crosshead was connected to a screw and an electric motor, which caused it to move up or down, at a prescribed speed selected by the user.

3 Results and Discussion

3.1 *Quality of Prepared Sample*

Blended Paraffin-HTPB based solid fuel density was measured for all compositions and results are summarized in Table 2. The blended solid fuel samples do not show any considerable change in density because both the fuels are having almost same density. The porosity value decides the quality of blending of fuel samples. Lower porosity range of 0.6–1.5% was observed for the prepared solid fuel samples, indicating good quality of blending of paraffin and HTPB. During hybrid rocket combustion, solid fuel melts, vaporizes and undergoes pyrolysis due to heat flux transferred from flame zone to fuel surface. Though voids and cracks present in the solid fuel are

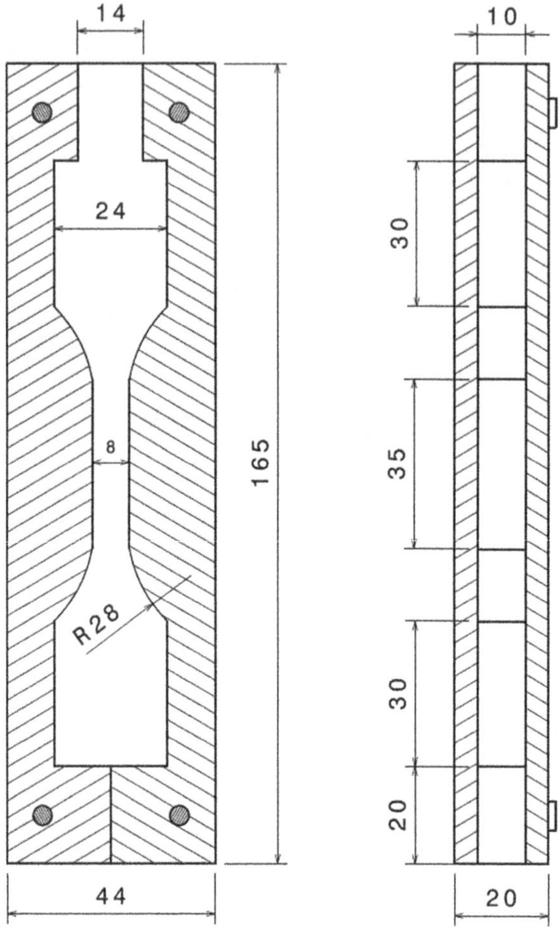

Fig. 3 Dog bone shape die—ASTM standard D412

Table 2 Density variations of paraffin-HTPB based solid fuel compositions

Sample name	Experimental density (kg/m^3)	Theoretical density (kg/m^3)	Porosity (%)
100P	894.0	900.0	0.7
90P	891.0	903.3	1.4
80P	901.2	906.6	0.6
70P	900.7	909.8	1.0
60P	904.0	913.1	1.0
50P	910.4	916.4	0.7
40P	911.1	919.7	0.9
25P	911.1	924.6	1.5
0P	920.3	932.9	1.4

not influencing fuel regression rate during combustion, lower porosity increases the volumetric loading of fuel in hybrid rocket motor [16].

3.2 Morphology of Blended Paraffin-HTPB Solid Fuel Compositions

To acquire better understanding about paraffin wax and HTPB blends SEM images are obtained. SEM images of 100P, 90P, 80P, 70P, 60P, 50P, 40P, 25P and 0P fuel blends are presented in Fig. 4. It is found that paraffin wax got distributed quite uniformly in all fuel samples. In solid fuel samples of 60P, 50P and 40P we found paraffin and HTPB mixed more homogenously compared to other fuel samples. In some fuel samples porosity is found, this may be because of samples are cured for one day in atmospheric conditions. It will not affect the combustion performance, as during combustion the surface temperature is well above 600 °C. At this temperature solid fuel gets melted, vaporize and undergo pyrolysis process with the result of heat flux transferred from flame zone to fuel surface.

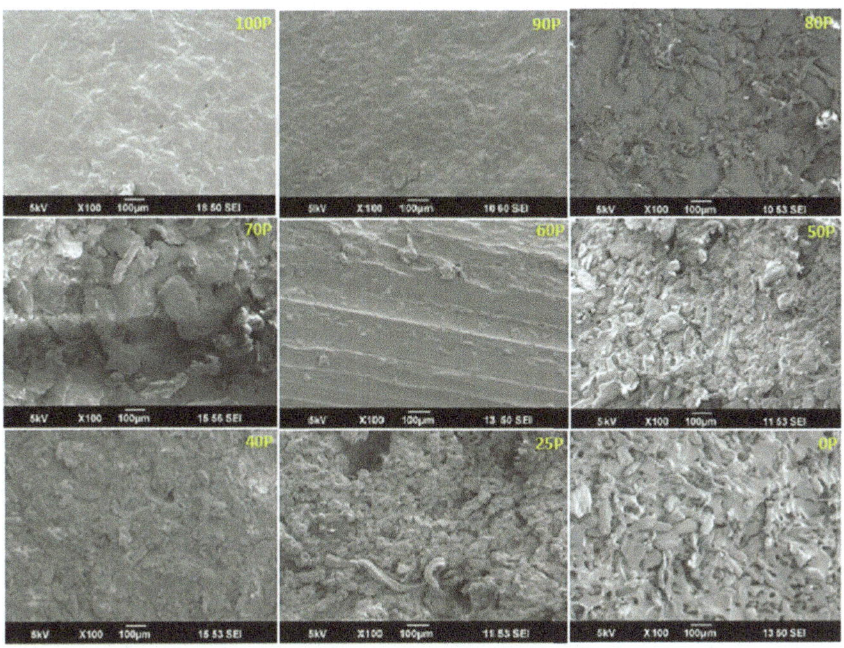

Fig. 4 SEM image of 100P, 90P, 80P, 70P, 60P, 50P, 40P, 25P and 0P solid fuel compositions

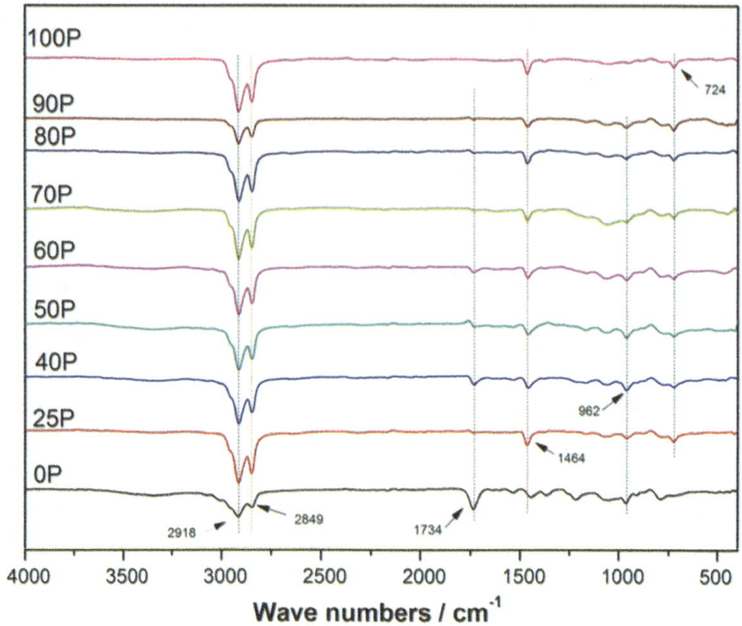

Fig. 5 FTIR spectra of 100P, 90P, 80P, 70P, 60P, 50P, 40P, 25P and 0P solid fuel compositions

3.3 Chemical Composition of Blended Paraffin-HTPB Solid Fuel Compositions

The FTIR spectra of 100P, 90P, 80P, 70P, 60P, 50P, 40P, 25P and 0P solid fuel compositions are presented in Fig. 5. It is observed that all fuel samples having peak at 2918 and 2849 cm^{-1} were attributed to the aliphatic C–H stretching vibration of the $-CH_3$ and $-CH_2$ groups [17] respectively. For HTPB, the peak at 1734 cm^{-1} belonged to the urethane groups and the absorption peaks at 962 cm^{-1} were corresponding to the bending vibrations of the C=C band [18, 19]. These results are confirming presence of paraffin wax and HTPB blends physically without any chemical reaction.

3.4 Tensile Behaviour of Blended Paraffin Wax and HTPB Solid Fuels

Paraffin wax has higher regression rate compared to polymeric fuel but it suffers with poor mechanical properties like low percentage of elongation. HTPB has low regression rate compared to paraffin wax and it suffers with low tensile strength but it has higher percentage of elongation. To overcome these problems HTPB is blended with paraffin wax. Nine different paraffin wax and HTPB compositions were prepared

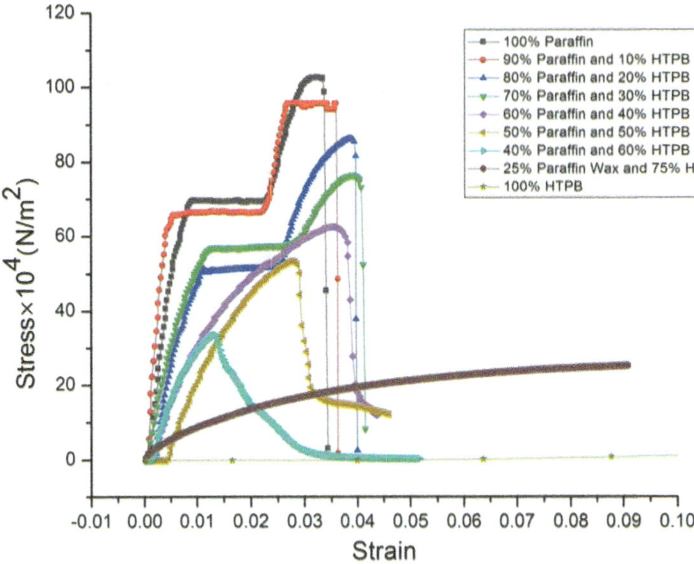

Fig. 6 Stress versus strain curve for nine different solid fuel compositions

as per the ASTM standards and tested in UTM machine of 9000 Series to measure the tensile strength and percentage of elongation. The tensile test was carried out with the moving head velocity as 5 mm/min. For the repeatability, three tensile samples were tested for each composition and the stress versus strain values were recorded. Figure 6 shows stress-strain traces of nine different fuel compositions tested at room temperature. Initially the stress increases linearly with strain till it reaches to the elastic modulus and then plastic deformation takes place and finally reaches to failure.

3.5 *Effect of Blending of Paraffin Wax with HTPB on Maximum Stress, Strain and Young's Modulus*

From Fig. 7, It is observed that pure paraffin wax sample having maximum stress of 102.311×10^4 N/m^2 and pure HTPB sample having maximum stress of 17.644×10^4 N/m^2. It is confirmed that increasing paraffin wax weight percentage in HTPB-paraffin wax blends, the maximum stress is increasing in the range of 16–74%. From Fig. 8. It is observed that pure paraffin wax sample having percentage of elongation of 0.03342 and pure HTPB sample having percentage of elongation of 1.6585. It is confirmed that increasing paraffin wax weight percentage in HTPB-paraffin wax blends, the percentage of elongation is decreasing in the range of 98–57%. From Fig. 9. It is observed that pure paraffin wax sample having Young's modulus of

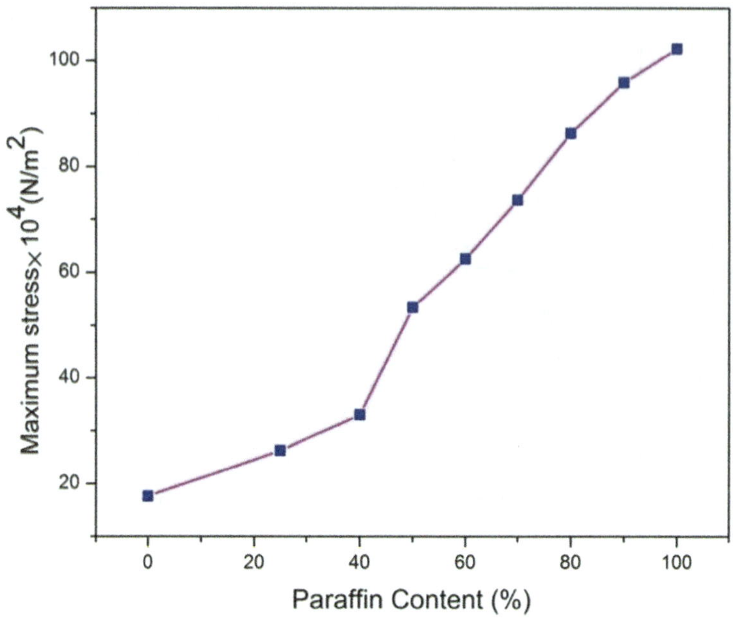

Fig. 7 Effect of blending of HTPB with paraffin wax on maximum stress

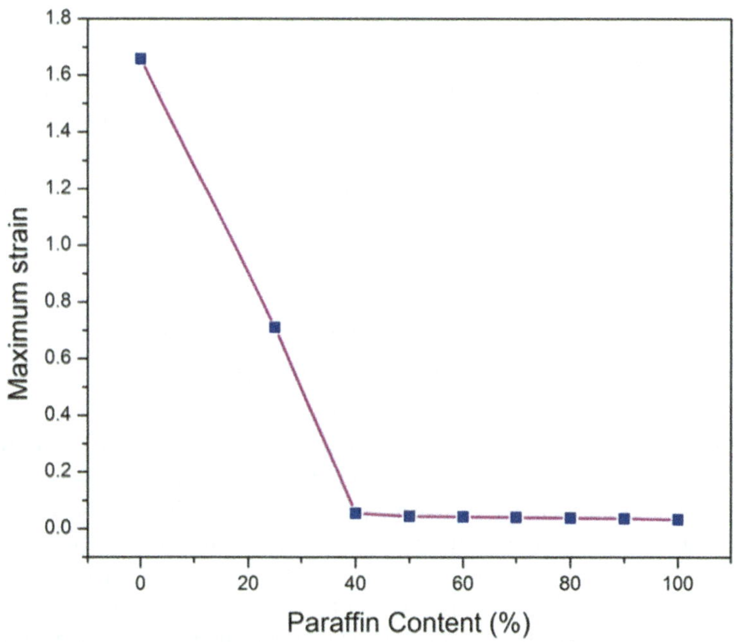

Fig. 8 Effect of blending of HTPB with paraffin wax on maximum strain

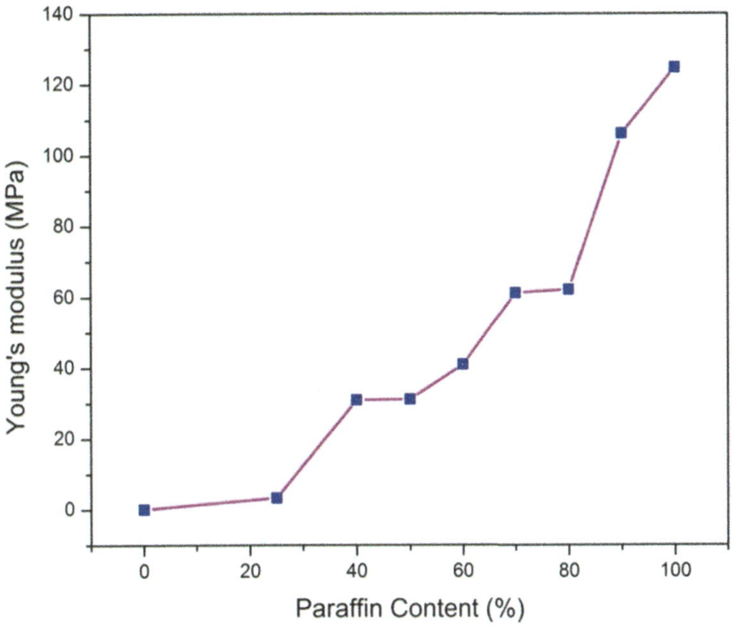

Fig. 9 Effect of blending of HTPB with paraffin wax on Young's modulus

124.7 MPa and pure HTPB sample having Young's modulus of 0.145 MPa. It is conformed that increasing paraffin wax weight percentage in HTPB-paraffin wax blends, Young's modulus is increasing in the range of 15–97%.

4 Conclusion

In this study HTPB-paraffin wax blended solid fuel compositions were evaluated experimentally for their density, surface morphology, chemical composition, tensile strength, percentage of elongation and Young's Modulus.

- The FTIR spectra test confirmed the physical blend of paraffin wax and HTPB without any chemical reaction. Surface morphology studies confirmed paraffin and HTPB are mixed more homogenously. Lower porosity observed in range of 0.6–1.5% for the prepared fuel sample indicates good quality of blending of HTPB and paraffin wax.
- Pure paraffin wax fuel sample had shown higher tensile strength and Young's modulus than pure HTPB and pure HTPB fuel sample had shown higher percentage of elongation than pure paraffin wax.
- With increasing Paraffin wax weight percentage in solid fuel composition, the percentage of elongation is decreasing in the range of 98–57%.

- Increase in Paraffin wax weight percentage in solid fuel composition, increases the maximum stress and Young's modulus in the range of 16–74% and 15–97% respectively.

Acknowledgements This work was supported by the Aeronautics Research and Development Board-DRDO.

References

1. Sutton GP, Biblarz O (2016) Rocket propulsion elements. John Wiley & Sons, New York
2. Altman D Hybrid rocket development history. In: 27th joint propulsion conference. American Institute of Aeronautics and Astronautics
3. George P, Krishnan S, Varkey PM, Ravindran M, Ramachandran L (2001) Fuel regression rate in hydroxyl-terminated-polybutadiene/gaseous-oxygen hybrid rocket motors. J Propuls Power 17:35–42
4. Chiaverini MJ, Serin N, Johnson DK, Lu Y-C, Kuo KK, Risha GA (2000) Regression rate behavior of hybrid rocket solid fuels. J Propuls Power 16:125–132
5. Jayapal SNM, Rafnaz M, Kadiresh PN (2018) Spray characteristics of plug type swirl injector. In: 2018 advances in science and engineering technology international conferences (ASET). IEEE, pp 1–6
6. Karabeyoglu A, Zilliac G, Cantwell BJ, DeZilwa S, Castellucci P (2004) Scale-up tests of high regression rate paraffin-based hybrid rocket fuels. J Propuls Power 20:1037–1045
7. Zilliac G, Karabeyoglu M Hybrid rocket fuel regression rate data and modeling. In: 42nd AIAA/ASME/SAE/ASEE joint propulsion conference & exhibit. American Institute of Aeronautics and Astronautics
8. Lee T-S, Tsai H-L (2008) Combustion characteristics of a paraffin-based fuel hybrid rocket. In: 9th Asia-Pacific international symposium on combustion and energy utilization, pp 2–6
9. Mahottamananda SN, Joshi PC (2016) Combustion study of HTPB–sugar hybrid fuel with gaseous oxygen. J Therm Anal Calorim 123:1927–1934
10. Sinha YK, Sridhar BTN, Kishnakumar R (2016) Study of thermo-mechanical properties of HTPB–Paraffin solid fuel. Arab J Sci Eng. 41:4683–4690
11. Sinha YK, Sridhar BTN, Santhosh M (2015) Thermal decomposition study of HTPB solid fuel in the presence of activated charcoal and paraffin. J Therm Anal Calorim 119:557–565
12. Maruyama S, Ishiguro T, Shinohara K, Nakagawa I Study on mechanical characteristics of paraffin-based fuel. In: 47th AIAA/ASME/SAE/ASEE joint propulsion conference & exhibit. American Institute of Aeronautics and Astronautics
13. Kumar R, Ramakrishna PA (2016) Studies on EVA-based wax fuel for launch vehicle applications. Propellants Explos Pyrotech 41:295–303
14. Kim S, Moon H, Kim J, Cho J (2015) Evaluation of paraffin-polyethylene blends as novel solid fuel for hybrid rockets. J Propuls Power 31:1750–1760
15. DeSain J, Brady B, Metzler K, Curtiss T, Albright T Tensile tests of paraffin wax for hybrid rocket fuel grains. In: 45th AIAA/ASME/SAE/ASEE joint propulsion conference & exhibit. American Institute of Aeronautics and Astronautics
16. Sakote R, Yadav N, Karmakar S, Joshi PC, Chatterjee AK (2014) Regression rate studies of paraffin wax-HTPB hybrid fuels using swirl injectors. Propellants Explos Pyrotech 39:859–865
17. Sarı A, Alkan C, Bilgin C (2014) Micro/nano encapsulation of some paraffin eutectic mixtures with poly(methyl methacrylate) shell: preparation, characterization and latent heat thermal energy storage properties. Appl Energy 136:217–227

18. You M, Wang X, Zhang X, Zhang L, Wang J (2011) Microencapsulated n-Octadecane with styrene-divinybenzene co-polymer shells. J Polym Res 18:49–58
19. Lv X, Zha M, Ma Z, Zhao F, Xu S, Xu H (2017) Fabrication, characterization, and combustion performance of Al/HTPB composite particles. Combust Sci Technol 189:312–321

Aerodynamic Performance of Shark Skin Shape Vortex Generator

Nur Faraihan Zulkefli, Mohamad Asmidzam Ahamat, Nurul Fatihah Mohd Safri, Nurhayati Mohd Nur, and Azmin Syakrine Mohd Rafie

Abstract The performance of NACA 4415 airfoils with counter rotating triangular or shark skin shape sub-boundary layer vortex generators were evaluated. This is relevant to the improvement of lift-to-drag ratio of the airfoil which will contribute to development of a more efficient airfoil. The lift and drag coefficients of NACA 4415 airfoils without vortex generator, with triangular or shark skin shape vortex generators were obtained from wind tunnel tests. For the airfoil with shark skin shape vortex generator, its performance was compared with the results from computational fluid dynamics simulation. For all tests and simulations, the location of vortex generators was maintained at the 50% of chord length. The Reynolds number was maintained at 2.5×10^5. The only varied parameter other than the shape of vortex generator was the angle of attack which was varied between 0° and 20°, with 2° increment. In some conditions, the performance of airfoil with shark skin shape vortex generator was better than the airfoil with triangular vortex generator. The simulated lift and drag coefficients for airfoil with shark skin shape vortex generator are comparable with the results from experiments. These findings are important for future development of highly efficient airfoil with innovative shape of vortex generator.

Keywords Sub-boundary layer · Flow control · Counter-rotating · Lift coefficient

---

N. F. Zulkefli (✉) · N. F. M. Safri · N. M. Nur
Universiti Kuala Lumpur Malaysian Institute of Aviation Technology, Sepang, Selangor, Malaysia
e-mail: nurfaraihan@unikl.edu.my

M. A. Ahamat
Universiti Kuala Lumpur Malaysia France Institute, Bandar Baru Bangi, Selangor, Malaysia

A. S. M. Rafie
Universiti Putra Malaysia, Seri Kembangan, Selangor, Malaysia

P. Rajendran et al. (eds.), *Proceedings of International Conference of Aerospace and Mechanical Engineering 2019*, Lecture Notes in Mechanical Engineering,
https://doi.org/10.1007/978-981-15-4756-0_10

1 Introduction

1.1 Flow Separation Control

Vortex generators are commonly used in the commercial and military aircraft to control the flow pattern of air on the top surface of its wings. Some of the effects of vortex generator on the air flow are delaying transition, increase lift coefficient, enhance turbulent, postpone flow separation and reduce drag coefficient. Vortex generators also increase the lift-to-drag ratio which contributes to lower fuel consumption, reducing stall speed and reduce the takeoff and landing distances.

Passive vortex generators are devices that firmly fixed to the surface of an airfoil, less expensive, and less complicated to be implemented. The sub-boundary layer vortex generator is also known as low-profile vortex generator or micro-vortex generator that has the height(h)-over-boundary layer thickness (δ) of much less than 1. It has a better performance in induce flow mixing, thickened the turbulent velocity profiles and eliminate flow separation from the surface of airfoil. Some of the researchers also refer this sub-boundary layer vortex generator as sub-merged vortex generator due to direct mixing in the lowest part of boundary layer.

The triangular sub-boundary layer vortex generator with spaces vanes are more effective than joined vanes in avoiding adverse pressure gradient [1]. Counter-rotating vortex generator is more effective compared to the co-rotating vortex generator [2]. The optimal angle of attack (β) is 18° for triangular vortex generator, which give 20% reduction in drag compared to rectangular vortex generator [3]. When the sub-boundary layer vortex generator is positioned at 30% of airfoil chord at the wind turbine blade, the maximum increase in lift force is 44% while drag force decrease at 0.2% at pre-stall [4].

Biomimetic technique has been implemented as passive control devices by modifying the leading and trailing edge of airfoil, and the surface of airfoil [5]. Recent studies concern on biomimetic flow control show that reduction in drag force can improve aerodynamic performance of airfoil. Leading edge bumps inspired from humpback whale flippers is one of the devices which can reduce drag force by almost 40% [6]. The bristled of shark skin structure have potential to control the boundary layer thickness and decrease the overall drag by a careful control of its surface roughness characteristics [7]. Skin friction drag also can be reduced by imitating the surface profile of shark skin [8]. The denticle shape inspired from shortfin of Mako shark skin is arranged in the wide range on NACA 0012 airfoil at 26% chord, trapped the boundary layer and generate short reattaching separation bubble along the chord. This leads to a drag reduction and prevent loss of lift at higher angle of attack [9]. Bechert et al. [10] used turbulent oil tunnel to measure the shear stress over flat plate and it was observed that skin friction decreased by approximately 3%. Meanwhile, Domel et al. [11] found that the smallest sharkskin denticles were able to reduce power consumption and drag force compared to smooth airfoil (without denticles). However, there is no attempt has been made by researchers to develop vortex generator with the shape that is inspired from nature.

In this paper, the lift and drag coefficients of NACA 4415 airfoil without vortex generator, with passive vortex generators with triangular or shark skin shape were evaluated by experiments and computational fluid dynamics simulation. The performance of NACA 4415 airfoil with triangular or shark skin shape vortex generator was compared with the airfoil with no vortex generator. The lift to drag ratio was evaluated at the angle of attack between 0° and 20° at the Reynolds number of 2.5×10^5.

2 Experimental Setup

2.1 Fabrication and Specification of Airfoil

In this paper, the performance of NACA 4415 airfoil (Fig. 1) with the chord length of 130 mm and spanwise of 300 mm (equal to width of the test section of wind tunnel) was evaluated. The airfoil was made from ABS and fabricated using 3D printing technique. A slot was made on the top surface of the airfoil at the 50%C. This slot can accommodate three different surfaces which were the smooth surface, surface with triangular vortex generator (Fig. 2a) and surface with shark skin shape vortex generator (Fig. 2b). Figure 3 shows the schematic diagram of arrangement of these passive vortex generators. The height (h) of the vortex generator is 4 mm with a length (L) of 12 mm or 3 h. The distance between the pair of vortex generator (D = 24 mm) is equal to h, and the spacing between one pair to another pair of vortex

Fig. 1 The NACA 4415 airfoil with slot for vortex generator at 50%C

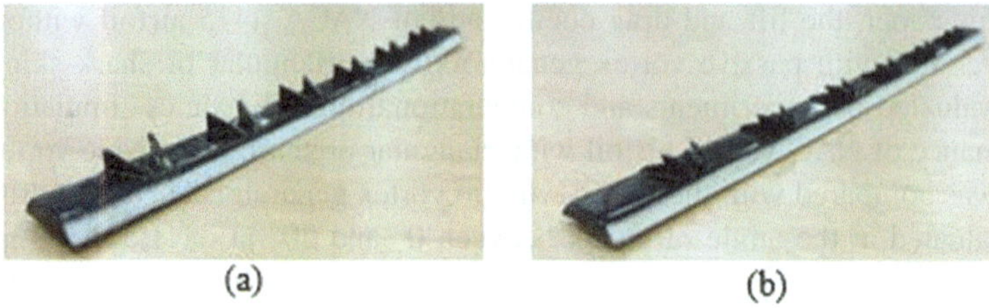

Fig. 2 (**a**) Triangular counter-rotating vortex generator, (**b**) shark skin shape counter-rotating vortex generator

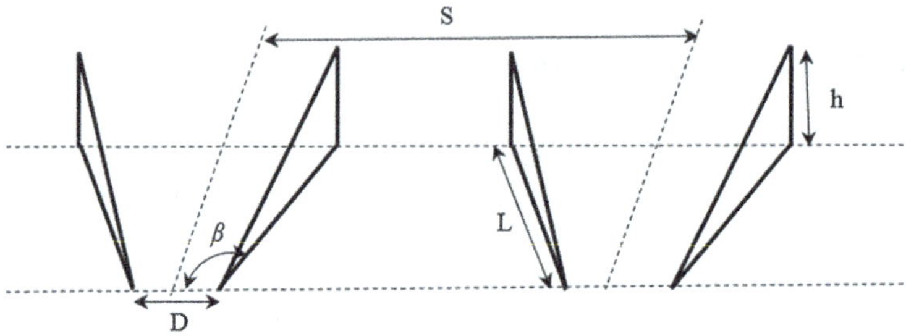

Fig. 3 Schematic design of passive vortex generator (h = 4 mm)

generator (S = 28 mm) is equal to 7 h. Both triangular and shark skin shape counter-rotating vortex generators has the same dimension with angle of incident (β) of 20° and located at 50% of chord length (C).

2.2 *Experiments to Determine Lift and Drag Coefficients*

The experimental work covered three cases which are airfoil with no vortex generator, airfoil with counter-rotating triangular vortex generator and airfoil with counter-rotating shark skin shape vortex generator. The airfoil with no vortex generator served as the baseline case.

The experiments were conducted in an open circuit wind tunnel as in Fig. 4. The test section has the dimension of 300 mm × 300 mm × 600 mm. The maximum velocity of air in the wind tunnel is 50 m/s at 1400 rpm. The air entered the wind tunnel through a settling chamber containing a honeycomb and screen before it passes through the contraction cone. Honeycomb structure straighten the flow by reducing lateral velocities while the screen reduces the axial turbulences.

The airfoil was attached to a structure that is connected to axis balance. The velocity of air was set to 28 m/s which corresponded to Reynolds number of 2.5 × 10^5. Experiments were conducted at the angle of attack from 0° to 20°, with the increment of 2° for all airfoils.

Fig. 4 Open circuit subsonic wind tunnel for the NACA 4415 airfoils test

2.3 Computational Domain and Simulation Set-up

The lift and drag coefficients of NACA 4415 airfoil was simulated at the Reynolds number of 2.5×10^5. The computational domain was extended to 11.5 length of chord (C) for upstream, 21C for downstream, 12.5C for upper and lower regions, and 0.01C in the direction of spanwise of the airfoil (Fig. 5). The hexahedral meshes were used

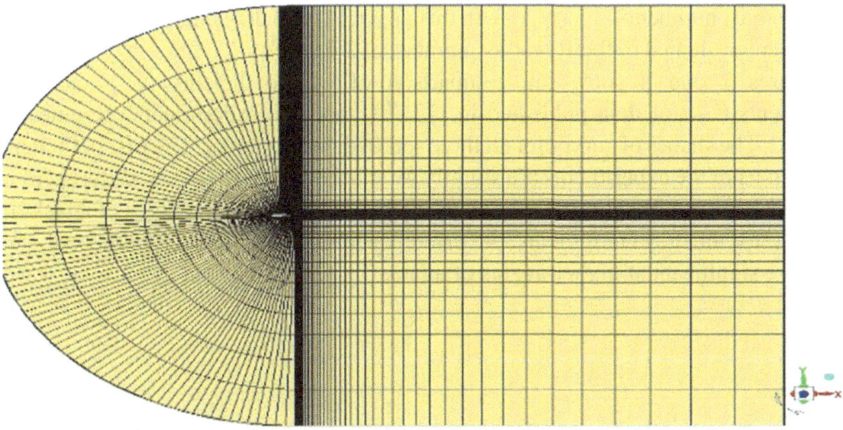

Fig. 5 Computational domain of the simulation of NACA 4415 airfoil

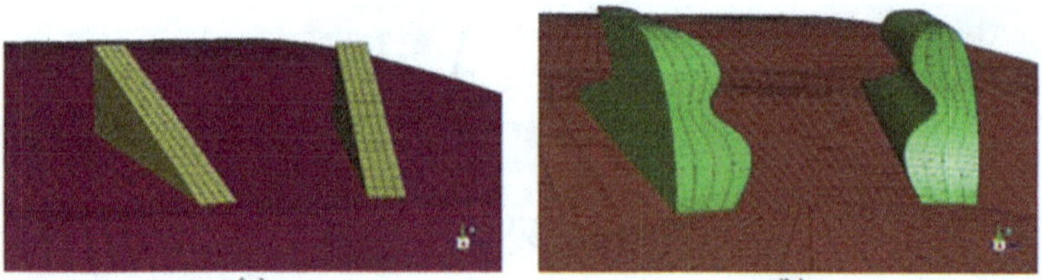

Fig. 6 Hexahedral meshes of (**a**) counter-rotating triangular vortex generator and (**b**) counter-rotating shark skin vortex generator

for both counter-rotating triangular and shark skin shape vortex generator (Fig. 6). The simulation was executed in ANSYS 17.2 Commercial simulation software using steady state analysis. The transition SST turbulent model was adopted because this model has better accuracy with less computational time. The simulation was run in pressure-based and the SIMPLE pressure-velocity coupling for both first and second order scheme with constant value of density, viscosity and temperature.

3 Results

3.1 Airfoil Without Vortex Generator

The results obtained from experiments were compared with the data available in published papers. The data in Foutih et al. [12] is on the lift and drag coefficients of NACA 4415 airfoil obtained through wind tunnel tests at the Reynolds number of 2.0×10^5. In addition, the result published by Zhen et al. [13] is for NACA 4415 at the Reynolds number of 2.13×10^5. In this paper, lift and drag coefficients of NACA 4415 were evaluated at the Reynolds number of 2.5×10^5.

Figures 7 and 8 show the comparison of lift coefficients (C_L) and drag coefficients(C_D) for data published in [12, 13], and the experiments presented in Sect. 2.2. The shapes of curves for all three sets of data are similar. The lift coefficients obtained from the experiments are slightly differs from [12, 13] due to differences in Reynolds number. The discrepancy is more pronounced at lower angles of attack. In contrast, for the drag coefficients, the discrepancies between these three sets of data are more significant at higher angles of attack.

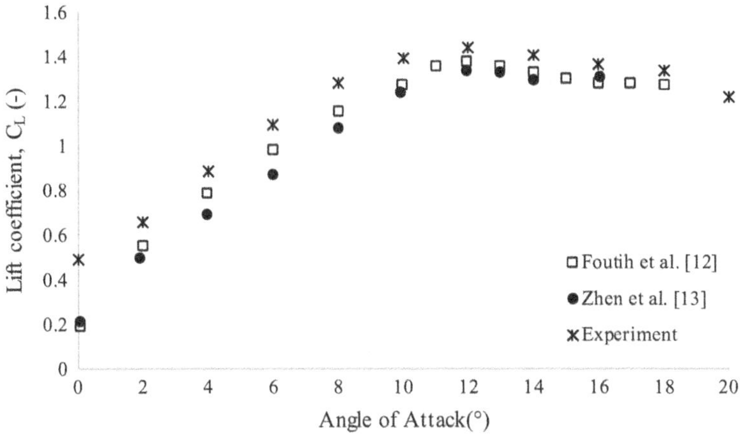

Fig. 7 Lift coefficients at various angles of attack for NACA 4415 airfoil without vortex generator

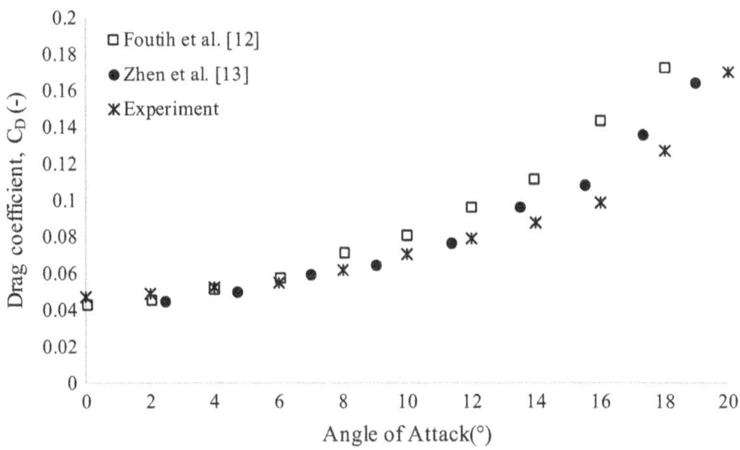

Fig. 8 Drag coefficients at various angles of attack for NACA 4415 airfoil without vortex generator

3.2 *Lift and Drag Coefficients for NACA 4415 Airfoils with Counter-Rotating Triangular or Shark Skin Shape Vortex Generators*

The comparison of lift coefficients for NACA 4415 airfoil without vortex generator, NACA 4415 airfoil with counter-rotating triangular vortex generator and NACA 4415 airfoil with counter-rotating shark skin shape vortex generator are presented in Fig. 9. It can be seen clearly that NACA 4415 airfoils with counter-rotating triangular or shark skin shape vortex generators have higher lift coefficients compared to the airfoil without vortex generator especially when the angle of attack is between 10°

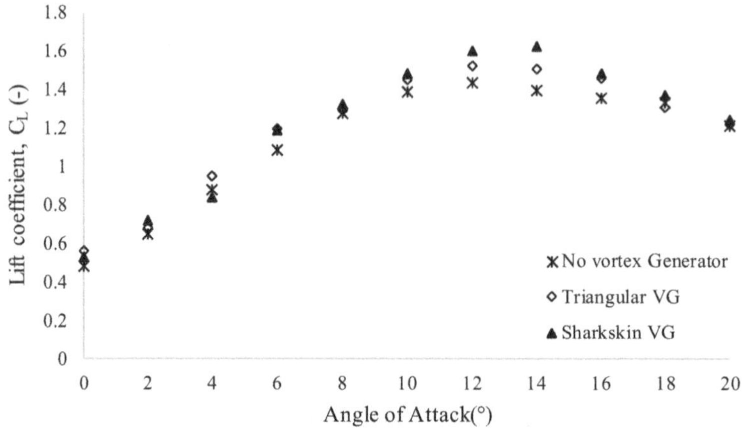

Fig. 9 Lift coefficients at various angles of attack for NACA 4415 airfoil without vortex generator, and airfoils with counter-rotating triangular or shark skin shape vortex generators

and 16°. At the angle of attack of 14°, the difference in lift coefficients between airfoil with counter-rotating triangular and shark skin shape vortex generator is about 5%.

For the comparison of drag coefficients (Fig. 10), airfoil with no vortex generator has the highest drag coefficients for all range of angles of attack. The drag coefficients for the airfoils with counter-rotating triangular or shark skin shape vortex generators are comparable. The maximum difference in drag coefficient is 50%, at the lower angle of attack.

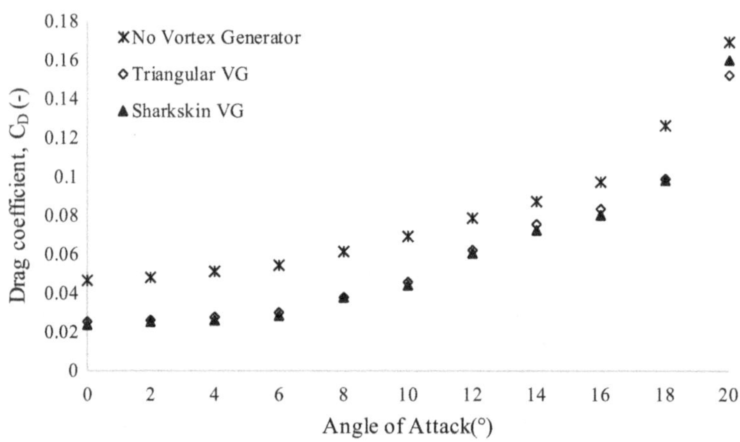

Fig. 10 Drag coefficients at various angles of attack for NACA 4415 airfoil without vortex generator, and airfoils with counter-rotating triangular or shark skin shape vortex generators

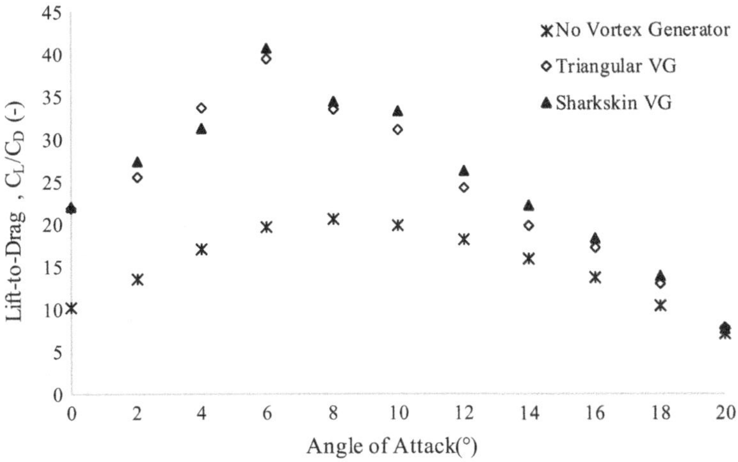

Fig. 11 The lift-to-drag ratio at various angles of attack for NACA 4415 airfoil without vortex generator, and airfoils with counter-rotating triangular or shark skin shape vortex generators

The lift-to-drag ratio (C_L/C_D) for airfoils with no vortex generator, airfoil with counter-rotating triangular vortex generator and airfoil with counter-rotating shark skin shape vortex generator are presented in Fig. 11. The maximum difference in lift-to-drag-ratio between the airfoils with vortex generators and the airfoil with no vortex generator is 50%, occurred at the angle of attack of 6°. There is no difference in lift-to-drag-ratio for the three airfoils at the angle of attack of 20°.

3.3 Performance of NACA 4415 Airfoil with Shark Skin Shape Vortex Generator: Comparison Between Wind Tunnel Tests and Computational Fluid Dynamics Simulation

Figure 12 shows the comparison of lift coefficients for the NACA 4415 airfoil with shark skin vortex generator. The discrepancies between data obtained from experiments and simulation are within 10%. For the comparison of drag coefficient (Fig. 13), significant differences can be observed when the angle of attack is higher than 12°. The difference in lift-to-drag ratio (Fig. 14) is at its highest at the angle of attack of 6°.

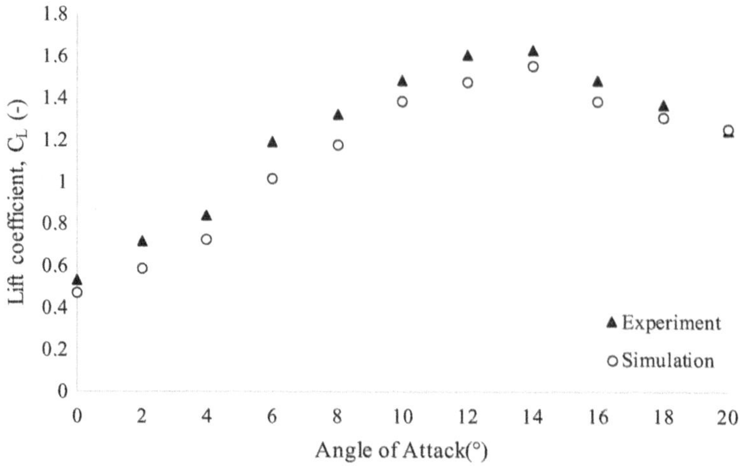

Fig. 12 Lift coefficients of NACA 4415 with counter-rotating skin shark shape vortex generator at various angles of attack

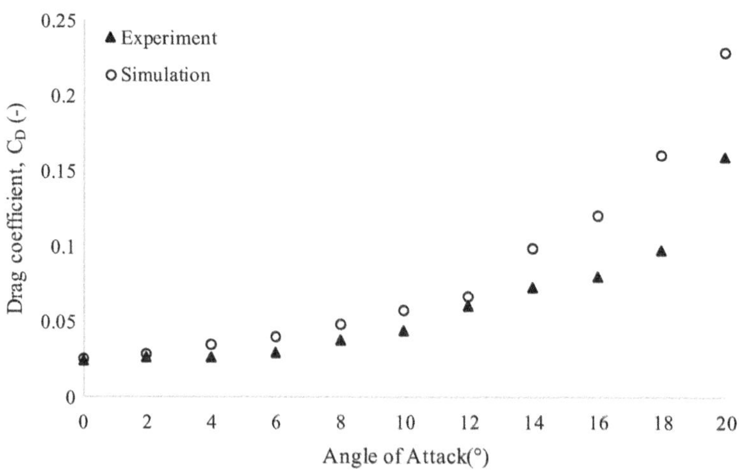

Fig. 13 Drag coefficients of NACA 4415 with counter-rotating skin shark shape vortex generator at various angles of attack

4 Discussion

This paper is part of the efforts to improve the aerodynamic performance of airfoil through biomimetic strategy. One of the challenges in mimicking the shape of shark skin on vortex generator is to determine the suitable fabrication method. In this paper, the vortex generator was fabricated using 3D printing technique. One of the limitations of this method is the smallest features that can be printed is depending on

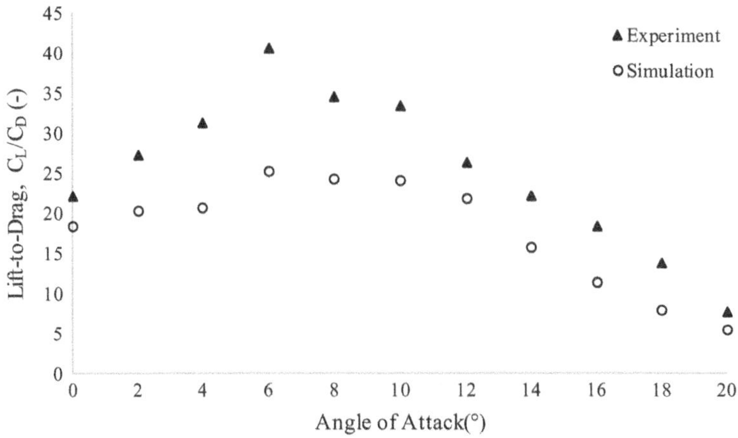

Fig. 14 Lift-to-drag ratio of NACA 4415 with counter-rotating shark skin vortex generator at various angles of attack

the capability of the 3D printer. Another challenge is to precisely replicate the shape of the denticles. Arguably, the denticles found on the surface of shark skin can have variations in size and shape. Further investigations to determine the effect of these variation may be required in the future.

The results presented in this paper is only at one Reynolds number. Thus, the validity of the performance of counter-rotating shark skin shape vortex generator is limited to this Reynolds number only. The results are applicable for the development of small aircraft that is relevant to military or surveillance applications. For the airfoil with longer chord line, the estimated relative velocity may be suitable for the operation of wind turbine. In future, the performance of airfoil with shark skin shape vortex generator at higher Reynolds number is worth to be investigated.

5 Conclusion

The performance of NACA 4415 airfoils without vortex generator, with counter-rotating triangular or shark skin shape vortex generators were evaluated. It is obvious that the effect of vortex generators significantly improves the lift-to-drag ratio of airfoils. The difference in lift-to-drag ratio for airfoils with counter-rotating triangular vortex generator or shark skin shape vortex generators are comparable. However, the difference in lift coefficient of airfoil with shark skin shape vortex generator is 5% higher at the angle of attack of ~14°. Further investigation on the performance of these airfoils at higher Reynolds number may reveal significant differences on their performance. The findings presented in this paper is relevant to the development of efficient wind turbine blades or wing of a small aircraft that can be used for surveillance systems.

Acknowledgements The authors would like to acknowledge the financial support from Ministry of Higher Education (MOHE) through Fundamental Research Grant Scheme (FRGS/1/2018/TK08/UNIKL/02/1).

References

1. Ashill PR, Fulker JL, Hackett KC (2002) Studies of flows induced by sub boundary layer vortex generators (SBVGs). In: 40th AIAA aerospace sciences meeting & exhibit, vol 968, p 13
2. Godard G, Stanislas M (2006) Control of a decelerating boundary layer. Part 1: optimization of passive vortex generators. Aerosp Sci Technol 10(3):181–191
3. Kumar KR, Morab S, Shekar S, Mahalingam A (2016) A CFD study of the performance characteristics of vented cylinders as vortex generators. IOSR J Mech Civ Eng 13(1):46–52
4. Manolesos M, Voutsinas SG (2015) Experimental investigation of the flow past passive vortex generators on an airfoil experiencing three-dimensional separation. J Wind Eng Ind Aerodyn 142:130–148
5. Choi H, Park H, Sagong W, Lee S (2012) Biomimetic flow control based on morphological features of living creatures. Phys Fluids 24(12):121302
6. Anders JB (2000) Biomimetic flow control. In: AIAA-2000–2543. AIAA, Denver, pp 1–28
7. Dean B, Bhushan B (2010) Shark-skin surfaces for fluid-drag reduction in turbulent flow: a review. Philos Trans R Soc A Math Phys Eng Sci 368(193):5737–5737
8. Raayai-Ardakani S, McKinley GH (2017) Drag reduction using wrinkled surfaces in high Reynolds number laminar boundary layer flows. Phys Fluids 29(9):093605
9. Domel AG, Saadat M, Weaver JC, Haj-Hariri H, Bertoldi K, Lauder GV (2018) Shark skin-inspired designs that improve aerodynamic performance. J Royal Soc Interface 15(139):1–9
10. Bechert DW, Bruse M, Hage W, Meyer R (2000) Fluid mechanics of biological surfaces and their technological application. Naturwissenschaften 87(4):157–171
11. Domel AG, Domel G, Weaver JC, Saadat M, Bertoldi K, Lauder GV (2018) Hydrodynamic properties of biomimetic shark skin: effect of denticle size and swimming speed. Bioinspir Biomim 13(5):1–46
12. Fouatih OM, Medale M, Imine O, Imine B (2016) Design optimization of the aerodynamic passive flow control on NACA 4415 airfoil using vortex generators. Eur J Mech B/Fluids 56:82–96
13. Zhen TK, Zubair M, Ahmad KA (2011) Experimental and numerical investigation of the effects of passive vortex generators on Aludra UAV performance. Chin J Aeronaut 24(5):577–583

A Prototype Development and Evaluation of Electrochemical Device for Heavy Metal Measurement

Siti Nur Hanisah Umar, Elmi Abu Bakar, Noorfazreena Mohammad Kamaruddin, and Naoki Uchiyama

Abstract Heavy metals are highly toxic, carcinogenic, and non-biodegradable and may cause harmful effects on human health and the environment. To reduce or eliminate heavy metal contamination, it is necessary to accurately determine the amount of heavy metal in the environment. Traditional detection techniques are expensive, bulky, and not suitable for in-site applications. The electrochemical technique using a potentiostat instrument is a reliable alternative as its potential to fabricate on a small circuit for in-site application. "Black boxes" nature of most commercial potentiostat leads to various development of lab-build potentiostat. Most lab-build potentiostat was designed for general purpose used. This paper focused on development and evaluation of potentiostat specific for heavy metal detection. The development consists of a readout circuit of electronic components together with NI myRIO-1900 as a controller system. Linearity and accuracy of the readout circuit were evaluated. Single and simultaneous detection of copper (Cu), cadmium (Cd) and lead (Pb) were tested. The results show that the proposed design has good repeatability comparable to a commercial potentiostat.

Keywords Potentiostat · myRIO · LabVIEW · Electrochemical · Square wave voltammetry · Operational amplifier · Heavy metal

1 Introduction

Heavy metals are defined as metals with atomic weights between 63.5 and 200.6 g mol^{-1} and a specific gravity greater than 5 g cm^{-3} [1]. The source of heavy metal in the environment can be both natural and anthropogenic origin [2]. Weathering of soil and rocks are the main source of natural origin. The main anthropogenic source such as agriculture, industrial and metallurgical activities, energy production

S. N. H. Umar (✉) · E. Abu Bakar · N. Mohammad Kamaruddin
School of Aerospace Engineering, Universiti Sains Malaysia, Nibong Tebal, Malaysia
e-mail: snhanisahumar@gmail.com

N. Uchiyama
Department of Mechanical Engineering, Toyohashi University of Technology, Aichi, Japan

P. Rajendran et al. (eds.), *Proceedings of International Conference of Aerospace and Mechanical Engineering 2019*, Lecture Notes in Mechanical Engineering,
https://doi.org/10.1007/978-981-15-4756-0_11

and transportation influence the arising of contamination of most rivers and estuaries [3]. The heavy metals are highly toxic, carcinogenic, non-biodegradable and lead to a greater risk to human health and environment [1]. To reduce heavy metal pollution, it is necessary to accurately determine the concentration [4].

Traditionally quantitative techniques, such as inductive-coupled plasma atomic emission spectroscopy (ICP-AES), inductive-coupled plasma mass spectrometry (ICP-MS), and graphite furnace atomic adsorption spectrometry (GFAAS), require a large set-up, high cost, and sophisticated instrument-equipped laboratories for sample analysis. Thus, cannot be applied to develop real-time and on-site monitoring devices [5]. An electrochemical technique is a reliable alternative to traditional technique and has gained more popularity as it is more affordable, user-friendly, allow simple procedures and suitable to fabricated on a small circuit in the portable form for in-site applications [6].

Unfortunately, most commercial electrochemical instruments, the potentiostat, function as "black boxes" cause limited information of their circuitry and behaviour, thus lead to significant constraints for system integration for new measurement and application [7]. Recently, a few lab-build potentiostats have been reported. Most of the potentiostat is designed for general-laboratory purposes [7, 8]. Some are designed as a multidisciplinary teaching tools experiment for exploring both electrochemistry and electronic aspects [9, 10].

The objective of this study is to design the I-HMstat, an electrochemical device specific for detection of heavy metal elements in the water. In order to meet this objective, a prototype will be developed. In the previous design, the prototype is based on an open-source microcontroller adjoined with nine electronic components as a readout circuit [6]. While this paper will explain the development and evaluation of a simpler design by minimized the component in the readout circuit into five electronic components. The controller, comprised of NI myRIO-1900 together with a virtual instrument (NI LabView) was used as control, acquisition and processing signal. The application of this virtual instrument allows for later signal processing for automatic quantification and identification of heavy metal elements [11].

2 Methodology

The overall design of the I-HMstat electrochemical heavy metal detection system as shown in Fig. 1. The design can be categories into a readout circuit of a basic potentiostat component and a controller system component. Each stage in the basic potentiostat component is a combination of active and passive electronic parts.

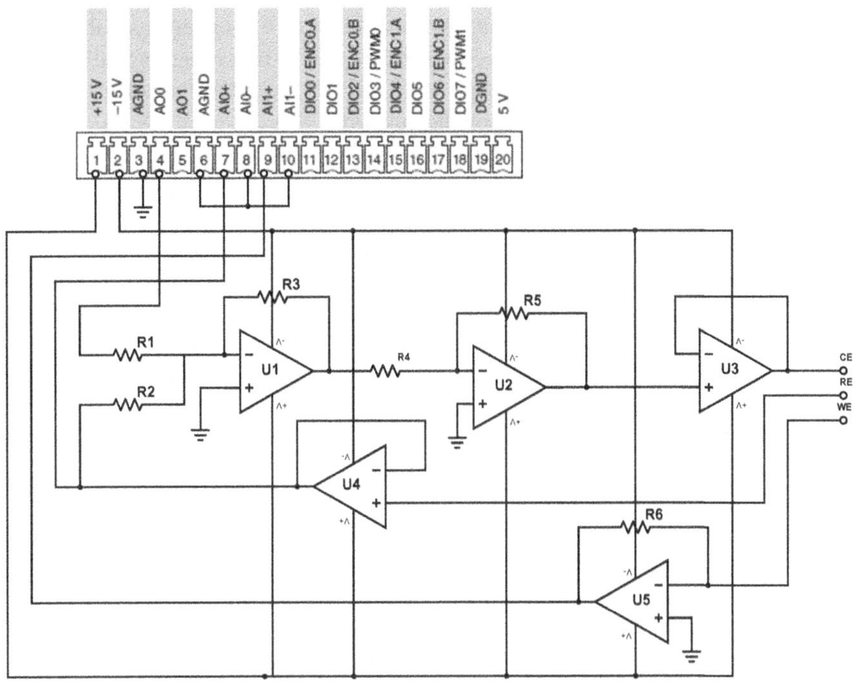

Fig. 1 Circuit diagram of basic potentiostat connected to the input (AI0 and AI1) and output (AO0) of controller part

2.1 Basic Potentiostat Design

The potentiostat readout circuit comprised of the potential control part and the current measurement part. The function of the potential control part is to control the interfacial potential at the working electrode (WE) with referral to reference electrode (RE) in which the voltage is applied at the counter electrode (CE) [7]. The circuit was designed in several stages. The first stage is the summing inverting amplifier (U1) considering the feedback component from RE through a unity gain amplifier (U4). Then, another inverting amplifier (U2) to rectify the signal to initial polarity, followed by a voltage follower (U3).

The potential input to the electrochemical cell through CE is express as,

$$V_{out} = \frac{R_4}{R_5}\left(\frac{R_3}{R_1}V_{in_1} + \frac{R_3}{R_2}V_{in_2}\right) \tag{1}$$

where, V_{in_1} is an output potential signal applied from the controller through analog output 0 (AO0) and V_{in_2} is an output potential form U4. The resistor value for this design is set as $R_1 = R_2 = R_3 = R_4 = R_5 = 10k\Omega$.

For the current measurement part, the function was to convert a small current change from redox reaction in the electrochemical cells to the voltage signal [12]. For this purpose, the trans-impedance amplifier (TIA) (U5) was designed. The amplifier output can be determined as the following equation,

$$V_{in} = I_i R_6 \tag{2}$$

where I_i is the measured current and R_6 is a feedback resistor of the transimpedance amplifier. The output of current measurement can be altered to the specific range of current value by changing the R_6 value [12].

2.2 Signal Control and Process

The controller was used to generate an algorithm for signal function, signal processing, and graphical user interfaced (GUI). The program development environment used a graphical programming language which is in the form of blocks diagram [13]. In this design, the output potential signal from NI myRIO and the input signal to be processed by NI myRIO was connected to the circuit readout through mini system port (MSP) connector C. The MSP connector C has two analog output channels (AO0 and AO1) and two analog input channels (AI0 and AI1) which can be used to generate and measure signals up to ± 10 V respectively. The analog output channel has a dedicated digital-to-analog converter (DAC), controlled by two serial communication buses from field-programmable gate array (FPGA). While analog input equipped with an analog-to-digital converter (ADC) of 12 bits resolution [14]. The build-in DAC and ADC in the controller system reduced the component used in circuit readout potentiostat compared to previous design [6]. The generated potential signal by the controller can be directly applied to the CE through the AO0 channel. While the output current measurement can be directly acquired to the controller through the AI0 channel for signal processing.

2.3 Detection of Heavy Metal Using Square Wave Anodic Stripping Voltammetry

Figure 2 shows the experimental setup of the controller, readout circuit, electrochemical cell and, GUI. The Square wave anodic stripping voltammetry (SWASV) method was used to detect heavy metal element due to its ability to minimize the contribution of non-faradaic current, usually provide more precise current measurement, better separation of interfering voltammetric peak and lowers levels of quantification [12]. The main step in SWASV is the deposition step followed by the stripping step. The deposition is where the negative potential is applied to the WE. The purpose is to

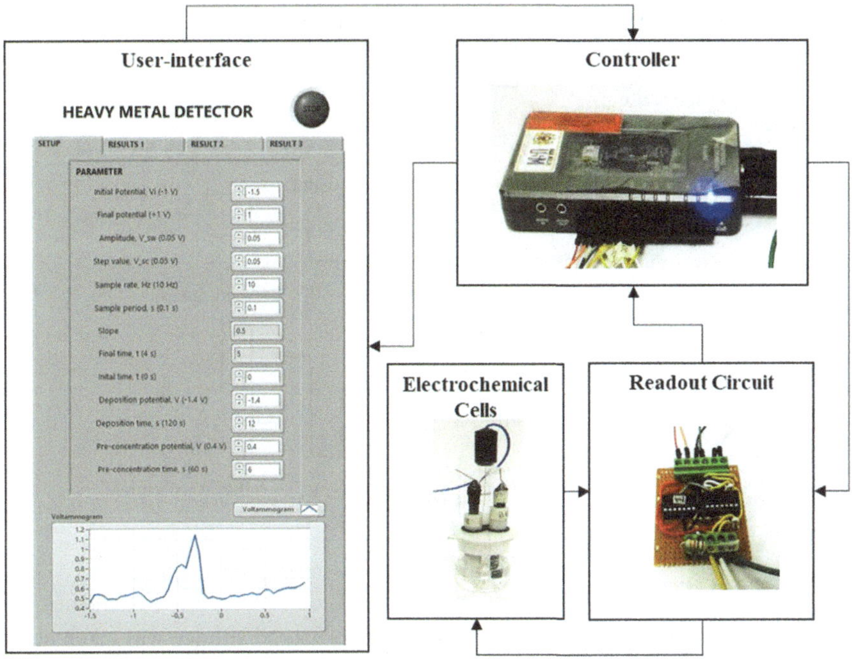

Fig. 2 Overall component in electrochemical heavy metal detection system

reduce the metal ion in the solution onto the electrode surface. The reduction will occur if the applied potential is more negative than the reduction potential of the metal [15]. Subsequent, the stripping step with a potential scan of square wave voltammetry (SWV) caused the metal deposited on the electrode surface to re-oxidized and dissolve into the solution. The oxidation will occur when potential matches the oxidation potential of the metal [15]. The applied signal of SWV was implemented by superimposed a staircase signal on a symmetrical square wave element as illustrated in Fig. 3.

3 Results and Discussions

3.1 Linear Calibration Test

Two bench test was designed and conducted. The first test was conducted to evaluate the linearity and accuracy of the input/output signal at the control part readout circuit. The second test was to examine the capability of the current measurement readout circuit to convert current received from WE to the voltage signal. To do so, the output of CE was connected to 10 kΩ resistor and 11 samples of input voltage were supplied

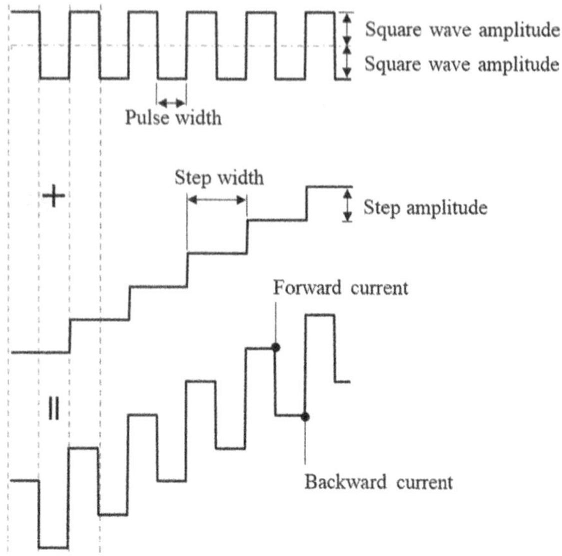

Fig. 3 The potential scan of SWASV is the sum of simple symmetrical staircase and square wave of same phase and frequency

range from −1.0 to +1.0 V. In the first test, the measured voltage was taken at the output voltage of CE using a digital voltmeter. An average error of the measured voltage was 0.29% compared to the ideal case calculated using Eq. 1. For the second test, the resultant current was determined based on voltage measurement from the previous test and resistive load measured using a digital ohmmeter. The measured voltage of converted current by readout circuit was taken at the output of TIA. The voltage was compared to the ideal case calculated using Eq. 2 with an average error of 0.22%. Referring to Fig. 4, both tests indicated a regression line with a correlation

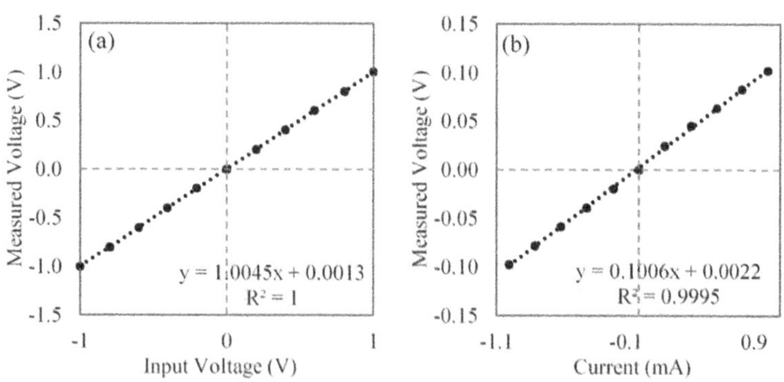

Fig. 4 Linear calibration result showing **a** capability of output circuit readout to produce variety of output voltages and **b** the accuracy of the circuit readout to convert current received at the WE

coefficient of perfectly fit and nearly fit for control part and current measurement readout circuit respectively.

3.2 *Detection of Various Metal in Solution*

Detection of copper (Cu), cadmium (Cd) and lead (Pb) were carried out using SWASV. Deionized water was used to dilute the stock solution of Cu, Cd, and Pb into the desired concentration. The experiment was conducted using pencil graphite electrode (PGE) as WE and CE, and Ag/AgCl as RE. During the detection of Cu, −1.4 V was applied to the WE for 120 s, caused the Cu ion in the solution to reduced and deposited onto the electrode surface. Then, the deposited ion was stripped back into solution when potential scan of SWV were applied from −1.4 to +1.5 V, with square wave amplitude of 0.05 V, staircase amplitude of 0.05 V and sample rate of 10 Hz. The stripping (oxidation) of Cu ion occurred when scan potential matches the oxidation potential of Cu. The same procedure also applied for the detection of Cd and Pb. Referring to Fig. 5, the stripping of Cu element in the solution happened at the potential of +0.25 V. While Cd and Pb were detected at −0.7 and −0.3 V respectively. This showed, SWASV method employed differences oxidation potential for the detection of the metal element. For simultaneous detection, Cu, Cd, and Pb were detected at the potential of +0.05, −0.75 and −0.45 V respectively. It is important to consider the shifted detection potential of each element during individual and simultaneous detection especially when an automated system to be designed.

The current measurement provided information of each heavy metal concentration. The current signal was sampled at the end of each pulse called forward current and backward current as depicted in Fig. 3. The differences between forward and backward current indicated the peak current as seen in Fig. 5. Most elements were able to produce a sharp distinct peak except for the detection of Cu element using the commercial instrument in Fig. 5b. The peak current was blunter and round. This is due to, the difference between forward and backward current increased and reduced slightly before and after it reached a peak current of 0.45 mA respectively. Each measurement was repeated several times. Repeatability of the proposed system shown in Table 1 was comparable to a commercial potentiostat (Metrohm Autolab) under the same procedure and condition.

4 Conclusion

This paper outlined the design development of a simpler readout circuit equipped with NI myRIO-1900 as a controller. The accuracy of the readout circuit potential control and current measurement part were compared to an ideal case with an average error of 0.29% and 0.22% respectively, these were within the 5% tolerance range

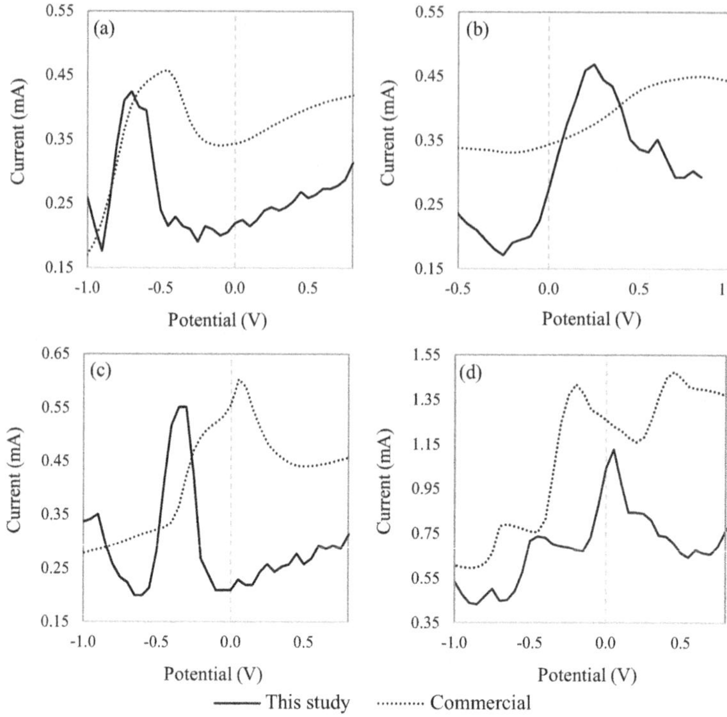

Fig. 5 Detection of heavy metal using proposed device and commercial instrument (Autolab) for **a** 10 ppm Cd, **b** 10 ppm Cu, **c** 10 ppm Pb and **d** simultaneous detection of 10 ppm Cd, 10 ppm Cu and 10 ppm Pb

Table 1 Repeatability test for peak current measurement

Element		$\bar{y}$	σ	$\bar{\sigma}$
Cd	This study	0.43	0.0153	0.0108
	Commercial	0.47	0.0115	0.0082
Cu	This study	0.47	0.0058	0.0041
	Commercial	0.43	0.0208	0.0147
Pb	This study	0.55	0.0058	0.0041
	Commercial	0.57	0.0300	0.0212

$\bar{y}$ is mean value
σ is standard deviation
$\bar{\sigma}$ is standard deviation of mean

of components used in fabrication. The SWASV method was used to detect Cu, Cd, and Pb. Individually, Cu, Cd, and Pb were detected at potential +0.25 V, −0.7 V and −0.3 V respectively. Detection potential was shifted to more negative area when simultaneous detection we conducted. Shifting phenomena of detection potential is a crucial element to consider especially when an automatic analytical measurement system to be designed.

Acknowledgements The authors would like to acknowledge Universiti Sains Malaysia (USM) and Ministry of Education (MOE) for providing the financial assistance under RU Top-Down research grant (1001/PAERO/6740041) and awarding her USM Academic Staff Training Scheme (ASTS).

References

1. Gumpu MB, Sethuraman S, Krishnan UM, Rayappan JBB (2016) A review on detection of heavy metal ions in water—An electrochemical approach. Sensors Actuators, B Chem 213(2015):515–533
2. Griswold W, Martin S (2009) Human health effects of heavy metals. Environ Sci Technol 15:1–6
3. Alkarkhi AFM, Ahmad A, Ismail N, Easa AM (2008) Multivariate analysis of heavy metals concentrations in river estuary. Environ Monit Assess 143(1–3):179–186
4. Lu Y, Liang X, Niyungeko C, Zhou J, Xu J, Tian G (2018) A review of the identification and detection of heavy metal ions in the environment by voltammetry. Talanta 178(2018):324–338
5. Biyani M et al (2016) DEP-on-go for simultaneous sensing of multiple heavy metals pollutants in environmental samples. Sensors 17(12):45
6. Umar SNH, Bakar EA, Kamaruddin NM, Uchiyama N (2018) A low cost potentiostat device for monitoring aqueous solution. MATEC Web Conf 217(04001):8
7. Dryden MDMM, Wheeler AR (2015) DStat: a versatile, open-source potentiostat for electroanalysis and integration. PLoS ONE 10(10):1–17
8. Rowe AA et al (2011) CheapStat: an open-source, 'Do-It-Yourself' potentiostat for analytical and educational applications. PLoS ONE 6(9):1–7
9. Li YC et al (2018) An easily fabricated low-cost potentiostat coupled with user-friendly software for introducing students to electrochemical reactions and electroanalytical techniques. J Chem Educ 95(9):1658–1661
10. Meloni GN (2016) Building a microcontroller based potentiostat: a inexpensive and versatile platform for teaching electrochemistry and instrumentation. J Chem Educ 93(7):1320–1322
11. Jakubowska M (2011) Signal processing in electrochemistry. Electroanalysis 23(3):553–572
12. Adams S, Doeven EH, Quayle K, Kouzani A (2019) MiniStat: development and evaluation of a mini-potentiostat for electrochemical measurements. IEEE Access 7:31903–31912
13. Abdul-Kadir NA, Noi S, Che Harun FK (2017) The evaluation of potentiostats: electrochemical detection devices. J Telecommun Electron Comput Eng 9(3–9):7–14 (2017)
14. National Instruments, User Guide and Specification NI myRIO-1900
15. Nemiroski A et al (2014) Universal mobile electrochemical detector designed for use in resource-limited applications. Proc Natl Acad Sci 111(33):11984–11989

Analysis of Pressure Coefficient Around Three Airfoils Operating at Different Reynolds Number Using CFD and XFOIL

Aravind Seeni and Parvathy Rajendran

Abstract In this paper, CFD is used to perform analyses of 3 different airfoils at varied angles of attack. The objective is to determine a suitable CFD model and find a single validated simulation setting applicable for different airfoils. The pressure coefficient around an airfoil is analyzed. A single Reynolds number condition is assigned for testing of each airfoil over different angles of attack. The CFD results are compared with experimental and XFOIL data. The solver used is the commercially available ANSYS v16.0 CFX. From this study the following results are derived: It is found that the simulation results closely match experimental results at Reynolds number in the range of 3 million. At low Reynolds number, the CFD approach struggles to reach the higher values of pressure coefficient achieved experimentally. It is also found that XFOIL provided better results compared to CFD and also converged at a faster rate. The above results are discussed in this paper.

Keywords Airfoils · Pressure coefficient · CFD · ANSYS CFX · XFOIL · NACA0012 · NACA0015 · ClarkY

1 Introduction

There are several approaches to perform aerodynamic tests on an airfoil in fluid flow. Some of the tests are based on numerical approach such as CFD, XFOIL while others are based on experimental testing in a wind tunnel. Before the advancement of computer technology, all analyses were carried out in wind tunnel and those tests were widely reported in literature. Also a numerical method known as XFOIL was developed in the 1980s as a rapid tool to perform airfoil analysis at low Reynolds

A. Seeni · P. Rajendran (✉)
School of Aerospace Engineering, Universiti Sains Malaysia, Engineering Campus, 14300 Nibong Tebal, Penang, Malaysia
e-mail: aeparvathy@usm.my

A. Seeni
e-mail: aravindseeni@gmail.com

P. Rajendran et al. (eds.), *Proceedings of International Conference of Aerospace and Mechanical Engineering 2019*, Lecture Notes in Mechanical Engineering,
https://doi.org/10.1007/978-981-15-4756-0_12

number (*Re*) [1]. The tool uses the panel method with a linear vorticity stream-function formulation. The recent advancement in semiconductors has given rise to Computational Fluid Dynamics (CFD) as an alternate powerful tool. The use of CFD has been applied in many areas of engineering and across different fields [2–7]. CFD uses the Finite element Method for discretization of a body into smaller elements before performing numerical integration to solve for Reynolds Averaged Navier Stokes Equations in fluid dynamic analysis. The advantage of using CFD is that it can be used to perform any analysis at a reasonable computational cost. As a result, it is extensively used in aircraft structural analysis and design. For instance, it has been used in different studies for fluid dynamic analysis of an aircraft airfoil [8–12].

Since CFD follows a numerical approach the accuracy of the results is highly dependent on the modelling technique as well as simulation settings. The present research focuses on determining a suitable CFD model that can be used to analyze a variety of airfoils through a single validated simulation setting. Validation of the model will need to be performed through comparison with experimental data. For performing this, a survey of experimental results of different airfoils is undertaken that provides the pressure coefficient (*Cp*) distribution plots at different α. Different α are chosen and a comparison of experimental results with CFD and XFOIL data is subsequently performed. The modeling technique and simulation setting proposed is the output of this research. A limitation of this work is that no attempt has been made to capture the intricate flow physics, boundary layer separation and reattachment. No discussion has been made to that effect. The focus is only on obtaining closest *Cp* values to experimental results.

2 Survey of Experimental Data for Different Airfoils

2.1 Study by Gregory and O'Reilly (1970)

The study by Gregory and O'Reilly [13] is an experimental test on NACA0012 airfoil for *Re* flows of 2.88×10^6. The tests were conducted on a 4×2.7 m low speed wind tunnel at N.P.L, UK. The *Re* of flow tests were carried out in the range of 3 million i.e. 2.88×10^6 for chord lengths (c) of 0.76 m and 55 m/s flow velocity (V_∞). The results of the test were characterized by laminar separation near the leading edge without *Re*-attachment as well as turbulent boundary layer separation at the trailing edge. Undesired 3D flow behavior across the section was characterized in this 2D test. A description of the study is provided in Table 1.

Table 1 Description of study by Gregory and O'Reilly [13]

Property	Description
Dataset	Gregory and O'Reilly [13]
Location of test	Aerodynamics division, N.P.L., UK
Airfoil	NACA0012
Chord length	0.76 m
Reynolds number	2.88×10^6
Flow velocity	55 m/s

Table 2 Description of study by Marchman and Werme [14]

Property	Description
Dataset	Marchman and Werme [14]
Location of test	Virginia Polytechnic Institute and State university, USA
Airfoil	ClarkY
Chord length	0.15 m
Reynolds number	75,000
Flow velocity	–

2.2 *Study by Marchman and Werme (1984)*

The study of Marchman and Werme [14] focused on studying ClarkY airfoil at very low *Re*. Data were obtained on pressure distributions on the upper and lower surface of airfoil section using a 6 × 7 ft wind tunnel. While the *Re* of flow investigated was in the range of 50,000 to 200,000, 75,000 is particularly investigated for studying *Re* effects for separation. The flow turbulence level was measured at 0.05%. Flow separation was observed at $\alpha = 14°$. The study was conducted for α ranging from 0 to 28° in steps of 1° forward and back, to study hysteresis effect if present. A description of the study is provided in Table 2.

2.3 *Study by Miller (2008)*

The study by Miller [15] provides pressure distribution data around a NACA0015 airfoil for flow velocities of 17 m/s and *Re* of 232,940. Four α namely, 0°, 5°, 10° and 15° were investigated. The study was performed to study the lift, drag and moment characteristics of the symmetrical airfoil as well as study the hysteresis effect during stall. A description of the study is provided in Table 3.

Table 3 Description of study by Miller [15]

Property	Description
Dataset	Miller [15]
Location of test	Ohio State university, USA
Airfoil	NACA0015
Chord	0.2 m
Reynolds number	232,940
Flow velocity	17 m/s

3 Methodology

3.1 Numerical Approach

The implementation of CFD analysis is on ANSYS CFX v16.0, a general purpose CFD simulation code. The XFOIL is chosen as an alternate numerical approach which uses the Panel Method.

The airfoil coordinates are collected from standard airfoil databases. The ordinates are loaded to the software from which the points are connected together to form lines and 3D surface. A thickness of 0.01 m is assumed for enabling a 3D analysis in ANSYS.

3.2 CFD Computational Domain

The domain for analysis is sketched as a C-mesh. A spacing of 12.5c is allowed between the airfoil and the wall and 12.5c between the airfoil trailing edge and pressure outlet. This spacing is sufficient enough to prevent the development of flow affecting the result convergence. A gauge pressure of 0 is set at the outlet. The no slip condition is enabled at the airfoil surface. Similarly a zero shear condition is enabled at the outer walls.

For the grid generation, an unstructured mesh strategy is followed for grid generation of flow domain (see Fig. 1a). The region near the airfoil wall is packed with finer elements while the remaining regions consisted of coarse elements (see Fig. 1b). The region surrounding the wall is designed with wedge elements using "inflation" property in ANSYS to capture the effects associated with thin boundary layer. Thereafter, boundary conditions are applied and simulation settings are applied.

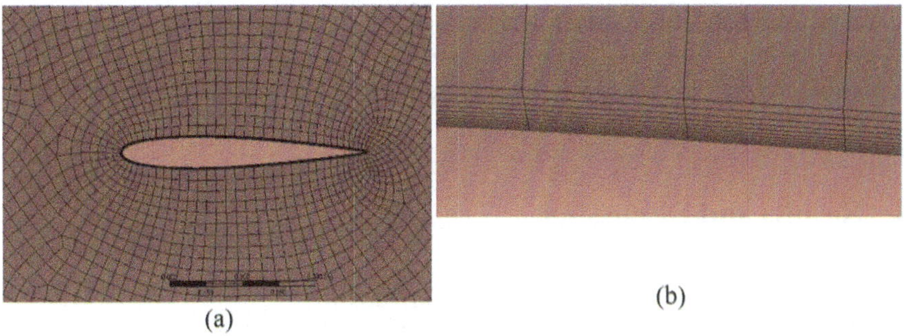

Fig. 1 **a** Unstructured mesh generation for flow domain, **b** Mesh close-up near airfoil wall for capturing viscous, thin boundary layer effects

3.3 XFOIL Domain

Similarly, the ordinates are loaded in JBLADE, an integrated XFOIL/XFLR5 interface supported by an intuitive GUI [16]. The distribution of points are "refined" further to form smooth, unobtrusive surface with the sharp curvatures at leading edge are corrected for roundness. A total of 250 points are created to form a 2D airfoil spline.

In JBLADE XFOIL, the refined points are simulated under similar conditions as suggested in experimental tests. A similar approach is followed for boundary conditions as in CFD and the results are extracted for further analysis.

3.4 Boundary Conditions and Simulation Settings

In ANSYS CFX, the simulation settings are chosen independent of any airfoil as all the simulation should be run under uniform settings. The boundary conditions are

Table 4 Domain and boundary conditions

Airfoil	NACA0012	ClarkY	NACA0015
Re	2,880,000	75,000	232,940
ρ	1.225	1.225	1.225
c	0.76	0.15	0.2032
μ	1.7894E-05	1.7894E-05	1.7894E-05
$V\infty$	55.35	7.189	16.75
Mach	0.1667	0.0217	0.0504

however varied such that velocity, chord and *Re* are maintained similar to experimental data. These are listed in Table 4. These conditions are set identical in XFOIL. A convergence criteria set at 10^{-6} is sufficient for reasonable accuracy of results.

4 Results and Discussion

4.1 Benchmark Data and Error Estimation

For establishing a comparison between numerical and experimental data, a benchmark dataset is needed. For NACA0012, data provided by Gregory and O'Reilly (1970), for ClarkY, data provided by Marchman and Werme (1984) and for NACA0015, data provided by Miller (2008) will be fixed as benchmark data. This will be assumed to have zero error and provide reference data for this work.

The datasets obtained from CFD and XFOIL are compared with benchmark data. The comparison is performed by finding the difference between obtained CFD and benchmark values for each chord length percent. These are considered as error estimates. The difference between CFD and benchmark value ($Error_1$) as well as XFOIL and benchmark value ($Error_2$) are found using the following expressions for each α as,

$$Error_1 = Cp_{CFD} - Cp_{bm} \tag{1}$$

$$Error_2 = Cp_{XFOIL} - Cp_{bm} \tag{2}$$

where Cp_{CFD}, Cp_{XFOIL} and Cp_{bm} are pressure coefficients of CFD, XFOIL and benchmark (experimental) respectively.

The *Cp* around the airfoil for different α is shown in Fig. 2 for NACA0012, Fig. 3 for ClarkY and Fig. 4 for NACA0015.

4.2 NACA0012

The error for 0°, 10° and 15° are found to be −0.1, 1.2 and 0.3 respectively (see Tables 5 and 6). From this it can be seen that the CFD values have achieved a good quality of results at low and high α. At mid α, the quality suffered due to peaks at leading edge.

In the case of XFOIL, the results are found to be very close to experimental values. There is no significant difference between experimental and XFOIL at all α and chord lengths. Thus, the values of XFOIL are found to have better quality than observed for CFD in case of NACA0012 airfoil.

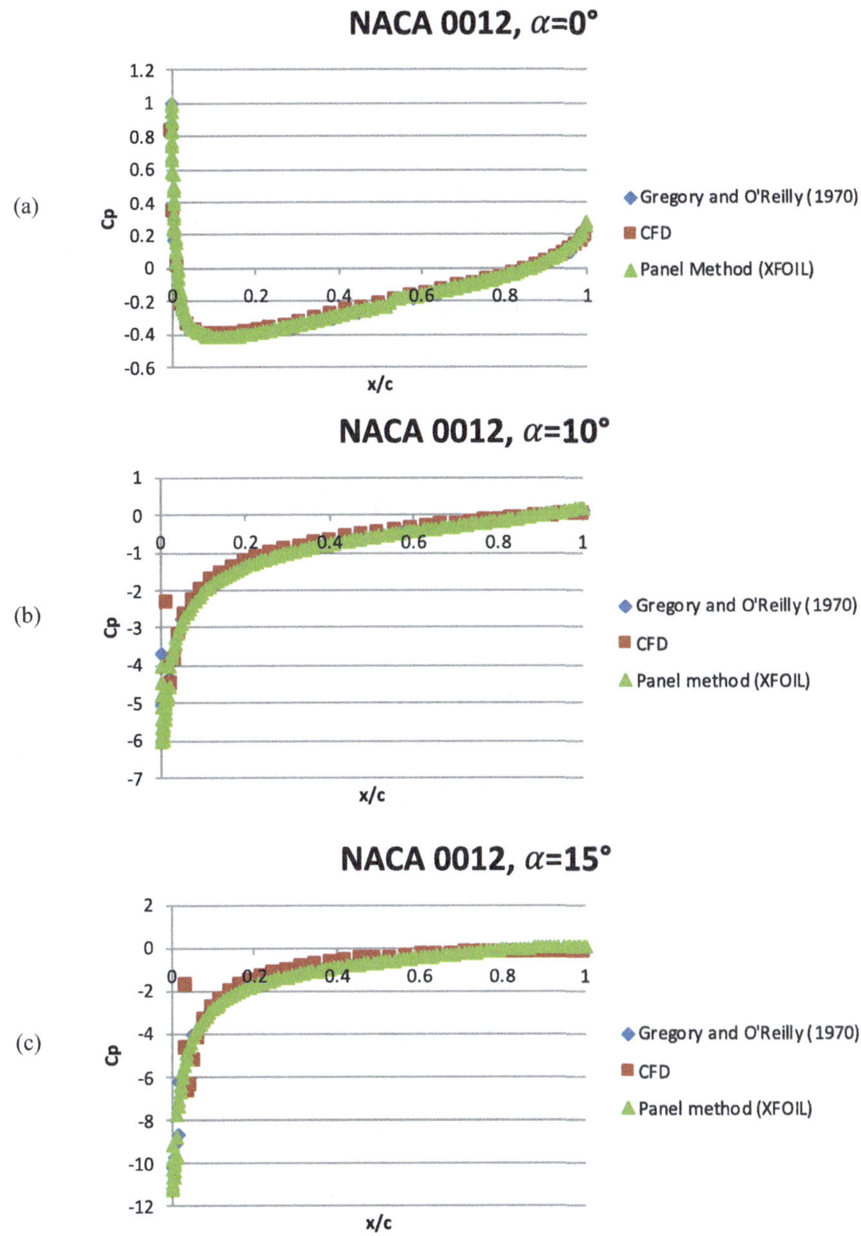

Fig. 2 Pressure coefficient plots of NACA0012 airfoil at **a** $\alpha = 0°$, **b** $\alpha = 10°$ and **c** $\alpha = 15°$

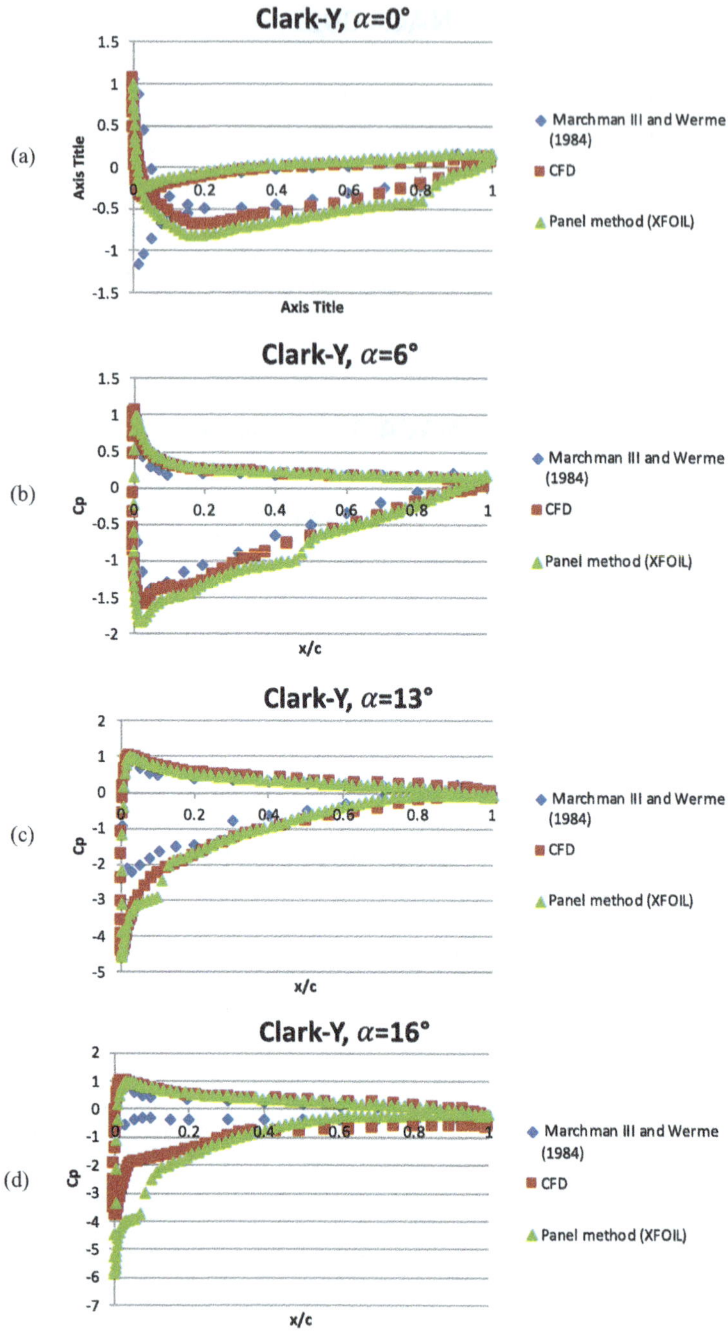

Fig. 3 Pressure coefficient plots of ClarkY airfoil at **a** $\alpha = 0°$, **b** $\alpha = 6°$, **c** $\alpha = 13°$ and **d** $\alpha = 16°$

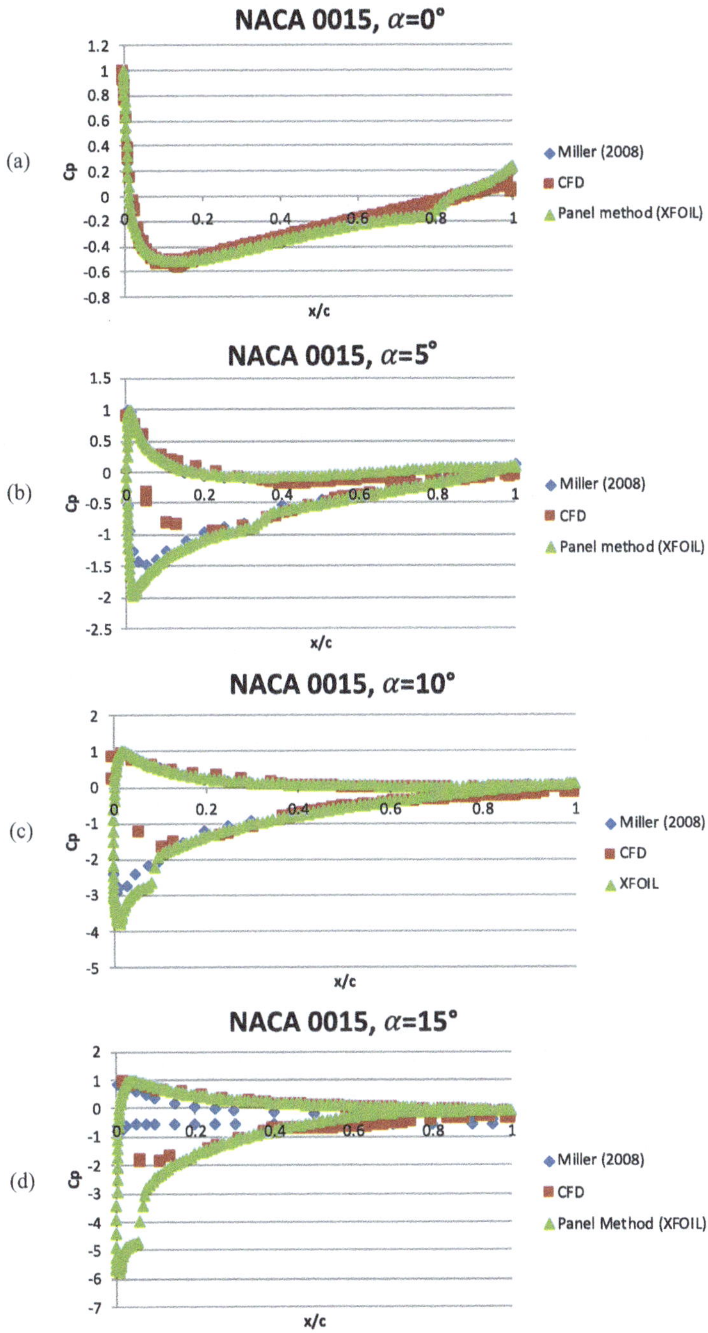

Fig. 4 NACA0015 simulation and experimental results comparison. **a** $\alpha = 0°$, **b** $\alpha = 5°$, **c** $\alpha = 10°$ and **d** $\alpha = 15°$

Table 5 Error estimates for different angles of attack from 0° to 10°

Airfoil	$\alpha = 0°$		$\alpha = 5°$		$\alpha = 6°$		$\alpha = 10°$	
	$Error_1$	$Error_2$	$Error_1$	$Error_2$	$Error_1$	$Error_2$	$Error_1$	$Error_2$
NACA0012	−0.1	−0.1	–	–	–	–	1.2	0.1
ClarkY	0	−0.1	–	–	−0.1	−0.1	–	–
NACA0015	0	0	0.5	0.1	–	–	0.1	0

Table 6 Error estimates for different angles of attack from 13° to 16°

Airfoil	$\alpha = 13°$		$\alpha = 15°$		$\alpha = 16°$	
	$Error_1$	$Error_2$	$Error_1$	$Error_2$	$Error_1$	$Error_2$
NACA0012	–	–	0.3	0	–	–
ClarkY	−0.3	−0.1	–	–	−0.5	−0.3
NACA0015	–	–	0.5	−0.1	–	–

4.3 *ClarkY*

The error estimate shows that all the CFD values have performed uniformly similar to all α,—0°, 6°, 12° and 14° (see Tables 5 and 6). At all α, the leading edge suction peak occurs. The maximum error is found to be at higher α which implies that the results are obtained as expected. At 14° α, the maximum error of −0.5 is obtained.

The error in XFOIL is found to be lower than CFD with no leading edge peaks. The maximum error observed is −0.3 for 14° α. For other α, the error is observed to be relatively less at −0.1 which explains the better results output of XFOIL.

4.4 *NACA0015*

The error estimate in the case of NACA0115 for CFD is close to 0 at 0° α. Leading edge peaks are observed at higher α the overall error average is found to be maximum for 5° and 15° and lowest at 0° (see Tables 5 and 6).

The error results obtained show than XFOIL has produced better results compared to CFD for all α.

5 Conclusion

A comparative analysis between CFD, XFOIL and Experimental has been carried out. The results show that out of two numerical methods, XFOIL had closer predictions to benchmark values for all 3 airfoils. It was also found that the CFD simulation in airfoils is applicable only for *Re* of at least 75,000. The ANSYS CFX solver with simulation settings proposed has yielded satisfactory results.

References

1. Drela, M (1989) XFOIL: an analysis and design system for low Reynolds number airfoils. In: Conference on low reynolds number airfoil aerodynamics, vol 54, pp 1–12
2. Addepalli SK, Mallikarjuna JM (2016) Parametric analysis of a 4-stroke GDI engine using CFD. Alexandria Eng J 57(1):23–34
3. Babich F, Cook M, Loveday D, Rawal R, Shukla Y (2017) Transient three-dimensional CFD modelling of ceiling fans. Build Environ 123:37–49
4. Bottarelli M, Bortoloni M, Zannoni G, Allen R, Cherry N (2017) CFD analysis of roof tile coverings. Energy 137:391–398
5. Močilan M, Žmindák M, Pecháč P, Weis P (2017) CFD simulation of hydraulic tank. Proc Eng 192:609–614
6. Prachař A, Hospodář P, Vrchota P (2018) Gust alleviation of aeroelastic aircraft using CFD simulation. Transp Res Proc 29:366–375
7. Vittorio C, Aldo F, Karim AS, Vincenzo B, Emanuele R, Marco M (2018) Preliminary transonic CFD analyses of a PrandtlPlane transport aircraft. Transp Res Proc 29:82–91
8. Nordanger K, Holdahl R, Kvamsdal T, Kvarving AM, Rasheed A (2015) Simulation of airflow past a 2D NACA0015 airfoil using an isogeometric incompressible Navier-Stokes solver with the Spalart-Allmaras turbulence model. Comput Methods Appl Mech Eng 290:183–208
9. Nordanger K, Holdahl R, Kvarving AM, Rasheed A, Kvamsdal T (2015) Implementation and comparison of three isogeometric Navier-Stokes solvers applied to simulation of flow past a fixed 2D NACA0012 airfoil at high Reynolds number. Comput Methods Appl Mech Eng 284:664–688
10. Bragg MB, Broeren AP, Blumenthal LA (2005) Iced-airfoil aerodynamics. Prog Aerosp Sci 41(5):323–362
11. Morgado J, Vizinho R, Silvestre MAR, Páscoa JC (2016) XFOIL vs CFD performance predictions for high lift low Reynolds number airfoils. Aerosp Sci Technol 52:207–214
12. Kapsalis PS, Voutsinas S, Vlachos NS (2016) Comparing the effect of three transition models on the CFD predictions of a NACA0012 airfoil aerodynamics. J Wind Eng Ind Aerodyn 157:158–170
13. Gregory N, O'Reilly CL (1970) Low-speed aerodynamic characteristics of NACA 0012 aerofoil section, including the effects of upper-surface roughness simulating hoar frost. Reports Memo 3726(3726):35
14. Marchman JF, Werme TD (1984) ClarkY airfoil performance at low reynolds numbers. In: AIAA 22nd aerospace sciences meeting
15. Miller SD (2008) Lift, dragand moment of a NACA 0015 airfoil, p 25 (2008)
16. Morgado J, Silvestre MÂR, Páscoa JC (2015) Validation of new formulations for propeller analysis. J Propuls Power 31(1):467–477

CFD Analysis of a Novel Propeller Design Operating at Low Reynolds Number

Aravind Seeni and Parvathy Rajendran

Abstract The aerodynamic performance of a propeller is characterized by the thrust produced, torque and efficiency. In this paper, novel blade shapes are tested using the technique Computational Fluid Dynamics (CFD). The blade shapes are based on a passive slotted design in which slots are present on the suction side i.e. on the upstream front side of the blade. Two slotted design models are tested using the commercial CFD solver ANSYS Fluent. The slotted designs are identical in terms of slot geometry (0.1 mm width, 0.2 mm depth) but differ in the location of slot along chord from leading edge (0.206c and 0.382c). In order to test the modified designs, a validation study of the baseline model of a propeller is conducted. Then slotted designs comprising two models are tested and compared to baseline design. The propeller model considered is the APC10x7 Slow Flyer and the Reynolds number of flow analyzed is approximately 68,500 (estimated at 75% radial distance) The results showed that the presence of slots have altered the performance. The presence of slots has reduced the thrust performance for the entire operational range of advance ratios. However, the power performance has increased due to decrease in torque or rotational resistance of the propeller. The efficiency is thereby increased for one model for a specific range of advance ratios. In the case of another model, for most advance ratios the efficiency was found to be decreased.

Keywords Slotted design · Propeller performance · Computational fluid dynamics · Validation · Thrust · Power · Efficiency

---

A. Seeni · P. Rajendran (✉)
School of Aerospace Engineering, Universiti Sains Malaysia, Engineering Campus, 14300 Nibong Tebal, Penang, Malaysia
e-mail: aeparvathy@usm.my

A. Seeni
e-mail: aravindseeni@gmail.com

P. Rajendran et al. (eds.), *Proceedings of International Conference of Aerospace and Mechanical Engineering 2019*, Lecture Notes in Mechanical Engineering,
https://doi.org/10.1007/978-981-15-4756-0_13

1 Introduction

Propellers have so far been employed in the design of aerospace vehicles to provide mobility in both terrestrial and extra-terrestrial applications. While the use of propellers as a thrust generation device for terrestrial aircraft mobility is long-standing, several NASA studies employing propellers and/or rotors in an electric-powered propulsion system for UAVs or drones have been successfully conceptualized and applied in space missions (Fig. 1). Such space missions include those that are designed to explore Venus [1–3], Earth, Mars [4] and Titan [5]. The speed in which these machines operate, correspond to a propeller operating in the very low Reynolds number (Re) regime [6].

Propellers are aerodynamic devices that are designed to provide thrust force. New novel propeller shapes are of constant and widespread interest in the scientific community in order to design propellers with improved aerodynamic performance and efficiency. The aerodynamic performance of a propeller is characterized by the thrust produced, torque and efficiency of rotation.

Recent development of a novel design of propeller is the slotted design. The design is inspired by the usage of grooves along the surface. Prior implementation of

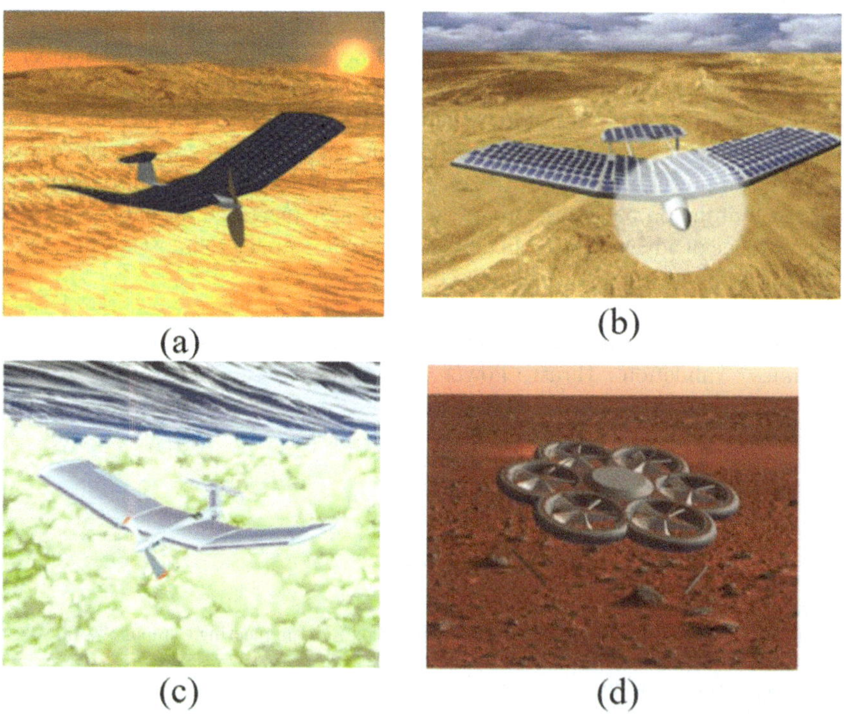

Fig. 1 Examples of space missions with propeller-based mobility. **a** Venus solar airplane [1], **b** Solar Venus drone [3], **c** Venus space drone [2], **d** Mars Hexacopter concept [4]

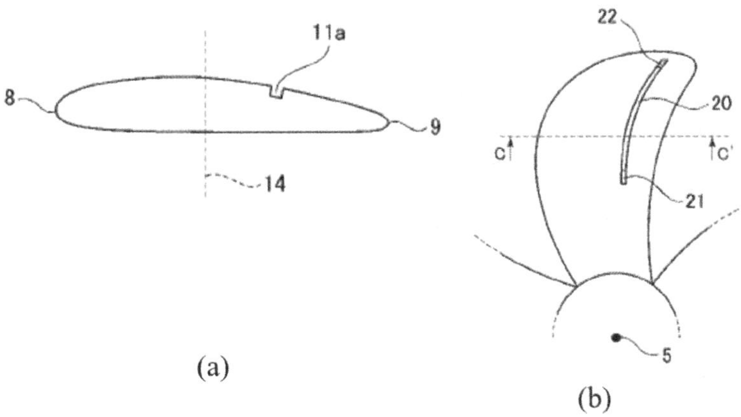

Fig. 2 Kei Sato's invention pertaining to slots and cavitation effects on marine propeller **a** slot geometry, **b** full scale blade model with slot [7]

grooves on circular cylinder and airfoil surface and their performance enhancements can be found in the following studies [8–13]. Sato [13] proposed the first concept of slotted propeller suited for marine application. In the design, grooves or slots are introduced on the suction side of a propeller as shown in Fig. 2. The grooves extend in the radial direction from the hub to tip. As per the invention, the grooves suppress cavitation. The propeller design is intended for marine application.

Noorazyze [14] in her work, studied the aerodynamic benefits of using slots on a 5 m diameter wind turbine operating at low wind speeds. As per the work, the introduction of slots reduced the material weight, contributing to low rotational inertia. This has led to yield better energy performance. Two types of slotted designs were studied namely horizontal and vertical. The results were obtained by performing a CFD analysis using ANSYS software. The results also showed that implementing slots increased the lift coefficient for varied angles of attack.

Similarly, the work of Didane et al. [15] studied the effects of different slot geometry on a low wind speed wind turbine blade. Three types of slot geometry namely circular, horizontal rectangular, vertical rectangular were studied. The result showed that circular slots with diameter of 5 mm would be the best configuration with improved C_l/C_d ratios. Also the overall weight of the blade is reduced.

So far, the effect of slots on aerodynamic performance in terms of lift, drag coefficients and ratios have been deduced. Despite the similarity in operation compared to wind turbine, no studies exist on studying the aerodynamic performance of a propeller in terms of thrust, torque and efficiency. Only the effects of cavitation on marine propellers have so far been studied.

In this paper, the aerodynamic characterization of slotted propeller operating at low Re will be investigated. The concept of using slots in UAV propellers is novel. So far, no reported works have been found in literature that has studied the aerodynamic performance characterization of slotted propellers. A first attempt will be made on the study of slotted design for a UAV propeller.

2 Methodology

The Computational Fluid Dynamics (CFD) technique will be used to test propeller models. Initially, CAD models of propeller, both baseline and modified are modelled in Catia v5 software. The models are then exported to finite volume based solver namely ANSYS Fluent. Fluent is a code that has been reported in several research works in the past [16–20]. This explains the capability of the code to generate accurate results.

2.1 *Governing Equations*

The governing equations consist of the continuity and Reynolds-averaged momentum equation which can be written as follows:

$$\frac{\partial \rho}{\partial t} + \frac{\partial}{\partial x_j}(\rho u_i) = 0 \tag{1}$$

$$\frac{\partial(\rho \overline{u_i})}{\partial t} + \frac{\partial\left(\rho \overline{u_i u_j}\right)}{\partial x_j} = -\frac{\partial \bar{p}}{\partial x_i} + \frac{\partial}{\partial x_j}\left[\mu\left(\frac{\partial \overline{u_i}}{\partial x_j} + \frac{\partial \overline{u_j}}{\partial x_i} - \frac{2}{3}\delta_{ij}\frac{\partial \overline{u_m}}{\partial x_m}\right)\right] + \frac{\partial}{\partial x_j}\left(-\rho\overline{u_i^{'} u_j^{'}}\right) \tag{2}$$

where δ_{ij} is the Kronecker delta and $-\rho\overline{u_i^{'} u_j^{'}}$ is the Reynolds stresses. The Reynolds stresses could be linked to mean rates of deformation as proposed by Boussinesq hypothesis as,

$$-\rho\overline{u_i^{'} u_j^{'}} = \mu_t\left(\frac{\partial u_i}{\partial x_j} + \frac{\partial u_j}{\partial x_i}\right) - \frac{2}{3}\left(\rho k + \mu_t \frac{\partial u_i}{\partial x_i}\right)\delta_{ij} \tag{3}$$

where μ_t is the turbulent viscosity. μ_t will be estimated by turbulence model equations.

The flow around the propeller can be modelled by steady, RANS equations for an incompressible flow. The inclusion of turbulence models generally makes the flow unsteady. RANS turbulence models provide closure to Reynolds Stress tensor that represents the effect of turbulent fluctuations in the mean flow. This allows steady state simulations of turbulent flow being performed in ANSYS Fluent [21].

The turbulence model considered is the one-equation Spalart-Allmaras (S-A). S-A is specifically designed for aerodynamics flows [22]. It is also applicable for flows with low Re [23]. The work of Zhang et al. [24] have also implemented S-A model in their investigation of a ducted propeller.

2.2 Thrust, Power and Efficiency

The thrust coefficient, power coefficient and efficiency can be expressed as follows:

$$K_T = \frac{T}{\rho n^2 D^4} \tag{4}$$

$$K_P = \frac{P}{\rho n^3 D^5} \tag{5}$$

$$\eta = J \frac{K_T}{K_P} \tag{6}$$

where ρ is the density, n is the rotational speed, D is the diameter and J is the advance ratio. Advance ratio can be expressed by the following relationship.

$$J = \frac{V}{nD} \tag{7}$$

where V is the freestream velocity.

2.3 Propeller Geometry

The model chosen for slotted design study is the Applied Precision Composites (APC) 10 × 7 Slow Flyer (SF) propeller designed with a diameter (D) of 0.254 m and pitch of 0.1778 m. The propeller is composed of Eppler 63 and Clark-Y airfoils near the hub and tip respectively. This propeller is widely used in applications operating in low Re such as quadcopters, small educational UAVs and the like.

2.4 Computational Domain

A three-dimensional computational grid is implemented in which the velocity components u, v and w and pressure component p at the center of the control volumes is solved. The computational domain is modelled based on Multiple Reference Frame. Two reference frames, stationary and rotational are considered. The propeller is enclosed by the rotational reference frame which moves or rotates at a speed of 3008 rpm. It is designed to have a cylindrical geometry with a diameter of 1.1D. The stationary reference frame is assigned with a cubic geometry with sides of 8D in length. The dimensions are set sufficient enough for the flow to reach steady state and vanish downstream of the propeller. The advantage of this modelling approach is the simplicity of domain construction.

2.5 *Mesh*

The domain is meshed into multiple numbers of small fluid control volumes. The mesh is made up of tetrahedron elements in both stationary and rotational domain. An unstructured mesh methodology is followed due to the complexity of the geometry. While the use of structured mesh is followed for its refined topology, however, the use of unstructured mesh does not decrease the accuracy of the results [25–27].

2.6 *Boundary Conditions*

A no slip condition is assumed at the wall which is the rotating domain. Inlet velocities are specified from 2.4449 to 10.1744 m/s for various corresponding advance ratio conditions of 0.192–0.799. The velocities are assumed based on the experiments by [28]. The flow Re is approximately 68,500 that is estimated based on chord distance at 75% radial distance. The outlet is specified with a pressure gauge of 0 Pa. A Semi-implicit Method for Pressure Linked Equations (SIMPLE) is assumed as pressure-velocity coupling scheme. The second-order upwind scheme is assumed for pressure, momentum and modified turbulent viscosity. A least squares cell-based algorithm is assumed for gradient calculations. The fluid is assumed to be air at a temperature of 25 °C with a density of 1.225 kg/m^3 and dynamic viscosity of 1.7894 $\times$ 10^{-5} kgm^{-1}s^{-1}.

3 Results and Discussion

3.1 *Verification and Validation of Baseline Propeller in Viscous Flow*

Verification

A grid independency test is performed for baseline propeller model in order to verify the solution. The Grid resolution method is adopted. Five meshes of varying mesh resolutions comprising 200 thousand (coarsest), 800 thousand (coarse), 1.1 million (medium), 1.4 million (mid-fine) and 3.8 million (fine) elements are used to obtain the thrust and power performance for one advance ratio, J = 0.486. The results showed that the value of thrust and power do not change significantly when the mesh size in increased from medium to fine. This is illustrated in Fig. 3a for thrust coefficient and Fig. 3b for power coefficient. The medium mesh resolution is sufficient enough to provide a grid independent solution.

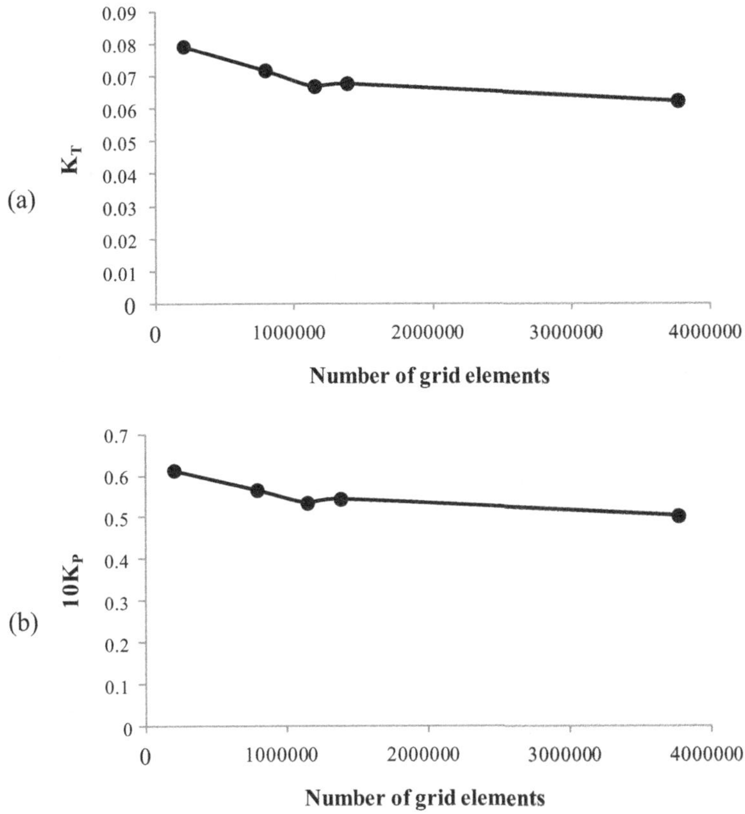

Fig. 3 Grid independence study for **a** Thrust coefficient, **b** Power coefficient for J = 0.486

Validation

It can be found from Fig. 4a that as the advance ratios increases, the thrust generated by the propeller decreases. The increase in velocity of flow causes the propeller to generate decreasing thrust with ρ and n remaining constant. A comparison of results between CFD and experiment reveals that the thrust results generated using the numerical method matches closely to that of experimental results across all advance ratios.

The power coefficient is a very miniscule quantity. Therefore, the obtained power coefficient is multiplied by a factor of 10 and plotted. It can be seen in Fig. 4b that as the advance ratio increases, the power coefficient decreases gradually. It can be added that the torque moment begins to decrease as fluid velocity increases. A comparison of experimental and numerical results reveal that the CFD results matches experimental results satisfactorily, although experiment produced slightly higher results.

The obtained CFD results are compared with experimental measurements for the chosen range of advance ratios. It can be seen in Fig. 4c that the efficiency matches experimental results satisfactorily. Thus the CFD results are validated.

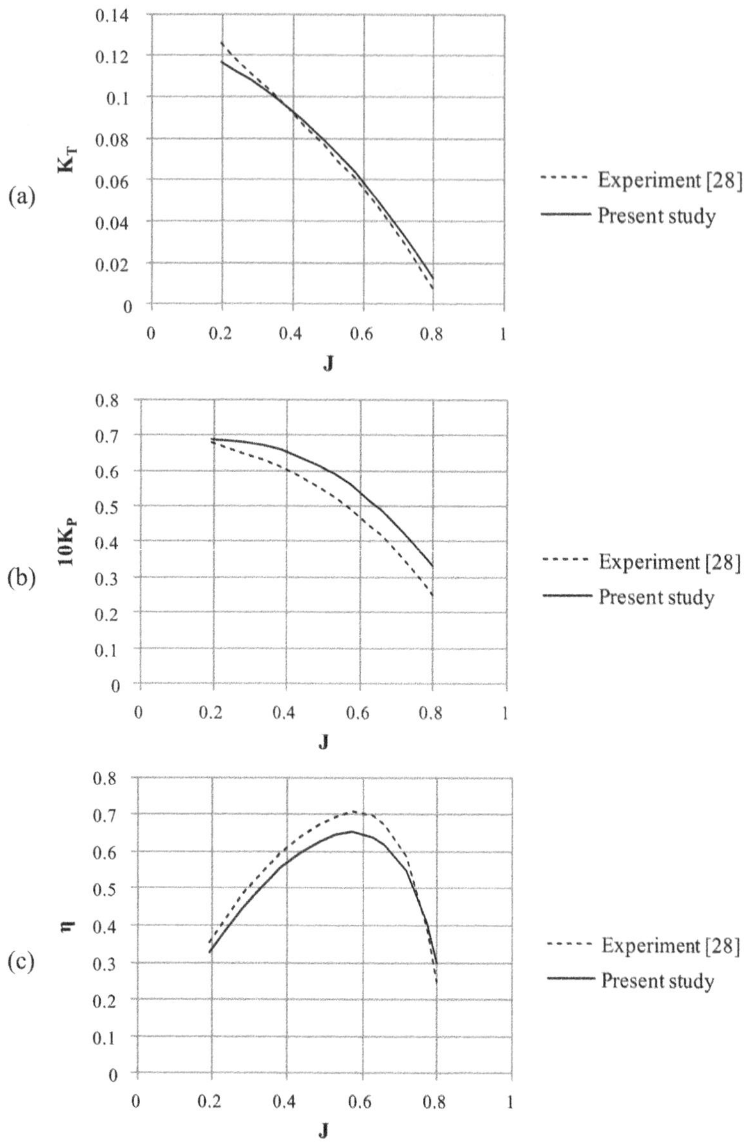

Fig. 4 Validation of **a** Thrust coefficient, **b** Power coefficient and **c** Efficiency

3.2 Aerodynamic Performance Characterization of Slotted Propeller Designs

Now slotted designs with implementations on the baseline model APC10x7SF propeller are tested to analyze the aerodynamic performance characteristics. Two new slotted designs are tested. The design models have identical slot geometry but have slots placed in different chordwise positions on the blade. Model1 has a dimension of 0.1 mm width, 0.2 mm depth and located at 0.206c measured from leading edge. Model2 has identical dimensions with slot located at 0.382c measured from leading edge. The slots stretch along the entire section of the blade from root to tip and form a groove with constant dimensions and location.

Figure 5a provides the results of thrust coefficient variation for baseline and the two slotted design models. As per the results given in Fig. 5, it can be found that the thrust coefficient of the slotted design models is slightly lower than the baseline model. The thrust coefficient has decreased by −6.4 to −16.5% for Model1. It has decreased by −4.2 to −65.2% for Model2. Among the design models, Model2 produced slightly higher thrust for lower to mid advance ratios of 0.192–0.486. For higher advance ratios, from 0.527 to 0.799, Model1 produced higher thrust compared to Model2.

Figure 5b provides the results of power coefficient variation for baseline and slotted designs. As far as power coefficient is concerned, the slotted designs performed better. Lower torque is produced by the modified design models, which would mean lower resisting force to rotation in the presence of slots. The torque produced by Model1 is lower than Model2 from advance ratio of 0.192–0.717. At advance ratio of 0.773 and 0.799, the power coefficient of Model2 is relatively lower.

Figure 5c provides an illustration of the efficiency variation of baseline and slotted configurations. It can be found that Model1 has higher propeller efficiency than baseline for advance ratios from 0.192 to 0.659. For advance ratios from 0.717 to 0.799 the efficiency has decreased. The propeller efficiency of Model2 is lower than baseline for all advance ratios except for $J = 0.192$, 0.236 and 0.334. The efficiency of Model2 is also relatively lower than Model1.

Based on the efficiency of Model1, it can be argued that the implementation of slots on APC10x7SF baseline could be used to sustain a flight with improved efficiency but at the cost of reduced thrust levels. This flight will have a higher efficiency than baseline model for advance ratios from 0.192 to 0.659. This operational range correspond to inflow freestream velocities of 2.4449–8.3916 m/s. Model2 can be operated at higher efficiency only at two low advance ratios. The desired operational range for this design is significantly small.

4 Conclusion

The following conclusions have been drawn from this study:

1. The implementation of slots on a low Re propeller showed the effects on thrust and power performance as well as efficiency. It showed that slots have reduced the thrust performance for the entire operational range of advance ratios. This was found in both cases of slotted design models.
2. However, the power performance of the propeller has increased with decrease in rotational resistance to the propeller. The impact of torque on the overall

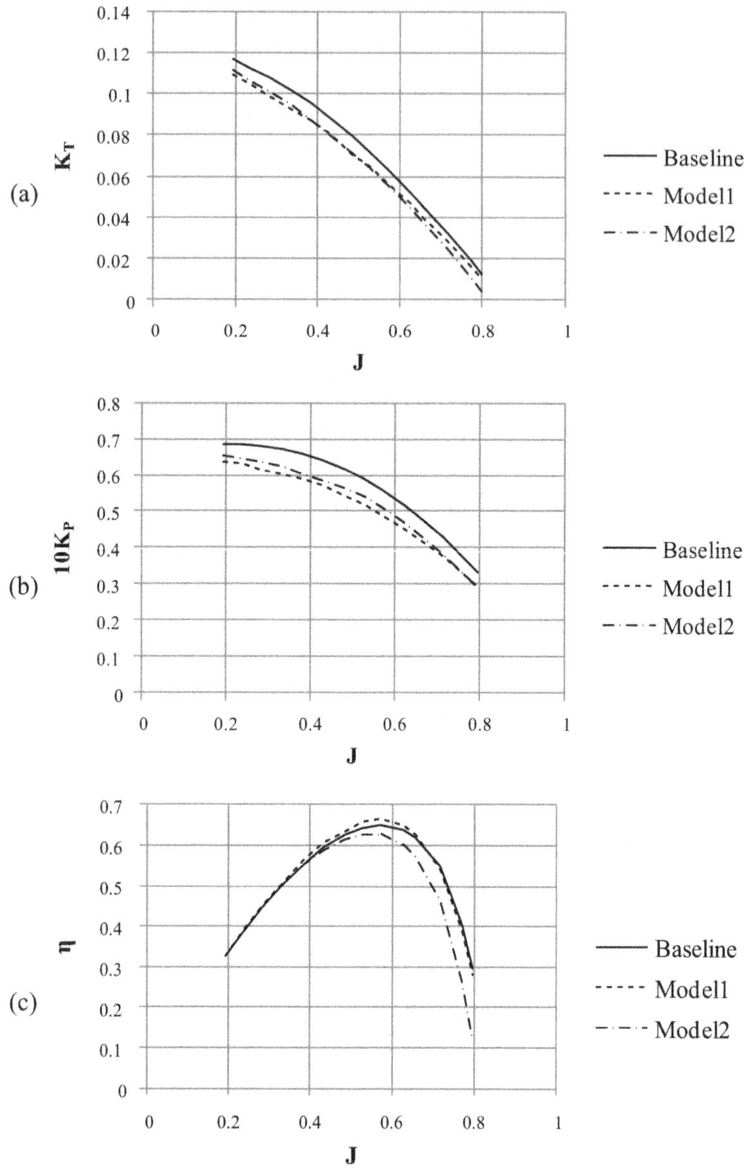

Fig. 5 **a** Thrust coefficient, **b** power coefficient and **c** efficiency comparison of baseline and slotted designs

performance was found to be dependent on the location of slot. At Model1 location, the slot produced relatively lesser torque compared to slot at Model2.
3. The efficiency of the propeller was found to be increased with the implementation of slots for one model for a specific range of advance ratio. For other design model, the efficiency was found to be lesser than baseline for most advance ratios.

References

1. Xiongfeng Z, Zheng G, Zhongxi H (2015) Sun-seeking eternal flight solar-powered airplane for venus exploration. J Aerosp Eng 28(5):1–7
2. Landis GA, Colozza A, Lamarre CM (2003) Almospheric flight on venus: a conceptual design. J Spacecr Rockets 40(5):672
3. Landis GA (2006) Robotic exploration of the surface and atmosphere of Venus. Acta Astronaut 59(7):570–579
4. Pergola P, Cipolla V (2016) Mission architecture for Mars exploration based on small satellites and planetary drones. Int J Intell Unmanned Syst 4(3), 142–162
5. Hassanalian M, Rice D, Abdelke A (2018) Evolution of space drones for planetary exploration: a review. Prog Aerosp Sci 97(February):61–105
6. Colozza AJ (2001) Overview of propulsion systems for a Mars aircraft, no. NASA/TM–2001-210575
7. Sato K (2014) Propeller JP2014169017A
8. Yokoi Y, Igarashi T, Hirao K (2011) The study about drag reduction of a circular cylinder with grooves. J Fluid Sci Technol 6(4):637–650
9. Yamagishi Y, Kimura S, Oki M (2013) Study on drag reduction of a circular cylinder with grooves. Trans Japan Soc Mech Eng Ser B 79(805):1742–1751
10. Mohammadi A, Floryan JM (2013) Groove optimization for drag reduction. Phys Fluids 25:113601
11. Seo SH, Nam CD, Han J-Y, Hong C-H (2013) Drag reduction of a bluff body by grooves laid out by design of experiment. J Fluids Struct 135:111202
12. Seo S, Hong C (2016) Performance improvement of airfoils for wind blade with the groove. Int J Green Energy 13(1):34–39
13. Song X, Qi Y, Zhang M, Zhang G, Zhan W (2019) Application and optimization of drag reduction characteristics on the flow around a partial grooved cylinder by using the response surface method. Eng Appl Comput Fluid Mech 13(1):158–176
14. Noorazyze SN (2014) Performance evaluation of slotted and continuous types wind turbine blade. Universiti Tun Hussein Onn Malaysia
15. Didane DH, Mohd S, Subari Z, Rosly N, Ghafir MFA, Masrom MFM (2016) An aerodynamic performance analysis of a perforated wind turbine blade. In: IOP conference series materials science and engineering 160(1)
16. Kaidi S, Smaoui H, Sergent P (2018) CFD investigation of mutual interaction between hull, propellers, and rudders for an inland container ship in deep, very deep, shallow, and very shallow waters. J Waterw Port Coastal Ocean Eng 144(6)
17. Paik K, Hwang S, Jung J, Lee T, Lee Y, Ahn H (2015) Investigation on the wake evolution of contra-rotating propeller using RANS computation and SPIV measurement. Int J Nav Archit Ocean Eng 7(3):595–609
18. Shora MM, Ghassemi H, Nowruzi H (2018) Using computational fluid dynamic and artificial neural networks to predict the performance and cavitation volume of a propeller under different geometrical and physical characteristics. J Mar Eng Technol 17(2):59–84
19. Villalpando F, Reggio M, Ilinca A (2011) Assessment of turbulence models for flow simulation around a wind turbine airfoil. Model Simul Eng 2011

20. Wang C, Sun S, Sun S, Li L (2017) Numerical analysis of propeller exciting force in oblique flow. J Mar Sci Technol 22(4):602–619
21. Bell B (2012) The most accurate and advanced turbulence capabilities. In: Confidence by design workshop (2012)
22. Spalart S, Philippe RA (1994) A one equation turbulence model for aerodynamic flows. La Rech Aerosp 1:5–21
23. Liu F, Cokljat D (2003) Contribution by fluent. In: Davidson L, Cokljat D, Fröhlich J, Leschziner MA, Mellen C, Wolfgang R (eds) LESFOIL: large eddy simulation of flow around a high lift airfoil. Springer-Verlag, Berlin, Heidelberg, New York, pp 145
24. Zhang Q et al (2018) Investigation on the performance of a ducted propeller in oblique flow. J Offshore Mech Arct, Eng
25. Alakashi AM, Basuno IB (2014) Comparison between structured and unstructured grid generation on two dimensional flows based on finite volume method (FVM). Int J Mining Metall Mech Eng 2(2):97–103 (2014)
26. Molnar B (2014) Comparison of structured and unstructured meshes for the computations of an H-type darrieus wind turbine. Otto von Guericke University Magdeburg, Germany
27. Husaini M, Samad Z, Arshad MR (2011) Autonomous underwater vehicle propeller. In: Computational fluid dynamics technologies and applications, no. July, Intech
28. Brandt J, Deters R, Ananda G, Selig M (2010) Small-scale propeller performance at low speeds—online database. [Online]. http://www.ae.illinois.edu/m-selig/props/propDB.html. Accessed 1 May 2018

Anisotropic Median Bilateral Filtering

Kai Yit Kok and Parvathy Rajendran

Abstract In this study, a new image filter—Anisotropic Median Bilateral filter (AM-Bilateral) is proposed for image noise reduction. Unlike the Bilateral filter, AM-Bilateral filter is an iterative non-linear filter and decreases image noise continuously in every iteration. Performance evaluation of AM-Bilateral filter is done and compared with Bilateral filter, Median filter and Arithmetic Mean filter using various images with Gaussian noise. The simulation results have shown that AM-Bilateral filter outperforms other filters with higher Peak Signal to Noise Ratio (PSNR) and Structural Similarity (SSIM) on average.

Keywords Bilateral filter · Anisotropic median · Image filter · Gaussian noise · PSNR · SSIM

1 Introduction

In the image processing field, one of the main challenges is solving image noise [1]. An image can be defined as matrix data with a certain intensity level, and any abnormal intensity variations in the matrix data are considered as image noise [2]. The presence of image noise affects image quality and lower performance of image processing applications [3]. Hence, many image denoising techniques are developed to detect and eliminate image noise without affecting image details ultimately.

K. Y. Kok · P. Rajendran (✉)
School of Aerospace Engineering, Universiti Sains Malaysia, Engineering Campus, 14300 Nibong Tebal, Penang, Malaysia
e-mail: aeparvathy@usm.my

K. Y. Kok
e-mail: kok901221kaiyit@student.usm.my

P. Rajendran et al. (eds.), *Proceedings of International Conference of Aerospace and Mechanical Engineering 2019*, Lecture Notes in Mechanical Engineering,
https://doi.org/10.1007/978-981-15-4756-0_14

Generally, there are 2 types of image filters: linear or non-linear filter [4, 5]. A linear filter such as Arithmetic Mean filter usually removes image noise through replacing with average pixel value within local area [6] but will lead to blurring issue. As the size of local area increases, the blurring problem will be more terrible. Anyway, it is straightforward and can be implemented easily with simple mathematics.

A non-linear filter is designed to improve on linear filter especially in minimizing signal-dependent noise. A median filter is a widely used non-linear filter for denoising images. This filter is based on pixel value ranking [7] whereby each pixel of image is compared within local region, and all pixels from the region will be sorted in ascending order. The 'median' value in the list is then chosen to replace previous value. It is excellent in eliminating salt and pepper noise since the highest and lowest pixel values will be ignored mostly during the pixel value selection. Although it can perform well in significant cases [8], the performance can be affected drastically when image noise is severe.

Besides, the Bilateral filter is another widely used non-linear filter for image noise reduction [9–11]. In this filter, weight is assigned to neighborhood pixels around center pixel. The weight varies according to distance from center pixel and intensity difference to ensure preserving of edge features [12]. Still, large degree of smoothing is required to remove high level of image noise, but edge features are no longer preserved under this condition [13].

Hence, in this study, we propose a new iterative image filter for image noise reduction by combining two image filters: Anisotropic Median filter and Bilateral filter. Anisotropic Median filter plays the role of eliminating abnormal pixels while Bilateral filter smoothens image and further reduces the image noise. Moreover, the edge features can be preserved due to the characteristics of these two filters. With this combination, primary 'noise' pixels can be taken out, and the rest 'noise' pixels can be adjusted closer to original image.

2 AM-Bilateral Filter

In this proposed method, a high degree of smoothing setting is avoided to ensure noise image can be recovered, and sharp edges are preserved at the same time. In order to do this, an iterative means of new image filter—AM-Bilateral filter is designed whereby the input image is filtered progressively with different range parameter in intensity kernel (from low to high) in every iteration. The process flow of the AM-Bilateral filter is presented in Fig. 1.

AM-Bilateral filter starts from the Anisotropic Median scheme. The idea of the Anisotropic Median approach was introduced to overcome the limitation of Median filter by only considering the neighboring pixels that have certain similarity of center pixels [14]. With this condition, most of the 'noise' pixels and irrelevant pixels will be ignored, remaining 'good' and relevant pixels for pixel value selection. This method can find the correct pixel with higher success rate compared to conventional Median

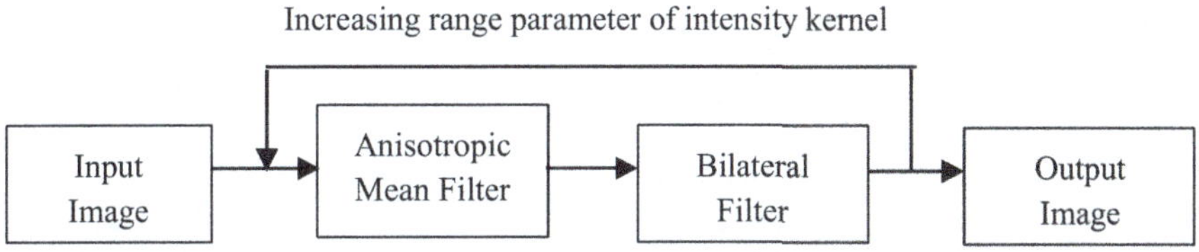

Fig. 1 The framework of AM-Bilateral filter

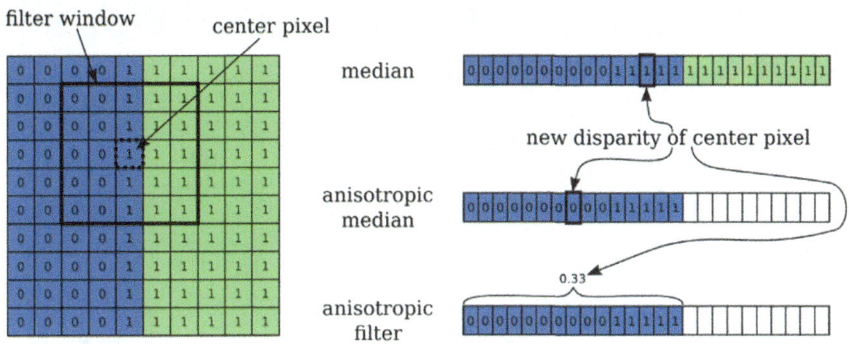

Fig. 2 Illustration of Anisotropic Median Filter [14]

filter. Figure 2 presents the illustration of how Anisotropic Median filter works and the difference compared to Median filter and Anisotropic Mean filter.

Next, the image will be processed using a Bilateral filter. Bilateral filter was first introduced by Tomasi et al. [15]. It was developed to smoothen images and preserve edges at the same time. This is done by using weighted average of neighborhood pixels whereby the weight is depending on intensity distance and spatial distance [16]. For instance, the output at x pixel location using Bilateral filter can be calculated through Eq. (1) [17, 18]:

$$\tilde{I}(x) = \frac{1}{C} \sum_{y \in N(x)} e^{\frac{-|y-x|^2}{2\sigma_d^2}} e^{\frac{-|I(y)-I(x)|^2}{2\sigma_r^2}} I(y) \qquad (1)$$

where $N(x)$ is the local region, $e^{\frac{-|y-x|^2}{2\sigma_d^2}} e^{\frac{-|I(y)-I(x)|^2}{2\sigma_r^2}}$ is the Bilateral filter kernel, and C is the normalization constant.

$$C = \sum_{y \in N(x)} e^{\frac{-|y-x|^2}{2\sigma_d^2}} e^{\frac{-|I(y)-I(x)|^2}{2\sigma_r^2}} \qquad (2)$$

In AM-Bilateral filter, the domain parameter for spatial kernel remains constant while the range parameter for intensity kernel is increasing every iteration, starting from small value in the first iteration. Through this, most of the 'noise' pixels can

Fig. 3 The output of AM-Bilateral filter in different iteration **a** Noise image, **b** 1st iteration, **c** 2nd iteration, **d** 3rd iteration

be eliminated eventually, and the image can be smoothened without affecting the details of the sharp edge. After trials and errors, optimum number of iteration for AM-Bilateral filter can be set from 2 to 4, depending on range parameter for intensity kernel setting. Figure 3 illustrates the output image of AM-Bilateral filter in different iteration stage.

3 Results and Discussion

In order to evaluate the performance of AM-Bilateral filter, four noise images: Lena, Boy, Boat, and Barbara, are used throughout the study as shown in Fig. 4. Further-

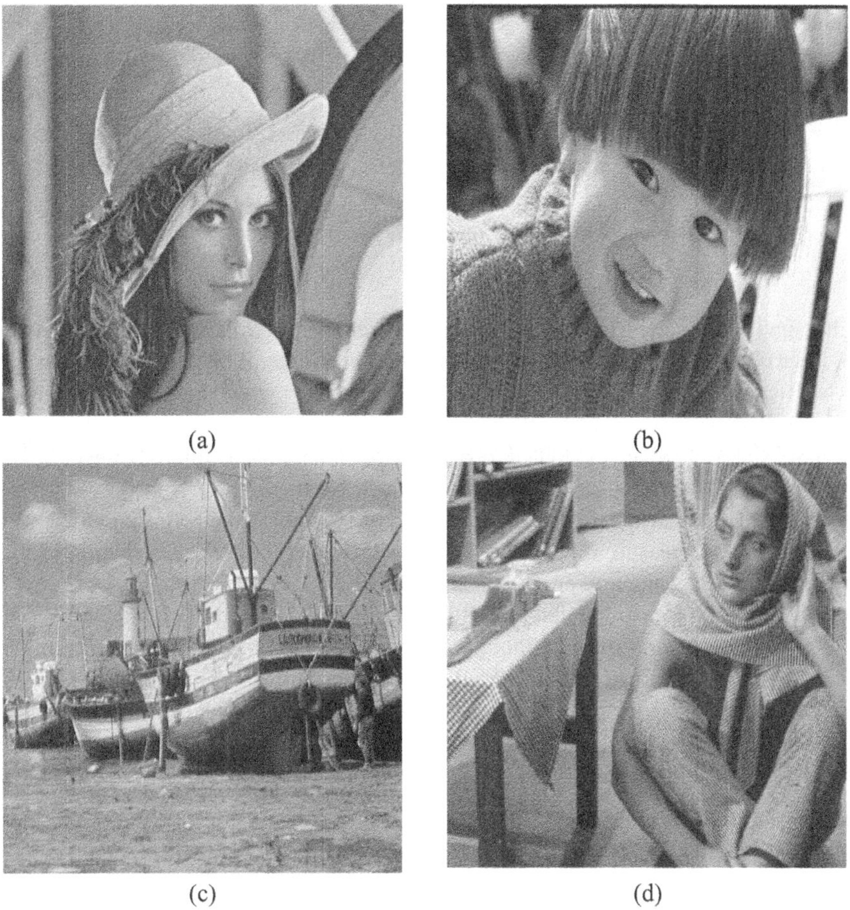

Fig. 4 Standard images used for image analysis **a** Lena, **b** Boy, **c** Boat, **d** Barbara

more, it will be compared with Bilateral filter, Median filter, and Arithmetic Mean filter.

The noise images are generated using MATLAB built-in Gaussian noise function with setting (mean = 0 and variance = 0.01). Mask with a size of 3×3 is used for all the image filters in this analysis. The number of iteration for Bilateral filter, Median

Table 1 Performance of different image filters on various noise images

Image filter	Lena		Boy		Boat		Barbara	
	PSNR	SSIM	PSNR	SSIM	PSNR	SSIM	PSNR	SSIM
Median	26.75	0.564	26.36	0.565	25.73	0.578	23.22	0.542
Arithmetic mean	28.15	0.643	27.70	0.646	26.58	0.642	23.96	0.604
Bilateral	27.51	0.599	27.41	0.629	26.84	0.637	25.32	0.641
AM-Bilateral	29.55	0.782	28.66	0.729	27.47	0.725	24.51	0.686

filter, and Arithmetic Mean filter is set as 1, whereas for AM-Bilateral filter is set as 3. In terms of range parameter for intensity kernel, Bilateral filter is set to 0.3, while AM-Bilateral filter is set to [0.1, 0.2, 0.3] whereby the pixel values of the image are within [0, 1].

PSNR and SSIM are estimated to validate the quality of the filtered image. PSNR parameter is a widely used traditional approach for image filter analysis, in which higher value indicates better quality of image restoration while SSIM parameter is a new method to measure the similarity between two images. The simulation results of performance evaluation are tabulated in Table 1.

The results have shown that AM-Bilateral filter has improved significantly compared to the Bilateral filter. Although Bilateral filter is slightly higher in terms of PSNR in Barbara noise image, AM-Bilateral filter still has better SSIM value for the same image. Figure 5 demonstrates the Lena restored image using various image filters. In this figure, it is evident that AM-Bilateral filter can remove image noise much better than the others without significant blurring effect.

4 Conclusion

In this study, the performance of the proposed method is validated using different noise images. The results have shown that AM-Bilateral filter can recover images with higher quality compared to conventional Bilateral filters and other image filters. Yet, the drawback of this filter is computational load issue since it is iterative and combined from two image filters. Therefore, further development in solving this problem is required in future studies so that it can be implemented in real-time applications.

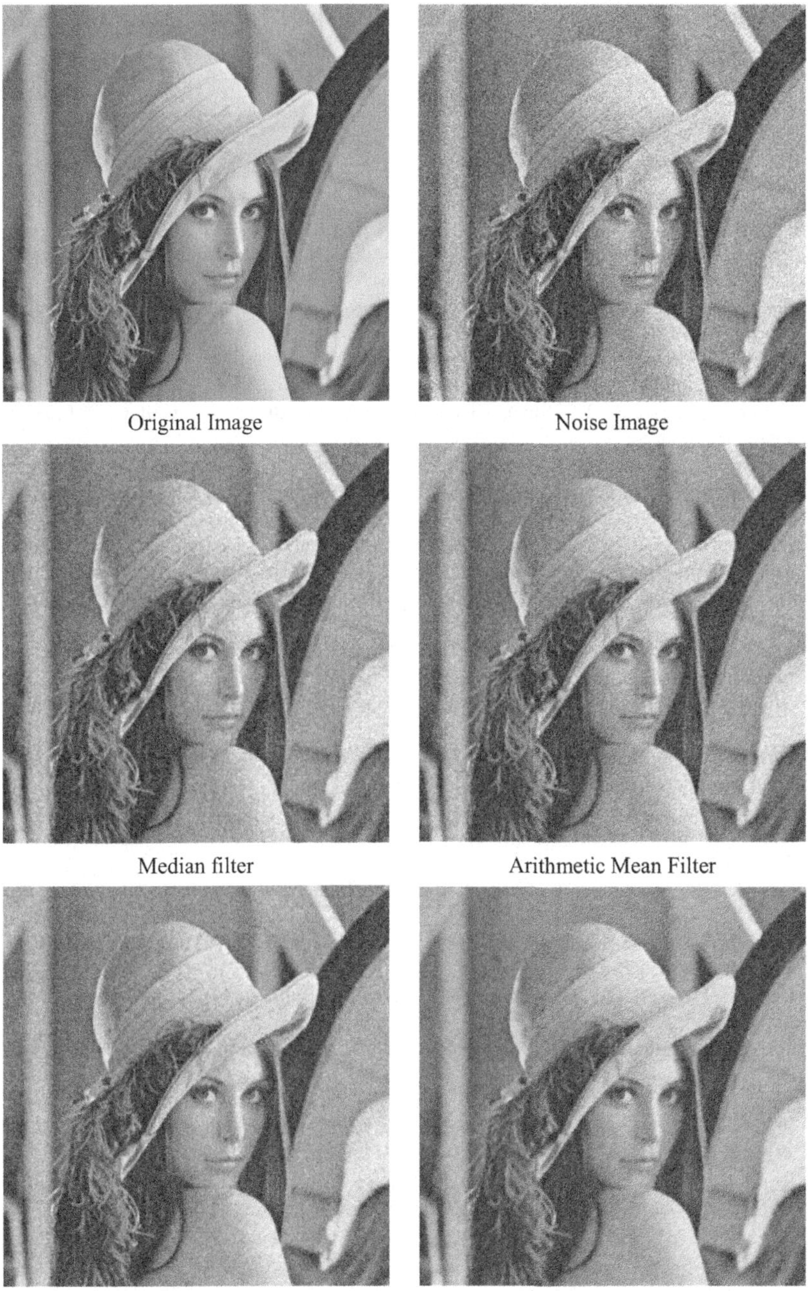

Fig. 5 Lena original image, noise image and restored image using different image filters

Acknowledgements This publication was supported by FRGS Kementerian Pendidikan Malaysia Grant No. 203/PAERO/6071437 and USM Fellowship.

References

1. Verma AK, Saini BS (2017) Forward-backward processing technique for image denoising using FDZP 2D filter. J Appl Res Technol 15(6):583–592
2. Selvin S, Ajay S, Gowri BG, Sowmya V, Soman K (2016) $\ell 1$ trend filter for image denoising. Procedia Comput Sci 93:495–502
3. Agrawal P, Verma JS (2013) A survey of linear and non-linear filters for noise reduction. Int J Adv Res Comput Sci Manage Stud 1(3)
4. Karami A, Tafakori L (2017) Image denoising using generalised Cauchy filter. IET Image Proc 11(9):767–776
5. Kumar SV, Nagaraju C (2018) Support vector neural network based fuzzy hybrid filter for impulse noise identification and removal from gray-scale image. J King Saud Univ-Comput Inf Sci
6. Kaur A, Chopra V (2012) A comparative study and analysis of image restoration techniques using different images formats. Int J Sci Emerg Technol Latest Trends 2(1):7–14
7. Gonzalez RC, Woods RE (2012) Digital image processing. Prentice Hall, Upper Saddle River, NJ
8. Somasundaram K, Kalavathi P (1888) Medical image denoising using Non-linear spatial mean filters for edge detection. Image 7(4816):2063
9. Ghosh S, Nair P, Chaudhury KN (2018) Optimized Fourier bilateral filtering. IEEE Signal Proc Lett 25(10):1555–1559
10. Paris S, Kornprobst P, Tumblin J, Durand F (2009) Bilateral filtering: theory and applications. Found Trends® Comput Graph Vis 4(1):1–73
11. Liu W, Zhang P, Chen X, Shen C, Huang X, Yang J (2018) Embedding bilateral filter in least squares for efficient edge-preserving image smoothing. IEEE Trans Circuits Syst Video Technol
12. Gavaskar RG, Chaudhury KN (2018) Fast adaptive bilateral filtering. IEEE Trans Image Proces 28(2):779–790
13. Reich S, Wörgötter F, Dellen B (2018) A real-time edge-preserving denoising filter. In: VISIGRAPP (4: VISAPP), pp 85–94
14. Einecke N, Eggert J (2013) Anisotropic median filtering for stereo disparity map refinement. In: VISAPP (2), pp 189–198
15. Tomasi C, Manduchi R (1998) Bilateral filtering for gray and color images. In: ICCV, vol 1, p 2
16. Weiss B (2006) Fast median and bilateral filtering. ACM Trans Graph (TOG) 25(3):519–526
17. Zhang M, Gunturk BK (2008) Multiresolution bilateral filtering for image denoising. IEEE Trans Image Proces 17(12):2324–2333
18. Riji R, Rajan J, Sijbers J, Nair MS (2015) Iterative bilateral filter for Rician noise reduction in MR images. Sig Image Video Proce 9(7):1543–1548

Aircraft Wing Aerodynamic Efficiency Improvement Using Longitudinal Spanwise Grooves

Samuel Merryisha and Parvathy Rajendran

Abstract Passive flow control techniques are the advanced boundary surface modifiers to improve the aerodynamic performance by keeping the flow attached. The overall aim of this study is to implement semi-circular grooves over the NACA 2412 wing at various orientations, to keep the flow attached along with the relative aerodynamic performance changes of baseline and groove wing. Numerically simulation over a Low Aspect Ratio wing of order 1 with variant angle of attack, at an operating velocity of 30 m/s, and Re of 4.4×105 is presented. Grooves over a baseline wing disturb the flow behavior, thereby creates turbulence by producing rotating small eddies. These rotating vortices keep the flow attached for longer time by re-attaching the separated flow with enhancement in L/D characteristics. Variant of longitudinal grooves has been employed over the wingspan at different x/c location (1) near leading edge (0.2C), (2) near trailing edge (0.8C), (3) mid-span (0.5C), (4) triplet location (0.2C, 0.5C, 0.8C). The models are designed using CATIA V5R20, and ANSYS Fluent helps to simulate the flow behavior, vortex development and aerodynamic performance difference between models. From the simulation study, it is clearly understood; the baseline wing shows massive boundary flow separation on the suction side of the wing beyond 14° AOA with immediate stalling. The presence of these longitudinal grooves enhances the stalling characteristics by keeping the flow attached up to 18° AOA. In all the optimized wing models, the L/D shows at least 0.05% improvement compared to baseline wing. However, the aerodynamic characteristics show pronounced results in the case of the triplet groove wing. It proves that creating roughness over smooth wing shifts the stalling angle and also improves the lift in that region with minor turbulence drag reduction.

Keywords Airfoil · Aerodynamic performance · Boundary layer · Groove · Stalling · Turbulence

S. Merryisha · P. Rajendran (✉)
School of Aerospace Engineering, Universiti Sains Malaysia, Engineering Campus, 14300 Nibong Tebal, Penang, Malaysia
e-mail: aeparvathy@usm.my

S. Merryisha
e-mail: merryishagracia@gmail.com

P. Rajendran et al. (eds.), *Proceedings of International Conference of Aerospace and Mechanical Engineering 2019*, Lecture Notes in Mechanical Engineering,
https://doi.org/10.1007/978-981-15-4756-0_15

1 Introduction

Over decades scientists are inquisitive about the optimized lift generation designs to enhance the aircraft characteristics. Significant efforts were shown in decreasing the surface shear stress which affects the boundary layer. Challengers have to be faced to improve the performance of lift incorporated with minimal drag formation and delayed flow separation, particularly at higher AOA. As in the case of 0° AOA, the pressure gradient maintains to be zero, and at higher AOA the adverse pressure gradient reaches maximum value. The physical flow behavior around an object throws out a better understanding of engineering disciplinaries of existing aerodynamic design and encourages a leap to improve surface argumentation. When an object pear through fluid, a boundary layer is enclosed around the object; the performance of these objects will, therefore, depend on the boundary layer deviation. When the flow is initialized, the flow maintained to be laminar as long as the surface is smooth; this will eventually lead to flow separation [1]. This flow separation is mainly due to the non-linear breakdown of flow over the smooth surface, generating shear layer, which decays the aerodynamic performance [2, 3]. When the turbulence is introduced through the shear layer the detached flow re-attaches the surface due to its natural adhesive property [1, 4]. Researchers have showcased that the early detachment of flow is the primary source to improve drag and degrade lift control. Hence, creating roughness on the surface of the wing has attracted the attention of researchers in recent trends [3, 5].

Separation control or decay shows vast improvement in the performance of airfoil/wing [1, 6]. Recent competitive flow control techniques [7, 8] are: (1) Active modifiers: process involves improving the aerodynamic characteristics by generating additional Kinetic Energy (K.E) to the flow-through blowing or suction process, by retrofitting control devices over surface. (2) Passive modifier: modifying the wing surface to disturb the pressure distribution, thereby delays/prevents the flow separation. Delaying the flow separation improves the maximum cross-sectional load by cutting down aerodynamic drag [9]. Reattachment flow is achieved by introducing grooves/bumps, which partially eliminates the boundary layer separation up to significant extent. The separation flow deviation has been minimized by 80% under immersed boundary layer condition [4]. Passive flow control techniques have been considered in-order to degrade wall-bounded drag with simultaneous reduction in turbulence drag [6]. Flow over active and passive VG's has been numerically simulated [4] with 3D Navier stokes equation to determine the average natural frequency flow separation along the spanwise direction.

Results from past studied highlights [10] that placing modifiers over the wing surface interacts the near-wall flow. This leads to degradation in intensity of streamwise turbulence along with 10–15% drop in Reynolds shear stress. NASA study [11] declares that drag formation is proportional to number of surface irregularities and the inter distance between them. Flow over the riblet surface [12] shows 25% reduction in surface drag compared to flat plate. Deep concerning the delay in boundary flow separation by many researchers investigation [9] has shown vast increase in velocity fluctuations over the modified surface, which may result in consecutive shear layer instability. The characteristic of the incoming boundary layer flow is mainly based on modified geometry. These fluctuations thicken the boundary layer in the order of cavity orientation.

The ultimate objective of the current study is to determine the flexibility of flow and its detachment point by using variably oriented grooves along x/c location. The study focuses on the control flow separation techniques with an enhancement in aerodynamic performance by understanding the stalling characteristics and healing the flow. The sensitivity of each groove orientation to that of the facing flow along with transverse boundary layer separation has been deeply discussed. Direct numerical simulation over a Low Aspect Ratio NACA 2412 wing of order 1 has been enhanced to study the aerodynamic behavior at various AOA 0°–20°, respectively. The flow control mechanism is enhanced in this study by attaching longitudinal groove (spanwise). The longitudinal groove of width 2.3 mm and depth 1.15 mm has been indented over the wing surface at 0.2C, 0.5C, 0.8C, and at all the three locations. We anticipate that the flow field will be energized by interacting with the smooth flow past the leading edge of the wing, which restricts the chordwise vortex formation with control over multiple flow effects.

2 Flow Separation Mechanism of Rough Surface and Positioning

Analysis of flow over an aircraft wing has recently shown interest in rough surface aerodynamics to improve the flow efficiency, which is well realized. Flow separation positioning of the baseline wing has to be well studied to confirm the orientation of the surface modifier [11]. Creating roughness over the wing surface not only translates flow from laminar to turbulence but also shows significant contribution to Reynolds number beyond the critical limit [13]. Implementing roughness over the path of the flow changes the aerodynamic characteristics by shifting the separation bubbles downstream. Alternatively, the performance seems to degrade to the ratio of the roughness elements [13]. It is well known that aerodynamic performance is mainly degraded due to flow separation [1].

Two main attempts can be indulged to improve the aerodynamic property [10] at higher AOA (1) delayed transition of the boundary layer due to laminar-turbulent, (2) redirecting the existing boundary layer turbulent structure. Creating turbulence in the free stream flow path stimulate flow transition by degrading the pressure between adjacent indentation leading to attached flow [3]. According to NASA study [11], scientists claim that placing surface modifiers at the forward part of the wing will create pre-mature transition effect thereby reduced drag effect by one-third. The concept of riblets [3, 10] (micro protruded surface) throughout wing has the potential of shifting the transition point separation with reduced viscous drag. The flow over spanwise streak spacing [12] has been improved by 40% compared to a flat plate.

2.1 Indented/Grooved Surface

Net surface drag reduction, along with the positive confliction results has been observed by intending grooves/riblets over the flow surface. A passive strategy study has been carried over boat-tailed bluff body with counter transverse grooves over a R_e of 9.6×10^4 [9]. The flow over grooved surface project out significant delay of flow separation with productive 9.7% reduction in drag. The effect of suction over the suction surface of the NACA 0012 [8] wing was numerically studied. This study was based on optimizing the length of the suction jet. Upon analyzing, the results show that increasing the length of the suction jet improves L/D by 43% and also delays the separation flow.

Longitudinal triangular grooves were indented over a non-dimensional flat plate to study the heat transfer and drag characteristics. From the study [14] it is clear that the flow becomes turbulent at $y^+ = 30$ for a flat base plate. Flow over the model varies from 15 to 40 m/s, which shows 4 to 5% reduction in drag characteristics in the case of V-groove type configuration. A similar structural attempt was made by Bacher et al. [12], and their results reveal that riblets positioned at $y^+ = 15$ show a more significant effect in lateral spreading behavior with increased boundary layer momentum thickness. The flow control surface modifiers should have a variable ratio of depth to the location to reduce momentum losses [9].

2.2 Analysis of Flow Field Around Modified Surface

A study proved that the presence of roughness over the wing surface thickness the boundary layer in the order of x/c, where x terms to be the function of boundary layer thickness [15]. The thickening factor was observed keenly and concluded that it is mainly due to transition promotion at growing viscous layer. Visible internal flow delay over laminar and turbulent layer has been shown with appropriate positioning of transverse grooves. This flow visualization scheme is enhanced to determine the effect of the recirculation zone [9, 16]. The flow characteristic effect is concerned with flow pressure distribution over the transverse airfoil corrugation. According to some researchers' work [16], it is clear that trapping of vortices at each stage of corrugation leads to laminar reattachment with delayed stall and evidential lift increment. The vortices which get trapped inside the grooves generate high momentum boundary layer [9]. The increase in drag magnitude depends on the ratio of surface roughness height to pitch [11]. Studies clearly show that the smaller surface roughness improves drag formation due to the massive generation of absolute pressure gradient.

3 Research Methodology

3.1 Theoretical and Numerical Study

Incompressible unsteady fluid flow is considered with turbulence and viscous nature. The flow is governed by a partial differential equation, such as the law of conservation of mass and momentum, as stated in Eqs. 1 and 2. The turbulence generation in the flow can be well established through Reynolds-Averaged Navier Stokes equation (RANS). By taking time average components into account, the following equations are incorporated.

$$\frac{\partial \rho}{\partial t} + \frac{\partial}{\partial x_i}(\rho u_i) = 0 \tag{1}$$

$$\frac{\partial}{\partial t}(\rho u_i) + \frac{\partial}{\partial x_j}(\rho u_i u_j) = -\frac{\partial_\rho}{\partial x_i} + \frac{\partial}{\partial x_j}\left[\mu\left(\frac{\partial u_i}{\partial x_j} + \frac{\partial u_j}{\partial x_i} - \frac{2}{3}\delta_{ij}\frac{\partial u_i}{\partial x_i}\right)\right] + \frac{\partial}{\partial x_j}\left(-\rho \underset{u_i}{\rightarrow} \underset{u_j}{\rightarrow}\right) \tag{2}$$

where $\left(-\rho \underset{u_i}{\rightarrow} \underset{u_j}{\rightarrow}\right)$ is Reynolds stresses, here the turbulence shear stress is related to linear strain according to Boussinesq approximation [17]. The momentum equation about Boussinesq approximation can be written as follow.

$$\frac{\partial}{\partial t}(\rho u_i) + \frac{\partial}{\partial x_j}(\rho u_i u_j) = -\frac{\partial p}{\partial x_i} + \frac{\partial}{\partial x_j}\left((\mu + \mu_j)\frac{\partial u_i}{\partial x_j}\right) + \frac{\partial}{\partial x_j}\left((\mu + \mu_j)\frac{\partial u_j}{\partial x_i}\right) \tag{3}$$

The turbulence characteristics of the flow are achieved by Spalart-Allmaras one-equation turbulence model. The Spalart-Allmaras transport equation in the case of non-conservative form can be written as follows,

$$\frac{\partial}{\partial t}(\rho \tilde{v}) + \frac{\partial}{\partial x_i}(\rho \tilde{v} u_i) = G_v + \frac{1}{\sigma \tilde{v}}\left[\frac{\partial}{\partial x_j}\left\{(\mu + \rho \tilde{v})\frac{\partial \tilde{v}}{\partial x_j}\right\} + C_{b2}\rho\left(\frac{\partial \tilde{v}}{\partial x_j}\right)^2\right] - Y_v + S_{\tilde{v}} \tag{4}$$

where,

G_v	production of turbulent viscosity
$u_i u_j$	mean velocity components
v	molecular kinematic viscosity
$\tilde{v}$	modified turbulent viscosity
$S_{\tilde{v}}$	user-defined source term
Y_v	destruction of turbulent viscosity

Wingspan: 230mm
Aspect Ratio: 1
Root airfoil thickness: 15mm
Tip airfoil thickness: 15mm
Root chord: 230mm
Tip chord: 230mm
Incidence angle: 0°
Sweep angle: 0°

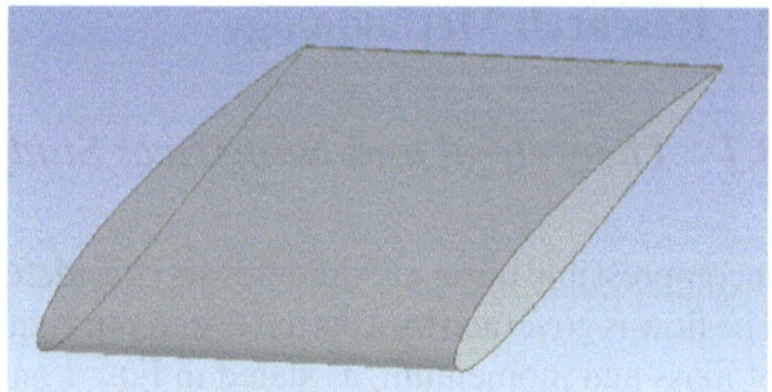

Fig. 1 Baseline wing (model 1)

$\sigma_{\tilde{\upsilon}}$, C_{b2} constants
ρ density

The main objective of this study is to predict the flow disturbance difference between baseline wing and grooved wing. The evaluation of this simulation is to determine the dissipation of aerodynamic characteristics associated with boundary layer behavior. The complete study has been carried out with NACA 2412 airfoil with the following specification. The wing model is modeled using CATIA V5R20, as shown in Fig. 1 and analyzed with the help of ANSYS 18.1.

3.2 Groove Geometric Description

In order to accomplish the primary focus of this investigation, simplified groves were created over the wing surface, which runs from root chord to tip chord of wing. The groove of width 2.3 mm and depth 1.15 mm are considered (shown in Fig. 2), where depth and width are maintained constant throughout all models. Longitudinal grooves are placed over the suction side of the wing with four different slot distribution (i.e.) (1) Groove near leading edge (0.2C), (2) Groove near trailing edge (0.8C), (3) mid-span (0.5C), (4) triplet location (0.2C, 0.5C, 0.8C) as shown in Fig. 3 the diversion of flow separation and the aerodynamic characteristics changes depends on the location of groove, inter-spacing between grooves, and the ratio of length to height of groove.

3.3 Boundary Condition and Grid Generation

C-block semi-bullet domain is chosen for this study, where the domain stretches from −y to +y, −z to +z about 10C each apart from the model. Where the model is fixed on the +x axis with a lateral spreading domain along −x-axis with 10C boundary from the root of the wing. The computational domain was chosen to be bigger and farther (far away from the model) (as shown in Fig. 4) to prevent the interaction of

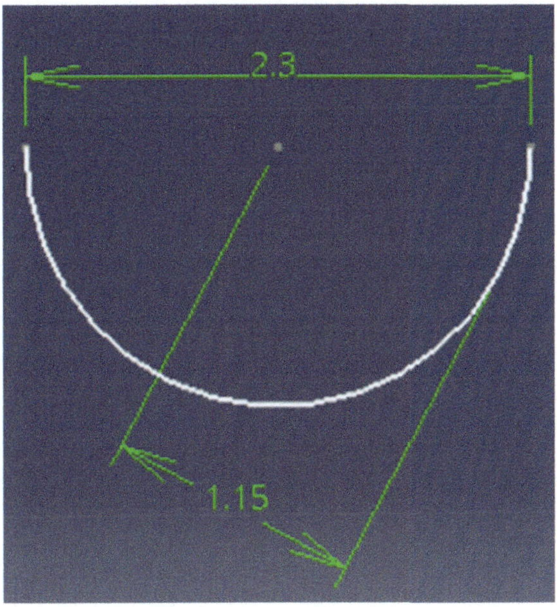

Fig. 2 Groove dimensions (dimensions are in mm)

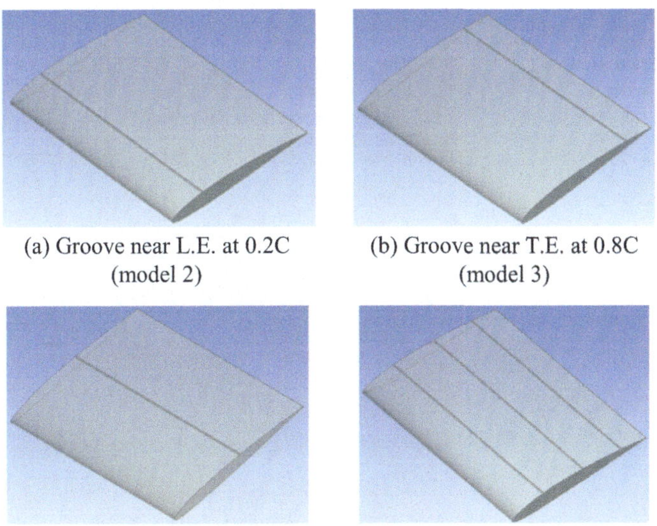

(a) Groove near L.E. at 0.2C (model 2)

(b) Groove near T.E. at 0.8C (model 3)

(c) Mid span groove at 0.5C (model 4)

(d) Triplet groove located at 0.2C, 0.5C, 0.8C (model 5)

Fig. 3 Groove model orientations

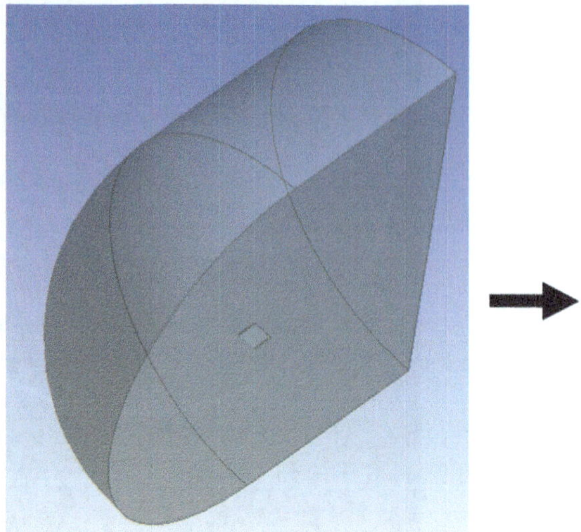

Fig. 4 Isometric view of a semi-bullet domain

domain boundary layer flow to the flow field around the airfoil [8]. As in the case of subsonic flow over a wing surface, parallel computational has been adopted to perform linear scaling process with projected domain boundaries Fig. 5.

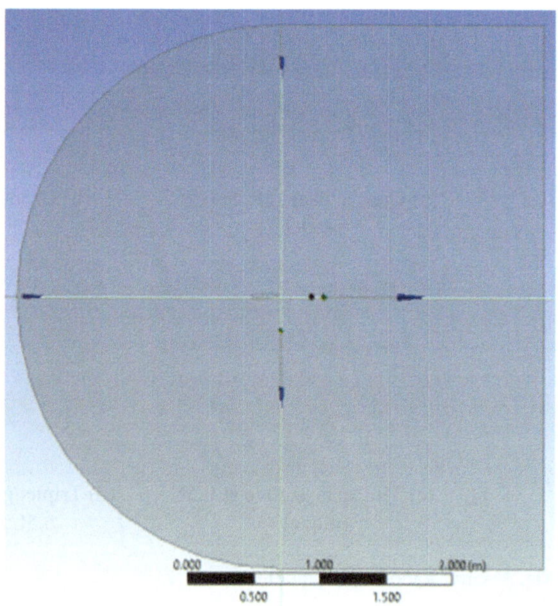

Fig. 5 Projected view of domain with wing

Adiabatic non-slip condition is given to the airfoil and also for the domain walls. The upper and lower boundaries are assigned to be far-field. The meshed view is as Figs. 6 and 7 and the mesh settings are as follows.

Size function: Curvature
Relevance and span angle center: Fine
Maximum face size: 20 mm
Growth rate: 1.20

In this study, the velocity of 30 m/s has been adopted with the incompressible flow. As the Reynolds number is in million Re—4.4 × 105 turbulent boundary flow condition has been considered. To make the simulation easier, the impact over the boundary layer has been minimized by applying non-slip conditions over the

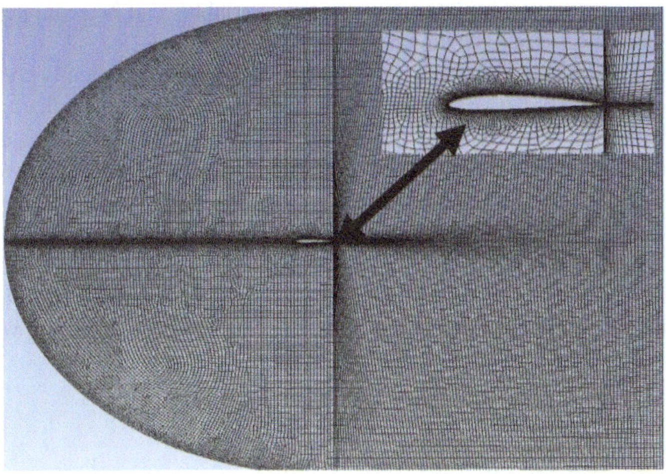

Fig. 6 Domain grid generation

Fig. 7 Fine grid wing

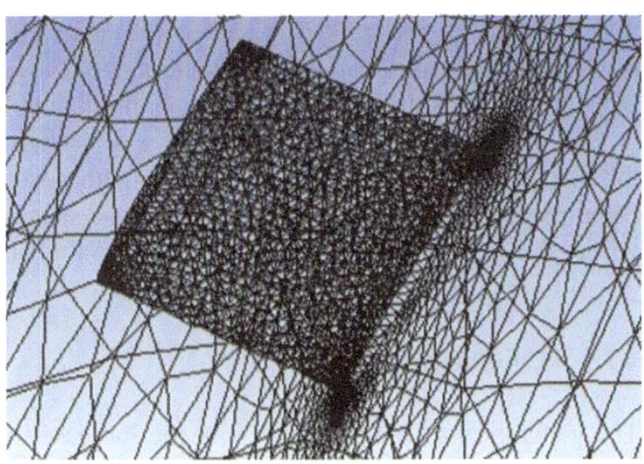

Table 1 Boundary conditions

Boundary conditions	Values
Flow velocity	30 m/s
Turbulence kinetic energy	4.184×10^{-7} m^2 s^2
Density of air	1.225 kg/m^3
Dynamic viscosity	1.7894×10^{-5} kg/m s
Density	1.2043 kg/m^3
Reference length	0.23 m
Gauge pressure	0 Pa
Operating pressure	101,325 Pa

airfoil and also for the boundary walls. The interconnected interior nodes calculate the outgoing characteristics. The characteristic variables can be regulated through boundary condition setup are shown in Table 1.

The viscous sublayer [18] of the Spalart Allmaras was automatically computed by using $\frac{u}{u_\tau} = \frac{\rho u_\tau y}{\mu}$, where, u—velocity parallel to the wall, u_τ—shear velocity and Y—distance from wall. Among all the different turbulence models simple one-equational Spalart Allmaras model represents better behavior for this numerical investigation as it shows beneficial results overflow separation and counter-rotating turbulence. A similar comparative study has been done [19] over simple 1D SA-Model to solve transport equation for turbulent K.E. Turbulence intensity were maintained at 1–5% [20, 21] under RANS model coupled with Spalart Allmaras model has been implemented to match the free stream turbulence level. Velocity inlet was set for inlet boundary condition and pressure outlet for outlet boundary.

4 Results and Discussion

4.1 Aerodynamic Characteristics

As the angle of attack varies, the co-efficient of lift and drag values change accordingly for each model. This study reveals the lift and drag coefficient for different groove orientation, as shown in Figs. 8 and 9 when the angle of attack increases, the lift, and drag increase simultaneously with development of boundary vortices. By implementing grooves over the wing surface degrades the strength of the surface vortices, thereby keeps the flow attached on to the wing for greater angle of attack.

4.1.1 Co-efficient of Lift

The generation of lift increases with the increase in the angle of attack and then decreases with a further increase in AOA. As in the case of 0° AOA, the baseline

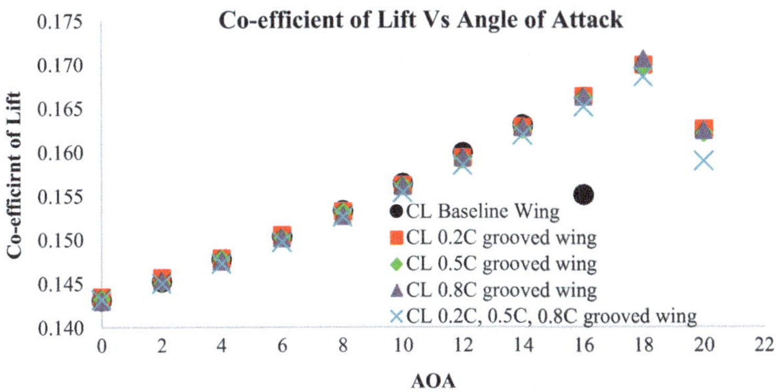

Fig. 8 Cl versus AOA Graphical representation for different airfoil Models

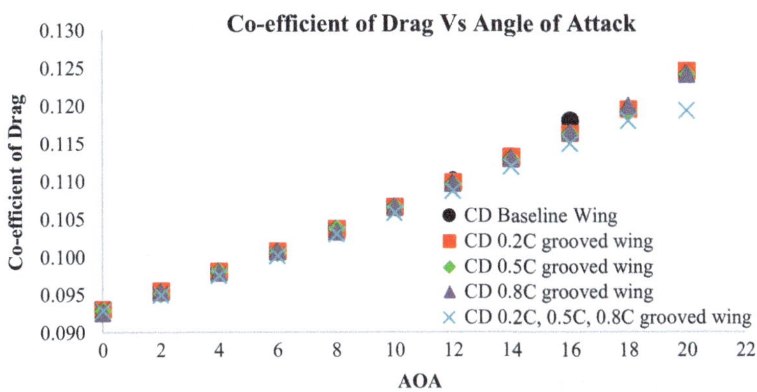

Fig. 9 Cd versus AOA Graphical representation for different airfoil Models

wing generates lift of 0.14313, whereas the grooved wing models generated lift between 0.14303 and 0.14353. Baseline wing generates maximum lift of 0.16310 at 14° AOA, whereas the grooved wing models generated maximum lift between 0.16854 to 0.17060 at 18° AOA. From Fig. 8 it can be seen that baseline wing stalls at 14° AOA, whereas grooved wings stalls at 18° AOA. From the graph it can be assured that grooved wing fits the best performance when considering stalling and behavior.

4.1.2 Co-efficient of Drag

The influence of the wing surface modifier can be observed by the co-efficient of drag plotted against the angle of attack. As the AOA increases, the drag also increases slowly up to stalling AOA, beyond which there is a rapid rise in drag generation.

Table 2 Percentage improvement in lift to drag (L/D) ratio

Angle of attack	Model 2	Model 3	Model 4	L/D Model 5
0°	0.05	0.18	0.07	0.01
2°	0.08	0.02	0.07	0.23
4°	0.15	0.12	0.12	0.36
6°	0.14	−0.04	0.13	0.19
8°	0.03	0.04	−0.06	0.25
10°	0.09	0.05	0.10	0.32
12°	0.13	0.25	0.23	0.62
14°	−0.05	0.13	0.11	0.56
16°	8.79	8.65	8.72	9.32

This rapid generation of drag is mainly due to the region of separated flow over the wing surface, leading to more significant pressure drag formation. From Fig. 9 its visualized that the triplet grooved wing (model 5) shows better performance in drag reduction, with a maximum of 2.63% in drag reduction compared to baseline wing. All the remaining grooved wing models (model 2—model 4) also show better drag reduction compared to baseline wing. The following are the average percentage of drag reduction compared to baseline wing (average of all AOA): model 2—0.25%, model 3—0.41%, model 4—0.40%, model 5—1.01%. This condition is probably due to the relaxed re-circulation region within the groove by decaying the momentum losses.

4.1.3 Lift to Drag Ratio

The performance of the wing is considerably noted with the outcome of the lift to drag ratio. This performance shows better results when the wing shows higher lift and reduced drag. From the study it is noted that lift to drag ratio increases with the increase in AOA. From Table 2 it's clearly understood that the vortex intensity is proportional to lift as the vortex formation is related to the L/D ratio. The triplet grooved wing (model 5) shows out the best performance compared to all other grooved wing and the baseline wing.

4.2 *Pressure Contour*

The pressure transformation shows its divergence in leading edge and at the point of zero pressure gradient. As the angle of attack increases, the pressure gradient also increases, as shown in Fig. 10 which leads to detachment of boundary layer as shown in velocity contour.

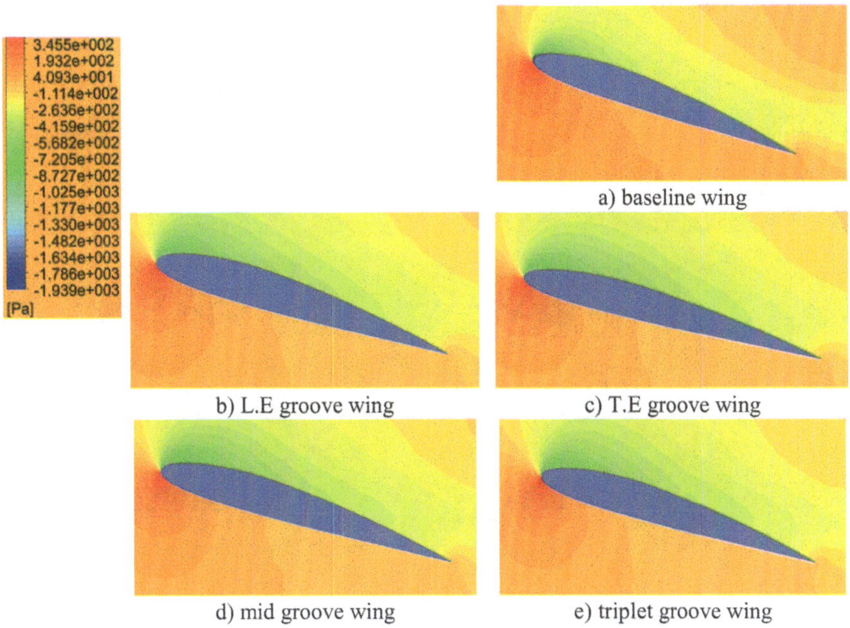

Fig. 10 Pressure distribution of wing models at 16° AOA

4.3 Velocity Contour

From the visual image of velocity contour the development of boundary layer is clearly seen, which starts from the L.E stagnation point until the T.E. Figure 11 clearly differentials the boundary layer different (i.e.) the baseline wing has a thick boundary layer (means the deflection of boundary flow), whereas the grooved wing shows thin boundary layer throughout the flow. At 16° angle of attack, the figures differentiate the delayed separation point between the baseline wing and the grooved wings, with degradation in wake region. It can be generally observed, locating the groove at adverse pressure gradient region diplomatically pushes the stagnation point farther.

4.4 Development of Vortex Due to Turbulence Factor

The flow is resolved incompressible by implementing the Navier Stokes equation. One-equational Spalart Allmaras model has been implemented, which diffuses the molecules leaving behind energized flow with turbulent kinetic energy. This turbulence kinetic energy generates eddies that contribute the vortex. From Fig. 12 it is observed that the grooved wing dissipates the vortex at higher angle of attack, where

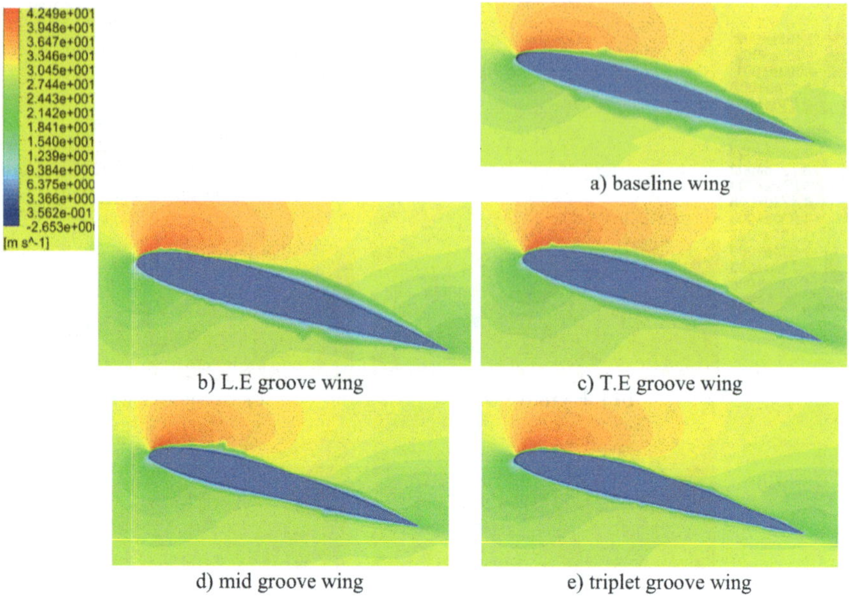

Fig. 11 Velocity distribution of wing models at 16° AOA

the kinetic energy is reduced. Generation of streamwise vortices due to irregularities on the wing surface act as a vortex generator to push the separation point forward, as observed by Ogawa et al. [22].

4.5 General Discussion

The flow behaviors of each model has its own pace of flow dynamics and are not similar for all angle of attack. The groove region reveals a mixing of flow and thus shows out more significant velocity gradient with lesser magnitude. Beyond the groove the strength of the rotating vortices decreases. Similar visualization is observed by Puri [23] while simulating a slot near leading edge, which showed ideal re-attachment due to mixing of flows. Laminar re-attachment is observed above the separated flow region at re-circulation zone, same as observed by Buckholz [16]. Re-attachment of flow occurs beyond the separation bubbles. The flow past the lateral position of the separation point develops wider wakes, and these wakes diffuses the re-attachment of flows. Reducing the wake width makes the flow stabilized to re-attach back.

At 16° AOA its visualized the baseline wing shows the divergence of flow, whereas the grooved wing shows attached flow due to the high-velocity deviation on the wing surface. The attached flow is mainly due to the re-circulation of flow inside the concave groove cavity. These re-circulations raise the zone of velocity difference along with velocity gradient re-enhancement.

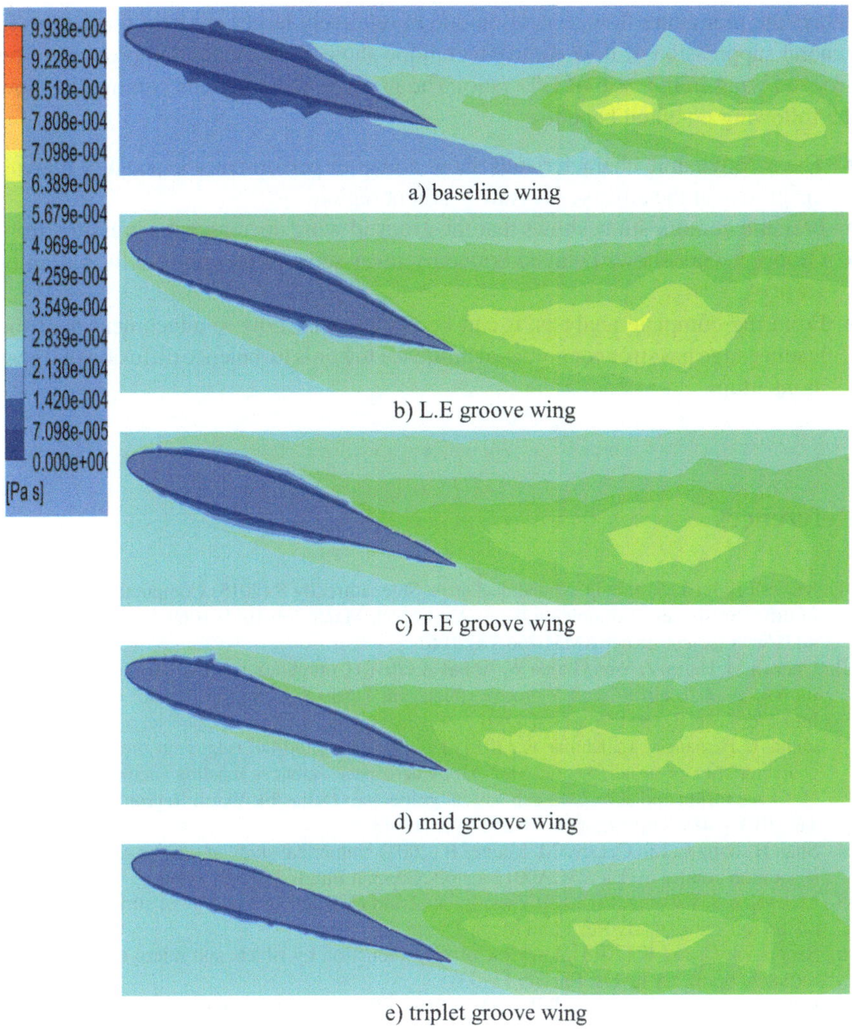

Fig. 12 Vortex flow visualization of different models

5 Conclusion

The current investigation has shown the effect of different groove orientation and its significant influence. The complete characteristic of flow is altered due to surface roughness. The thickness of the boundary layer is even without scattered flow in the case of grooved wing, which resolves the complete flow characteristic. Results confirm that placing grooves at various x/c location perform better than the baseline wing. Comparatively in the case of baseline wing stalling angle grooved wing shows 7.21% enhancement in lift with 9.32% improvement in L/D ratio. Triplet groove delays the premature flow separation due to relatively larger gradient formation. In general suppression of flow control technique showed a beneficial improvement on grooved wing in the case of L/D ratio. The grooved wing models commonly show the following potential benefits.

- The aerodynamic characteristics show a leap in performance especially placing the groove at the adverse pressure gradient region.
- Pressure contour study shows that the grooved wing decreases the pressure drop.
- Grooved wing shows delayed boundary layer separation with less formation of vortices.
- From the complete analysis, it can be evident that flow re-attachment near the leading edge has stronger potential influence towards turbulence diffusion of linear momentum.

References

1. Mahesh Babu T, Sairaja D, Chandrasekar A, Sreenathreddy S (2015) Computational analysis of different shapes of dimple on wing. Appl Mech Mater 766:1061–1069. https://doi.org/10.4028/www.scientific.net/AMM.766-767.1061
2. Ghazali M, Harun Z, Wan Ghopa W, Abbas A (2016) Computational fluid dynamic simulation on NACA 0026 airfoil with V-groove riblets. Int J Adv Sci Eng Inf Technol 6(4):529–533. https://doi.org/10.18517/ijaseit.6.4.901
3. Guha T, Fernandez E, Kumar R (2013) Effect of longitudinal ridges on the aerodynamic characteristics of an airfoil. In: 51st AIAA aerospace sciences meeting including the New Horizons forum and aerospace exposition, grapevine (Dallas/Ft. Worth Region), Texas, 7–10 Jan 2013. p 489. https://doi.org/10.2514/6.2013-489
4. Shan H, Jiang L, Liu C, Love M, Maines B (2008) Numerical study of passive and active flow separation control over a NACA0012 airfoil. Comput Fluids 37(8):975–992
5. Merryisha S, Parvathy R (2019) Experimental and CFD analysis survey of surface modifiers on aircraft wing. J Adv Res Fluid Mech Therm Sci
6. Saravi SS, Cheng K (2013) A review of drag reduction by riblets and micro-textures in the turbulent boundary layers. Eur Sci J 9(33)
7. Ganesh N, Arunvinthan S, Pillai SN (2019) Effect of surface blowing on aerodynamic characteristics of tubercled straight wing. Chin J Aeronaut 32(5):1111–1120
8. Yousefi K, Saleh R (2015) Three-dimensional suction flow control and suction jet length optimization of NACA 0012 wing. Meccanica 50(6):1481–1494. https://doi.org/10.1007/s11012-015-0100-9
9. Mariotti A, Buresti G, Salvetti MV (2018) Flow separation control and drag reduction for a two-dimensional boat-tailed bluff body through transverse grooves. In: ASME 2018 5th joint US-European fluids engineering division summer meeting, Montreal, Quebec, Canada, 15–20 July 2018. American society of mechanical engineers, pp V001T008A007–V001T008A007
10. Viswanath P (2002) Aircraft viscous drag reduction using riblets. Prog Aerosp Sci 38(6–7):571–600
11. Hood MJ (1939) The effects of surface waviness and of rib stitching on wing drag. National Advisory Committee for Aeronautics. Langley Manorial Aeronautical Laboratory, Washington
12. Bacher EV, Smith CA (1985) Combined visualization-anemometry study of the turbulent drag reducing mechanisms of triangular micro-groove surface modifications. In: Shear flow control conference, Boulder, Colorado. American Institute of Aeronautics and Astronaoutics, p 548, 12–14 Mar 1985

13. Bloch D, Mueller T (1986) Effects of distributed grit roughness on separation and transition on an airfoil at low Reynolds numbers. In: 4th applied aerodynamics conference, San Diego, CA, 9–11 June 1986. p 1788. https://doi.org/10.2514/6.1986-1788
14. Walsh M, Weinstein L (1979) Drag and heat-transfer characteristics of small longitudinally ribbed surfaces. AIAA J 17(7):770–771. https://doi.org/10.2514/3.61216
15. Chakroun W, Al-Mesri I, Al-Fahad S (2004) Effect of surface roughness on the aerodynamic characteristics of a symmetrical airfoil. Wind Eng 28(5):547–564
16. Buckholz R (1986) The functional role of wing corrugations in living systems. J Fluids Eng 108(1):93–97
17. Seror S, Rubin T, Peigin S, Epstein B (2005) Implementation and validation of the Spalart-Allmaras turbulence model in parallel environment. J Aircraft 42(1):179–188. https://doi.org/10.2514/1.4009
18. Islam MT, Arefin AM, Masud M, Mourshed M (2018) The effect of Reynolds number on the performance of a modified NACA 2412 Airfoil. In: AIP conference proceedings, December 2018, vol 1. AIP Publishing, p 040015. https://doi.org/10.1063/1.504432
19. Zakaria MY, Ibrahim MM, Ragab S, Hajj MR (2018) A computational study of vortex shedding from a NACA-0012 airfoil at high angles of attack. Int J Aerodyn 6(1):1–17. https://doi.org/10.1504/IJAD.2018.10010814
20. Manni L, Nishino T, Delafin P-L (2016) Numerical study of airfoil stall cells using a very wide computational domain. Comput Fluids 140:260–269
21. Lopes A (2016) A 2D software system for expedite analysis of CFD problems in complex geometries. Comput Appl Eng Educ 24(1):27–38. https://doi.org/10.1002/cae.21668
22. Ogawa H, Babinsky H, Pätzold M, Lutz T (2008) Shock-wave/boundary-layer interaction control using three-dimensional bumps for transonic wings. AIAA J 46
23. Puri K, Laufer M, Müller-Vahl H, Greenblatt D, Frankel SH (2018) Computations of active flow control via steady blowing over a NACA-0018 airfoil: implicit LES and RANS validated against experimental data. In: 2018 AIAA aerospace sciences meeting, 8–12 Jan 2018. p 0792. https://doi.org/10.2514/6.2018-0792

Generalized Eshelby Integral Formula for Multiple Inclusion Composite Materials

Abdul Rauf Abdul Raif, Muhammed Fadzli Ismail, and Mulia Minhat

Abstract Eshelby integral formula, which was initially derived for a single inclusion embedded in all matrix system, plays a fundamental role in the micromechanics of composite or heterogeneous materials. The integral formula is remarked to be readily generalized to the case of multiple inclusions. However, the proof for such case has never been established. Herein, the integral formula is reformulated within the frameworks of generalized self-consistent method for the case of multiple inclusion composite system and thus, establishing the proof of its generalized version.

Keywords Eshelby integral formula · Generalized self-consistent method · Multiple inclusion · Multiphase composites

List of Abbreviation and Symbols

GSC	Generalized self-consistent
RVE	Representative volume element
eff	Effective composite
C_{ijkl}	4th-order stiffness tensor
σ_{ij}	2nd-order stress tensor
ε_{ij}	2nd-order strain tensor
u_i	Displacement vector
n_j	Unit vector normal to surface
U	Elastic energy
W	Strain energy density
N	Maximum number of phases in a composite system
i	Phase numbering, which represents inclusion's number in a composite, e.g. $i = 1, 2, \ldots, N$

A. R. A. Raif · M. F. Ismail · M. Minhat (✉)
Aerospace Section, Universiti Kuala Lumpur Malaysian Institute of Aviation Technology, Jalan Jenderam Hulu, 43800 Sepang, Selangor DE, Malaysia
e-mail: mulia@unikl.edu.my

P. Rajendran et al. (eds.), *Proceedings of International Conference of Aerospace and Mechanical Engineering 2019*, Lecture Notes in Mechanical Engineering,
https://doi.org/10.1007/978-981-15-4756-0_16

r	Radius of a phase
$N+1$	The outermost phase in a multiphase composite based on the representative volume element defined in generalized self-consistent method. The phase is also referred to as an equivalent homogeneous medium
s, S, Σ	Surface of a body or contact surface between two bodies
v, V	Volume of a body
$\langle(\cdot)\rangle$	Average value of a field variable per unit volume, e.g. $\langle\sigma\rangle = \frac{1}{V}\int_V (\sigma)dV$
$(\cdot)^0$	Field variable for medium without inclusions under linear or uniform boundary conditions, i.e. U^0
$(\cdot)^{RVE}$	Field variable for medium with the inclusions under linear or uniform boundary conditions, e.g. W^{RVE}
$(\cdot)^{eff}$	Field variable for effective composite medium under linear or uniform boundary conditions, e.g. W^{eff}
$(\cdot)^\wedge$	Field variable for effective composite medium containing a set of uniform body forces under linear or uniform boundary condition, e.g. $W^\wedge$
$(\cdot)'$	Field variable for effective composite medium containing a set of uniform body forces without boundary conditions, e.g. W'

1 Introduction

The concepts of interaction energy between stress systems and the reduction of volume integration to a particular type of surface integration using divergence theorem have been great utilities in micromechanics of heterogeneous body since they were first developed and applied respectively by Eshelby in his study of a body containing inhomogeneity [1]. Christensen and Lo [2, 3] in their attempts to correctly estimate the effective shear modulus of a two-phase composite system had applied those concepts, from which useful Eshelby's formula derived for a single inclusion embedded in all matrix medium was obtained (see Fig. 1a). The formula was derived by differentiating the states of elastic energy systems of microscale mediums with and without the inclusion, and is given as follow:

$$U = U^0 + \frac{1}{2}\int\limits_s \left(\sigma_{ij}^0 u_i + \sigma_{ij} u_i^0\right) n_j dS \tag{1}$$

where U is the elastic energy of a medium with inclusion, i.e. a medium consists of matrix and inclusion, U^0 is the elastic energy of a medium without the inclusion, i.e. medium composed entirely of matrix material only, σ_{ij} and u_i are the field variables of a medium that surrounds the inclusion, and σ_{ij}^0 and u_i^0 are the fields of the same medium without the inclusion, n_j is a unit vector normal to the surface s, and s is the contact surface between matrix and inclusion.

The reduced version of the Eshelby's formula was first proposed by Christensen and Lo [2–4] in their three-phase model. The model is commonly known as the

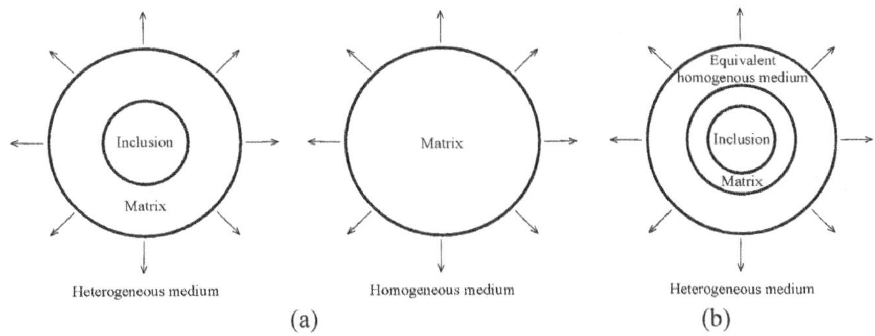

Fig. 1 **a** Comparison in the elastic energy of a medium with inclusion and a medium without the inclusion that leads to Eshelby's formula, and **b** three-phase model for the two-phase spherical and cylindrical composite materials, which employs Eshelby integral formula

generalized self-consistent (GSC) method. In their microscale model, the single inclusion is first surrounded by matrix material, which is then embedded in an infinite equivalent homogeneous medium such as shown in Fig. 1b. The presence of a single inclusion in matrix material allows the model to capture the effect of inclusion's volume fraction on the effective properties of composite material. The existence of equivalent homogeneous medium is justified considering the assumption that all other inclusions and matrix in a composite except for that single inclusion and its surrounding matrix material under consideration, are homogenized all together and subsequently, become that equivalent medium. Its properties are unknown but actually represent the effective properties of composite material that we seek to solve for. The assumption of that medium being unbounded is also valid if were to consider the size of that single inclusion and its surrounding matrix material in comparison to the overall size of a composite structure. According to Christensen and Lo [2], the micromechanics problem in seeking those unknown properties is solved when the elastic energies of mediums with and without both the inclusion and matrix are assumed to be identical, i.e. $U = U^0$. Thus, from Eq. (1) we have:

$$\frac{1}{2}\int_S (\sigma_{ij}^0 u_i + \sigma_{ij} u_i^0) n_j dS = 0. \tag{2}$$

The reduced form of the Eshelby's formula shown in Eq. (2) is famously known as the Eshelby integral formula. It is one of the essential equations in GSC method as it allows for the method to produce closed-form accurate analytical solutions of effective properties up to a typically large volume fraction of inclusion in composite materials [5–9]. According to Christensen [3], both formula in Eqs. (1) and (2) can readily be generalized to the case of multiple inclusions, or more accurately in the modeling context based on the GSC method, can readily be generalized to the case of multiphase composite system having concentric circular shape inclusions such as shown in Fig. 1a, b. Hereinafter, such multiphase composite system is referred to whenever the subject of multiple inclusions composite materials is discussed. In fact, the Christensen's three-phase model, which is used to predict the effective properties of composites is the classic example of a multiphase composite system, e.g. fiber (single inclusion) and matrix are respectively treated as the first two concentric circular phases embedded in the third phase of an equivalent homogeneous medium. Nevertheless, the proof for the case of multiple inclusions was never established by original authors and to our knowledge, to date has never been proven satisfactorily despite the success history of the GSC method employed in various micromechanics studies since 1980s and more recently, in the micromechanics study of nanostructure hybrid fiber reinforced composite materials [8–12]. In fact, many solutions obtained from the GSC method for composites having multiple inclusions found in the literature were the result of direct application of the original but not the generalized formulae. Such approach may suffer from significant inconsistencies. Thus, this triggers our motivation to prove the generalized version of Eshelby integral formula and in this paper, for the first time the correct formulations of the Eshelby's equations for composites having multiple inclusions within the frameworks of the GSC model will be derived and illustrated.

2 Generalized Eshelby's and Eshelby Integral Formula

In this section, we reformulate the Eshelby's and its integral formula for the N-phase composite system representing the special type of multiple inclusion composite material based on the representative volume element (RVE) defined by the GSC model such as shown in Fig. 2a. Note that N is the number of phases representing the multiple inclusions considered in the RVE and the $N + 1$-th phase represents the infinite equivalent homogenous medium that surrounds the inclusions. The derivations given

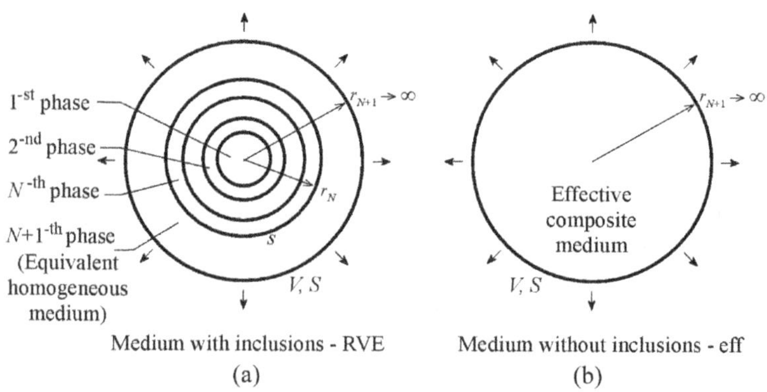

Fig. 2 Comparison in the energy system of an N-phase composite medium in an unbounded equivalent homogeneous medium or an $N + 1$-th phase with its equivalent effective composite medium

here closely follow the procedures proposed by Christensen [3]. However, it differs in several features, which exhibit the novelty of this paper. First, it is obvious that the formulation here is for the case of multiple inclusions or specifically, multiphase composite system based on the GSC model. Second, the formulae are derived using the homogeneous boundary condition of applied linear displacement vector instead of uniform surface traction vector. Third, the energy system used in the formulations is based on the average strain energy density system $\langle W \rangle$ rather than the elastic energy system U. Finally, an unbounded medium without the inclusions that has the same size and boundary conditions as the RVE is represented by an effective composite medium instead of an equivalent homogeneous medium (see Fig. 2b). It must be emphasized that effective composite medium is a medium that is resulted from when the medium with inclusions or the RVE defined earlier is altogether homogenized. Thus, its properties are also unknown but are actually about the same as the unknown properties of an unbounded equivalent homogeneous medium, i.e. $C_{ijkl}^{eff} \cong C_{ijkl}^{N+1}$ as the outer radii of both mediums reach to infinity.

Based on Fig. 2, the average strain energy density of a medium with inclusions $\langle W^{RVE} \rangle$ can be represented as

$$\left\langle W^{RVE} \right\rangle = \frac{1}{2V} \int_V \sigma_{ij}^{RVE} \varepsilon_{ij}^{RVE} dV \tag{3}$$

and for medium without the inclusions $\langle W^{eff} \rangle$ is given by

$$\left\langle W^{eff} \right\rangle = \frac{1}{2V} \int_V \sigma_{ij}^{eff} \varepsilon_{ij}^{eff} dV \tag{4}$$

where σ_{ij}^{RVE} and ε_{ij}^{RVE} are the stress and strain fields of a medium with inclusions respectively, σ_{ij}^{eff} and ε_{ij}^{eff} are the respective stress and strain fields of the same medium without the inclusions, and V is the volume of RVE. Using Eqs. (3) and (4), the difference in average strain density between the two mediums can be written as

$$\left\langle W^{RVE} \right\rangle = \left\langle W^{eff} \right\rangle + \frac{1}{2V} \int_V \left(\sigma_{ij}^{RVE} \varepsilon_{ij}^{RVE} - \sigma_{ij}^{eff} \varepsilon_{ij}^{eff} \right) dV. \tag{5}$$

Applying divergence theorem in conjunction with equilibrium equations where $\sigma_{ij,j}^{RVE} = \sigma_{ij,j}^{eff} = 0$, Eq. (5) can be transformed into the following form

$$\left\langle W^{RVE} \right\rangle = \left\langle W^{eff} \right\rangle + \frac{1}{2V} \int_S \left(\sigma_{ij}^{RVE} u_i^{RVE} - \sigma_{ij}^{eff} u_i^{eff} \right) n_j dS \tag{6}$$

where u_i^{RVE} and u_i^{eff} are the displacement vector fields for medium with and without inclusions respectively, and S is to be taken as the outer surface of RVE. Notice that the volume integral in Eq. (5) has been transformed into surface integral shown in Eq. (6) (see appendix section for details). Since the surfaces of both mediums are imposed with the same surface displacement boundary conditions, i.e. $u_i^{RVE} = u_i^{eff}$, Eq. (6) can now be rewritten as follow

$$\left\langle W^{RVE}\right\rangle = \left\langle W^{eff}\right\rangle + \frac{1}{2V}\int\limits_S \left(\sigma_{ij}^{RVE} - \sigma_{ij}^{eff}\right)u_i^{eff} n_j dS. \tag{7}$$

At this point, we intend to get a different form of the integral term in Eq. (7) and to achieve this, we analyze a different but equivalent set of problems related to Fig. 2a.

Let us consider the same RVE but is composed entirely from the material of effective composite medium. To account for the effects of the N inclusions, which is experienced by the $N+1$-th phase in a medium with N-inclusions shown in Fig. 2a, several particular sets of uniform or non-fluctuated body forces are applied over the fictitious surfaces s_i at where would be the outer radius of each or i-th phase supposed to be, i.e. at $r = r_i$ where $i = 1, 2, \ldots, N$ such as shown in Fig. 3a. Consequently, with these effects of body forces, the field variables in the region outside that of the outer radius of N-th phase, i.e. at $r \geq r_N$ will be the same state of field variables experienced by the $N+1$-th phase shown in Fig. 2a. Subsequently, based on the superposition method we can further decompose this equivalent problem 3.a into the two problems shown in Fig. 3b, c. We shall now assign different notations to differentiate the field variables in each problem, e.g. $\sigma_{ij}^{\wedge}$, $\varepsilon_{ij}^{\wedge}$, $u_i^{\wedge}$ for the equivalent problem 3.a, σ_{ij}^{eff}, ε_{ij}^{eff}, u_i^{eff} for the problem 3.b and σ_{ij}', ε_{ij}', u_i' for the problem 3.c.

It is obvious that the problem Fig. 3b is basically the same as that shown in Fig. 2b for the effective composite medium. Furthermore, outside the region enclosed by body forces shown in Fig. 3a, the following relations where $\sigma_{ij}^{\wedge} = \sigma_{ij}^{N+1}$, $\varepsilon_{ij}^{\wedge} = \varepsilon_{ij}^{N+1}$ and $u_i^{\wedge} = u_i^{N+1}$ are valid and will be made of use later. In addition, the obvious difference between Fig. 3a, c is the absence of applied external boundary conditions in the latter. Based on the decomposition scheme illustrated in Fig. 3, the field variables in equivalent problem 3.a can now be written as

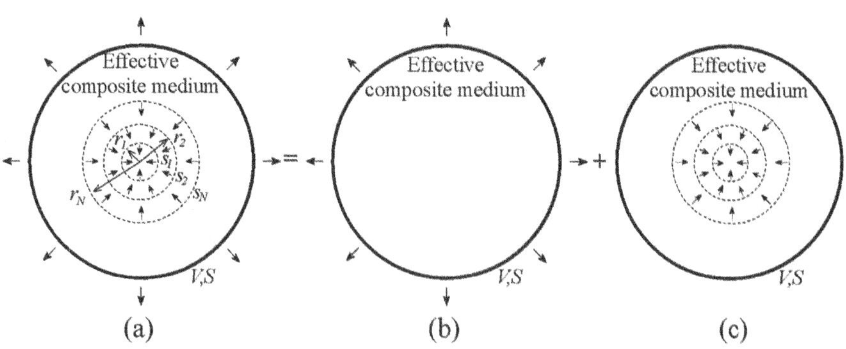

Fig. 3 Decomposition of the equivalent problem

$$\sigma_{ij}^{\wedge} = \sigma_{ij}^{eff} + \sigma_{ij}',$$
$$\varepsilon_{ij}^{\wedge} = \varepsilon_{ij}^{eff} + \varepsilon_{ij}',$$
$$u_i^{\wedge} = u_i^{eff} + u_i'. \tag{8}$$

With Eq. (8), the average strain energy density of the equivalent problem 3.a can be written in the following form

$$\langle W^{\wedge} \rangle = \frac{1}{2V} \int_V \left(\sigma_{ij}^{eff} + \sigma_{ij}' \right) \left(\varepsilon_{ij}^{eff} + \varepsilon_{ij}' \right) dV. \tag{9}$$

The terms inside the integral of Eq. (9) can be expanded giving us

$$\langle W^{\wedge} \rangle = \langle W^{eff} \rangle + \langle W' \rangle + \langle W^{INT} \rangle \tag{10}$$

where

$$\langle W' \rangle = \frac{1}{2V} \int_V \sigma_{ij}' \varepsilon_{ij}' dV, \tag{11}$$

$$\langle W^{INT} \rangle = \frac{1}{2V} \int_V \left(\sigma_{ij}^{eff} \varepsilon_{ij}' + \sigma_{ij}' \varepsilon_{ij}^{eff} \right) dV, \tag{12}$$

and the expression $\langle W^{eff} \rangle$ is already given by Eq. (4). As mentioned by Christensen [4], the term $\langle W^{INT} \rangle$ describes the effect of interaction energies between the two stress states shown in problems 3.b and c.

The expression $\langle W^{INT} \rangle$ shown in Eq. (12) can be represented in a more convenient form. We now consider the first term inside the integral of Eq. (12) and by using the stress-strain relation, we have

$$\sigma_{ij}^{eff} \varepsilon_{ij}' = C_{ijkl}^{eff} \varepsilon_{kl}^{eff} \varepsilon_{ij}'. \tag{13}$$

Using the symmetry condition for the elastic stiffness tensors of an effective composite medium, e.g. $C_{ijkl}^{eff} = C_{klij}^{eff} = C_{klij}' = C_{ijkl}'$, Eq. (13) can be rearranged into

$$\sigma_{ij}^{eff} \varepsilon_{ij}' = C_{ijkl}' \varepsilon_{kl}' \varepsilon_{ij}^{eff} = \sigma_{ij}' \varepsilon_{ij}^{eff}. \tag{14}$$

Thus, substituting the first term in the integral of Eq. (12) with Eq. (14) gives us

$$\langle W^{INT} \rangle = \frac{1}{V} \int_V \sigma_{ij}' \varepsilon_{ij}^{eff} dV \tag{15}$$

and applying divergence theorem in conjunction with equilibrium equations, we obtain

$$\left\langle W^{INT}\right\rangle = \frac{1}{V}\int\limits_{S} \sigma'_{ij}u_i^{eff}n_j dS. \tag{16}$$

We now have obtained the convenient form of the average strain interaction energy density.

Next we wish to relate the interaction energy expression shown in Eq. (16) with the integral term in Eq. (7) so that full form of the Eshelby formula can be obtained. First, we must evaluate in detail the average strain energy density of the RVE or medium with inclusions shown in Fig. 2a, which is given in Eq. (5). Expanding the first term inside the integral of Eq. (5) gives us

$$\begin{aligned}\left\langle W^{RVE}\right\rangle &= \left\langle W^{eff}\right\rangle \\ &+ \frac{1}{2V}\int\limits_{V}\left(\sigma_{ij}^{(1)}\varepsilon_{ij}^{(1)} + \sigma_{ij}^{(2)}\varepsilon_{ij}^{(2)} + .. + \sigma_{ij}^{(N)}\varepsilon_{ij}^{(N)} + s_{ij}^{N+1}\varepsilon_{ij}^{N+1}\right)dV \\ &- \frac{1}{2V}\int\limits_{V}\sigma_{ij}^{eff}\varepsilon_{ij}^{eff}dV.\end{aligned} \tag{17}$$

Rearranging the energy contribution from the $N+1$-th phase from the first integral into the second integral of Eq. (17), we have

$$\begin{aligned}\left\langle W^{RVE}\right\rangle &= \left\langle W^{eff}\right\rangle + \left\langle W^{(i)}\right\rangle \\ &+ \frac{1}{2V}\int\limits_{V}\left(\sigma_{ij}^{N+1}\varepsilon_{ij}^{N+1} - \sigma_{ij}^{eff}\varepsilon_{ij}^{eff}\right)dV\end{aligned} \tag{18}$$

where $\left\langle W^{(i)}\right\rangle = \frac{1}{2v}\int_{V(i)}\left(\sigma_{ij}^{(i)}\varepsilon_{ij}^{(i)}\right)dV$ and $i = 1, 2, \ldots, N$. Let us now evaluate the last integral term of Eq. (18). By using divergence theorem together with equations of equilibrium and by considering the fact that both surfaces S of mediums shown in Fig. 2 experience the same displacement vector, i.e. $u_i^{N+1} = u_i^{eff}$, Eq. (18) can be transformed into

$$\left\langle W^{RVE}\right\rangle = \left\langle W^{eff}\right\rangle + \left\langle W^{(i)}\right\rangle + \frac{1}{2V}\int\limits_{S}\left(\sigma_{ij}^{N+1} - \varepsilon_{ij}^{eff}\right)u_i^{eff}n_j dS. \tag{19}$$

With the new version of Eq. (7) shown in Eq. (19), it is now possible for us to relate Eq. (16) with the second integral term in Eq. (19). To achieve this, we employ the first row of Eq. (8) to rewrite Eq. (19) as

$$\langle W^{RVE} \rangle = \langle W^{eff} \rangle + \langle W^{(i)} \rangle + \frac{1}{2V} \int_S \sigma'_{ij} u_i^{eff} n_j dS \tag{20}$$

where the relation $\sigma_{ij}^{N+1} = \sigma_{ij}^{\wedge}$ has been used to obtain the above expression. As mentioned earlier, that relation is valid because the field variables in region outside the boundary surface s_N and in the $N + 1$-th phase are in fact, identical. Thus, with Eq. (16) it is possible to rewrite Eq. (20) as

$$\langle W^{RVE} \rangle = \langle W^{eff} \rangle + \langle W^{(i)} \rangle + \frac{1}{2} \langle W^{INT} \rangle. \tag{21}$$

Here, we argue that as the radius of equivalent homogeneous media in the RVE shown in Fig. 2a reaches to the infinity, the average strain energy density contribution from the i-th inclusions where $i = 1, 2, \ldots, N$ are vanishingly small and thus, can safely be ignored. As a result, Eq. (21) is reduced to

$$\langle W^{RVE} \rangle = \langle W^{eff} \rangle + \frac{1}{2} \langle W^{INT} \rangle. \tag{22}$$

At this point, we are about to get the final form of Eshelby's formula and also its associated integral formula in relation to Eq. (22). To do this, we return to the interaction energy form $\langle W^{INT} \rangle$ given in Eq. (15) and we focus on the equivalent problem 3.a with particular attention to the region outside the boundary surface s_N since the average energy interaction contribution from this region will be of a great significant due to its infinite size assumption.

We basically deconstruct the region of volume integration in problem 3.a into two parts such as shown in Fig. 4. The two parts or regions are separated by surface Σ, which is taken outside the region where body forces are applied. As such, Eq. (15) takes the following form

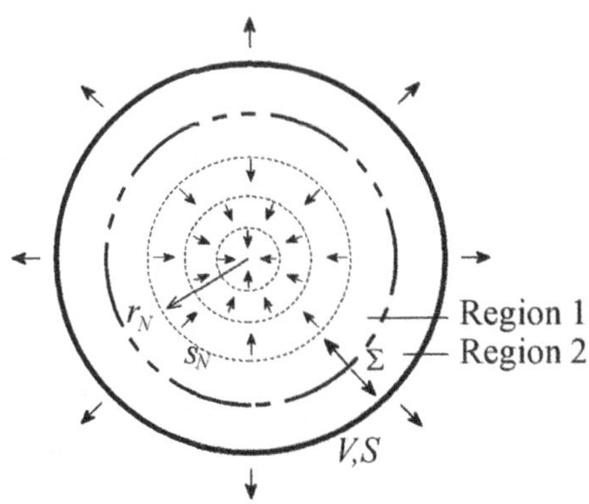

Fig. 4 Surface integral location of the deconstructed problem 3.a

$$\left\langle W^{INT}\right\rangle = \frac{1}{V}\int_{V1}\sigma'_{ij}\varepsilon^{eff}_{ij}\,dV + \frac{1}{V}\int_{V2}\sigma'_{ij}\varepsilon^{eff}_{ij}\,dV. \tag{23}$$

Replacing the last integral term of Eq. (23) with the left-hand term of Eq. (14), we have

$$\left\langle W^{INT}\right\rangle = \frac{1}{V}\int_{V1}\sigma'_{ij}\varepsilon^{eff}_{ij}\,dV + \frac{1}{V}\int_{V2}\sigma^{eff}_{ij}\,\varepsilon'_{ij}\,dV. \tag{24}$$

Applying divergence theorem and the condition of equilibrium equations in the volumetric region 1 and 2, Eq. (24) becomes

$$\begin{aligned}\left\langle W^{INT}\right\rangle = &\frac{1}{V}\int_{\Sigma}\sigma'_{ij}u^{eff}_{i}n_j\,dS - \frac{1}{V}\int_{\Sigma}\sigma^{eff}_{ij}\,u'_i n_j\,dS\\ &+\frac{1}{V}\int_{S}\sigma^{eff}_{ij}\,u'_i n_j\,dS\end{aligned} \tag{25}$$

where the appearance of negative sign in Eq. (25) is due to the proposition of where the positive direction of a unit vector normal to the surface Σ is taken to be outward. It is interesting to note that in problem 3.c, u'_i equals zero on surface S. As a result, Eq. (25) is reduced to

$$\left\langle W^{INT}\right\rangle = \frac{1}{V}\int_{\Sigma}\left(\sigma'_{ij}u^{eff}_{i} - \sigma^{eff}_{ij}\,u'_i\right)n_j\,dS. \tag{26}$$

Using the relations in Eqs. (8), (26) can be transformed into

$$\left\langle W^{INT}\right\rangle = \frac{1}{V}\int_{\Sigma}\left(\sigma^{\wedge}_{ij}u^{eff}_{i} - \sigma^{eff}_{ij}\,u^{\wedge}_i\right)n_j\,dS. \tag{27}$$

Next, with particular attention to the region outside the surface s_N and recalling the fact that $\sigma^{\wedge}_{ij} = \sigma^{N+1}_{ij}$ and $u^{\wedge}_i = u^{N+1}_i$ are valid for that region, Eq. (27) can be further transformed into

$$\left\langle W^{INT}\right\rangle = \frac{1}{V}\int_{\Sigma}\left(\sigma^{N+1}_{ij}u^{eff}_{i} - \sigma^{eff}_{ij}\,u^{N+1}_i\right)n_j\,dS. \tag{28}$$

Here, we now have obtained the desired expression for our interaction effects between these two average strain energy density systems. Finally, by substituting Eq. (28) into Eq. (22) gives us the desired form of the Eshelby's formula, which is written as

$$\langle W^{RVE} \rangle = \langle W^{eff} \rangle + \frac{1}{2V} \int\limits_{s_N} \left(\sigma_{ij}^{N+1} u_i^{eff} - \sigma_{ij}^{eff} u_i^{N+1} \right) n_j dS \tag{29}$$

where the surface Σ has been taken over by the surface s_N at the outer radius of N-th phase or at $r = r_N$. It must be remarked that the Eshelby's formula shown in Eq. (29) is obtained when a homogenous displacement vector is applied at the external boundary of RVE. It is easy to show that when a uniform surface traction vector is enforced, the corresponding formula shall have the following form

$$\langle W^{RVE} \rangle = \langle W^{eff} \rangle + \frac{1}{2V} \int\limits_{s_N} \left(\sigma_{ij}^{eff} u_i^{N+1} - \sigma_{ij}^{N+1} u_i^{eff} \right) n_j dS. \tag{30}$$

Next, it is in our main interest to obtain the generalized Eshelby integral formula and this is achieved by equivalencing the average strain energy density of the two mediums with and without the inclusions shown in Fig. 2, e.g. $\langle W^{RVE} \rangle = \langle W^{eff} \rangle$. Consequently, Eqs. (29) and (30) respectively become

$$\begin{aligned} &\frac{1}{2V} \int\limits_{s_N} \left(\sigma_{ij}^{N+1} u_i^{eff} - \sigma_{ij}^{eff} u_i^{N+1} \right) n_j dS \\ &= \frac{1}{2V} \int\limits_{s_N} \left(\sigma_{ij}^{eff} u_i^{N+1} - \sigma_{ij}^{N+1} u_i^{eff} \right) n_j dS = 0. \end{aligned} \tag{31}$$

Notice that the generalized Eshelby's and Eshelby integral formulae shown in Eqs. (29), (30) and (31) are consistent with their counterpart expressions given in Eqs. (1) and (2), which were originally developed for a single inclusion embedded in all matrix medium. In addition, one can intuitively see from the appearance of those equations that the number of inclusions considered in the problem is immaterial since the formulations rely on the basis where the equivalent homogeneous medium and the effective medium are treated as an infinite or unbounded medium. To finish, the generalization of the original formulae is now proven.

3 Conclusion

In this paper, the generalized Eshelby's and Eshelby integral formulae have been formulated based on the frameworks of generalized self-consistent method through a series of solid micromechanics analysis. Initially, the energy system of a medium with multiple inclusions are compared with the energy of its effective medium obtained when all inclusions including the equivalent homogenous material are homogenized together. Next, the interaction energy between those two mediums was thoroughly analyzed using different sets of equivalent problems. Through those problems and energy conversion procedures, which reduce the usual volume integral expression into much simpler surface integral form, useful interaction energy expression was obtained. This obtained expression was then substituted back into the main energy comparison equations, resulting the final desired form of the generalized Eshelby's formula. Finally, the generalized Eshelby integral formula was received when the energy systems of the two mediums were equalized. With these results, the proof to where the original formulae can readily be generalized to the case of multiple inclusions is established.

Acknowledgements The authors are indebted to UniKL Center/Section of Research and Innovation for the publication grant and Dr. Sergey Lurie and MAI for continuous support.

Appendix

Herein, the mathematical treatments based on the continuum tensor mechanics in multiscale theory of solids [13–17], converting the energy expression from the volume integral shown in Eq. (5) to a surface integral form given in Eq. (6) of the manuscript, are illustrated and explained for the sake of clarity in understanding the conversion procedures. Focusing on the first term in the integral expression of Eq. (5) without specifically referring to any of the mediums, we have

$$\frac{1}{2V}\int\limits_{V}\sigma_{ij}\varepsilon_{ij}dV \tag{32}$$

where σ_{ij} and ε_{ij} are the respective stress and strain tensor fields of a microscale medium under uniform boundary conditions. Next, we wish to get a different expression for the strain tensor field shown in Eq. (32). Under the assumption of where the displacement gradient fields in the RVE are small compared to the unity, the displacement gradients $u_{i,j}$ can be represented as

$$\langle u_{i,j}\rangle = \langle\varepsilon_{ij}\rangle + \langle\omega_{ij}\rangle \tag{33}$$

where comma denotes derivative, ε_{ij} is the infinitesimal strain tensor and ω_{ij} is the infinitesimal rigid body rotation tensor. In micromechanics analysis, the small rigid body rotation tensor in microscale is typically taken to be zero, i.e. $\omega_{ij} = 0$. As such, with the help of Eqs. (33), (32) can now be rewritten as

$$\frac{1}{2V}\int\limits_{V}\sigma_{ij}u_{i,j}dV. \tag{34}$$

Next, we wish to have a different form of Eq. (34) so that it will be convenient for us to apply divergence theorem to that equation. Recalling the full form of equilibrium equations, we have

$$\langle \sigma_{ij,j} \rangle + \langle b_i \rangle = \langle a_i \rangle \tag{35}$$

where σ_{ij} are components of Cauchy stress, b_i is a body force, ρ is mass density and a_i is an acceleration. Under static condition where $a_i = 0$, the microscale inertia force shown on the right-hand side of Eq. (35) vanishes. Furthermore, by considering no microscale body forces exist in the RVE, we have $b_i = 0$. Even if exist such as in the case of Fig. 3c, these body forces are considered 'invisible' in the RVE equilibrium equations since they are non-fluctuated body forces [17]. Moreover, one can treat the problems of the body forces negligible in comparison to the boundary conditions especially in the context of the GSC method where its RVE has an infinite radius or size. Hence, the equilibrium equations of Eq. (35) become

$$\langle \sigma_{ij,j} \rangle = 0. \tag{36}$$

With Eqs. (36), (34) is implied to have the following form

$$\frac{1}{2V} \int_V \left(\sigma_{ij,j} u_i + \sigma_{ij} u_{i,j} \right) dV. \tag{37}$$

Meaningfully, Eq. (37) can be rewritten as

$$\frac{1}{2V} \int_V \left(\sigma_{ij} u_i \right)_{,j} dV. \tag{38}$$

Equation (38) is now ready for the application of divergence theorem, which is also known as the Gauss's or Ostrogradsky's theorem, or sometimes referred to as the Gauss-Ostrogradsky's theorem. In general, the theorem mathematically states that

$$\int_V \left(t_{ij\ldots k} \right)_{,q} dV = \int_S t_{ij\ldots k} n_q dS \tag{39}$$

where $t_{ij\ldots k}$ is an arbitrarily order of continuous differentiable tensor field, n_q is a unit vector normal to surface S, which bounds the volume V. Basically, it is a relation, which relates an integral over a closed volume to an integral over its bounding surface. Thus, by using divergence theorem exemplified in Eq. (39), the volume integral of Eq. (38) can now be desirably transformed to its equivalent surface integral form, which gives us

$$\frac{1}{2V}\int_S \sigma_{ij}u_i n_j dS. \tag{40}$$

As we can see, Eq. (40) exhibits the typical average energy expression in surface integral form whenever the divergence theorem along with the equations of equilibrium are invoked and applied to its equivalent volume integral form given in Eq. (32) earlier.

References

1. Eshelby JD (1956) The continuum theory of lattice defects. Solid States Phys 3:79–144
2. Christensen RM, Lo KH (1979) Solutions for effective shear effective properties in three phase sphere and cylinder models. J Mech Phys Solids 27(4):315–330
3. Christensen RM, Lo KH (1986) Erratum: Solutions for effective shear effective properties in three phase sphere and cylinder models. J Mech Phys Solids 34:369
4. Christensen RM (2005) Mechanics of composite materials, 2nd edn. Dover, New York
5. Benveniste Y (2008) Revisiting the generalized self-consistent scheme: clarification of some aspects and a new formulation. J Mech Phys Solids 56(10):2984–3002
6. Christensen RM (1990) A critical evaluations of a class of micromechanics models. J Mech Phys Solids 38(3):379–404
7. Gusev AA, Lurie SA (2009) Loss amplification effect in multiphase materials with viscoelastic interfaces. Macromolecules 42:5372–5377
8. Lurie S, Minhat M (2014) Application of generalized self-consistent method to predict the effective properties of bristled fiber composites. Composites B—Eng 61:26–40
9. Lurie S, Minhat M, Tuchkova N (2015) Estimations of effective dynamics properties of bristled fiber composite materials based on a self-consistent Eshelby method. J Eng Mech 95(1):7–29
10. Lurie S, Minhat M, Tuchkova N, Soliaev J (2014) On remarkable loss amplification mechanism in fiber reinforced laminated composite materials. Appl Compos Mater 21(1):179–196
11. Lurie SA, Volkov-Bogorodskiy DB, Menshykov O, Solyaev YO, Aifantis EC (2018) Modeling the effective mechanical properties of "fuzzy fiber" composites across scales length. Composites B—Eng 142:24–35
12. Lurie SA, Volkov-Bogorodskii DB, Kriven GI, Rabinsky LN (2018) On estimating structural stresses in composites with whiskered fibers. Int J Civil Eng Technol 9(6):294–308
13. Malvern LE (1969) Introduction to the mechanics of continuous medium. Prentice Hall, New Jersey
14. Mase GT, Smelser RE, Mase GE (2010) Continuum mechanics for engineers, 3rd edn. CRC Press, Florida
15. Aboudi J, Arnold SM, Bednarcyk BA (2013) Micromechanics of composite materials. Butterworth-Heinenmann, Amsterdam
16. Hill R (1972) On constitutive macro-variables for heterogeneous solids at finite strain. Proc R Soc Lond A 326:131–147
17. De Souta Neto EA, Blanco PJ, Sanchez PJ, Feijoo RA (2015) An RVE-based multiscale theory of solids with micro-scale inertia and body force effects. Mech Mater 80:136–144

Applicability of Rule of Mixtures to Estimate Effective Properties of Nanocomposite Materials

Muhammad Lutfi Mat Rodzi, Muhammed Fadzli Ismail, and Mulia Minhat

Abstract A rule of mixtures is employed and modified to examine the effective elastic properties of a unidirectional composite lamina reinforced with nanostructure-hybrid fibers. Such fiber system is designated when nanostructure such as nanowires or carbon nanotubes are radially grown on the surface of primary fiber. When combines with matrix, a complex three-phase composite with enhanced elastic properties is expected. Herein, the applicability of this simple micromechanics method to reliably estimate the effective properties of such advanced novel composite material is assessed. The results demonstrated that the method is capable of modeling the effects on elastic properties of a composite due to the presence of nanostructure. However, in light of published experimental data and other micromechanics results, the proposed method is found to be at best, applicable for a composite that has a very low fiber volume fraction only except when an axial Young' modulus is predicted.

Keywords Micromechanics · Rule of mixtures · Nanocomposites

1 Introduction

Over the last decade, there has been a great success in growing carbon nanotubes or nanowires on the surface of carbon fiber, which leads to the development of novel nanostructure-hybrid fiber system [1–6]. Embedded in a matrix material, simultaneous enhancements in several structural functions of composite materials are observed [7–9]. In terms of fiber-matrix interface or interphase characteristics, the presence of fully covered nanostructure on primary fibre surface enhances the interfacial strength of composites due to the greater surface area for load transfer and mechanical interlocking mechanism exists between fiber and matrix. As a result, composites with greater stiffness and strength can be achieved. Depending on the types of nanostructure, other enhancements in damping, fracture, toughness, thermal, electrical

M. L. M. Rodzi · M. F. Ismail (✉) · M. Minhat
Aerospace Section, Universiti Kuala Lumpur Malaysian Institute of Aviation Technology, Jalan Jenderam Hulu, 43800 Sepang, Selangor dE, Malaysia
e-mail: mfadzli@unikl.edu.my

P. Rajendran et al. (eds.), *Proceedings of International Conference of Aerospace and Mechanical Engineering 2019*, Lecture Notes in Mechanical Engineering,
https://doi.org/10.1007/978-981-15-4756-0_17

and electromagnetic properties may simultaneously be obtained. Such simultaneous improvements are very attractive for engineering applications. For example, fuzzy fiber, a nickname for fiber that has carbon nanotubes radially grown on its surface, has shown to be a promising material for the next generation strong lightweight heat-resistant composite structures for rockets and engines as well as for structural health monitoring sensors [5, 9, 10]. Nanowire also offers unique and valuable properties, e.g. nanowire-coated fiber composites have great potentials to be the next generation shock-resistant structures, energy harvester and piezoelectric materials [6, 7, 11, 12]. In light of these manifested technological importance, a micromechanics method based on the rule of mixtures (ROM) is proposed to examine the overall elastic behaviors of such advanced multifunctional materials. This method is adopted due to the fact that it appears in every standard literature of composite mechanics [13–17, 21] and above all, due to its modeling simplicity that yields simple mathematical expressions of effective properties.

ROM is considered as a first-order model where it ignores the shape and distribution of the constituents and only include their aggregates [18]. Despite this limitation, ROM is still a valuable tool for certain applications such as when determining the tensile properties of conventional unidirectional fiber reinforced plastic composites or when upper and lower bounds are the chief interest in evaluating effective properties of composite materials. In literature, majority of ROM applications are for the two-phase composite system with both constituents having isotopic properties. In our case, the nanocomposite exhibits complex microstructure with various material properties (see Fig. 1). For example, the composite is assumed to have three phases, which the first and the third phase represents fiber and matrix respectively while the second phase can be considered as a cylindrically orthotropic nanocomposite interphase layer comprising nanostructure and matrix material. Thus, unlike previous typical ROM applications, the model will be modified accordingly to take into account those three phases with each phase being an orthotropic material. Here, we note that growing nanostructure to a certain extent often reduces the properties of

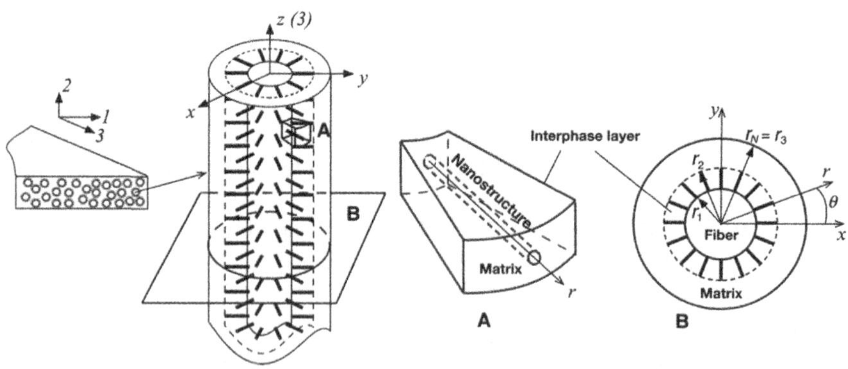

Fig. 1 Schematics of nanostructure-hybrid fiber composite lamina and its representative volume elements of the composite and interphase layer in 2-D and 3-D

base fiber but such condition is ignored in our proposed model although accounting for it is possible. With these in mind, our goal is now to examine on whether such modified ROM can correctly model the presence of nanostructure and at the same time, gives reliable estimations on the effective properties of such nanocomposite materials. In other words, its applicability is evaluated.

2 Modeling Approach and Mathematical Models

Let us consider a unidirectional lamina shown in Fig. 1 subjected to in-plane normal and shear stresses. The axial principal axis of fiber is parallel to *3*-axis. Understanding that ROM is a first order model as mentioned earlier, it is possible to have the lamina be presented as a system of strips [18] such as shown in Fig. 2. Basically, the fiber is designated by the dark-shaded strips, interphase layer by light-shaded strips and pure matrix material by un-shaded areas. The volume fractions of constituents are given as $\upsilon_{f,i,m} = w_{f,i,m}/w$ such that $\upsilon_f + \upsilon_i + \upsilon_m = \upsilon_* = \upsilon = 1$ where υ is the total volume fraction, *w* is the total width of every constituent and with subscript *, *f*, *i* and *m*, they represent the quantity for lamina, fiber, interphase and matrix phase respectively.

Since lamina is a thin ply structure, a plane stress state can be assumed, i.e. zero stresses are considered in the direction of 2-axis. Furthermore, the composite is considered orthotropic if the normal stresses are assumed not to cause an in-plane shear deformation and the shear stresses do not elongate the ply in both axial and transverse directions. Correspondingly, the constitutive equations for such composite and its constituents can be written as follows:

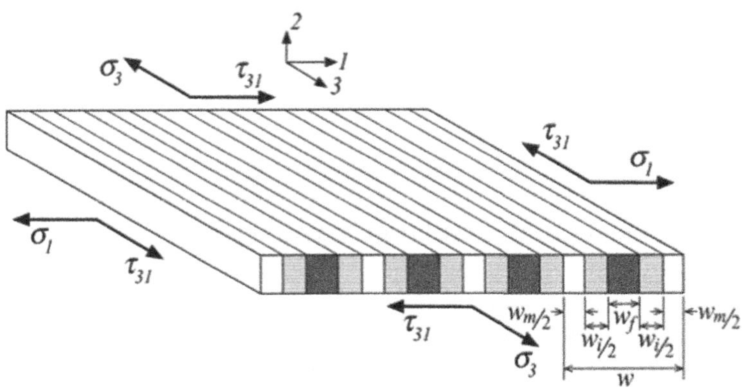

Fig. 2 A first-order model of a unidirectional composite lamina

$$\begin{aligned}\varepsilon_3^{*,f,i,m} &= (\sigma_3^{*,f,i,m} - \upsilon_{31}^{*,f,i,m}\sigma_1^{*,f,i,m})/E_3^{*,f,i,m};\ \varepsilon_1^{*,f,i,m}\\ &= (\sigma_1^{*,f,i,m} - \upsilon_{13}^{*,f,i,m}\sigma_3^{*,f,i,m})/E_1^{*,f,i,m};\ \gamma_{31}^{*,f,i,m}\\ &= \tau_{31}^{*,f,i,m}/G_{31}^{*,f,i,m}\end{aligned} \quad (1)$$

where E_3, E_1, and $G_{31} = G_{13}$ are axial Young's modulus, transverse Young's modulus and shear modulus respectively, $\sigma_{3,1}$ and $\varepsilon_{3,1}$ are the respective normal stresses and strains while the τ_{31} and γ_{31} are the shear stress and strain respectively, and ν_{31} and ν_{13} are the respective major and minor Poisson's ratio, only one of which is independent based on the following symmetry, $E_3\nu_{13} = E_1\nu_{31}$.

By considering the iso-strain condition in the axial loading as well as the iso-stress state in both transverse and shear loading conditions, it is easy to show that relations in Eq. (1) can be manipulated in such a way giving us the following effective properties (see appendix for details):

$$\begin{aligned}E_3^* = E_3^f\upsilon_f + E_3^i\upsilon_i + E_3^m\upsilon_m;\ 1/E_1^* = \upsilon_f/E_1^f + \upsilon_i/E_1^i + \upsilon_m/E_1^m;\\ 1/G_{31}^* = \upsilon_f/G_{31}^f + \upsilon_i/G_{31}^i + \upsilon_m/G_{31}^m;\ \upsilon_{31}^* = \upsilon_{31}^f\upsilon_f + \upsilon_{31}^i\upsilon_i + \upsilon_{31}^m\upsilon_m\end{aligned} \quad (2)$$

Dropping the terms associated to interphase layer, the ROM expressions shown in Eq. (2) become the typical two-phase composite formulae. It must be noted that Eq. (2) are derived for orthotropic phase but in many modeling efforts, at most transversely isotropic property is used. In our modeling, the fiber is considered transversely isotropic with axis of symmetry parallel to fiber's major axis or 3-axis while for matrix, is isotropic material. The effective properties of interphase layer must be obtained first by homogenizing the nanostructure and matrix material and it assumes the transversely isotropic property with axis of symmetry parallel to 1-axis. The volume fraction of nanostructure in the interphase layer is defined as $\upsilon_b = M_b^2 d_b^2/(4\pi(l_b + D)D)$ [25, 26] where d_b and l_b are the respective average diameter and length of nanostructure, D is the diameter of base fiber, and M_b is the quantity of nanostructure grown along the circumference of base fiber and for 100% density of grown nanostructure, $M_b = \pi D/d_b$.

3 Methodologies and Simulation Parameters

To achieve the goal of this paper, micromechanics numerical computations are simulated. In order to verify the ability of the proposed ROM model to correctly capture the presence of nanostructure, we shall briefly analyze the *estimation trends*, which will be simulated for the two- and three-phase fiber composite, i.e. composite with and without the nanostructure respectively. Here, an estimation trend is defined as the trend of changes in the effective property, e.g. increase or decrease, when fiber volume fraction is varied. Next, the *enhancement trends* are also examined as to evaluate the effects on the composite effective properties due to the presence of nanostructure at fiber-matrix interface. In this case, the enhancement trend describes the relative changes in effective property when the estimation trends between the two- and three-phase composites are compared. All these are to ensure the estimates by ROM are consistent with the physical principles associated to the rules of micromechanics of composite materials so that the presence of nanostructure on the elastic properties can be correctly interpreted and reasoned. In this way, concrete conclusions on the effectiveness of ROM in capturing the effects of nanowire on composite elastic properties can be made.

To further verify the capability of the proposed model, its enhancement trends are also benchmarked against the enhancement trends produced by generalized self-consistent (GSC) model [19–23]. Such method has a strong reputation in providing good reliable estimates on the effective properties of such fiber composite materials [24–27]. In this case, its estimates for the two- and three-phase fiber composites are also simulated and presented for comparison purposes. If some consistency is observed between these two enhancement trends, then the ability of the proposed ROM model can be further concluded. Apart from correlating with the GSC method, some of the effective properties are compared to published experimental data [28]. In this way, the ability to provide accurate estimation and limitation of rule of mixtures can also be evaluated.

For such purposes stated above, we analyze a nanowire-hybrid fiber composite, which consists of IM7 carbon fibers coated with vertically aligned zinc oxide nanowires and EPON 862 epoxy resin. The primary carbon fibers have average diameters of 5.2 μm. The surfaces of carbon fibers are fully-grown with nanowires, i.e. 100% growth density, which translates into approximately 71% volume fraction of nanowire in the interphase layer. In addition, the nanowire have average diameters and lengths of 50 nm and 500 nm respectively. Note that the growing process of nanowire did not significantly degrade the properties of primary carbon fiber as claimed by original authors [2]. The properties of these constituents are given in Table 1.

Table 1 Constituent properties of nanowire-hybrid fiber composites [25]

Material	Properties
IM7 carbon fiber	Axial Young's modulus—256.76 GPa Transverse Young's modulus—25.51 GPa Axial shear modulus—22.06 Transverse shear modulus—9.25 GPa Major Poisson's ratio—0.289
Zinc oxide nanowire	Young's modulus—140 GPa Poisson's ratio—0.35
EPON 862 epoxy	Young's modulus—2.55 GPa Poisson's ratio—0.35

4 Simulation Results and Discussion

The results for all effective properties, which consist of axial Young's modulus E_3, transverse Young's modulus E_1, axial shear modulus G_{13} and major Poisson's ratio ν_{31}, obtained by the ROM model as well as by the GSC model for the two- and three-phase composite respectively are shown in Fig. 3.

First, we now analyze the estimation trends provided by the rules of mixtures model for the two- and three-phase composites, i.e. as mentioned earlier, the composite with and without the nanostructure. As stated earlier, the *estimation trend* is referred to as the trend of changes in the effective property as volume fraction of fiber varies. Looking at the numerical results of the two-phase fiber composite, which are produced by the original rule of mixtures, typical estimation trends are observed. For example, in the case of axial Young' modulus, a familiar linear relation with fiber volume fraction is illustrated where its value increases when fiber volume fraction is increased (see Fig. 3a). The same can be said about major Poisson's ratio (Fig. 3d) where a linear relationship is observed but instead of increasing, its value is decreasing. Finally, the usual nonlinear curve showing increasing value of an effective transverse Young's modulus and axial shear modulus can be seen when fiber volume fraction increases (Fig. 3b, c). Basically, these estimation trends of effective properties for the two-phase composite can be explained from the physical principles associated to micromechanics rules. Those reasons can easily be found in many

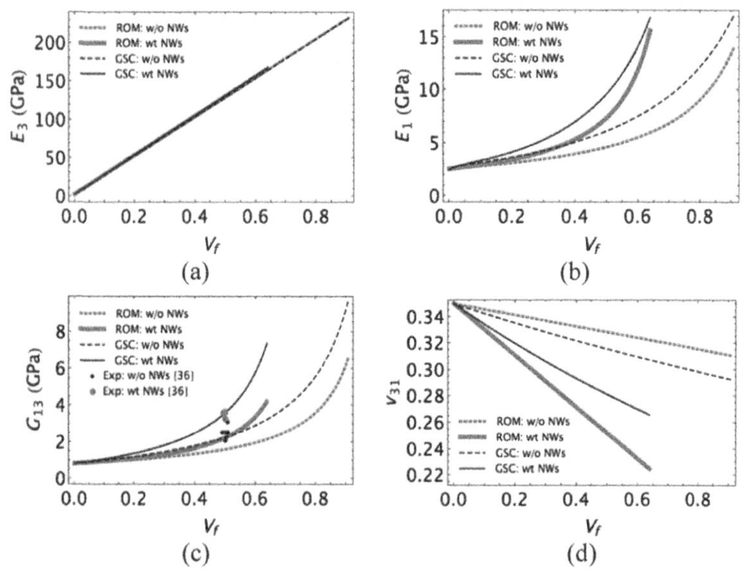

Fig. 3 Comparison of effective properties for **a** axial Young's modulus E_3, **b** transverse Young's moduli E_1, **c** shear modulus G_{13}, and **d** Poisson's ratio ν_{31} for the two- and three-phase fiber composites (IM7 carbon fiber epoxy composite with and without zinc oxide nanowire) obtained by the ROM, GSC method and published experimental data (Exp)

standard composite micromechanics literatures and as such, we do not wish to dwell on them further.

More importantly, the estimation trends of three-phase composite simulated by our ROM model must be evaluated. As we can see from Fig. 3, the estimation trends for all effective properties are consistent with the trends exhibited by the two-phase composite system. For example, similar linear relations for an effective axial Young's modulus and major Poisson's ratio are observed, which saw the value of the former increases while the latter decreases when volume fraction of fiber is increased. In addition, the same nonlinear relations are observed for transverse Young's modulus and axial shear modulus, in which their values increase as fiber content increases. Correspondingly, this means that the estimates of proposed rule of mixture do not contradict with the physical principles of micromechanics of composite materials when it comes to presence of additional phase in a composite. Thus, it is possible for us to suggest that the modified ROM model has so far correctly capture the presence of an additional interphase layer containing the nanostructure. However, at this point we must stress that such ability is not fully confirmed yet and is subjected to further investigation.

Before we proceed further, notice that the volume fraction of fiber in the three-phase composites is limited to ~64%. This value is actually the theoretical limit of fiber volume fraction in a composite and is directly related to the length of nanostructure. Consequently, if exceeds, the interphase layers in a composite will cross each other. Although this may happen in real composite (in fact, the intersect of interphase layers may also occur at volume fraction less than the theoretical limit), such condition is ignored in our proposed model. Next, the enhancement trends will be investigated, and this may provide further information on the actual capability of the proposed ROM model.

As defined earlier, *enhancement trend* describes the trend of enhancement in the effective property due to the presence of nanostructure and is interpreted by comparing the estimation trends between the two- and three-phase composites. We now proceed with the evaluation on the enhancement trend for each effective property and provide some explanations on why such trends occur from the viewpoint of physical principles, so that some concrete conclusions can be made from the observations. At this point, we are still analyzing the results produced by ROM model only. Now, looking at axial Young's modulus E_3 specifically (see Fig. 3a), it seems that the presence of nanowire surrounding the base fiber in a composite offers no significant advantage over the conventional two-phase fiber composite system. For example, at 50% fiber volume fraction, less than 1% increase is observed. The reason to this trend can be traced back from our assumption on the physical growth of nanostructure. In the proposed model, we assume nanowires are vertically aligned nanostructures grown radially on the surface of fiber. As such, they mainly provide reinforcement in transverse direction of the fiber and as a result, minimal improvement in that elastic modulus is observed even when fiber content is increased. In fact, the minimal improvement shown by the curve is actually contributed to the lesser amount of matrix in a composite due to the presence of interphase layer when evaluated at constant volume fraction of fiber. However, such observation is invalid in the case of transverse Young's modulus E_1 and axial shear modulus G_{13}. Upon closer examinations on those effective properties (see Fig. 3b, c), the enhancements can noticeably be seen when fiber volume of fraction reaches to 20% and are increasingly significant as volume fraction approaches to its allowable limit. For example, at 50% fiber concentration, transverse Young's modulus and axial shear modulus is increased approximately by 60% and 50% respectively. Such observations are expected considering the transverse-direction reinforcement provided by nanowire as mentioned earlier. For instance, when volume fraction of fiber increases, the matrix volume fraction decreases and simultaneously, the relative content of nanowire in the interphase layer and composite increases. As such, better and greater transverse-related elastic properties are achievable. In the case of major Poisson's ratio ν_{31} shown in Fig. 3d, its value decreases as the volume fraction of fiber increases for both two- and three-phase fiber composite. However, the decrease in Poisson's ratio value is greater for the three-phase fiber composite. Again, similar reasoning can be justified, e.g. the transverse reinforcement provided by nanowire reduces the transverse strain when axial load is applied to the composite and as a result, the ratio of transverse strain over axial strain, which defines the Poisson's ratio gets smaller. In addition, the value of transverse strain becomes even smaller when volume fraction of fiber and interphase layer gets higher. To sum, the enhancement trends of effective properties obtained by ROM have been properly evaluated, and these trends can adequately be explained and justified with sound reasonings based on micromechanics physical principles. Evidently, all these observations and physically justifiable reasons strongly suggest that the presence of nanostructure is correctly captured by our proposed ROM model. Nevertheless, to validate further this finding, we shall compare these results with the estimates and trends provided by GSC model.

Based on Fig. 3, when comparing the enhancement trends of effective properties between ROM and GSC model, they are generally consistent to one another. For example, both models show that the presence of nanowire has minimal effect on the axial Young's modulus E_3. As for transverse Young's modulus E_1 and axial shear modulus G_{13}, the enhancements due to the presence of nanowire are also similar where their values are drastically increased as fiber volume fraction reaches to its theoretical limit. Specifically, on transverse Young's modulus E_1, both estimates on the three-phase composite illustrate that higher value of effective property can be obtained at the theoretical limit when compared with the value obtained at maximum fiber concentration of a two-phase composite. In the case of axial shear modulus G_{13}, both estimates point out that the maximum value of effective property shown by the two-phase system can never be reached by the three-phase composite. Hence, with these observations, we conclude that the proposed ROM model is undoubtedly capable of modeling the presence of nanostructure and its effects on the effective properties of nanostructure-hybrid fiber composite materials.

Next, we are interested with accuracy of the estimations and also, the limitation of the proposed model. These are analyzed by comparing the values of ROM estimates to the relevant experimental data [28]. As can be seen in Fig. 3c, only experimental data on the axial shear modulus G_{13} for both two- and three-phase composites are available. The comparison reveals that the estimates by ROM for the two- and three-phase composites are very conservative. For example, the average experimental value of axial shear modulus for the two- and three-phase composites measured at 50% fiber volume fraction are found to be 2.54 ± 0.35 GPa and 3.49 ± 0.21 GPa respectively, which saw an increase by 54%. On the other hand, the ROM gives the values about 1.6 GPa and 2.24 GPa respectively, which shows roughly 40% increment. Here, two conclusions can be made. First, the ROM as demonstrated earlier, indeed has the capability to capture the effects of nanowire presence in fiber composite. Second, the accuracy of estimations, however, is rather poor, e.g. the percentage difference between the experimental data and ROM estimates for both two- and three-phase composite is about 59% and 56% lower respectively. These large contrasts are somehow expected due to nature of ROM being a first-order model where as mentioned earlier, only fiber aggregate is taken into account while the shape and its distribution in a composite is neglected. Such assumption leads to a lamina being modeled as strips of constituents and under shear loading condition, each constituent is assumed to receive an equal amount of stress, which in reality is not the case. The stresses are not actually constant among the constituents and they are not even uniform throughout the same constituent especially at higher range of fiber volume fraction.

Notwithstanding the conservative or inaccuracy estimates by ROM for the shear loading case, we still wish to examine whether the same observation still applies to the remaining effective moduli. Unfortunately, this is rather a difficult task to accomplish because of the limited experimental data available. To overcome this, we use GSC estimates as the benchmark data since the method as remarked earlier, has been successfully employed to model the presence of such nanostructure [24–27]. For example, Fig. 3c shows that the GSC estimate on axial shear modulus is in close agreement with experimental data. The predicted values are found to be 2.55 GPa and 3.52 GPa for the two- and three-phase composite respectively and the percentage differences are respectively less than 0.5% and 1%. This implies that the GSC method may indeed be used as a benchmark data. With this in mind, the estimations on axial Young's modulus E_3 of the three-phase composite between ROM and GSC model are first compared and the result shows that they are in excellent agreement to each other. This suggests that ROM can actually give accurate estimate on that particular modulus as the GSC does. Unfortunately, the same cannot be said on the other transverse-related moduli, i.e. transverse Young's modulus E_1 and major Poisson's ratio ν_{31} where similar as before, ROM gives increasingly conservative estimates as volume fraction of fiber increases. Such observation demonstrates that ROM has some limitation in its predictive capability when estimating those transverse-related properties. Moreover, Fig. 3 illustrates that reliable predictions by ROM can only be achieved for a composite that has low fiber content only. For example, a less than 10% difference between ROM and GSC estimations can only be satisfied when the composite has roughly 10–20% volume fraction of fiber.

5 Conclusion

A micromechanics model based on the ROM model has been adopted and modified to investigate the overall elastic behaviors of nanostructure-hybrid fiber composite materials. The goal is to evaluate its applicability in estimating the effective properties of such fiber composite system. In another words, its capability, accuracy and limitation is methodologically examined. In relation to this, some modeling strategies were developed and the ROM mathematical expressions for all effective elastic properties were derived. Subsequently, several numerical simulations were run and the outputs were compared to the GSC estimates and experimental data.

From the comparisons, the ROM estimates have been found not to violate any micromechanics physical principles and are undeniably consistent with the GSC estimates and experimental data. Evidently, these findings verify the capability of the proposed model to estimate the effective properties of fiber composites when nanostructure exists at the interface of fiber and matrix. Nevertheless, upon deeper examinations, several distinct patterns on its estimates are conclusively identified. For example, the estimates are only accurate or reliable for a composite having low fiber volume fraction only, say about less than 20% and if more than that, its estimates are just as good as a rule-of-thumb for qualitative assessment only, except in the case of axial Young's modulus where the estimation is in excellent agreement with the GSC estimate across all possible range of fiber volume fraction.

Acknowledgements Authors are indebted to UniKL Center/Section of Research and Innovation for the publication grant and support. Special appreciation to Dr. Sergey Lurie and MAI for continuous support.

Appendix

The details of mathematical modeling of the ROM model modified for a three orthotropic phase lamina that yields final expressions for the four effective properties shown in Eq. (2) are derived here. Referring to Fig. 2, the constitutive equations for the effective composite or lamina [18] are given as

$$\begin{aligned}\varepsilon_3 &= \sigma_3/E_3^* - \upsilon_{13}^*\sigma_1/E_1^*,\\ \varepsilon_1 &= \sigma_1/E_1^* - \upsilon_{31}^*\sigma_3/E_3^*,\\ \gamma_{31} &= \tau_{31}/G_{31}^* \end{aligned} \tag{3}$$

where E_3^*—effective axial Young's modulus, E_1^*—effective transverse Young's modulus, $G_{31}^* = G_{31}^*$—effective shear modulus, $\sigma_{3,1}$—normal stresses in axial and transverse direction experienced by the lamina respectively, $\varepsilon_{3,1}$—normal strain in axial and transverse direction respectively, τ_{31}—shear stress, γ_{31}—shear strain, υ_{31}^*—effective major Poisson's ratio and υ_{13}^*—effective minor Poisson's ratio obeying the following general symmetry

$$E_3 \nu_{13} = E_1 \nu_{31}. \tag{4}$$

The volume fraction of every constituent is given as follows

$$\upsilon_{f,i,m} = w_{f,i,m}/w \tag{5}$$

and as before, $\upsilon_f + \upsilon_i + \upsilon_m = \upsilon = 1$ where υ designates total volume fraction and w is the total width of a lamina and with subscript *f*, *i* or *m*, they represent the quantity for fiber, interphase and matrix phase respectively. Finally, the constitutive equations for every constituent, which are similar to the relations given for lamina in Eq. (3) but have been rearranged with the aid of Eq. (4), can be written as

$$\begin{aligned} \varepsilon_3^{f,i,m} &= (\sigma_3^{f,i,m} - \upsilon_{31}^{f,i,m} \sigma_1^{f,i,m})/E_3^{f,i,m}, \\ \varepsilon_1^{f,i,m} &= (\sigma_1^{f,i,m} - \upsilon_{13}^{f,i,m} \sigma_3^{f,i,m})/E_1^{f,i,m}, \\ \gamma_{31}^{f,i,m} &= \tau_{31}^{f,i,m}/G_{31}^{f,i,m}. \end{aligned} \tag{6}$$

Next, we shall define the states of stresses and strains of the constituents under specific loading conditions applied to the lamina. When a lamina is axially loaded in tension, the iso-strain state condition requires every constituent or strip to experience uniform strain as the lamina, which we have

$$\varepsilon_3 = \varepsilon_3^f = \varepsilon_3^i = \varepsilon_3^m \tag{7}$$

and the apparent resultant stress of a lamina is distributed accordingly between the constituents, which is given as follow

$$\sigma_3 w = \sigma_3^f w_f + \sigma_3^i w_i + \sigma_3^m w_m. \tag{8}$$

In transverse loading condition, iso-stress condition requires all strips to experience equal amount of apparent stress experienced by the lamina, which gives us

$$\sigma_1 = \sigma_1^f = \sigma_1^i = \sigma_1^m. \tag{9}$$

Furthermore, the apparent elongation of a lamina is basically the sum of fiber, interphase and matrix strip's elongation, e.g. $\Delta w = \Delta w_f + \Delta w_i + \Delta w_m$. Introducing transverse strains for the lamina and also for every constituent: $\varepsilon_1 = \Delta w/w$, $\varepsilon_1^f = \Delta w^f/w^f$, $\varepsilon_1^i = \Delta w^i/w^i$ and $\varepsilon_1^m = \Delta w^m/w^m$, the apparent elongation of a lamina can now be written as

$$\varepsilon_1 w = \varepsilon_1^f w^f + \varepsilon_1^i w^i + \varepsilon_1^m w^m. \tag{10}$$

Finally, similar assumptions employed in the transverse loading condition are applicable to shear loading case, in which the shear stresses of lamina and its constituents are found to be

$$\tau_{31} = \tau_{31}^{f} = \tau_{31}^{i} = \tau_{31}^{m} \tag{11}$$

while the shear strain of a lamina are defined as

$$\gamma_{31} w = \gamma_{31}^{f} w^{f} + \gamma_{31}^{i} w^{i} + \gamma_{31}^{m} w^{m}. \tag{12}$$

Once the states of stresses and strains are properly defined, we now proceed to the determination of effective properties of the lamina. First, we wish to determine the expression for an effective axial Young's modulus. Considering Eqs. (7), (9) and (11), the constitutive equations for every constituent given in Eq. (6) can be reduced to the following form

$$\begin{aligned}
&\varepsilon_3 = (\sigma_3^f - \upsilon_{31}^f \sigma_1)/E_3^f,\ \varepsilon_3 = (\sigma_3^i - \upsilon_{31}^i \sigma_1)/E_3^i,\ \varepsilon_3 = (\sigma_3^m - \upsilon_{31}^m \sigma_1)/E_3^m,\\
&\varepsilon_1^f = (\sigma_1 - \upsilon_{13}^f \sigma_3^f)/E_1^f,\ \varepsilon_1^i = (\sigma_1 - \upsilon_{13}^i \sigma_3^i)/E_1^i,\ \varepsilon_1^m = (\sigma_1 - \upsilon_{13}^m \sigma_3^m)/E_1^m,\\
&\gamma_{31}^f = \tau_{31}/G_{31}^f,\ \gamma_{31}^i = \tau_{31}/G_{31}^i,\ \gamma_{31}^m = \tau_{31}/G_{31}^m.
\end{aligned} \tag{13}$$

The three relations in the first row of Eq. (13) can be rewritten to define the axial stress experienced by every strip, e.g.

$$\sigma_3^f = \varepsilon_3 E_3^f + \upsilon_{31}^f \sigma_1,\ \sigma_3^i = \varepsilon_3 E_3^i + \upsilon_{31}^i \sigma_1,\ \sigma_3^m = \varepsilon_3 E_3^m + \upsilon_{31}^m \sigma_1. \tag{14}$$

Next, Eq. (8) can be rewritten with the aid of Eq. (5) giving us

$$\sigma_3 = \sigma_3^f \upsilon_f + \sigma_3^i \upsilon_i + \sigma_3^m \upsilon_m. \tag{15}$$

It is in our intention to express ε_3 in terms of σ_3 and σ_1 and this is achieved by substituting Eq. (14) into (15), which we have

$$\begin{aligned}
\varepsilon_3 = {} & \sigma_3/\left(E_3^f \upsilon_f + E_3^i \upsilon_i + E_3^m \upsilon_m\right)\\
& - \left[\left(\upsilon_{31}^f \upsilon_f + \upsilon_{31}^i \upsilon_i + \upsilon_{31}^m \upsilon_m\right)/\left(E_3^f \upsilon_f + E_3^i \upsilon_i + E_3^m \upsilon_m\right)\right]\sigma_1.
\end{aligned} \tag{16}$$

Here, we are about to get the desired expression of an effective axial Young's modulus. Equating Eq. (16) with the first expression of Eq. (3), the following relations can be obtained

$$E_3^* = E_3^f \upsilon_f + E_3^i \upsilon_i + E_3^m \upsilon_m, \tag{17}$$

$$\upsilon_{13}^*/E_1^* = \left(\upsilon_{31}^f \upsilon_f + \upsilon_{31}^i \upsilon_i + \upsilon_{31}^m \upsilon_m\right)/\left(E_3^f \upsilon_f + E_3^i \upsilon_i + E_3^m \upsilon_m\right). \tag{18}$$

As we can see, Eq. (17) defines the effective axial Young's modulus for our lamina.

Next, we wish to determine the expression for an effective transverse Young's modulus. Using Eq. (5), we can rewrite Eq. (10) as

$$\varepsilon_1 = \varepsilon_1^f \upsilon_f + \varepsilon_1^i \upsilon_i + \varepsilon_1^m \upsilon_m. \tag{19}$$

Substituting the three equations shown in the second row of Eq. (13) into (19) and taking into account Eq. (14) together with the first row relations of Eq. (13) as well as both Eqs. (17) and (18), we will arrive to a relation that expresses ε_1 in terms of σ_1 and σ_3 and comparing that newly obtained relation with the second expression of Eq. (3), we receive

$$\begin{aligned} 1/E_1^* = {} & \upsilon_f/E_1^f + \upsilon_i/E_1^i + \upsilon_m/E_1^m \\ & - \left[\left(E_3^f \upsilon_{31}^i + E_3^i \upsilon_{31}^f\right)\left(E_1^f \upsilon_{13}^i + E_1^i \upsilon_{13}^f\right) E_1^m \upsilon_f \upsilon_i\right] / \left[E_1^f E_1^i E_1^m \left(E_3^f \upsilon_f + E_3^i \upsilon_i + E_3^m \upsilon_m\right)\right] \\ & - \left[\left(E_3^f \upsilon_{31}^m + E_3^m \upsilon_{31}^f\right)\left(E_1^f \upsilon_{13}^m + E_1^m \upsilon_{13}^f\right) E_1^i \upsilon_f \upsilon_m\right] / \left[E_1^f E_1^i E_1^m \left(E_3^f \upsilon_f + E_3^i \upsilon_i + E_3^m \upsilon_m\right)\right] \\ & - \left[\left(E_3^i \upsilon_{31}^m + E_3^m \upsilon_{31}^i\right)\left(E_1^i \upsilon_{13}^m + E_1^m \upsilon_{13}^i\right) E_1^f \upsilon_i \upsilon_m\right] / \left[E_1^f E_1^i E_1^m \left(E_3^f \upsilon_f + E_3^i \upsilon_i + E_3^m \upsilon_m\right)\right] \end{aligned} \tag{20}$$

and

$$\upsilon_{31}^*/E_3^* = \left(\upsilon_{31}^f \upsilon_f + \upsilon_{31}^i \upsilon_i + \upsilon_{31}^m \upsilon_m\right) / \left(E_3^f \upsilon_f + E_3^i \upsilon_i + E_3^m \upsilon_m\right). \tag{21}$$

Here, Eq. (20) represents the needed expression for an effective transverse Young's modulus. Based on our computations with entire possible range of fiber volume fraction, the results indicated that the last three terms on the right-hand side of Eq. (20) can safely be ignored and thus, giving us the simplified and familiar expression of the said modulus such as shown in Eq. (2) earlier.

Next, we shall define the expression for an effective major Poisson's ratio and this is easily received by substituting Eqs. (17) into (21), which we have

$$\upsilon_{31}^* = \upsilon_{31}^f \upsilon_f + \upsilon_{31}^i \upsilon_i + \upsilon_{31}^m \upsilon_m. \tag{22}$$

On the other hand, to determine the expression for effective shear modulus, we first substitute Eqs. (5) into (12) and obtain

$$\gamma_{31} = \gamma_{31}^f \upsilon_f + \gamma_{31}^i \upsilon_i + \gamma_{31}^m \upsilon_m. \tag{23}$$

Then, having the last relations of Eq. (13) substituted into (23) and subsequently equating it with last row expression of Eq. (3), we receive

$$1/G_{31}^* = \upsilon_f/G_{31}^f + \upsilon_i/G_{31}^i + \upsilon_m/G_{31}^m. \tag{24}$$

Thus, Eqs. (22) and (24) completes our tasks in determining the expressions for all effective moduli of the lamina.

References

1. Sager RJ, Klein PJ, Lagoudas DC, Zhang Q, Liu J, Dai L, Baur JW (2009) Effect of carbon nanotubes on the interfacial shear strength of T650 carbon in an epoxy matrix. Compos Sci Tech 69:898–9004
2. Lin Y, Ehlert G, Sodano HA (2009) Increased interface strength in carbon fiber composites through a ZnO nanowire interphase. Adv Funct Mater 19(16):2654–2660
3. Galan U, Lin Y, Ehlert GJ, Sodano HA (2011) Effect of ZnO nanowire morphology on the interfacial strength of nanowire coated carbon fibers. Compos Sci Tech 71(7):946–954
4. Steiner SA, Li R, Wardle BL (2013) Circumventing the mechanochemical origins of strength loss in synthesis of hierarchical carbon fibers. ACS Appl Mater Interfaces 5(11):4892–4903
5. Hart AHC, Koizumi R, Hamel J, Owuor PS, Ito Y, Ozden S (2017) Velcro-inspired SiC fuzzy fibers for aerospace applications. ACS Appl Mater Interfaces 9(15):17342–17350
6. Du Y, Fu C, Gao Y, Liu Y, Xing L, Zhao F (2017) Carbon fibers/ZnO nanowires hybrid nanogenerator based on an interface insulating barrier. RSC Advan 7:21452–21458
7. Nayarana KJ, Burela RG (2018) A review of recent research on multifunctional composite materials and structures with their applications. Mater Today Proc 5(2,1):5580–5590
8. Liu Y, Zhang C, Zhang X (2018) Design, fabrication and application of multi-scale, multi-functional nanostructured carbon fibers. In: Khanna, R., Cayumil, R. (eds.) Recent Developments in the field of carbon fibers. Intech Open, pp 33–49
9. Bezzon VDN, Montaheiro TLA, Menezes BRC, Ribas RG, Righetti VAN, Rodrigues KF, Thim GP (2019) Carbon nanostructure-based sensors: A brief review on recent advances. Adv Mater Sci Eng 1–21
10. Sebastian J, Schehl N, Bouchard M, Boehle M, Li L, Lagounov A, Kafdi K (2014) Health monitoring of structural composites with embedded carbon nanotubes coated glass fiber sensors. Carbon 66:191–200
11. Ray, M.C.: Smart fuzzy fiber-reinforced piezoelectric composites. In: Meguid, S.A (ed.) Advances in Nanocomposites: Modeling, characterization and applications, pp. 126–150. Springer, Heidelberg (2016)
12. Malakooti MH, Patterson BA, Hwang HS, Sodano HA (2016) ZnO nanowire interface for high strength multifunctional composites with embedded energy harvesting. Energy Environ Sci 9(2):634–643
13. Jones RM (1999) Mechanics of composite materials, 2nd edn. Taylor & Francis, Philadelphia
14. Daniel IM, Ishai O (2005) Engineering mechanics of composite materials, 2nd edn. Oxford, New York
15. Barberro EJ (2011) Introduction to composite materials design, 2nd edn. CRC Press, Florida
16. Aboudi J, Arnold S, Bednarcyk B (2012) Micromechanics of composite materials. Elsevier, Oxford
17. Gibson RF (2016) Principles of composite materials mechanics, 4th edn. CRC Press, Florida
18. Vasiliev VV, Morozov EV (2013) Advanced mechanics of composite materials and structural elements. Elsevier, Oxford
19. Eshelby JD (1956) The continuum theory of lattice defects. Solid States Phys 3:79–144
20. Christensen RM, Lo KH (1979) Solutions for effective shear effective properties in three phase sphere and cylinder models. J Mech Phys Solids 27(4):315–330
21. Christensen RM (2005) Mechanics of composite materials, 2nd edn. Dover, New York
22. Christensen RM (1990) A critical evaluations of a class of micromechanics models. J Mech Phys Solids 38(3):379–404
23. Benveniste Y (2008) Revisiting the generalized self-consistent scheme: clarification of some aspects and a new formulation. J Mech Phys Solids 56(10):2984–3002
24. Gusev AA, Lurie SA (2009) Loss amplification effect in multiphase materials with viscoelastic interfaces. Macromolecules 42:5372–5377
25. Lurie S, Minhat M (2014) Application of generalized self-consistent method to predict the effective properties of bristled fiber composites. Composites B 61:26–40

26. Lurie S, Minhat M, Tuchkova N (2015) Estimations of effective dynamics properties of bristled fiber composite materials based on a self-consistent Eshelby method. J Eng Mech 95(1):7–29
27. Lurie S, Minhat M, Tuchkova N, Soliaev J (2014) On remarkable loss amplification mechanism in fiber reinforced laminated composite materials. Appl Compos Mater 21(1):179–196
28. Ehlert GJ (2012) Development of a zinc oxide nanowire interphase for enhanced structural composites. Ph.D. thesis, University of Florida

Flow Field Measurement of Wake Generated by Gourami Fish Tail

Lai Hoong Chuin, Soh Ling Xin, N. A. Razak, Zarina Itam, and A. F. Osrin

Abstract Application of biomimetic systems has been an advancement in the robotic field. The development of underwater robot imitates the locomotion and flow around a swimming fish since its manoeuvrability and efficiency is excellent. The work presented in this manuscript aimed to visualize the flow of a real fishtail to aid the design and fabrication of a mechanical fishtail. The flow across the tail of a Gourami fish when it was visualized and observed in a water channel under static water condition. The tail kinematics of the real fish were extracted and applied on the designed mechanical fishtail. The designed mechanical fishtail was employed to undergo the flow visualization experiment using Particle Image Velocimetry (PIV). The experiment was conducted with the flow in the water channel set at a velocity of 0.2 m/s while the mechanical fishtail was set to flap with an amplitude of 20°. Measurements were repeated with the tail flapping amplitude of 40°. Chain of vortices was observed to be generated by both real fishtail and mechanical fishtail. Through comparison, it is noticed that the flow pattern generated by the mechanical fishtail flapping with the amplitude of 20° has the highest similarity with the actual Gourami fish compare to the flapping amplitude of 40°.

Keywords Mechanical fish · Particle image velocimetry · Biomimetic system

1 Introduction

Biomimetics is a field of science which attempts to design systems and synthesize materials through biomimicry [1]. It has a great application in both architecture and engineering field. For example, the Beijing National Stadium designed by Swiss

L. H. Chuin · S. L. Xin · N. A. Razak (✉) · A. F. Osrin
School of Aerospace Engineering, Engineering Campus, Universiti Sains Malaysia, 14300 Nibong Tebal, Penang, Malaysia
e-mail: norizham@usm.my

Z. Itam
Civil Engineering Department, College of Engineering, Universiti Tenaga Nasional, 43000 Kajang, Selangor, Malaysia

P. Rajendran et al. (eds.), *Proceedings of International Conference of Aerospace and Mechanical Engineering 2019*, Lecture Notes in Mechanical Engineering,
https://doi.org/10.1007/978-981-15-4756-0_18

Architects Herzog and de Meuron was described by Rogers et. al [2] as an excellent example of the use of biometrics in modern architecture. For robotics application, biomimetic systems have been the focus of research for a long time. The natural world provides a large body of biomimetic knowledge where engineers could replicate and apply the lesson learned on the mechanism of the robotic field. Pillsbury et. al. [3] discussed that the pneumatic artificial muscles (PAMs) with high specific work, simple design and long fatigue life are often used for their light-weight design and superior static performance.

Today, the development of the underwater robotic field is popular as it has a wide area of applications which include undersea operation, oceanic supervision, aquatic life-form observation, pollution search and military detection [4]. An underwater robot has been widely employed to replace human activities especially in the deep sea diving activities in the oil and gas industry or underwater observations purpose [5]. The creation and improvement of robot fish are being researched, where it could improve efficiency and bring benefits to humans in a different field.

Therefore, it is common to find studies related to the undulating motion of aquatic animals, especially swimming fish [6, 7]. For example, Cha et. al. [8] seek to understand the feasibility of active compliant material to harvest energy from the beating motion of a fishtail. Marut et. al. [9] on the other hand designed and developed a bioinspired jet propulsion mechanism for the development of unmanned undersea vehicles (UUVs) by mimicking jellyfish. With the kinematic study and flow field measurement and of live fish, it could lead to the design of efficient propulsion mechanism for an underwater robot.

Many research works were also conducted on the slender body theory of swimming fish to obtain the efficiency of fish propulsion in water [7, 10]. Vortices that were produced by the undulating swimming fishes could help them to save the energy required for propulsion. This indicates that the locomotion of live fishes undulatory motion generates the thrust required for fish to swim effectively. Theory of fish locomotion for forward motion and turning is important aspects of knowledge that could be applied to the underwater robot propulsion system.

With the fast development of technologies, many fish robots have been developed [11]. Among the pioneers is the RoboTuna developed by Massachusetts Institute of Technology (MIT) in 1994 [4, 12]. However, many of the fish robots developed are facing limitations in terms of technologies and cost. One of the greatest challenges in developing underwater fish robots is to improve its swimming efficiency and performance which hinders the usage of fish robots. Besides, robotic fish should act as a real fish underwater, where it could perform underwater missions without disturbing nature, i.e. fishes or other aquatic animals should not swim away from the robot fish [13] On the other hand, this also imposed the interest of researchers from different scientific realms to perform a further investigation on this field [4].

The focus of this work lies in the fish body locomotion and tail to generate the undulatory motion which propels them. This work aimed to visualize the flow across the Gourami fish's tail to aid the development of a mechanical fishtail for a fish-like robot. The flow across the Gourami's fishtail in terms of the wakes generated by the fish's tail was studied and investigated by particle image velocimetry (PIV) technique. This type of fish is chosen for its excellence swimming performance and high survivability in the fluctuated environment. To ensure the reliability of the developed mechanical fish-tail was evaluated in terms of its kinematics and wake characteristics. Comparisons were made to determine the similarities and functionalities of both Gouramis and developed mechanical fishtail.

2 Methodology

2.1 PIV Experiment on Flow Across Gourami's Tail

A water channel with an all-sided optically transparent test section of the cross-sectional area of 155 mm × 155 mm was employed to perform the PIV experiment. The entire flow visualization experimental setup is illustrated in Fig. 1. PIV technique is applied to visualized and extract the characteristics of the flow across the Gourami's tail. The laser was employed as the illumination source for the PIV setup. It is placed on the left side of the water channel. The light sheet illuminated perpendicularly to the fishtail while the CCD camera is placed on the other side. Since it is improper to suspend the CCD camera on the top of the test section to achieve the top view of the flow, a mirror is set to be 45° and placed underneath of the test section to reflect the illuminated flow field so that the images could be acquired by the CCD camera. As a precautions step, a piece of cardboard is placed to cover the right side of the test section to avoid direct exposure of the laser sheet into the CCD camera,

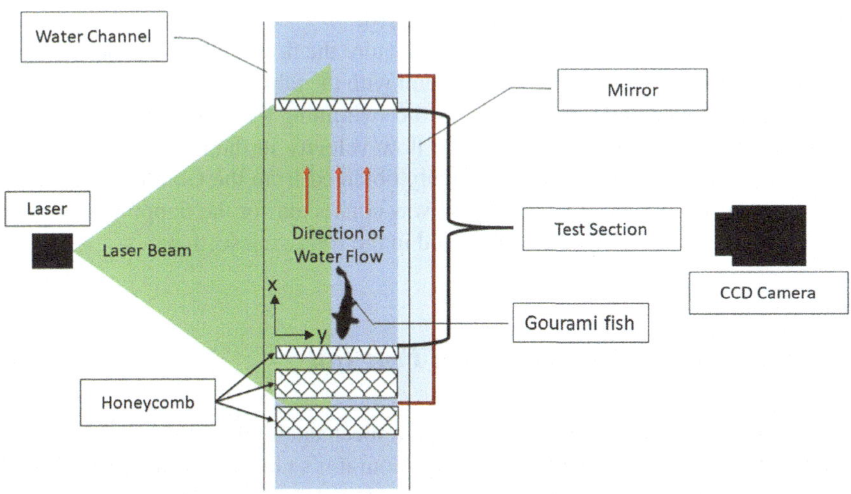

Fig. 1 Schematic illustration of the PIV experiment setup to visualize the flow across the Gourami's tail

causing damage to the CCD sensor. The acquired images were directly transferred to a computer for further PIV analysis (correlation).

In order to make the flow visible by the CCD camera, the flow in the water channel must be seeded with tiny seeding particles. The seeding particles introduced into the flow must be able to suspend in the fluid medium and tiny enough to follow the flow without causing additional drag. In this work, polyamide seeding particles (PSP) with a mean diameter of 50 μm is used as the seeding particles. PSP is chosen because its density is similar to water, so it is able to suspend freely in the water, following the flow. Besides, it also has good light scatter properties, which cause it to be the most suitable particle for this experimental application.

Image acquisition in this work is performed by a CCD camera which compatible with the PIV system. The PIV images set were acquired with double frame single pulse mode. The interval between pulse was set to be 5000 μs while the interval between images set was set to be 0.2 s. Data were extracted from acquired images by performing cross-correlation, a PIV analysis method that is commonly applied for this double frame single pulse mode of data acquisition. The cross-correlation process was performed by using PIVLab, a MATLAB-based PIV analysis software.

As shown in Fig. 1, the Gourami fish is placed in the test section of the water channel and allow to swim freely. This fish is first allowed to swim in the test section without any flow generated by the pump. A retort stand is used to hold and control the position of the light sheet from the illumination source. The light sheet is illuminated in the centre of the fish body. Since the position of the light sheet is not adjustable during the process of data acquisition, the motion of the fish in the z-direction (height of the test section) must be constrained. The limitation of the fish motion in the z-direction is achieved by manipulating the water level in the test section. In this work, the water level in the test section is set to be slightly higher than the height of the fish. The images of the flow generated by the Gourami's tail were acquired and the swimming velocity of the fish was extracted.

A similar experimental setup is used to acquire the flow field across the designed mechanical fishtail by replacing the Gourami with the fabricated mechanical fishtail. However, to simulate the environment of the swimming fish, the pump is turned on to generate flow in the test section. The flow velocity in the test section is set to be 0.2 m/s which is the swimming velocity obtained from the Gourami's fish. The measurement for the mechanical fishtail was conducted for the flapping amplitude of 20° and 40°. The results were presented in the following section.

2.2 Development of Mechanical Fish Tail

The second stage of this work is to design a mechanical fishtail. The shape of the mechanical fishtail was designed using the Gourami's tail as reference. The ratio of the tail was scaled based on the size of the actual Gourami's tail. The geometry of the mechanical fishtail is shown in Fig. 2 and was drawn using CATIA. The material used to fabricate the tail of the mechanical fishtail is a polypropylene sheet with a thickness

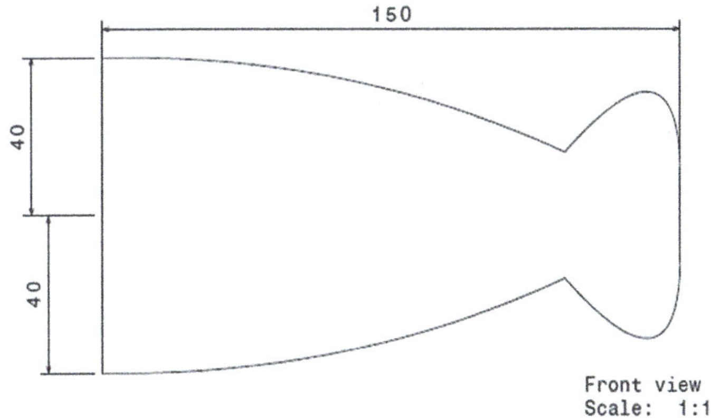

Fig. 2 The geometry and dimension of the mechanical fishtail

of 0.2 mm. This is because polypropylene has good flexibility and good resistance to fatigue. However, in order to ensure that the designed mechanical fishtail could withstand the force in the water when it is flapping, the stiffness of the mechanical fishtail should be slightly increased. This is achieved by increased the thickness of the mechanical fishtail to 0.4 mm by combining two sheets of polypropylene sheet together.

Vertebrates or the backbone of the fish are required to be installed on the mechanical dish to generate the flapping motion by the tail. It is made from a rectangle wooden stick with a length 40 mm and gradually decreased to 20 mm towards the tail with an interval of 5 mm. Each rectangle piece of the vertebra is slotted at the centre, where the vertebra could slot onto the fish body. Two holes with a diameter of 2 mm were drilled on both sides of the vertebrae, to allow the fishing line to pass through, generating the flapping motion of the mechanical tail. The fabricated mechanical fish is presented in Fig. 3. A simple 'fish head' is fabricated using Teflon to allocate the servo that provides the flapping motion on the tail.

The flapping motion of the mechanical fishtail was generated and controlled by a piece of Tower Pro SG90 servo. This type of servo is chosen because it is lightweight

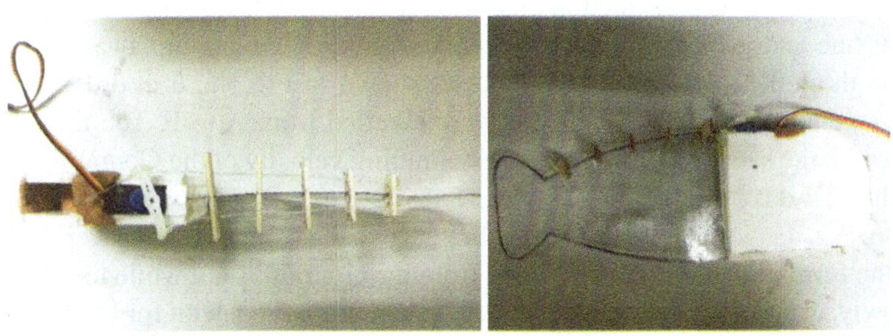

Fig. 3 Top view (left) and side view (right) of the mechanical fish

and has the required torque output to deflect the fishtail. Besides, it has a wide rotation angle ranges between ±90. The real Gourami fish flaps its tail using its muscle. Thus, for mechanical fish, two fishing lines act as the muscles of the fish. Each fishing lines is tied on each side of the servo arm and the last vertebrae, passing through the holes drilled on each side of the vertebrae. The pull and release of the fishing lines from the servo arm generate the mechanical fishtail flapping motion. The amplitude of the tail flapping is controlled by the degree of servo arm rotation.

The degree of servo arm rotation is controlled by using a microcontroller, Arduino Nano. With the pre-designed coding uploaded to the microcontroller, the mechanical fishtail flapping amplitude was set to be ±20°, ±40° and ±60°. The flow measurement across the mechanical fishtail for each flapping amplitude was conducted in the water channel as mentioned in Sect. 2.1. Comparisons of the flow field with the flow across the real Gourami's tail were made to determine the best flapping amplitude of the designed mechanical fishtail.

3 Results and Discussion

3.1 Flow Across Gourami's Tail

The data acquired from flow visualization focuses on the flow across the Gourami's tail. Other than the Gourami's tail flapping frequencies, sudden acceleration or ceased of the Gourami's will affects the flow field. However, the locomotion of the fish is impossible to be fully controlled. The evolution of the flow behind the Gourami fishtail is explained based on the behaviour of the vortex. Note that the scales on the Gourami fish are reflective, causing a large amount of the illuminated light to be reflected and dispersed, thus, the masking region for PIV analysis includes the reflected light region so that these regions are not taken into account for the correlation. This is because these regions are too bright for any particle to be clearly discerned. This led to an increase in spurious vectors.

The evolution of the vortices across the Gourami's tail is presented in Fig. 4. As mentioned previously, the locomotion of the fish is beyond controlled. Hence, the selected results to be presented in this work is where the fish swam with a relatively straight line and average velocity across the frame. The Gourami swam from right towards the left-hand side of the frame as shown in Fig. 4a, e in order while the vortices generated by Gourami's tail were labelled from A to F. The time interval between image sets was 0.2 s while the swimming velocity of the Gourami extracted from the PIV data was 0.2 m/s.

From Fig. 4a, vortex A started to generate when the Gourami's tail swept to its left. After the formation of vortex A, it shed from the Gourami's tail while the Gourami's tail slowly swept towards the right. As the fish continues to swim forward in Fig. 4b, the vortices A, B and C were formed. Vortices A and C rotates in anti-clockwise direction while Vortex B rotates in the clockwise direction. At this stage, vortex A

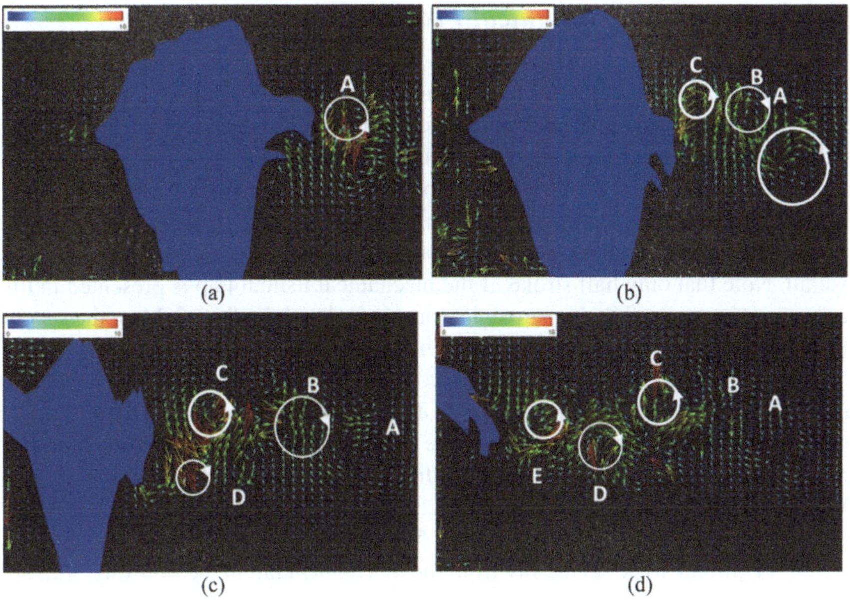

Fig. 4 Evolution of flow across the swimming Gourami's fishtail

had achieved the final expansion which gave up the energy and expands against the tail at the same time Vortices B and C continue to develop.

The growth of vortex D is observed in Fig. 4c, having a clockwise rotation when the Gourami continues to propel forward. Meanwhile, vortex A had achieved its quieting stage. This indicates that it had given up a large part of its kinetic and pressure energy. Vortex B had also reached its final expansion and start entering its quieting stage. Vortex C, on the other hand, continues to expand to achieve its highest energy state. As the Gourami continues to swim forward, new vortices were formed. Newly formed vortex E with rotating direction of anti-clockwise is observed from Fig. 4d. At this point, vortices A and B had achieved a quieting stage where it swirls perpendicular to the fish path while vortices.

Observing the flow field generated by the forward swimming Gourami in Fig. 4, it can be seen that the formation of vortices was initiated by the Gourami's body. The energy from the flow was transmitted to the tail and the vortices are newly formed. As the Gourami's tail continues to flap, the vortex sheds from the tail and move away from the Gourami's body while the fish propelled forward. New vortex was then initiated by the fish body and they cycle repeats until the fish stops swimming. The subsequent vortex formed will always rotate in the opposite direction. Thus, it is fair to infer that as the fish propelled forward, a series of vortices was formed to aid the propulsion of the fish.

3.2 Flow Across Mechanical Fish Tail

The flow visualization of mechanical fish in the water channel with a flow velocity of 0.2 m/s was conducted with two different flapping amplitudes, which are 20° and 40°. Evolution of flow behind the mechanical fishtail is shown in Figs. 5 and 6. Each vortex formed are labelled in alphabetical order and the mechanical tail in the images was masked. An unpretentious comparison was made with the flow across the Gourami's tail to determine the most suitable flapping amplitude of the mechanical fishtail. Note that only half stroke of the mechanical fishtail flap is presented in this manuscript since the flapping stroke is symmetrical and the flow field on the second half of the stroke is mirrored off the first half stroke.

Figure 5 presents the flow evolution across the mechanical fishtail flapping with an amplitude of 20°. The first image set was acquired when the mechanical tail flapped to its maximum amplitude on its left side as shown in Fig. 5a. At this instance, no clear vortex was observed in the flow. The first vortex in the flow field observed vortex A was formed when the tail was flapping towards its right side from the maximum left position as shown in Fig. 5b. Figure 5c shows as the flapping stroke continue, vortex A continue to be shed away from the mechanical tail in the flow direction and

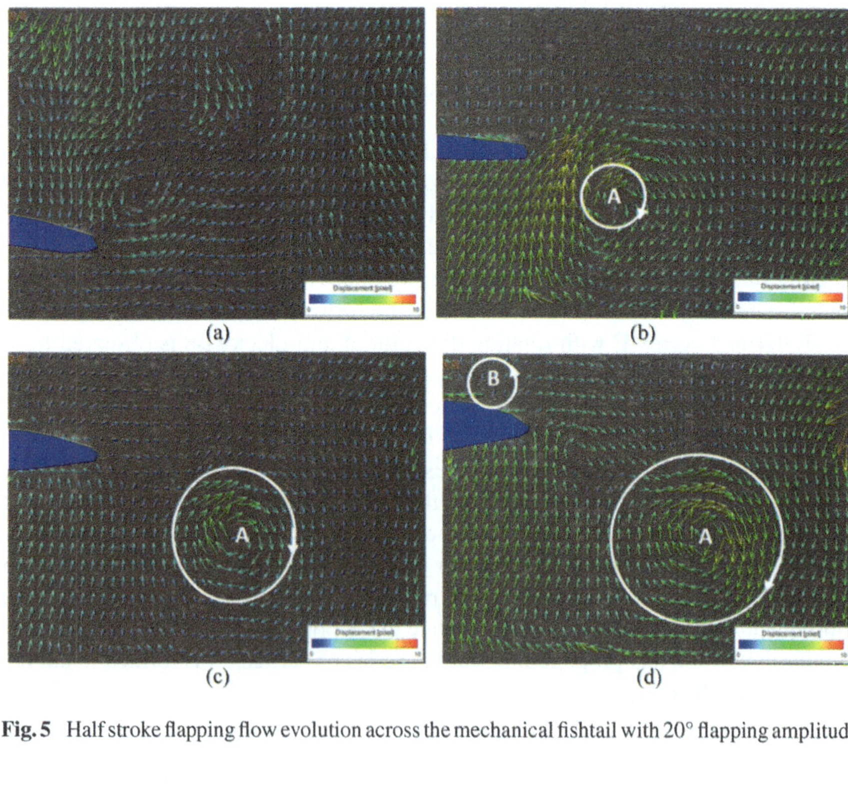

Fig. 5 Half stroke flapping flow evolution across the mechanical fishtail with 20° flapping amplitude

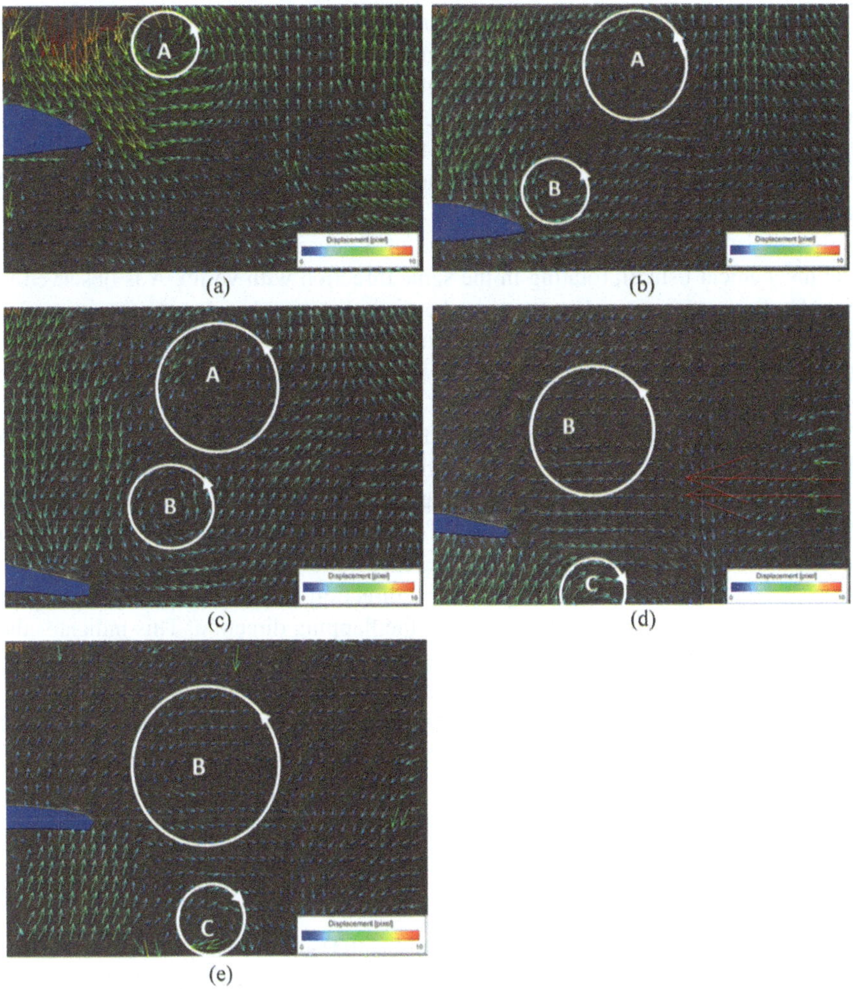

Fig. 6 Half stroke flapping flow evolution across the mechanical fishtail with 40° flapping amplitude

continue to grow as there is the obvious increase in the diameter of the vortex can be observed.

After the mechanical tail reached the maximum flapping amplitude on its right, it changed its flapping direction and proceed with the second half of the stroke. At this point, the flow in the tip of the mechanical fishtail started to experience changes in flow direction and increase in velocity as shown in Fig. 5d. This indicates that it is a potential for a new vortex rotating in the opposite direction of vortex A to be formed. At the same time, Vortex A still continues to grow and move away from the mechanical fishtail, following the flow direction of the water channel.

The half stroke flapping of the mechanical fishtail with a flapping amplitude of 40° is presented in Fig. 6. Firstly, it has an extra frame for the mechanical fishtail to complete the half stroke of the flapping as the flapping amplitude is higher. The acquisition of the first image started when mechanical fishtail is in a neutral position. In this measurement, the mechanical fishtail flapped towards its left. An anti-clockwise vortex A can be seen in Fig. 6a. This vortex is formed when the mechanical fishtail flapped from its maximum right position.

As the stroke continues, the formation of a new vortex, vortex B at the tip of the mechanical fishtail, rotating in the same direction with vortex A is observed in Fig. 6b. Both vortices were rotating in the same direction because the new vortex is formed when the tail is still flapping in the same direction of the stroke. At the same time, vortex A entered the final expansion stage, where it achieved its maximum diameter.

When the mechanical fishtail reached its maximum left position as shown in Fig. 6c, vortex B shed off from its tip while vortex A started to disintegrate. The stroke continued by changing the flapping direction, the mechanical fishtail started to flap towards its right. At this instance, vortex B had achieved its final expansion while vortex A had completely disappeared as shown in Fig. 6d. Similar to the previous measurement, the change in flow direction can be observed at the tip of the tail in when it experienced changes in the flapping direction. This indicates the potential of new vortex rotating in the opposite direction of vortex B to be formed. It is evidenced when newly formed clockwise vortex C was observed in Fig. 6e. Meanwhile, the mechanical tail had returned to its neutral position and vortex B starts to disintegrate.

4 Conclusion

Comparing the flow fields obtained from all the measurements, flow field across the mechanical fishtail flapped with the amplitude of 20° is highly similar to the flow field across the actual Gourami's tail. Each half stroke generates a vortex and the subsequent vortex is always rotating in the opposite direction. Increase in flapping amplitude could lead to an increase in the number of vortices formed in the same direction as the time required to complete the half stroke of flapping is longer. A new vortex rotating in the same direction is formed before the half stroke is completed. Hence, it is fair to infer that with 20° flapping amplitude, the designed mechanical fishtail is suitable to be applied to a robot fish and capable to propel the robot fish forward in a straight line.

Acknowledgments This work was supported by the UNITEN grant J510050834 and Universiti Sains Malaysia Bridging grant 6316310.

References

1. Bansode SS, Hiremath RB, Kolgiri S, Deshmukh RA (2016) Biomimetics and its applications-a review. Int J Emerg Technol Adv Eng 6(6):63–72
2. Rogers A, Yoon B, Malek C (2008) Beijing Olympic Stadium 2008 as biomimicry of a Bird's Nest, architectural structures, ARCH 251
3. Pillsbury TE, Robinson RM, Wereley NM (2013) Miniaturized pneumatic artificial muscles actuating a bio-Inspired robot hand. In: ASME 2013 smart materials, adaptive structures and intelligent systems, 2 Utah, USA
4. Masoomi SF, Gutschmidt S, Gaume N, Guillaume T, Eatwel C, Chen X, Sellier M (2015) Design and construction of a specialised biomimetic robot in multiple swimming gaits. Int J Adv Robotic Syst 12(11):168
5. Anderson JM, Chhabra NK (2002) Maneuvering and stability performance of a robotic tuna. Integ Comp Biol 42(1):118–126
6. Sfakiotakis M, Lane D, Davies J (1999) Review of fish swimming modes for aquatic locomotion. IEEE J Oceanic Eng 24(2):237–252
7. Shue TWH, Chen YH (2007) Numerical study of flow field induced by a locomotive fish in the moving meshes. Int J Num Meth Eng 69:2247–2263
8. Cha Y, Verotti M, Walcott H, Peterson SD, Porfiri M (2013) Energy harvesting from base excitation of a biomimetic fish tail hosting ionic polymer metal composites. In: ASME 2013 conference on smart materials, adaptive structures and intelligent systems, vol. 2. Utah, USA
9. Marut K, Stewart C, Villanueva A, Avirovik D, Priya S (2012) A biomimetic jellyfish-inspired jet propulsion system using an iris mechanism. In: ASME 2012 Conference on smart materials. adaptive structures and intelligent systems. Stone Mountain, Georgia, USA
10. Sakakibara J, Nakagawa M, Yoshida M (2004) Stereo-piv study of flow around a maneuvering fish. Exp Fluids 36(2):282–293
11. Razif M, Mohd FAA, Aimi I, Natarajan E, Yaakob O (2014) A review on development of robotic fish. J Trans Syst Eng 1(1):12–22
12. Yu J, Wang K, Tan M, Zhang J (2014) Design and control of an embedded vision guided robotic fish with multiple control surfaces. Sci World J 14:1–13
13. Rus D, Tolley MT (2015) Design, fabrication and control of soft robots. Nature 521:467–475

Design Factors of a Plasma Reactor with a Swirling Flow Field for Production of Rutile TiO_2 Nanoparticles

Byeongjun Jeon, Junghun Shin, and Donghoon Shin

Abstract This study presents the development of a plasma reactor producing rutile TiO_2 nanoparticles as a white pigment. The effects of reactor design factors are also considered based on experimental data. A Direct current thermal plasma using Ar as the plasma medium was applied to heat the reactor. $TiCl_4$ and O_2 were the reactants with TiO_2 and Cl_2 as the products. The advantage of this process is the production of titanium dioxide nanoparticles without the co-generation of HCl. The effects of design factors such as the position of the nozzles supplying the reactants, temperatures of the reactor and heater, residence time, and swirl number of the flow field on rutile phase generation were investigated. The collected titanium dioxide nanoparticles were examined using X-ray diffraction (XRD) and Scanning electron microscope (SEM). Although there was no significant effect of nozzle position, the temperatures of the reactor and heater showed strong correlations with the rutile phase contents, with temperatures of 1030 to 1060 °C for the reactor and 750 °C for the heater yielding the maximum rutile content of 79.8% at a residence time of 2.2 s. The gas flow pattern in the furnace also exhibited an effect on rutile content, i.e., as the swirl number in the reactor increased, the rutile phase content in the TiO_2 also increased. From experimental results and Computational fluid dynamics, the optimum swirl number was estimated to be between 0.7 and 0.8.

Keywords TiO_2 · Rutile · Nanoparticles · Thermal plasma · Reactor

1 Introduction

Titanium dioxide, also known as titania, is a white solid inorganic substance. It is used across a vast range of applications covering daily life to high-tech fields, e.g., paints, coatings, inks, plastics, paper, cosmetics, construction materials, ceramics, 3D-printing filaments, semi-conductor materials, and photo-catalysts. Titanium dioxide exhibits a chemically and physically stable structure, a high refractive index,

B. Jeon · J. Shin · D. Shin (✉)
Department of Mechanical Engineering, Graduate School, Kookmin University, Seoul, Korea
e-mail: d.shin@kookmin.ac.kr

P. Rajendran et al. (eds.), *Proceedings of International Conference of Aerospace and Mechanical Engineering 2019*, Lecture Notes in Mechanical Engineering,
https://doi.org/10.1007/978-981-15-4756-0_19

ultraviolet-cut ability, hiding power, and weathering resistance. The structure of titanium dioxide shows three phases: rutile, anatase, and brookite. Rutile has better absorbance properties and thermal resistance than the other phases. Owing to its excellent light-scattering properties, white opacity, and brightness, the most common application of rutile phase titanium dioxide is as a pigment [1–8].

Well-known methods for producing titanium dioxide are the sulfate process and chloride process. The sulfate process has also been used to generate the anatase phase. Owing to the batch-type operation of the sulfate process, washing and filtration are necessary. In addition, the sulfate process has the disadvantage of requiring the disposal of waste sulfuric acid, leading to environmental concerns. In contrast, the chloride process developed by Dupont in 1951 produces less and safer waste than the sulfate process. Owing to this advantage, the chloride process became more popular than the sulfate process after 1957 [9, 10]. Recently, a chloride process using thermal plasma has been studied. The heat sources used in previous chloride processes were hydrocarbon fuels, which cause corrosion issues owing to the HCl generated from the hydrogen of the hydrocarbon and the chlorine in $TiCl_4$ [11–14]. The thermal plasma process was developed to address this issue. The advantage of this process is continuous and single step reaction to produce titanium dioxide nanoparticles without pre-processing [15]. The chlorine of the products can be separated via a condensing process to be recycled to generate $TiCl_4$, the raw material.

The aim of this study was to optimize the design factors such as temperature, residence time, and swirl intensity of the flow field to develop a reactor using a DC thermal plasma as a part of the chloride process to produce rutile titania nanoparticles.

2 Experimental Apparatus and Procedure

2.1 Reactor Design

Figure 1 shows the basic design of the reactor. The main reactor was 60 mm in diameter and 500 mm in height. Swirl flow was produced by applying a tangential angle to the injector nozzles, thereby enhancing mixing of the raw material ($TiCl_4$) and oxidizer (O_2) and controlling flow uniformity. Insulation materials (Fibermax board 18R) were used to prevent heat loss. Inside the insulation materials, a SIC (silica carbide) tube was inserted to prevent damage to the insulation material and sedimentation of the produced TiO_2 nanoparticles. The reactor had two nozzles, one each for the oxidizer and raw material. The nozzles could be switched between upper nozzles at the level 1 (20 mm from the plasma torch) and lower nozzles at the level 2 (55 mm from the plasma torch), relative to the position of plasma torch. Swirl flow was produced by the nozzle angle (5°) to the radial direction. Swirl number(S) was calculated using formula (1) which means ratio of tangential momentum against axial momentum [16].

Fig. 1 Basic design of reactor

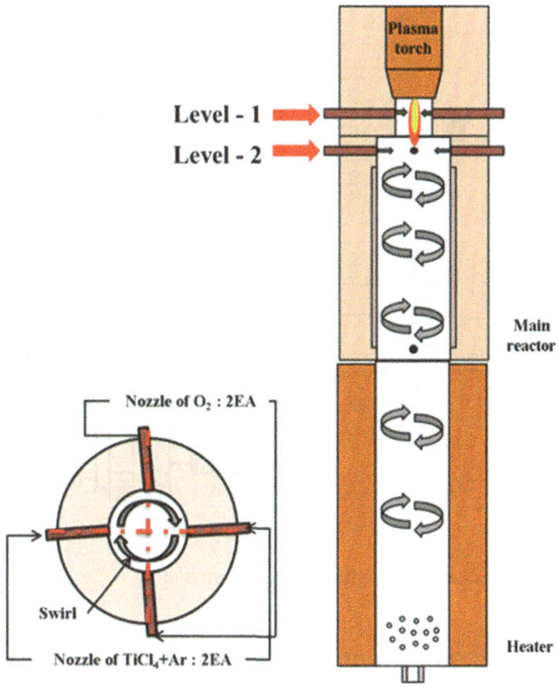

$$S = {}^{G_\theta}/{}_{G_x} \tag{1}$$

$$Where\, G_\theta = 2\pi \int_0^R \rho U W r^2 dr \tag{2}$$

$$G_x = 2\pi \int_0^R \rho U^2 r dr \tag{3}$$

where, G_θ is axial flux of the tangential momentum, G_x is axial flux of the axial momentum, ρ is density, r is radial position r, R is outer radius of the annulus, W is tangential velocity at corresponding radial position r, and U is axial velocity at corresponding radial position r.

A tube with halogen lamp heater was placed at the bottom of the reactor to extend the residence time while maintaining the reaction temperature above 750 °C which was 60 mm in diameter and 600 mm in height.

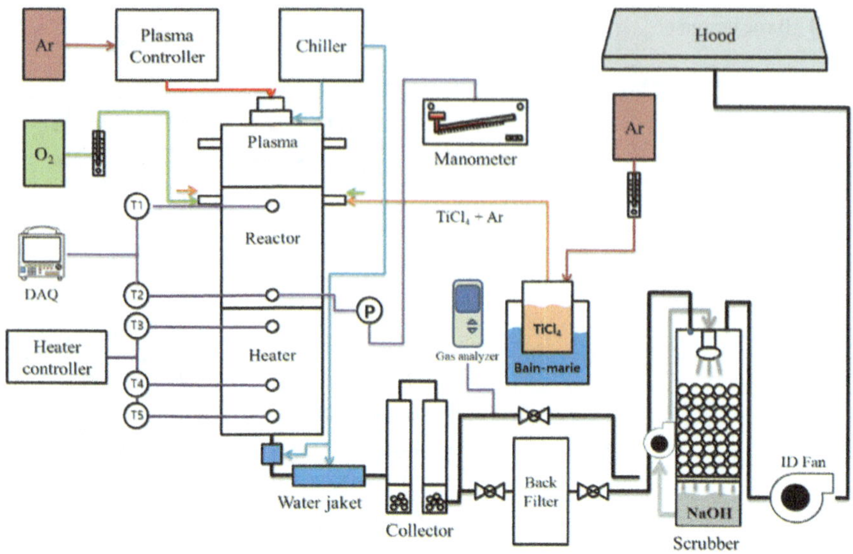

Fig. 2 Schematic diagram of the experiment

2.2 *Experimental Apparatus*

A schematic diagram of the experimental system is shown in Fig. 2. The plasma system consisted of a power supply and a plasma torch with a tungsten cathode (−) and copper anode (+). The power supply controlled the electric current and the plasma discharge gas (Ar). The chiller was used to cool the plasma torch and the off gas. The temperature in the reactor was measured at the upper (T1) and lower (T2) ends of the reactor. The temperature in the heater was measured at three points (T3–T5). A manometer was installed to monitor reactor pressure. A vessel of $TiCl_4$ (max. 3 L) was designed to act as a secure seal and was maintained at 120 °C by heated oil. A bubbler was installed in the vessel to evaporate liquid $TiCl_4$ and supply $TiCl_4$ to the reactor via the Ar carrier gas. Sodium hydroxide was used in the scrubber to remove Cl2 in the flue gases. The TiO_2 nanoparticles were sampled at the bag-filter for analysis. The O_2 concentration was measured using a Testo330 at the bypass-line to control the reaction stoichiometry. All flows of the oxidizer and carrier gas were controlled by the mass flow controllers (SMTEK, DFC-5000S).

2.3 *Experimental Procedure*

After heating of the reactor by the plasma and bottom heater for 1.5 h, the furnace reached a state of thermal equilibrium. Then, the raw material ($TiCl_4$) and oxidizer

(O_2) were supplied to the reactor. The reactor pressure was controlled at −5 mm H_2O. After steady state was achieved, the titanium dioxide was sampled in the bag-filter.

Table 1 shows the experimental conditions. Experiments to determine the influence of design factors on the content of rutile TiO_2 were conducted as follows. Case 1: the influence of the positions of the O_2 and $TiCl_4$ nozzles was tested; Case 2: the influence of heater temperature was tested; Case 3: the influence of reactor temperature was tested when the heater temperature was 750 °C; Case 4: the influence of residence time was tested by changing the flow rates of O_2 and Ar; Case 5: the influence of swirl number was tested.

The collected TiO_2 nanoparticles were analyzed using XRD (UltimaIV) and SEM (SSX550). XRD was used to analyze the phase of the TiO_2 and SEM was used to determine the shape and size of the collected TiO_2.

Table 1 Experimental conditions

Case		Nozzle position		Flow rate			ETC	
		Level-1	Level-2	Plasma gas Ar (LPM)	$TiCl_4$ Carrier gas Ar (LPM)	O_2 (LPM)	Heater (°C)	Swirl number
Case1	1	O_2	$TiCl_4$	25	5	8	NA	0.75
	2	$TiCl_4$, O_2	NA	25	5	8	NA	0.75
	3	NA	$TiCl_4$, O_2	25	5	8	NA	0.75
Case2	1	NA	$TiCl_4$, O_2	25	5	8	700	0.75
	2	NA	$TiCl_4$, O_2	25	5	8	750	0.75
	3	NA	$TiCl_4$, O_2	25	5	8	800	0.75
	4	NA	$TiCl_4$, O_2	25	5	8	850	0.75
Case3	1	NA	$TiCl_4$, O_2	25	5	8	750	0.75
Case4	1	NA	$TiCl_4$, O_2	25	5	4	750	0.65
	2	NA	$TiCl_4$, O_2	21	5	4	750	0.82
Case5	1	NA	$TiCl_4$, O_2	21	2	3.5	750	0.29
	2	NA	$TiCl_4$, O_2	21	4	6.5	750	0.72
	3	NA	$TiCl_4$, O_2	21	5.5	9	750	1.05

2.4 Computational Analysis

Computational fluid dynamics analysis was applied to elucidate the thermal fluid and reaction phenomena to maximize the mixing between $TiCl_4$ and O_2, and to give a uniform downward flow to increase residence time. Figure 3 presents the designed grid of the CFD model. The model divides the grid into many sections assigning a dense grid at the nozzles, and the main reaction zone. The grid contained 480,000 cells achieved using the cut-cell method. Steady-state flow of the reactor model was interpreted using ANSYS Fluent 17.0. The model used for turbulence was $Realizable\ k - \varepsilon$, which has advanced performance for swirling flow. Table 2 shows the main chemical reaction and reaction factors [17, 18].

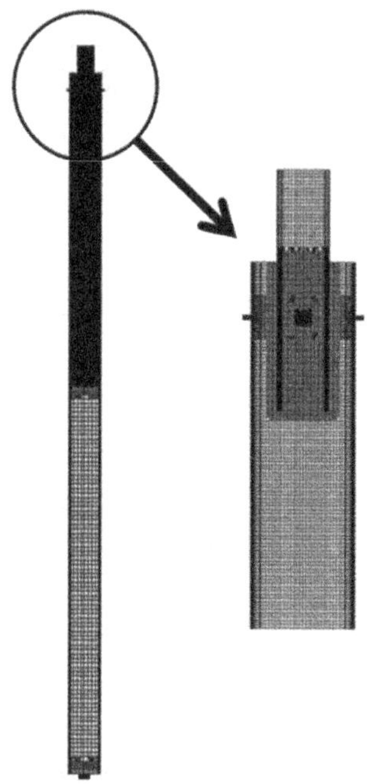

Fig. 3 Grid of 3D model for reactor simulation

Table 2 CFD conditions

Contents	Condition
Mesh	480,000 (cutcell)
Solver	Fluent 17.0
Model	realizable k-ε model
Chemical reaction model (1 step model)	$TiCl_4 + O_2 \rightarrow TiO_2 + 2Cl_2$ E_a: 25(kJ/mol), A: 6.8e + 11

3 Experimental Results and Discussion

3.1 *Effect of Nozzle Position on Rutile Content*

In Case 1, the effect of nozzle injection position was assessed. The reactor was not connected to the downstream heater so that the residence time was less than one sec. In Case 1-1, the O_2 nozzles were at level-1, i.e., close to the plasma flame, and those of $TiCl_4$ were at level-2, i.e., downstream, out of reach of the plasma flame. In Case 1-2, both the O_2 and $TiCl_4$ nozzles were at level-1. In Case 1-3, both the O_2 and $TiCl_4$ nozzles were located at level-2. The results (Fig. 4) show that the maximum rutile content of Case 1-2 was 44.7% at 1100 °C. In Case 1-3, the highest rutile content was 39.2% at 1150 °C. The average particle size of Cases 1-2 and 1-3 was in the range 149–320 nm, which was much larger than that of Case 1-1. The XRD results of the rutile contents are shown in Fig. 5. The SEM images used to measure the particle size are shown in Fig. 5. When $TiCl_4$ was injected at level-1, the produced particles easily blocked the plasma torch due to their closeness. Hence, the remaining experiments were carried out using nozzles at level-2 for O_2 and $TiCl_4$ injection.

Fig. 4 Percentage and particle size of rutile TiO_2 versus reactor temperature (Case 1)

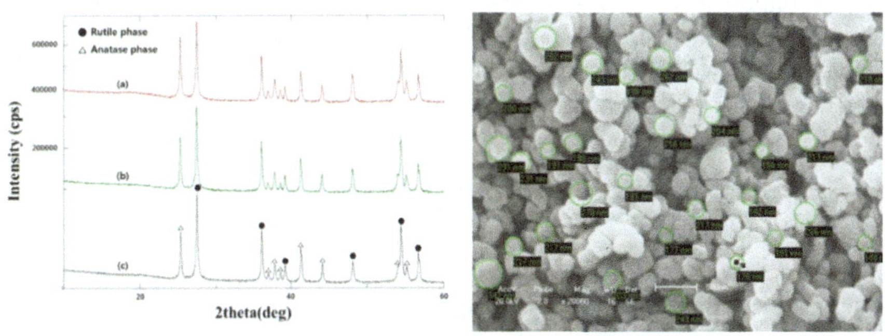

Fig. 5 XRD analysis results and SEM images of TiO_2 nanoparticles (Case 4)

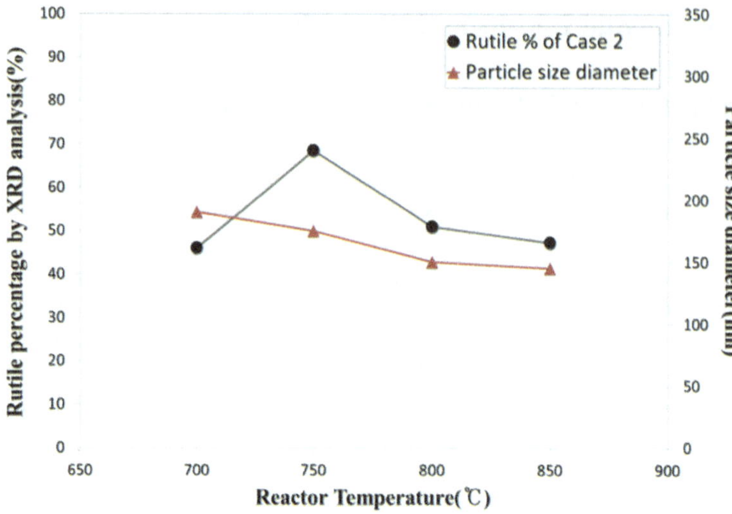

Fig. 6 Percentage of rutile TiO_2 and particle size diameter versus reactor temperature (Case 2)

3.2 *Effect of Heater Temperature on Rutile Content*

Case 2 considered the effect of the heater temperature. Figure 6 shows the rutile content and particle size in relation to mean heater temperature. In this case, the reactor temperatures were fixed at 1050 °C. The maximum rutile content was 68.5% at a heater temperature of 750 °C. Previous research has reported that 1000–1100 °C is the optimal temperature zone for maximizing rutile content; however, the present results suggest a different temperature [19]. The main reason for the difference is that the reactor configuration in the previous research used a separated heater; therefore, specimen temperature dropped before it reached the heater. In contrast, the present study used a directly-connected heater; thereby mitigating any temperature drop of the particles. The residence time in the reactor was around 0.8 s, and that in the connected heater was approximately 0.9 s. Owing to longer residence times in this case, the rutile contents were almost twice as compared to those of Case 1. The results also show that particle size decreased as heater temperature increased slightly (Fig. 6) which is possibly owing to shorten residence time as heater temperature increases.

3.3 *Effect of Reactor Temperature When Connected to the Heater*

Case 3 assessed the effect of reactor temperature when the heater was connected and the temperature was controlled at 750 °C. Figure 7 shows the rutile content in relation to reactor temperature as fitted with a second order polynomial regression. The rutile

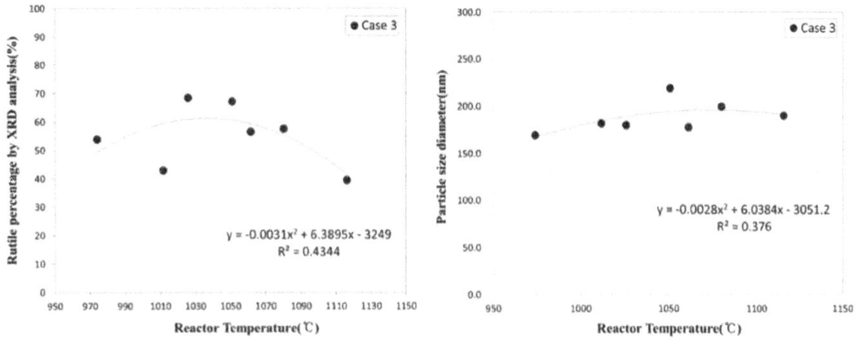

Fig. 7 Percentage and particle size of rutile TiO_2 versus reactor temperature (Case 3)

contents at reactor temperatures of 1030 and 1050 °C were more than 60%. However, the rutile content decreased as reactor temperature increased above 1050 °C. The particle sizes were in the range 169–219 nm, and showed slight increases as reactor temperature increased (Fig. 7). The increased temperature is expected to enhance the coalition between TiO_2 nanoparticles.

3.4 *Effect of Residence Time*

Case 4 determined the effect of flue gas residence time in the reactor. The residence time was controlled by the plasma gas flow rate. The temperature of both reactor and heater was fixed at 750 °C. Owing to limitations in controlling the plasma gas flow rate below 20 lpm, only two cases were examined. Figure 8 shows the rutile contents for Cases 4-1 and 4-2 in relation to reactor temperature. As residence time increased from 1.7 s (Case 3) to 1.9 s (Case 4-1), the maximum rutile content reached 69% at

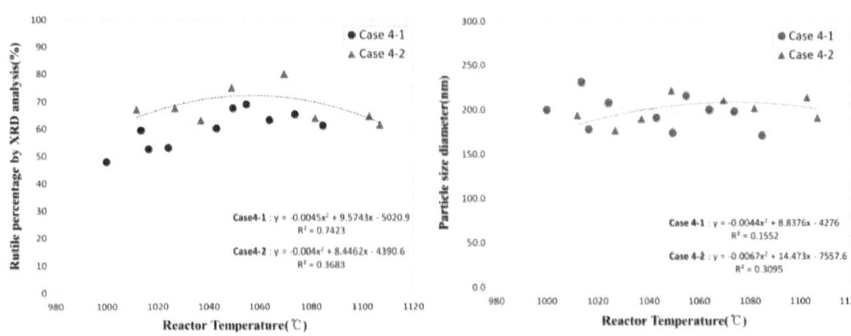

Fig. 8 Percentage and particle size of rutile TiO_2 versus reactor temperature for Case 4-1 (1.9 s) and Case 4-2 (2.2 s)

1050 °C. Particle sizes were in the range of 171–231 nm (Fig. 8). As residence time was increased to 2.2 s (Case 4-2), the rutile contents were greatest at 1050 °C (75%) and 1070 °C (79.8%), and they were always greater than 60%. Particle sizes were in the range of 177–222 nm which indicates that particle size did not differ greatly in relation to reactor temperature (Fig. 8). Comparing the results from Cases 3 and 4, it is evident that rutile percentage increased as residence time increased. As such, residence time appears to be one of the main parameters influencing rutile content.

3.5 *Effect of Swirl Intensity*

A comparison of Cases 3 and 5 was used to elucidate the effect of swirl intensity. The temperature of the reactor was controlled between 1020 and 1050 °C, and the temperature of the heater was fixed at 750 °C. Figure 9 shows rutile content in relation to swirl number. As the swirl number increased, the rutile content increased slightly. However, at swirl numbers greater than 0.8, there was no significant increase in the rutile content. Figure 9 shows that particle sizes were in the range of 195–218 nm which indicates that particle size increased depending on swirl number. The increased swirl number is related with turbulence in the reactor which enhances reaction and mixing so that the increase of rutile content as well as particle size appeared. Figure 10 shows the CFD results in relation to swirl number for Case 5. When the swirl number was 0.29, the mass fraction of TiO_2 was centrally concentrated, and its mixing in the reactor was relatively slow. Both temperature and Y-velocity (vertical velocity) showed a uniform distribution. Although uniform temperature and velocity profiles are generally necessary for an improved reaction, the mixing of reactants which can be controlled by swirl number is more important. For a swirl number of 0.72, the temperature and Y-velocity profiles are less uniform than those for the previous case. Also, the diamond shape at the center of the Y-velocity profile suggests a dead-zone, i.e., unused space. The TiO_2 concentrations are distributed more evenly

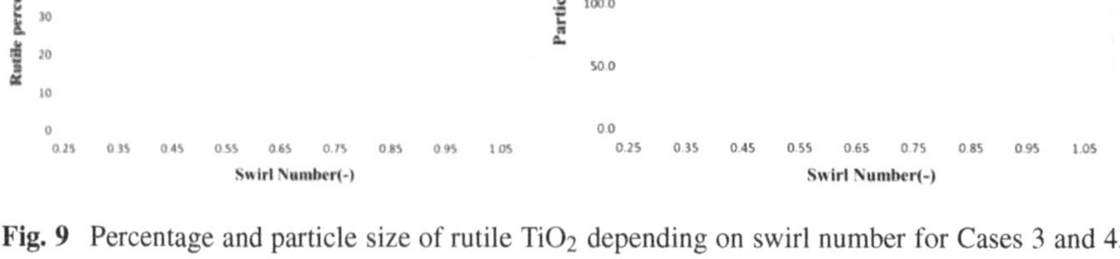

Fig. 9 Percentage and particle size of rutile TiO_2 depending on swirl number for Cases 3 and 4, each at 1020 and 1050 °C, and for Case 5

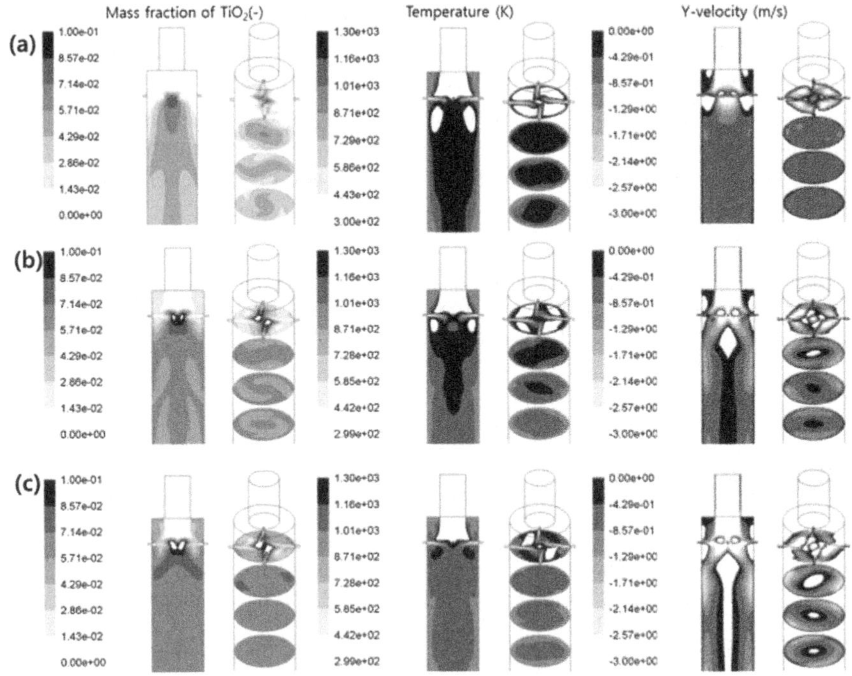

Fig. 10 CFD results of TiO_2 mass fraction [−], temperature, and Y-velocity magnitude (Case 5): **a** Case 5-1 (S 0.29) **b** Case 5-2 (S 0.72), **c** Case 5-3 (S 1.05)

downstream which indicates better mixing in the reactor and production of higher rutile contents. For a swirl number of 1.05, a larger dead-zone at the Y-velocity contour elucidate decreased actual residence time, although TiO_2 concentration and temperature exhibited even distributions, so that the rutile reaction occurred less.

4 Conclusion

Herein, we studied the design parameters of a reactor producing rutile TiO_2 nanoparticles using a series of experiments and computational analysis. The nozzle positions of the oxidizer and raw material, the temperatures of the reactor and the heater, the residence time, and the swirl number were selected as the main design parameters to be assessed. The conclusions obtained were as follows:

1. Nozzle position has a weak correlation with rutile content.
2. Reactor and heater temperatures have a significant impact on rutile content. It is recommended that the reactor be approximately 1050 °C and the heater 750 °C to maximize rutile contents.

3. The effects of residence time are also significant; longer residence times induce higher rutile contents. This relationship seems to be linear. However, further study is necessary to increase the rutile contents above 80%.
4. From the experimental results and CFD analysis it was found that rutile TiO_2 content increased according to swirl number. However, swirl numbers above 1.05 did not further increase rutile content because such high numbers can increase the size of the dead-zone and decrease the residence time.

Acknowledgements This work was partly supported by Institute for Information & communications Technology Promotion (IITP) (No. 10052730), the Basic Research Program through the National Research Foundation of Korea (NRF) (No. 2017M3A7B4042024).

References

1. Braun JH, Baidins A, Marganski RE (1992) TiO_2 pigment technology: a review. Prog Org Coat 20(2):105–138. https://doi.org/10.1016/0033-0655(92)80001-D
2. Hoffmann MR, Martin ST, Choi W, Bahnemann DW (1995) Environmental applications of semiconductor photocatalysis. Chem Rev 95(1):69–96
3. Linsebigler AL, Lu G, Yates JT Jr (1995) Photocatalysis on TiO_2 surfaces: principles, mechanisms, and selected results. Chem Rev 95(3):735–758
4. Chen X, Mao SS (2007) Titanium Dioxide Nanomaterials: synthesis, properties, modifications, and applications. Chem Rev 107(7):2891–2959. https://doi.org/10.1021/cr0500535
5. Reyes-Coronado D, Rodríguez-Gattorno G, Espinosa-Pesqueira M, Cab C, de Coss R.D, Oskam G (2008) Phase-pure TiO_2 nanoparticles: anatase, brookite and rutile. Nanotechnology 19(14):145605
6. Grant F (1959) Properties of rutile (titanium dioxide). Rev Modern Phys 31(3):646
7. Son HY, Park GY, Lee GI, Yang HS, Yu H (1989) S: Manufacture of TiO_2 by chloride process. Chem Ind Technol 7(3):336–346
8. Prevention IP (2007) Control. Reference document on best available techniques in the large volume inorganic chemicals, ammonia, acids and fertilizers industries. European Commission, Directorate-General JRC
9. Gázquez MJ, Bolívar JP, Garcia-Tenorio R, Vaca F (2014) A review of the production cycle of titanium dioxide pigment. Mater Sci Appl 5(07):441
10. Keegel M (2012) The changing dynamics of the TiO_2 value chain
11. Kim KH, Lee IH, Kim KS, Kim DJ (2004) Plasma equipments for nanotechnology process. Korea Institute of Science and Technology
12. Liao X (2011) The synthesis of nano TiO_2 particles using a DC transferred arc plasma reactor. McGill University Libraries
13. Park HK, Park KY (2015) Control of particle morphology and size in vapor-phase synthesis of titania, silica and alumina nanoparticles. KONA Powder Particle J 32:85–101
14. Park KY, Lee HY, Yang HS (1990) Pilot Scale Experiments on the Oxidation of $TiCl_4$. J Korean Ceramic Soc 27(6):817–823
15. Samal S (2017) Thermal plasma technology: the prospective future in material processing. J Cleaner Prod 142:3131–3150
16. Ishak MA, Jaafar MM, Eldrainy YA (2009) The effect of radial swirl generator on reducing emissions from bio-fuel burner system. Modern Appl Sci 3(6):45–51
17. Wu M, Windeler R, Steiner C, Börs T, Friedlander S (1993) Controlled synthesis of nanosized particles by aerosol processes. Aerosol Sci Technol 19(4):527–548

18. Shirley R, Liu Y, Totton TS, West RH, Kraft M (2009) First-principles thermochemistry for the combustion of a $TiCl_4$ and $AlCl_3$ mixture. J Phys Chem A 113(49):13790–13796
19. Morooka S, Yasutake T, Kobata A, Ikemizu K, Kato Y (1987) Production mechanism of TiO_2 fine particles by gas-phase reaction. Kagaku Kogaku Ronbunshu 13(2):159–165

Development of a Flameless Furnace with Swirling Reversed Air Injection Method

Namgyun Oh, Jaesam Shin, Byeongjun Jeon, Chonggun Choi, and Donghoon Shin

Abstract We developed a new flameless combustion called swirling-RAI (reversed air injection) and evaluated through experiments and numerical simulation. In a circular furnace which has a specific purpose to heat pellet bed, two pair of nozzle sets from the concept of RAI are installed in the inversed direction against each other. Due to the high momentum of angled air nozzles installed axisymmetric, the overall flow in the furnace is swirling. The experiments showed methane-air combustion without flame and NO_x emission was 3 ppm while CO was not detected at the exit. The numerical simulation showed detail of flow and reaction in the S-RAI furnace to understand its phenomena. The experimental and simulation results are also compared to see the reliability of the simulation. The complexity of flameless combustion makes the proper simulation difficult. In particularly, oxygen concentration at Port 2 showed a difference. However, the simulation gives enough information of flow field and insight of its effect on the flameless combustion, which shows two recirculating flow patterns in the furnace; the first recirculation is occurring at the top due to the ejection effect of nozzle jets, and the second recirculation is located at the reducing cone under the nozzles which is due to swirling flow.

Keywords Flameless combustion · Swirling flow

1 Introduction

Flameless combustion is an emerging technology to control NO_x emission via flue-gas internal recirculation and delayed mixing of fuel and oxidant. The enhanced recirculation of flue gas and low reaction speed owing to delayed mixing of reactants as well as dilution with flue gas make whole volumetric area of a furnace to participate in a mild reaction, which results in uniform temperature distribution to lower peak temperature so that thermal and prompt NO_x decreases significantly comparing to

N. Oh · J. Shin (✉) · B. Jeon · C. Choi · D. Shin
Department of Mechanical Engineering, Graduate School, Kookmin University, Seoul, South Korea
e-mail: sjs1673@kookmin.ac.kr

P. Rajendran et al. (eds.), *Proceedings of International Conference of Aerospace and Mechanical Engineering 2019*, Lecture Notes in Mechanical Engineering,
https://doi.org/10.1007/978-981-15-4756-0_20

conventional combustion technology. Meanwhile, unburned fuel is also not detected at exit when overall temperature of the furnace is high enough (> 850 °C) [1–3].

To reduce NO_x emission from combustion, the reburning concept was introduced by Wendt et al. [2]. Wunning et al., introduced flameless oxidation which induces flue-gas recirculation to increase inert contents in the oxidant so that the reaction rate decreases [3]. Furthermore, high speed oxidant flow generates high turbulent and mixing with flue-gas so that temperature profile in furnace becomes uniform. Both are the main mechanism of the reduced NO_x emission.

Meanwhile, the flameless combustion is sometimes called as HiTAC (High temperature air combustion) which means high temperature of oxidant is necessary for flameless combustion. Recently, Choi had performed experimental and numerical study to introduce a new flameless combustion called RAI (reversed air injection) flameless combustion which can be carried out with ambient temperature air and fuel [4]. Sorrentino et al., introduced a flameless furnace with circulating flow which has the temperature variation of 40 °C in the furnace [5, 6]. However, the oxidant gas (air) had to be heated to more than 1000 K to perform the flameless condition.

From the basic idea of Choi and Sorrentino, we generated a new flameless combustion called swirling-RAI (reversed air injection) and evaluated through experiments and numerical simulation herein. In a circular furnace which has a specific purpose to heat pellet bed, two pair of nozzle sets from the concept of RAI are installed in the inversed direction against each other. Due to the high momentum of angled air nozzles installed axisymmetric, the overall flow in the furnace is swirling. The experiments showed combustion without flame and NO_x emission was maintained at 3 ppm (@3%O_2) while CO was not detected at the exit. The numerical simulation showed the detail of flow and reaction in the S-RAI furnace to understand its phenomena. The experimental and simulation results are also compared to see the reliability of the simulation.

2 Concept of S-RAI Furnace

Figure 1 shows the conceptual drawing of the heating furnace by S-RAI, which has the top area of S-RAI combustion and the bottom area of the substance bed of pellet particles to be heated with flue-gas. The reducer zone connects the combustion zone and the heating zone.

The flow mechanism of S-RAI is shown in Fig. 2. The set of air and fuel nozzle confront each other which generates reverse injection against each other. Meanwhile, the momentum of air is much higher than the fuel, so the overall flow is governed by the air jet. The high speed of air jet makes the jet region to be lower pressure to induce the flue-gas nearby, so that the entrained flue-gas heats up the air jet stream and also lower the oxygen concentration which is the necessary condition for a flameless combustion. The two set of air and fuel nozzle were installed in an axisymmetric way

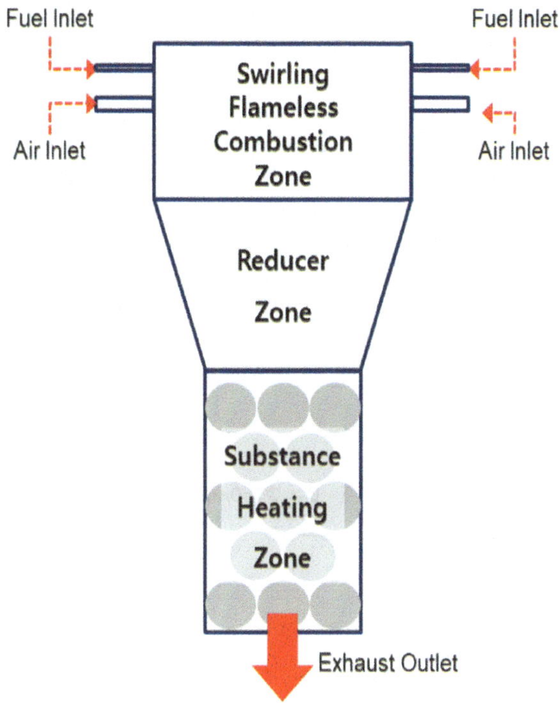

Fig. 1 Concept of swirling flameless combustion for heating substance

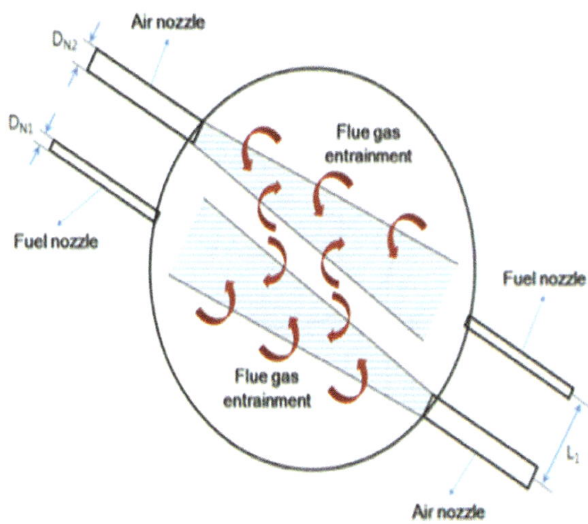

Fig. 2 Conceptual diagram of swirling flameless combustion

which has opponent nozzle direction of air and fuel, respectively. This configuration makes the overall flow to have swirl due to the unequal momentum between air and fuel.

Figure 3 shows the isometric view of the detailed internal geometry of the furnace for the case of ALFH (AL: air low position, FH: fuel high position). The top cylinder is for S-RAI combustion which includes axisymmetric two air nozzles at lower position and two fuel nozzles at higher position, respectively. The overall flow is counter clockwise in the combustion zone due to the axisymmetric angled air nozzles. At the reducer zone, the flow is expected to be uniform, and finally at the substance zone the flue-gas heats the bed uniformly and exhausts through the bottom. The other case of AHFL (AH: air high position, FL: fuel low position) is the same but the position exchange between the air nozzle and the fuel nozzle.

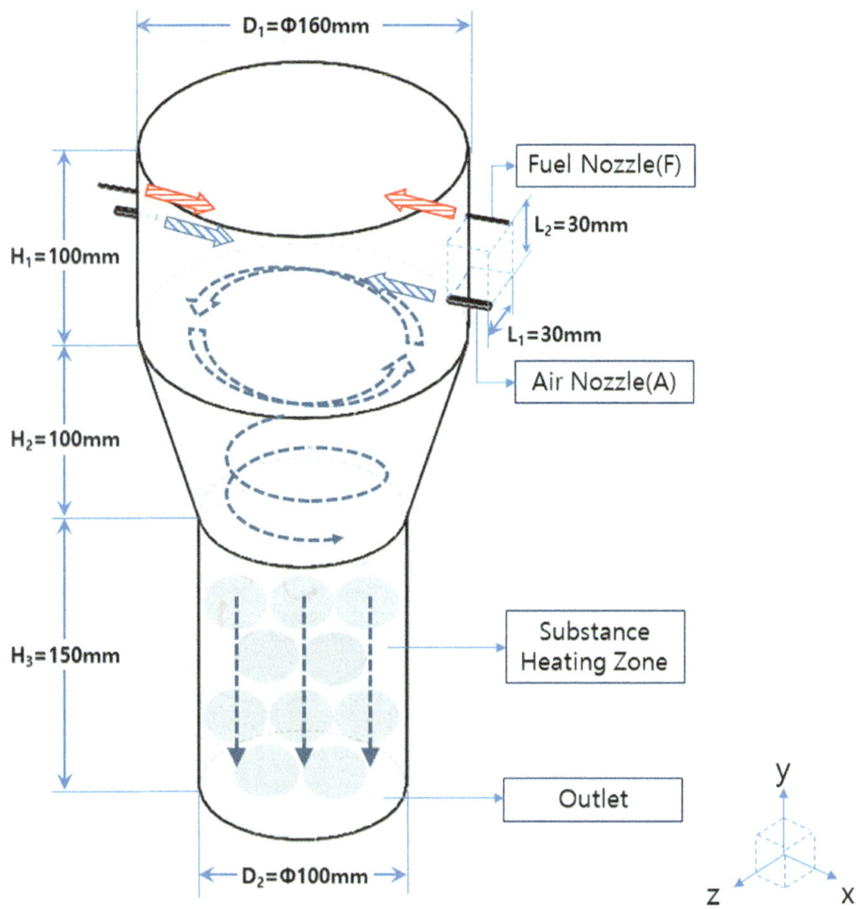

Fig. 3 Geometry of internal gas flow region

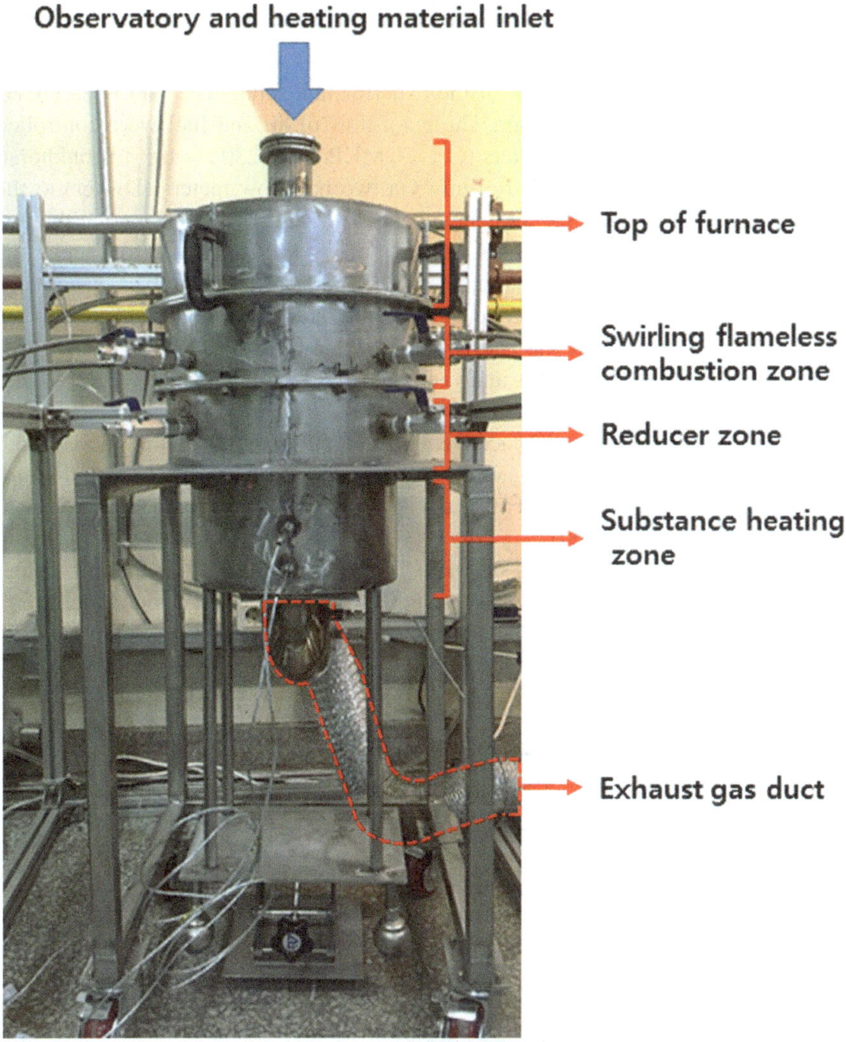

Fig. 4 Photo of swirling flameless combustion experiment apparatus

Figure 4 shows the photo of the experimental furnace, which is enclosed with ceramic wool board of 100 mm depth which has the average conductivity of 0.18 W/m K. The inlet of the pellet and the observatory port is installed at the top of the furnace. The air nozzle diameter is 1.3 mm to achieve the velocity higher than 100 m/s, and the fuel nozzle diameter is 1 mm which is the limit of the manufacturing. The distance between the air and fuel nozzle nearby is in diagonal of 30 mm × 30 mm.

3 Experimental Condition and Measurement

Figure 5 shows the flow diagram of the experimental setup. The fuel is pure CH_4 and the oxidant is compressed air. The main flow of air and fuel were controlled precisely with mass flow controllers (CH_4: < MKP: TSC-230 > , air: < Bronkhorst: F-202AV-M20 >) and distributed evenly via two ball flow meters (Dwyer) to the twin nozzles. The temperature and chemical distribution were measured at the two port installed at the exit of the S-RAI zone (port1, y = 320 mm, 30 mm under the last nozzle) and the middle of the reducer zone (port2, y = 290 mm), respectively. Temperature and gas sampling probe can travel in the radial direction at the ports to collect the gas to send to the gas analyzer (MRU: SWG200-1). The exhaust gas was measured with a combustion gas analyzer (TESTO-330-LL) to check the NO_x and CO emission with O_2 for operation control.

The measurements in the furnace were carried out at 9 points at the port1 and 7 points at the port2 for 10 min to get the steady value at each point. K-type thermocouple was used to measure the temperatures at 5 points (TC1 ~ 5) continuously with a DAQ (Yokogawa MV2000).

The experimental condition is shown in Table 1. The fuel, CH_4 is fed at 4.11 lpm to make the volumetric heat release to be 660 kW/m^3. In general, good combustion can be achieved at excess air ratio to be 10–20% and also the NO_x emission is usually high due to high adiabatic flame temperature. Hence the excess air ratio is fixed to

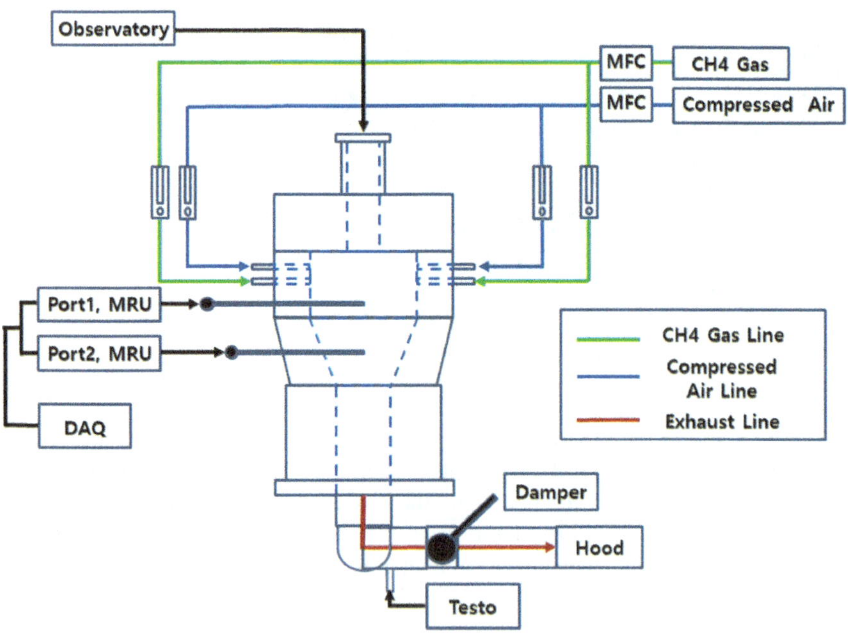

Fig. 5 Schematic diagram of swirling flameless combustion apparatus

Table 1 Experimental conditions according to nozzle position

Position of nozzle	Q/V	Air flow rate	Air velocity	Fuel flow rate	Fuel velocity	λ	Swirl number
	(kW/m³)	(lpm)	(m/s)	(lpm)	(m/s)	(-)	(-)
AHFL	606.6	45.0	132	4.11	44	1.15	55.32
ALFH							

be 15% so the air flow rate is 48.1 lpm. There are two nozzle configurations; the one is AHFL, and the other is ALFH. AHFL means the air nozzle position is high, and the fuel nozzle position is low. ALFH means the air nozzle position is low, and the fuel nozzle position is high.

The operation of the furnace was started with a conventional swirling diffusion burner positioned at the top observation port to heat the furnace to 800 °C. After the center temperature reached 800 °C, the furnace shifted to the S-RAI mode and the burner was shut down. After one hour of the S-RAI mode operation, steady condition could be achieved, then the measurement started at the ports.

4 Numerical Simulation Model and Condition

The grid for numerical simulation is shown in Fig. 6. The square nozzle is applied to simplify the grid generation and to improve the conversion speed. The wall side was concentrated with more grids because it is expected to have high profile change of numerical variables related with momentum and reaction. The total node number

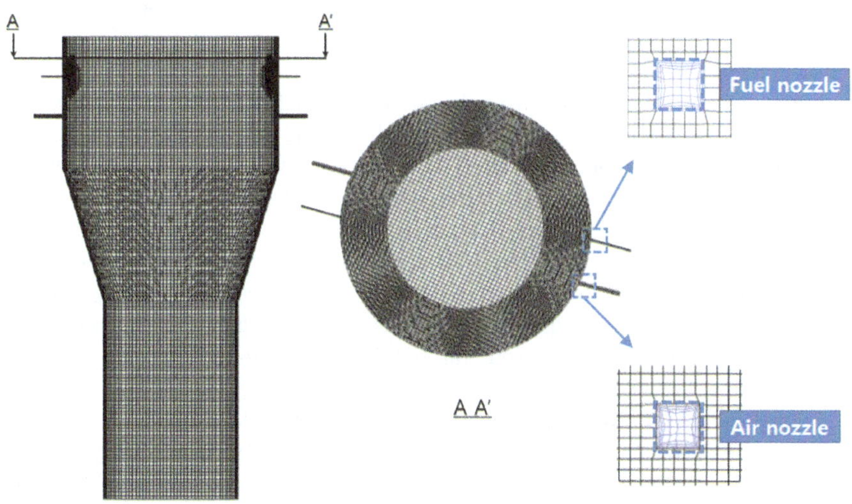

Fig. 6 Grid of swirling flameless combustion apparatus

is around 450,000 and the grid generation method is "Cutcell" which has benefit of economic grid generation as well as controlling grid density depending on the geometry. The substance zone is modeled as porous bed which has the particle of 40 mm diameter and the void fraction of 0.4. Standard k-e turbulence model, discrete ordinate radiation model, and Smooke's skeletal reaction model using 19 species and 46 steps were applied in FLUENT 17.0 [7].

5 Results and Discussion

Because the full measurement of variables such as temperature and chemical species in the furnace has technical limitation, the basic phenomena in the furnace is explained with CFD results herein. The Sect. 5.3 is the comparison between the simulation and the measurement at the port 1 and port 2.

5.1 Flow Characteristics

Figure 7 shows the photos of flame in the furnace according to the combustion mode changes. Figure 7a shows the moment after ignition to show the diffusion flame. Figure 7b shows the swirling flame while heating the furnace. Figure 7c shows the S-RAI mode at low temperature under 800 °C. Due to low temperature, it shows a faint flame. Figure 7d shows the S-RAI mode at high temperature above 800 °C which shows no flame at all and flameless combustion was accomplished. After the flameless condition accomplished and reached to steady state, the CO emission was 0 ppm and the NO_x emission was 3 ppm for all the cases, and the combustion zone temperatures were 1023 °C for the AHFL case and 1022 °C for the ALFH case. The flame noise also disappeared.

Figure 8 shows the path line from the air nozzle of CFD results. Figure 8a shows the results of the AHFL case and Fig. 8b of the ALFH case. The air stream of the AHFL is going down along the wall to the lower part of the furnace after it reaches the opponent wall, and then rise up through the middle with swirling to make an overall gas recirculation. The air stream of the ALFH case returns after it reaches the opponent wall, and goes down to the reducer part, and then rise up through the middle with swirling to make an overall gas circulation. Both cases show slightly different flow path of the air stream at the combustion section, however, after they come down along the wall, some portion rises up through the central furnace to develop the gas recirculation at the reducer zone.

Figure 9 shows the upward velocity profiles, and Fig. 10 shows the downward velocity profiles, which shows the circulation trend of each case. The gas recirculation at the middle is up to the air nozzle level, so the AHFL case shows the upward flow at the middle reaches higher level than the ALFH case. After the air streams of each

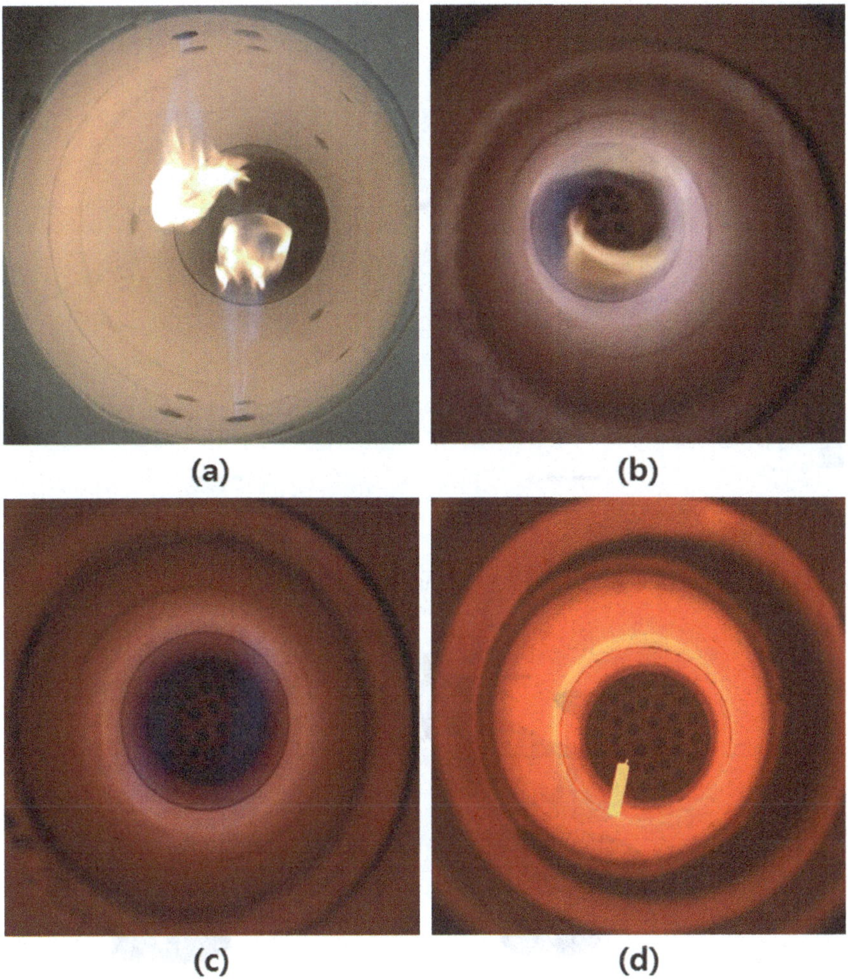

Fig. 7 A photograph of change of internal flame shape according to experimental time; **a** flame condition at before heat up, **b** swirling flame condition, **c** pre-flameless combustion condition, **d** flameless combustion condition

case reach the opponent wall, the stream split to the upward and the downward so that the high peak at the both side appear.

To summarize, the recirculation in the S-RAI furnace is two types; one is the horizontal gas entraining due to low pressure of the high speed air jet, and the other is overall recirculation which has downward flow at the wall and upward flow at the central furnace due to the swirling flow at the reducer zone. These recirculation modes contribute to generate the flameless combustion.

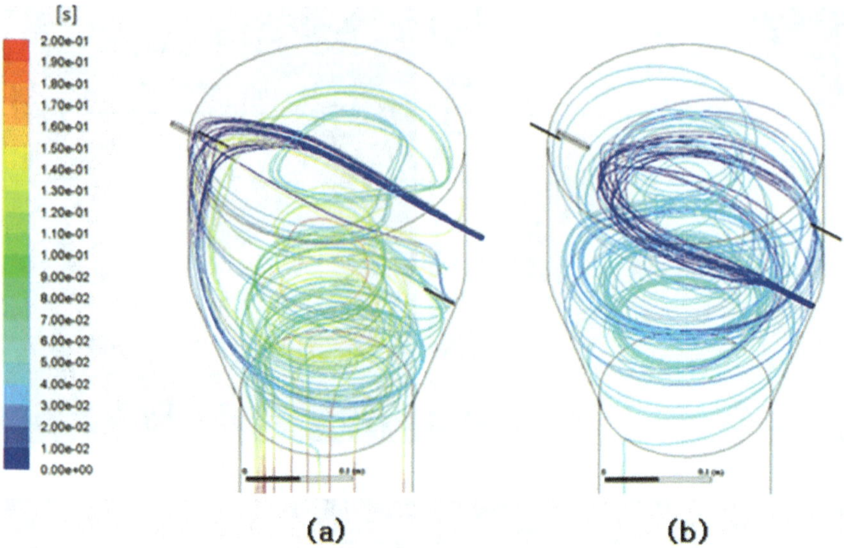

Fig. 8 Air flow path line of inside combustion furnace; **a** AHFL, **b** ALFH

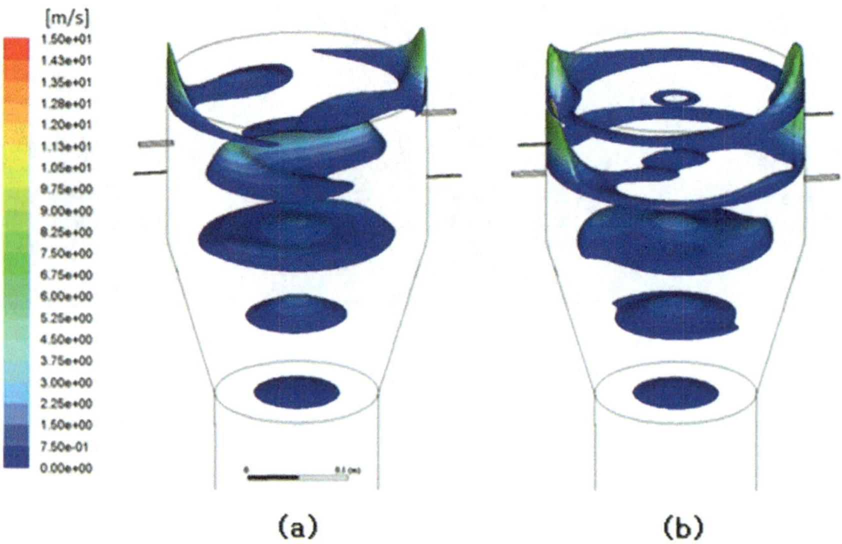

Fig. 9 Y-direction upward gas velocity distribution in horizontal cross sections depend on nozzle array; **a** AHFL, **b** ALFH

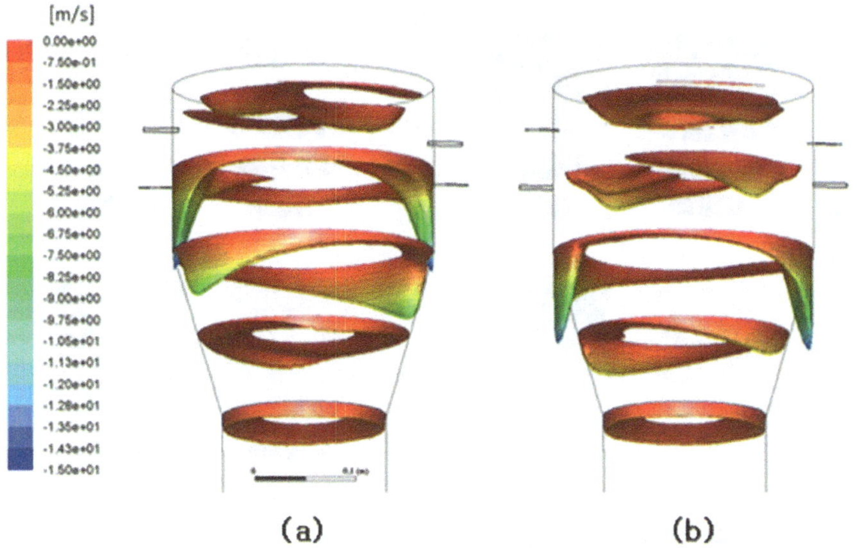

Fig. 10 Y-direction downward gas velocity distribution in horizontal cross sections depend on nozzle array; **a** AHFL, **b** ALFH.

5.2 Chemical Species and Temperature Distribution of CFD Results

Figure 11 shows the chemical species and temperature contours from CFD result at the different levels in the furnace. Based on the bottom of the substance zone, the lower nozzle level is at y = 290 mm, and the higher nozzle level is at y = 320 mm. Figure 11a show the CH_4 contours of the AHFL and ALFH case. The CH_4 does not react immediately after it enters the furnace, but its distribution generates a circulating long tail shape in the combustion zone which means volumetric slow degradation of CH_4 occurs. The O_2 contour in Fig. 11b shows the air penetration clearly at each level of air nozzle. Meanwhile the peak O_2 concentration at the end of the air jet is lower than 5% due to mixing with entraining nearby flue-gas. The top area of the AHFL has high O_2 concentration because the fuel is injected at the lower nozzle so that the fuel cannot reach enough to the top zone. Although the O_2 concentration varies significantly in the combustion zone, it becomes suddenly uniform in the reducer zone due to the overall flue-gas recirculation. The concentration of the intermediate fuel, H_2 (Fig. 11c) and CO (Fig. 11d), shows similar pattern for the AHFL and ALFH cases. The H_2 and CO of the AHFL case have high concentration at the level of fuel nozzle, and spread from the bottom of the combustion zone to the air nozzle level. Meanwhile, the H_2 and CO of the ALFH case have high concentration above the air nozzle level and still have high concentration at the middle of the reducer zone, which means the reaction is occurring in wider area than the AHFL case. The OH radical shown in Fig. 11e represents active reaction zone. The AHFL case has high OH

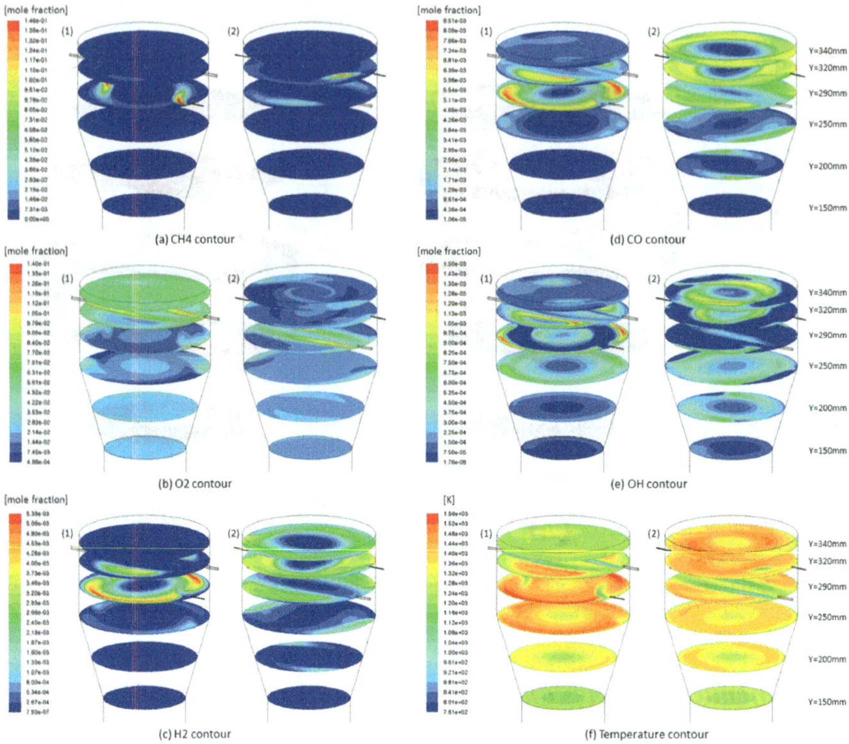

Fig. 11 Results of chemical species distribution and temperature distribution in horizontal section along height; 1 AHFL, 2 ALFH

concentration at the bottom of the combustion zone, while the ALFH case still has high OH concentration at the reducer zone. it shows again the reactions of the ALFH case is occurring in wider area than that of the AHFL case. The temperature shown in Fig. 11f is the results of flow and reactions. The AHFL case has high temperature zone in the range from the middle to the bottom of the combustion zone while the ALFH case has wider temperature zone which is from the top of the furnace to the middle of the reducer zone. Because the ALFH case has still reacting at the reducer zone, the temperature distribution at the bottom of the reducer zone of the ALFH case is less uniform than the AHFL case.

5.3 Comparison of Experimental Results with CFD Results

The measurement of temperature and chemical species was carried out at the port 1 (y = 280 mm) and port 2 (y = 200 mm) via inserting sampling probe into the furnace in depth, of which measurement error can be expected due to the radiation interference on the thermocouple and also reaction interference by the sampling probe. In spite of the uncertainty of the measurement technique, the trend of the measured values can show the characteristics of S-RAI combustion indirectly.

Figure 12a shows the temperature results of the CFD and measurement at the sampling line of the port 1. The center temperatures is low due to the overall flue gas recirculation and the wall side temperature is also low due to the diluted air is going down. The temperature between the centerline and the wall has symmetric dual high peaks for each case where active reaction is occurring according to OH radical distribution of Fig. 11e. The temperature distribution at the port 2 which is middle level of the reducer zone is shown in Fig. 12b. The minimum temperature appears at the center, and as the radial distance increases the temperature increases. This is also due to the secondary recirculation so that the cold exhaust gas is rising up through the center of the reducer zone. Although the CFD results has bigger variance than the measurements, their trends are the same to show that there is the flue gas recirculation. Meanwhile, from the Eq. (1) which calculates the estimated temperature after removing radiation effect, the actual temperature could be maximum 150 °C higher than the indicated by thermocouple. Hence, the simulation result is within reasonable range comparing to the temperature measurement.

$$Q_{rad} = Q_{cov} = \epsilon\sigma\left(T_1^4 - T_2^4\right) = h(T_1 - T_3) \quad (1)$$

Equation (1) represent the condition of radiation heat transfer is equal to the convection heat transfer at the bead of thermocouple. h is from laminar convection heat transfer equation (Nu = 64/Re) and, T1, T2 and T3 are the bead temperature, average wall temperature, and actual temperature of gas united in K, respectively. The average wall temperature is from the CFD result.

Figure 13 shows the O_2 and CO results of the CFD and measurement at the port 1. Due to the port 1 is located just under the lower nozzle, significant effect from the nozzle flow is expected. The AHFL case in Fig. 13a shows higher O_2 concentration as the radial distance increases which is due to the swirling flow of air so that the O_2

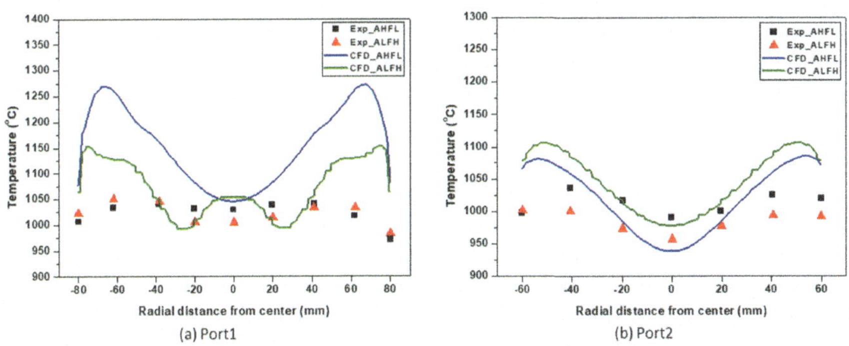

Fig. 12 Temperature distribution measured according to the radial depth of combustion furnace (x = 0, center)

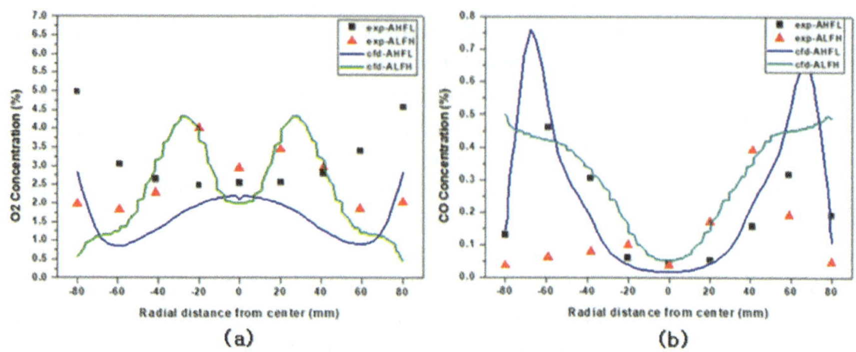

Fig. 13 O_2 and CO distribution measured according to the radial depth of combustion furnace at port 1 (x = 0 at the center); **a** O_2 distribution **b** CO distribution

is concentrated at the wall side. Meanwhile the ALFH case shows dual peaks at the radial distance of 20 mm and the O_2 concentration is low at the outside. Because the air nozzle is close to the port 1, the flow is disturbed significantly and the center and the outside is under the effect of flue gas recirculation and entraining flue gas to the air stream, respectively.

Figure 13b shows the CO results. The both case shows zero CO concentration at the center which is expected from the secondary gas recirculation. Meanwhile CO concentration increases at the wall side which means still active reaction during the coming down. The AHFL case shows low CO at the wall end which is due to enough O_2 is supplied as shown in Fig. 13a. The region of radial distance from 20 to 60 mm has high concentration of CO because it is a slow reacting zone.

6 Conclusion

A new axisymmetric flameless combustion system is proposed herein. Using the two pairs of simple air and fuel nozzles in an axisymmetric way and the air nozzle confronts the fuel nozzle, the swirling reversed air injection method is proved to have a flameless combustion characteristics using ambient temperature air and fuel without preheating. Two configurations of the air and fuel nozzle were studied to show the CO and NO_x emission of all the cases are 0 ppm and 3 ppm, respectively.

The flameless combustion condition is generated by two flow characteristics of S-RAI furnace; The first is entraining flue gas to the high speed air jet stream, and the second is overall recirculation through the center of the reducer zone.

Meanwhile, the air nozzle position appears as an important parameter because it splits into upward and downward flows after reaching the opponent wall. The ALFH case has wider reaction zone which is due to mixing between fuel and air is delayed longer than the AHFL case. The ALFH case has big recirculation at the upper space, while the AHFL case has rather plug flow pattern that air at the high level flows downward to meet the fuel rich flow. The temperature at the top area of the AHFL case is lower and the O_2 concentration is higher than the ALFH case. The OH radical distribution shows clearly that the ALFH case has wider reacting zone than the AHFL case.

Although there is numeric difference between the CFD and the measurement, they showed good agreement to express what is happening in the S-RAI furnace. Higher level of CFD simulation technology as well as non-contacting measurement technology will help to reduce the error of the CFD and the measurement. Meanwhile, the basic flow and mixing trend owing to the recirculation and nozzle jet flow is showed clearly with the conventional methods used herein.

Acknowledgements This work was partly supported by Institute for Information & communications Technology Promotion (IITP) (No. 10052730), the Basic Research Program through the National Research Foundation of Korea (NRF) (No. 2017M3A7B4042024).

References

1. Chung KY, Rhyu KS (2001) Technology trend of NO_x removal. J Korean Soc Marine Eng 25(1):57–64
2. Wendt JOL, Sternling C, Matovich MA (1974) Reduction of sulfur trioxide and nitrogen oxides by secondary fuel injection. Combust Symp 14:1085–1092
3. Wunning JA, Wunning JG (2000) Flameless oxidation to reduce thermal NO-formation. Progress Energy Combust Sci 23(1):81–94
4. Choi CG (2016) The combustion characteristics of the reversed air injection. Department of Mechanical Engineering Graduate School, Kookmin University, Seoul
5. Sorrentino G, Sabia P, Joannon M, Cavaliere A, Ragucci R (2015) Design and development of a lab-scale burner for MILD/flameless combustion. Chem Eng Trans 43:883–888
6. Sorrentino G, Goktolga U, de Joannon M, van Oijen J, Cavaliere A, de Goey P (2017) An experimental and numerical study of MILD combustion in a Cyclonic burner. Energy Procedia 120:649–656
7. Cho ES, Shin DH, Lu J, de Jong W, Roekaerts DJEM (2013) Configuration effects of natural gas fired multi-pair regenerative burners in a flameless oxidation furnace on efficiency and emissions. Appl Energy 107:25–32

Study on the Toluene Removal Characteristics of Glass Fiber Filter Coated by TiO_2 Photocatalyst

Fengyun Li, Jie Sun, and Donghoon Shin

Abstract This study analyzed the mechanism of photocatalytic reaction and described the toluene removal process through the equation of L-H reaction rate. Experiments were conducted by using the designed photocatalytic reactor to measure UV radiation, differential pressure in filter and the toluene removal efficiency as a function of air humidity, air velocity, toluene concentration of the reactor entrance and UV radiation. The computational analysis was performed to simulate the UV radiation on the filter surface, the flow field in the reactor and the toluene removal. As a result of the experiments, it was confirmed that the humidity had a large effect on the photocatalytic reaction efficiency. While the amount of the toluene removal increased, the toluene removal efficiency decreased in case of increasing the toluene concentration of the reactor entrance. The maximum and minimum toluene removal efficiency was at 28% and 3.1%, respectively. As a result of computational analysis, the uniformity of UV radiation on the filter surface was at 87%, and it was confirmed that the air flow in the reactor was uniform, and the L-H reaction model was considered as appropriate model for the photocatalytic reaction in this study.

Keywords Toluene · TiO_2 · Photocatalyst

1 Introduction

1.1 Research Background

Volatile organic compounds (VOCs) are one of indoor air pollutants, represented by toluene [1, 2]. Toluene (C_7H_8, molecular weight: 92.14) is a colorless, transparent, flammable liquid with a characteristic odor that dissolves approximately 535 mg/L in water [3–5]. Toluene can be released from indoor furniture and building materials, adhesives, paints, consumer products, smoking and burning processes, and exhaust

F. Li · J. Sun · D. Shin (✉)
School of Mechanical Engineering, Kookmin University, Seoul, Republic of Korea
e-mail: d.shin@kookmin.ac.kr

P. Rajendran et al. (eds.), *Proceedings of International Conference of Aerospace and Mechanical Engineering 2019*, Lecture Notes in Mechanical Engineering,
https://doi.org/10.1007/978-981-15-4756-0_21

gases from outdoor vehicles and factories can also release toluene through the ventilation system and into the room [5, 6]. Toluene is a neurotoxic agent. Short-term exposure to toluene can make serious effects on the human body, such as headache, dizziness, loss of consciousness, and even loss of coordination, cognitive dysfunction, hearing loss and loss of vision [5, 7–10]. In 2003, The report of Son et al. mentions that the toluene content in low-rise buildings is high, and the denser areas of the population have higher toluene pollution [11]. According to a study published by Kim et al. in 2005, the concentration of toluene in new buildings is greater than the safe value [3, 12]. According to a study by Anthwal, toluene is the highest concentration of BTEX in 2010 [13].

Recently, the method of treating VOCs through TiO_2 is becoming popular. Mo Summarized the reaction mechanism of VOCs removal with TiO_2 [1].

TiO_2 is used as a commonly used photocatalyst because of its stable chemical properties, fast reaction speed, harmless to human body, and low cost [14]. The main types of TiO_2 are rutile, anatase and brookite. According to the experiment of Muraswa [15] and Watanabe [16], the photolysis of anatase TiO_2 is stronger than that of rutile TiO_2. According to the results of Li et al., the mixed TiO_2 of rutile and anatase, anatase TiO_2 TiO_2 and rutile TiO_2 have successively reduced the COD removal effect [17]. Sun et al. found that in the phenol (C6H5OH) removal experiment, the photocatalytic activity of TiO_2 having a particle diameter of 10 nm was about 45% higher than that of TiO_2 having a particle diameter of 30 nm [18].

Visible light and UV light are the most common sources of light. In 1990, Dibble et al. Showed that the removal rate of TCE subjected to UV irradiation was higher than that without UV irradiation [19]. Lin et al. found that UV light is more efficient than visible light [20]. Cortes et al. found that the shorter the wavelength of UV light, the better the removal effect [21]. According to the study of Chen et al., Even though UV LED irradiation is weaker than UV lamp, the removal efficiency is higher [22]. However, due to the high price of UV LEDs, it is not commonly used. But with the development of UV LED production technology, UV LED will become a promising UV light source.

According to the TiO_2 fixing method, the reactor can be mainly classified into two types. The first type of photocatalytic reactor includes plate type and annular type. This reactor is TiO_2 coated on an internal constant surface, which has a simple structure, but has a low reaction efficiency due to a small UV irradiation area. Therefore, these reactors cannot be commercialized and are mainly used in laboratories to provide methods for studying kinetic models [23, 24]. Riffat et al. designed a mop fan reactor by coating TiO_2 on an optical fiber. The reactor has a large UV irradiation area, but the uneven thickness of the TiO_2 coating and the long fiber structure greatly affect the reaction, thus limiting the development of the fiber reactor [25]. The second photocatalytic reactor includes a packed bed type and a honeycomb type. Such reactors typically coat TiO_2 in an internal porous medium and therefore have a large reaction zone [23, 24]. Mehrvar et al. designed Packed-bed reactor with TiO_2-coated glass spheres. However, the surface activity of TiO_2 in this case is weak and the utilization efficiency of ultraviolet rays is low [26].

In 2003, Pareek et al. used the experiment and the first-order kinetic simulation to study the photodegradation performance of the reactor, and the similarity coefficient between the calculated results and the experimental results was 0.974 [27]. 2008, Mo et al. studied the effects of fin and Annular reactors on the photolysis of formaldehyde by theoretical analysis and computer simulation. The results show that the simulation results are consistent with the experimental results [28]. In 2013, Kumar used CFD to optimize the removal model of Rhodamine B, so that the photolysis efficiency of the theoretical model was 100%, the photolysis efficiency of the calculation and analysis was 94.3%, and the photolysis efficiency of the actual experiment was 89% [29].These examples show that computer simulations can yield structures that are similar to experiments. Therefore, it is feasible and useful to use software simulation to calculate the photolysis process.

In this study, a glass fiber filter coated with nano-TiO_2 and a UV LED were selected to meet the photocatalytic reaction conditions. And the experiment compared the difference in toluene removal effect according to humidity, toluene concentration, air flow rate and UV radiation intensity. In addition, the L-H reaction model will be used for simulation analysis and compared with the experimental results. The toluene removal characteristics of the designed reactor were studied by combining experimental and computer simulation results.

2 Experiment Apparatus and Method

2.1 Experimental Apparatus

Reactor design and filter coating. Figuers 1 and 2 is a photograph and schematic of the designed photocatalytic reactor. A perforated plate was installed near the reactor inlet to make even flow of air inside the reactor. There is a total of 260 holes with a diameter of 1 mm in the perforated plate, the diameter and thickness of the perforated plate are 50 mm and 5 mm, respectively, and the diameter of the hole is 1 mm. UV LED was installed at 200 mm from the porous plate. A glass fiber filter coated with TiO_2 was installed at 20 mm from the UV LED. The reactor was made of ABS material using a 3D printer.

The coating method of TiO_2 in the filter is as follows. (1) To remove the water resistance of the filter, the filter was washed with 0.08 mol./L sodium dodecyl sulfate (SDS) for 10 min and then dried at room temperature for 24 h. (2) In order to improve the adhesion of the coating, the filter was washed with 0.01 mol/L NaF for 5 min and then dried at room temperature for 24 h. (3) Then dipped with 2% TiO_2 suspension (+20% Polyethylene glycol 4000) for 10 min for TiO_2 coating and dried the filter at room temperature for 24 h. (4) The filter is baked at 50 °C for the last 30 min (Figs. 1 and 2).

Composition of Toluene Removal Experiment Apparatus. Figure 3 is a schematic of the experimental setup. The apparatus used for the experiment consisted

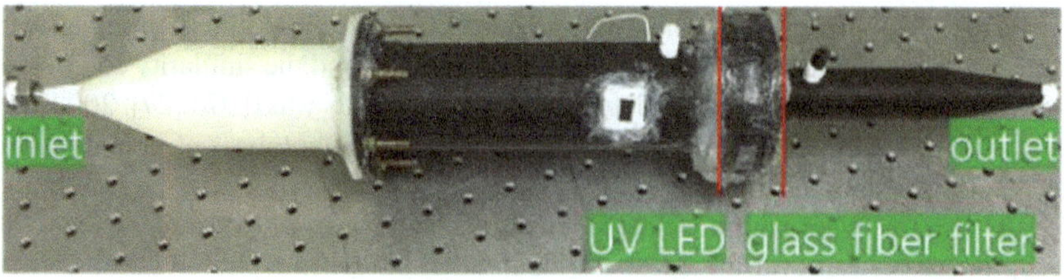

Fig. 1 Photograph of photocatalystic reactor

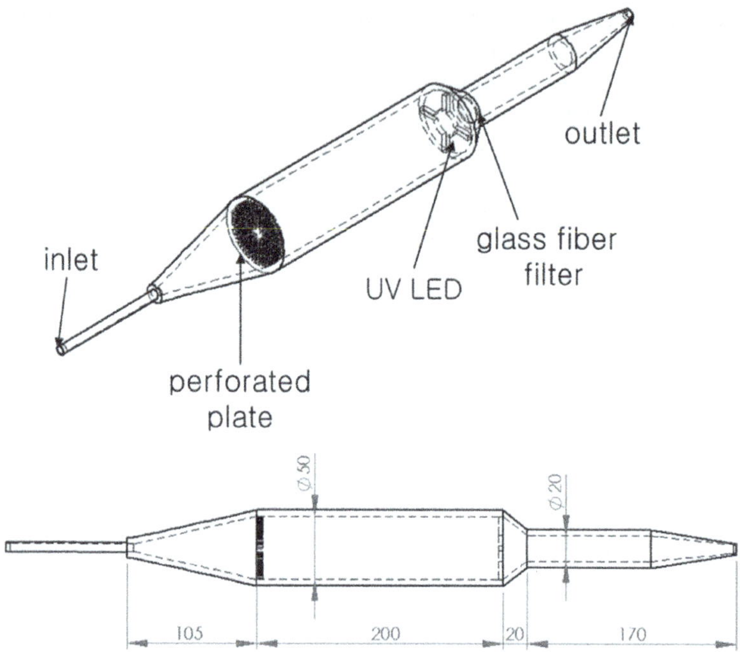

Fig. 2 Schematic diagram of photocatalystic reactor

of MFC, flow meter, bubbler, mixed glass bottle, reactor, power supply, manometer, temperature humidity meter and toluene meter. In the experiment, the compressed air dried by the air compressor was divided into three lines by adjusting the flow rate with MFC, and the upper line air was dilution air and flowed into the mixing glass bottle by adjusting the flow rate through the flow meter. The middle line air is a vapor carrier gas that flows through a flow meter to contain water vapor through a water bubbler and into the final mixed glass bottle. The bottom-line air was a toluene carrier gas that regulated the flow through a flow meter and contained toluene through a toluene bubbler and entered the final mixed glass bottle. In the mixed glass bottle, dilution air, air containing water vapor, and air containing toluene were mixed and

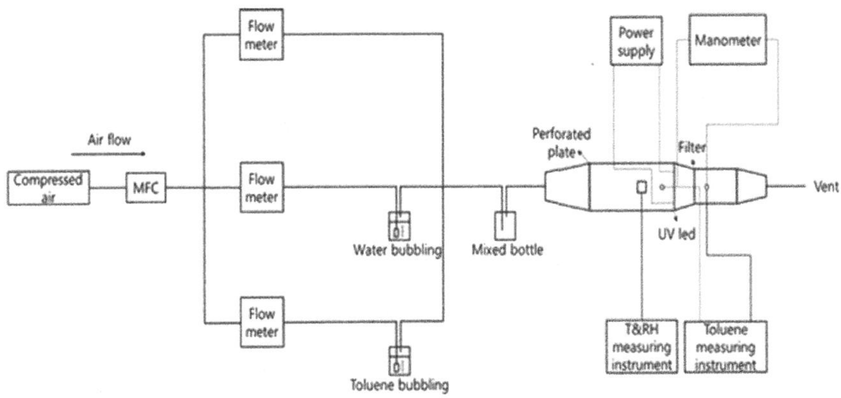

Fig. 3 Schematic diagram of experiment

introduced into the reactor. To supply the UV radiation needed for the photocatalytic reaction, a DC power supply was connected to the UV LEDs to provide the current.

Table 1 shows the device specifications used in the experiment.

2.2 *Experimental Conditions and Methods*

UV irradiation measurement experiment. The irradiation intensity of UV LEDs depends on the current supplied by the power supply and can also be affected by heat generated by itself. UV LED experiment was conducted to check the constant UV irradiation intensity according to air flow rate and UV LED current. The appearance of the UV LED tester is shown in Fig. 4. First, a constant flow rate of air was introduced into the reactor to regulate the UV LED current. Then, since the filter was installed at 2 cm distance from the UV LED, and the UV irradiation intensity was stabilized after 5 min, the UV irradiation intensity was measured at 0 cm distance and 2 cm distance from the UV LED. The illumination intensity at the 0 cm distance of the UV LED is as shown at the top of Table 2, and the illumination intensity at 2 cm is shown at the bottom of Table 2. As shown in Fig. 5, it was confirmed that the UV LED radiation was not significantly affected by the air flow rate. The radiation distribution of the UV LEDs obtained here can be used as data for toluene removal experiments and calculation models.

Toluene Removal Experiment. Figure 6 shows the state of the experimental apparatus, and the toluene removal efficiency was confirmed by adding air mixed with toluene to the reactor. First, only mixed air containing dilution air and water vapor was introduced into the reactor, and the air flow rate was stabilized, and then temperature and humidity were measured. Then, the toluene concentration value was recorded when the toluene concentration measured at the inlet was stabilized after introducing air containing toluene into the reactor. The toluene concentration value

Table 1 Instrument used in the experiment

No.	Type	Item	Specification	Quantity
1	System Devices	UV LED	Wavelength: 365 nm Current: 350 mA, Voltage: 3.2–3.5 V	1
2		TiO_2	Type: Anatase, Particle size: 20–30 nm	–
3		Glass fiber filter	Material: Borosilicate Glass Thickness: 675 m Pore size: 1 m	1
4		DC power supply	Constant Current: 0–3 A Current Resolution: 0.001 A Voltage: 0–30 V, Voltage Resolution: 0.1 V	1
5		Flow rate control	Bronkhorst F-202AV-M20, 0–200 LPM air Dwyer RMA-21-SSV, 1–10 LPM air Dwyer RMA-26-SSV, 0.5–5 LPM air Dwyer RMA-151-SSV, 5–50 ccm air	1
6	Measuring Instrument	Radiometer photometer	Wavelength Coverage: 250–675 nm Resolution: 1 $\mu m/cm^2$	1
7		Digital manometer	Range: 0–1.000″ w.c. (0.2491 kPa), with air velocity kit., Resolution: 0.001 kPa	1
8		Temperature and humidity Instrument	Temperature Ranage: −10 ~+60 °C, Resolution: 0.1 °C Humidity Ranage: 0 ~+100%RH, Resolution: 0.1%RH	1
9		Toluene measuring Instrument	Range: 0–205 mg/m^3, Resolution: 0.1 mg/m^3	1

Fig. 4 Picture of UV led experiment

Table 2 Experimental conditions and results of UV radiation measuring

Measuring location	Air flow rate (LPM)	Current (A)	Working time (min)	Radiation (w/m^2)
UV LED surface (0 cm)	1	0.04	5	43
		0.06	5	67
		0.08	5	105
		0.12	5	142
		0.18	5	176
	4	0.04	5	47
		0.06	5	67
		0.08	5	103
		0.12	5	138
		0.18	5	174
Filter surface (2 cm)	1	0.04	5	6
		0.06	5	10
		0.08	5	15
		0.12	5	20
		0.18	5	25
	4	0.04	5	6
		0.06	5	10
		0.08	5	15
		0.12	5	21
		0.18	5	26

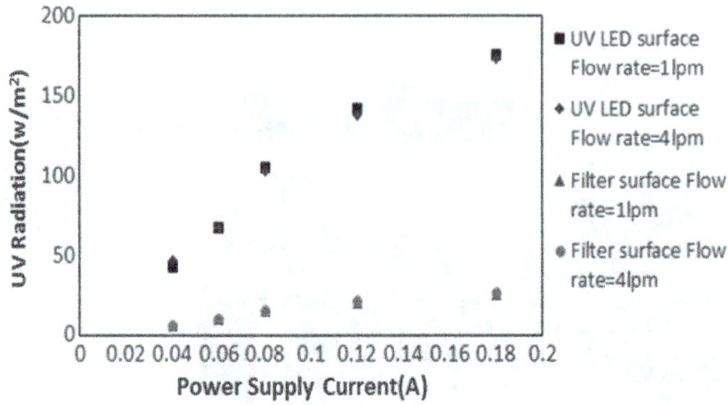

Fig. 5 UV radiation depending on power supply current

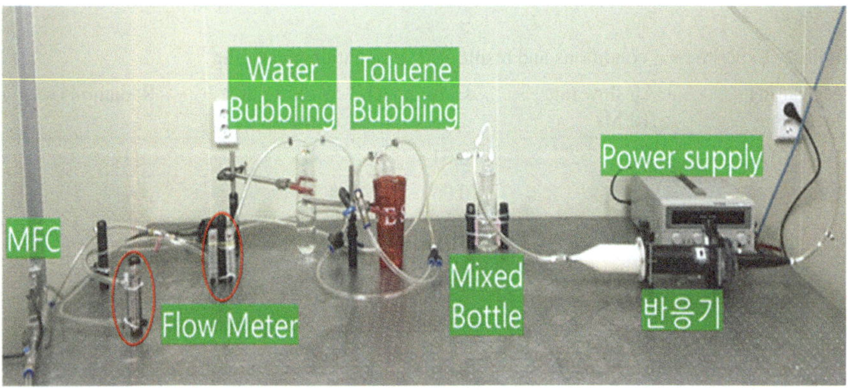

Fig. 6 Photograph of experiment system

was recorded when the toluene concentration measured at the outlet was stabilized after turning on the UV power supply. Finally, the power supply was turned off and the air supply was stopped. Purging was carried out with a large flow of dried compressed air for 15 min to remove toluene remaining inside the reactor.

Toluene removal experiments were carried out under the experimental conditions of Table 3. Toluene removal efficiency calculation method is as following Eq. (1). **ε** is the toluene removal efficiency, C_{in} is the reactor inlet toluene concentration, and C_{out} is the toluene concentration at the reactor outlet.

$$\varepsilon = \frac{C_{\text{in}} - C_{\text{out}}}{C_{\text{in}}} \tag{1}$$

Table 3 Variables and cases of toluene removal experiment

Case	Variables			
	Air	Air + water	Concentration of toluene	UV LED current
	Flow rate		(ppm)	(A)
	(LPM)			
1	3	0	19.4	0.04, 0.06, 0.08, 0.12, 0.18
	2	1		
	1.5	1.5		
	1	2		
	0	3		
2	0.5	0.5	15	0.04, 0.06, 0.08, 0.12, 0.18
	1	1		
	1.5	1.5		
	2	2		
3	1.5	1.5	16.5	0.04, 0.06, 0.08, 0.12, 0.18
			24.5	
			36.7	
			45.3	

3 Computational Analysis

In order to visualize the air flow in the reactor, the transmission of UV radiation and the removal of toluene, and to achieve a comparison of the experimental results, the photocatalytic reaction in the reactor was carried out by software ANSYS-Fluent 17.0 (the operating system is Windows 10 Enterprise Edition, the processor is AMD Ryzen 7 2700X Eight-core Processor, 32 GB RAM).

3.1 Mesh

The reactor was modeled using Solidworks 2015 and built using the Ansys Workbench grid (Fig. 7). It consists of approximately 500,000 grids, and the grid where the UV LEDs are located and where the photocatalytic reactions are most active is more densely meshed.

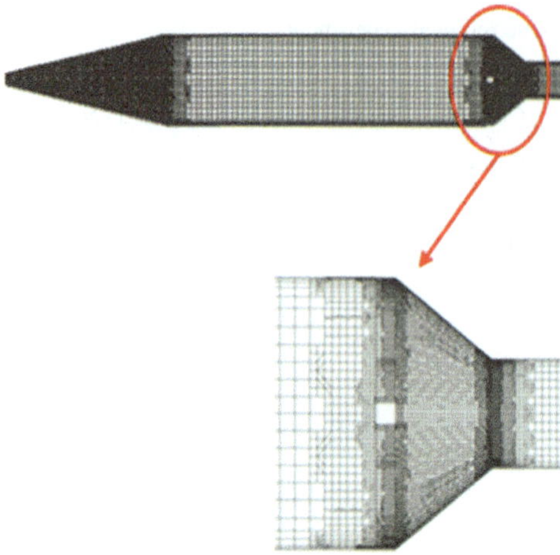

Fig. 7 Mesh of reactor

3.2 Analytical Models and Boundary Conditions

UV irradiation model. This study selected a DO model with a wide range of applications and high precision.

The governing equation for radiation in the emission, scattering, and absorption medium is shown in Eq. (2).

$$\vec{S} \cdot \nabla I = k - \beta I + \frac{\sigma s}{4\pi} \int 4\pi I(\vec{S}i, \vec{S}) d\Omega i \tag{2}$$

This equation represents the transfer of radiation (I) along the direction. k is the loss of photons due to absorption. βI is the loss of photons due to out-scattering. $\frac{\sigma s}{4\pi} \int 4\pi I(\vec{S}i, \vec{S}) d\Omega i$ denotes in-scattering of photons.

In clean air, there is no scattering and absorption, so the scattering and absorption coefficients are zero. The governing equation is shown in Eq. (3).

$$\vec{S} \cdot \nabla I = 0 \tag{3}$$

Due to the absorption of the filter, the amount of radiation inside the filter gradually decreases. Scattering assumes that the effect on Eq. (2) is small, and if omitted, the governing equation is as Eq. (4).

$$\vec{S} \cdot \nabla I = k \tag{4}$$

Absorption coefficient κ was calculated by Eq. (5) [30]. Dp and Φ were calculated by measuring the pressure difference between both filters.

$$k = -\frac{1}{Dp} In(\Phi) = -\frac{1}{0.003} In(0.052) = 895.91 \quad (5)$$

L-H Reaction Model. The reaction rate of a chemical reaction can be expressed as the rate of decrease of the mass of the reactant or the rate of increase of the mass of the product in unit time. By taking A → B C as an example, the reaction rate can be expressed by the following Eq. (6).

$$\mathrm{r} = \frac{d[A]}{dt} = \frac{d[B]}{dt} = \frac{d[C]}{dt} \quad (6)$$

According to the law of mass action proposed by Guldberg and Waage in 1867, If the temperature is constant, it is directly proportional to the reaction rate and the square of the number of each reactant concentration.

$$\mathrm{r} = \frac{d[A]}{dt} = k[A]^n \quad (7)$$

Hinshelwood first developed the Langmuir-Hinshelwood (L-H) reaction rate model by studying the reaction process between reactants and products through Langmuir Adsorption Isotherm. The L-H model is an ideal chemisorption model assuming that the surface of the adsorbent material is uniform, the energy values of each adsorption center are the same, and the adsorption efficiency and desorption efficiency are the same. Equation (8) is the equation of the L-H model. k is the response rate constant, K is the adsorption equilibrium constant, and Cs is the concentration of VOCs on the TiO_2 surface.

$$r = k\theta = \frac{kKCs}{1 + KCs} \quad (8)$$

The L-H reaction rate equation was chosen to simplify the study of gas phase and solid phase photocatalytic reactions. The L-H model is applied to remove indoor contaminants through photocatalytic reaction. Equation (9) is assumed because the amount of UV radiation received by TiO_2 and the number of adsorbed contaminants on the TiO_2 surface can affect the photocatalytic reaction rate. r is the reaction rate ($\mathrm{mg/m^3 \cdot s^{-1}}$) and C_3 is the reaction constant (mg/w) with respect to the UV radiation dose. I is the UV irradiation amount ($\mathrm{w/m^2}$). x is the distance (m) between two UV doses.

$$k = C_3 \frac{dI}{dx} \quad (9)$$

The simultaneous Eqs. (7), (8), and (9) yield (10). This equation is the L-H rate equation considering the UV dose. K is the adsorption equilibrium constant ($\mathrm{m^3/mg}$).

$$\mathrm{r} = \frac{d[A]}{dt} = \frac{C_3 dI}{dx} \cdot \frac{kCs}{1 + kCs} \quad (10)$$

Select the L-H model as the photocatalytic reaction model. Processing Eq. (10) yields Eq. (11). u is the speed (m/s), ΔC is the change in concentration of the reactants (mg/m^3), C_3 is the UV radiation dose and response coefficient (mg/w), ΔI is the amount of change in UV dose (w/m^2), K is the adsorption equilibrium constant (m^3/kg), Cs is the concentration of TiO_2 surface (mg/m^3).

$$u\Delta C = \frac{C_3 K \cdot Cs}{1 + K \cdot Cs} \cdot \Delta I \tag{11}$$

When the filter inlet irradiation amount I_0 is used to indicate the internal irradiation amount I of the filter, the Eq. (12) is obtained. x is the distance between the point in the illumination direction and the light source. Equation (11) is transformed to obtain Eq. (13).

$$I = I0e^{kapa \cdot x} \tag{12}$$

$$\frac{\Delta I}{u \cdot \Delta C} = \frac{1}{C_3 K \cdot Cs} + \frac{1}{C3} \tag{13}$$

If ΔC is the difference between the filter inlet and outlet concentrations and ΔI is the filter inlet UV radiation amount, then Cs is assumed to be the average of the filter inlet and toluene concentrations.

The data of $I_0 = 10\ w/m^2$ in case 3 is brought into Eqs. (12) and (13), resulting in a fitting equation as shown in Fig. 8. Therefore $\frac{1}{C_3 K} = 167.62$, $\frac{1}{C_3} = 3.6381$, so $C_3 = 0.275$, $K = 0.0217$.

According to Eq. (10), the semi-formal reaction rate of photocatalyst is Eq. (14). Δx is the distance between a point on the filter guide and a point on the filter surface.

$$r = 0.275(-896e^{-896\Delta x}) \cdot \frac{0.0217Cs}{1 + KCs} \tag{14}$$

r is made of UDF code and connected to Fluent.

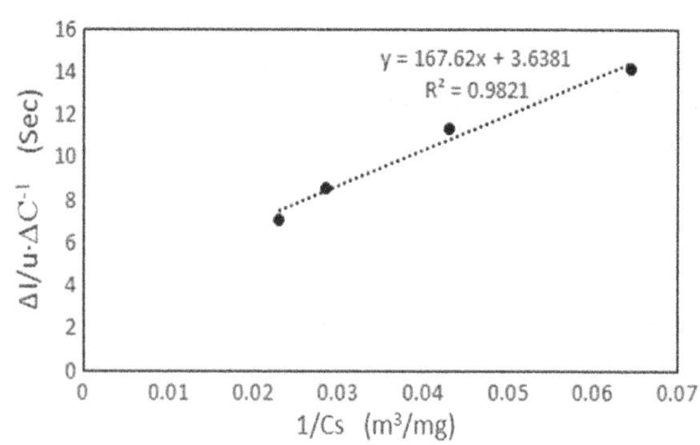

Fig. 8 Graph of 1/Cs-Δ I/u·Δ C-1

Table 4 Boundary conditions

Case	Inlet air velocity (m/s)	Outlet	Mass fraction of toluene (kg/m^3)	UV LED radiation (w/m^2)
4	1.58 (3LPM)	Pressure outlet (1 atm)	–	67
5	0.53 (1LPM)		–	–
	1.58 (3LPM)		–	–
6	1.58 (3LPM)		1.37E-05 (16.5 mg/m^3)	67

Boundary conditions. The boundary conditions for the computational analysis are summarized in Table 4. The inlet is set to a speed entry that allows for incompressible flow, the outlet pressure is set to atmospheric pressure, the radiation model of the UV LED uses a DO model that gives the initial amount of radiation and the wall is set to no slip.

4 Results and Discussion

4.1 Toluene Removal Experiment Results

Case 1 Experiment Results As shown in Fig. 9, the toluene removal efficiency was increased in the range of 13.3–64.7% of the relative air humidity, but the toluene removal efficiency was decreased in the range above 21 °C. The minimum removal efficiency of toluene is 3.13 when relative humidity is 13.3% and the UV irradiation amount is 6 w/m^2 (0.9 mg/m^3 removal), and the maximum removal efficiency of toluene is 64.7% when the relative humidity is 25 w/m^2. 19.6% (3.8 mg/m^3 removed).

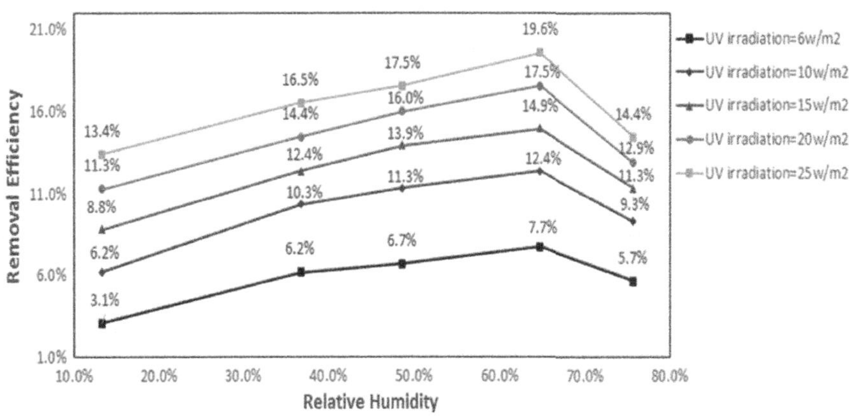

Fig. 9 Toluene removal efficiency depending on relative humidity

When the humidity is too low, the reflection efficiency increases with the humidity. When the humidity is too high, part of the water vapor will combine with TiO_2. The humidity control of the photocatalytic reaction is optimal at about 64.7%.

Case 2 Experiment Results. Figure 10 shows the effect of air flow rate on toluene removal efficiency. When the air flow rate is between 0.053 and 0.106 m/s, the toluene removal efficiency continues to increase to the highest, and then decreases. At the flow rate of 0.212 m/s, the toluene removal efficiency is the lowest. When the air flow rate is 0.106 m/s and the UV radiation is 25 w/m^2, the toluene removal efficiency is the highest, reaching 28% (removing two 4.2 mg/m^3).

Case 3 Experiment Results. Figure 11 shows the effect of toluene inlet concentration on toluene removal efficiency. When the toluene inlet concentration range is 16.5–45.3 mg/m^3, the toluene removal efficiency is lowered. When the toluene inlet concentration is 45.3 mg/m^3 and the UV radiation amount is 6 w/m^2, the toluene removal efficiency is the lowest, 4.9% (the removal amount is 2.2 mg/m^3). When

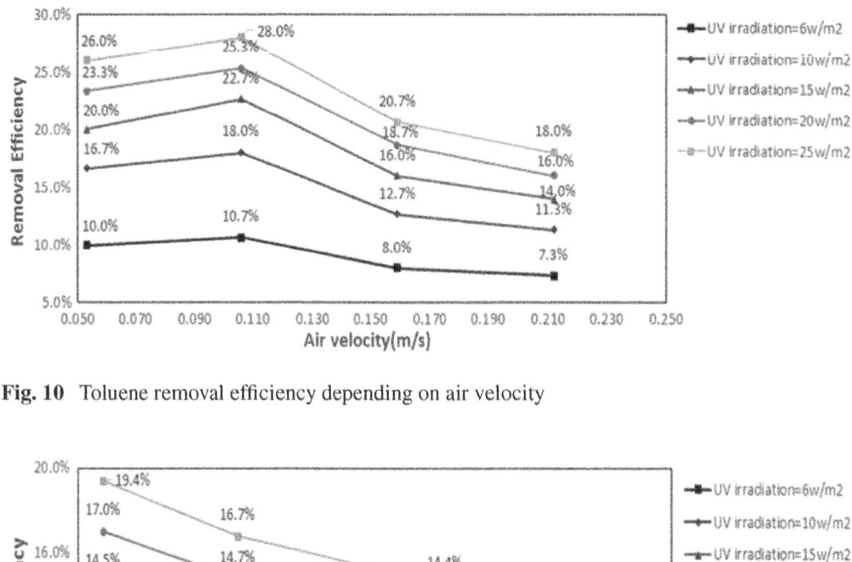

Fig. 10 Toluene removal efficiency depending on air velocity

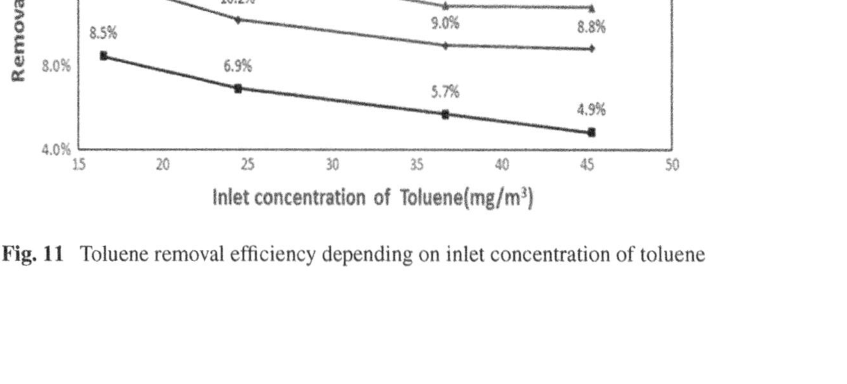

Fig. 11 Toluene removal efficiency depending on inlet concentration of toluene

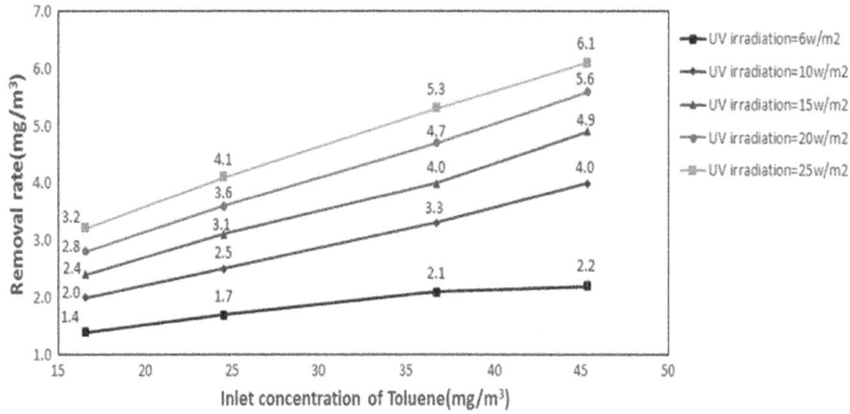

Fig. 12 Toluene removal rate depending on inlet of concentration of toluene

the inlet concentration is 16.5 mg/m^3, the UV irradiation amount is at 25 w/m^2, the toluene removal rate was 19.4% (the removal amount was 3.2 mg/m^3).

Figure 12 shows the effect of toluene inlet concentration on toluene removal. The removal of toluene from the toluene inlet concentration in the range of 16.5–45.3 mg/m^3 continued to increase. When the toluene inlet concentration is 16.5 mg/m^3 and the UV radiation amount is 6 w/m^2, the toluene removal amount is the smallest, 1.4 mg/m^3 (removal rate 8.5%), When the toluene inlet concentration was 45.3 mg/m^3 and the UV radiation amount was 25 w/m^2, the toluene removal amount was the largest, being 6.1 mg/m^3 (removal rate 13.5%).

According to the results of Case 1, Case 2 and Case 3, the toluene removal efficiency was increased in the UV irradiation range of 6–25 w/m^2, but the change was decreased as the UV irradiation amount increased.

4.2 Computational Analysis Results

After 8000 iterations, the calculation results basically converge, and the following simulation results are obtained.

Figure 13 shows the UV radiation distribution reaching the filter surface when the UV LED is calculated to reach 67 w/m^2. By calculation, the average value of the UV dose was 8.9 w/m^2, and the uniformity of UV irradiation was calculated to be 87% by the Eq. (15).

$$\text{UV radiation uniformity} = 1 - \frac{\text{UV radiation standard deviation}}{\text{UV radiation average}} \quad (15)$$

Figure 14a shows the velocity distribution in the reactor when the inlet air velocity is 0.53 m/s (flow rate = 1 LPM). The air flows through the perforated plate to obtain

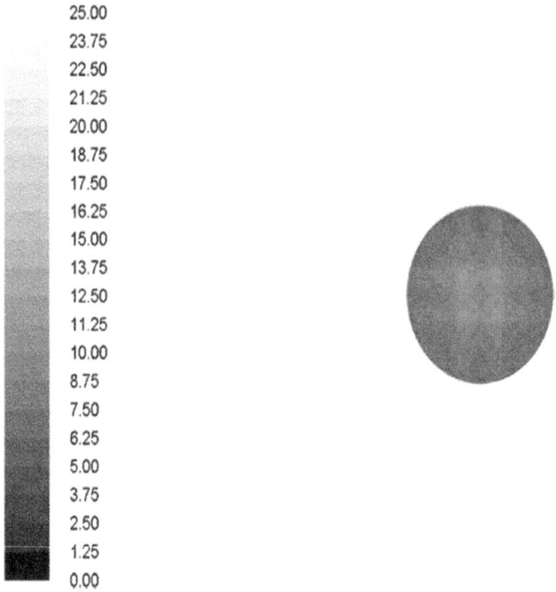

Fig. 13 Radiation on filter surface (w/m^2) UV LED radiation = 67 w/m^2

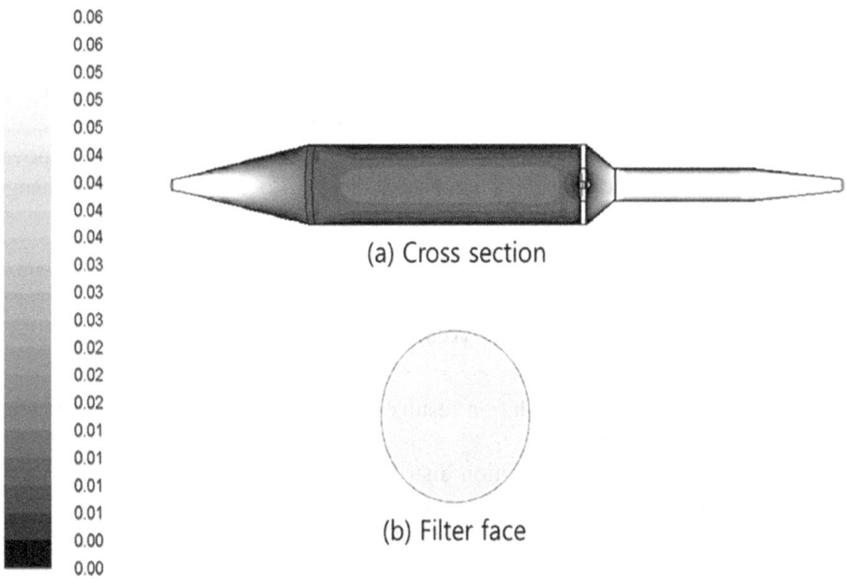

Fig. 14 Velocity distribution (m/s) (Inlet velocity = 0.53 m/s)

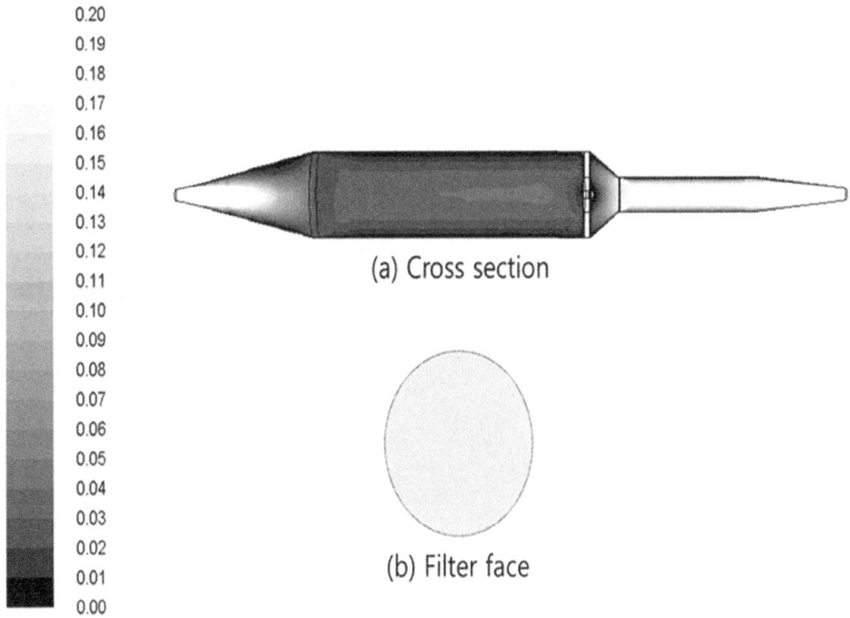

Fig. 15 Velocity distribution (m/s) (Inlet velocity = 1.58 m/s)

a uniform flow. Figure 15a shows the velocity distribution in the reactor when the inlet air velocity is 1.58 m/s (flow rate = 3 LPM). The air flows evenly after passing through the perforated plate. Comparing Figs. 14a and 15a, it can be concluded that the porous plate structure can achieve an approximately uniform flow even if the inlet air velocity is increased.

Figure 14b and 15b show the filter surface velocity distribution for an inlet air flow rate of 0.53 m/s and an inlet air flow rate of 1.58 m/s. The surface velocity distribution of the filter in Fig. 14b is uniform with an average velocity of 0.0504 m/s, which is 4.9% higher than the experimental data of 0.053 m/s. The surface velocity distribution of the filter in Fig. 15b is uniform with an average velocity of 0.154 m/s, compared with the experimental data of 0.159 m/s, the error is 3.1%.

When the inlet air flow rate is 0.053 m/s, the filter wind pressure is 59.89 Pa, which is 12% compared with the experimental data of 68.30 Pa. When the inlet air flow rate is 1.58 m/s, the pressure is 232.88 Pa, and the experimental data is 247.10 Pa, the error is 5.8%. Figure 16 shows the total pressure distribution in the reactor at an inlet velocity of 0.053 m/s.

The calculation process is completed by the computer, omitting the influence of the external environment and the reactor itself on the flow rate. Compared with the experimental results, there is a certain error. However, since the error is small, the flow field analysis can be performed by calculation.

As in case 3, the UV radiation on the filter surface is set to 10 w/m^2, the toluene removal rate at the inlet concentration of toluene of 16.5, 24.5, 36.7 and 45.3 mg/m^3

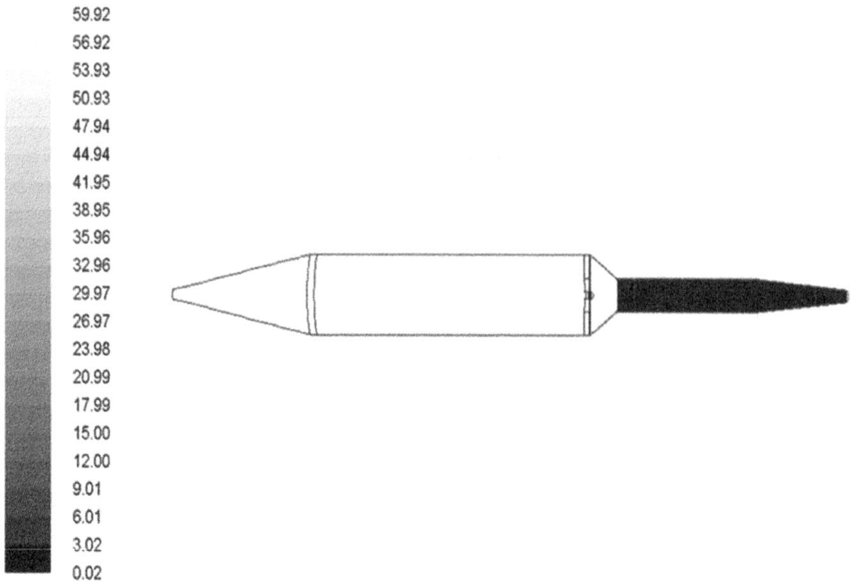

Fig. 16 Total pressure (Pa) (Inlet velocity = 0.53 m/s)

was calculated. Figure 17 shows that the distribution of toluene at a toluene inlet concentration of 16.5 mg/m^3 is uniform, and when toluene passes through the filter, toluene has a significant toluene concentration boundary in the reactor due to photocatalytic reaction.

Table 5 and Fig. 18 compare the experimental and calculated results. The experimental data and computer simulations of the toluene outlet mass fraction were compared with a minimum error of 0.17% and a maximum error of 0.62%. This is because computer simulation ignores the impact of the environment on the reaction. The toluene removal is reliable by computer simulation. In addition, the accuracy of the L-H reaction rate equation was verified.

5 Conclusion

In this study, the photocatalytic reaction removal characteristics of toluene were studied by experiments and computational analysis.

1. The experiment is divided into three parts: ultraviolet radiation measurement, filter paper partial pressure measurement and toluene removal.
   a. It is determined that the amount of UV radiation is not affected by the inlet air flow. The pressure distribution data inside the reactor was measured for computer simulation calculations.

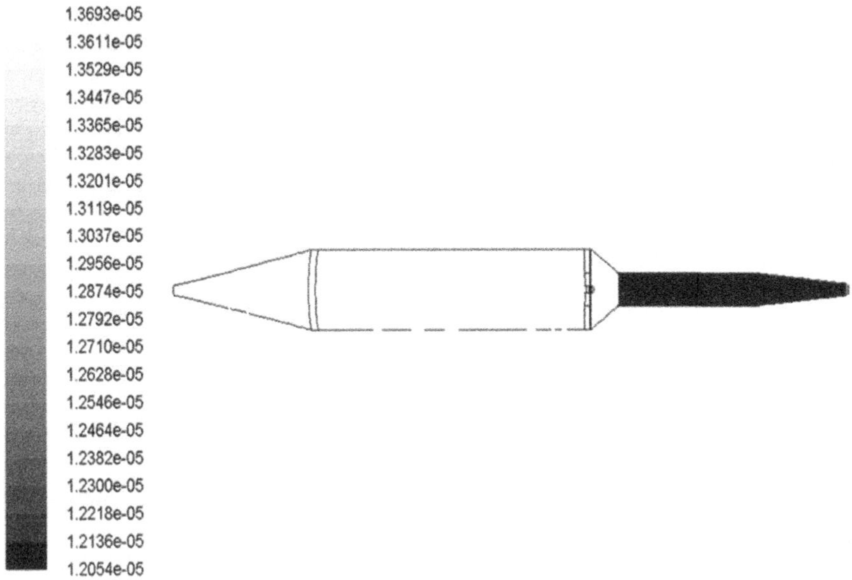

Fig. 17 Toluene mass fraction (Toluene concentration of the reactor entrance = 16.5 mg/m^3)

Table 5 Comparison of experiment and CFD results

Toluene concentration of the reactor entrance(mg/m^3)	Toluene mass fraction in experiment($\times 10^{-5}$)	Toluene mass fraction at CFD ($\times 10^{-5}$)	Error (%)
16.5	1.2033	1.2054	0.17
24.5	1.8257	1.8143	0.62
36.7	2.7718	2.7618	0.36
45.3	3.4274	3.4396	0.36

   b. Toluene removal experiments confirmed that humidity has a significant effect on the efficiency of photocatalytic reaction. As the inlet concentration of toluene increases, the removal of toluene increases, but the removal efficiency decreases. The minimum toluene removal efficiency in the experimental data was 3.1% and the maximum was 28%.

2. Use the experimental data to build a computer simulation model.

   a. Using the DO model, the uniformity of ultraviolet irradiation on the surface of the filter paper was 87%, and the error with the experimental result was 11%. It is inferred that the error is due to diffuse reflection inside the reactor during the experiment.
   b. Using the standard k-ε model and porous media model, the flow field inside the reactor was studied to confirm that the air flow inside the reactor was

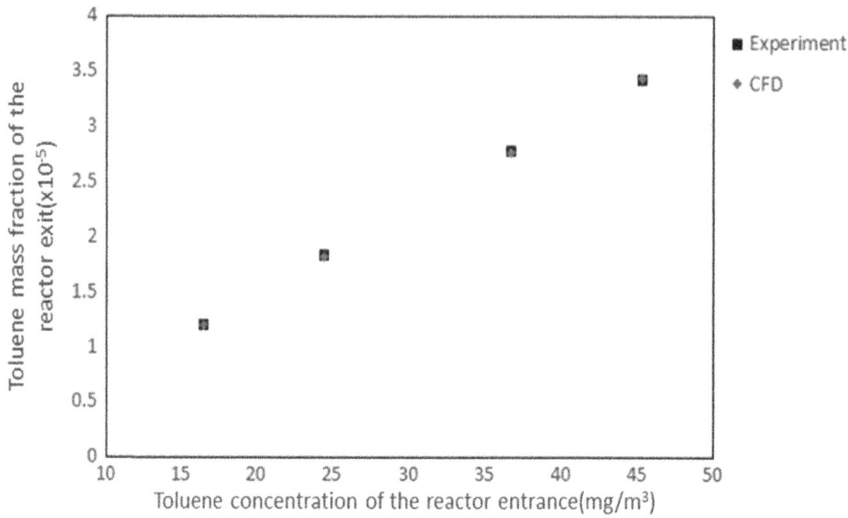

Fig. 18 Comparison of experiment and CFD results

uniform. According to the flow field conditions, computer simulations were performed, and the error between the simulation results and the experimental data was 5.8% and 12%. It is inferred that the error is caused by the influence of the ambient atmosphere and reactor materials on the experiment.

c. The computer simulation results and experimental data for toluene removal were approximately 0.16%. Therefore, the L-H response model is suitable for the photocatalytic reaction in this study.

In this study, only one toluene was studied, but various pollutants in the indoor environment coexisted and affected each other. Therefore, it is necessary to study the photocatalytic reaction on other pollutants and the study of the coexistence of various pollutants.

Acknowledgment This work was partly supported by Institute for Information & communications Technology Promotion (IITP) (No.10052730), the Basic Research Program through the National Research Foundation of Korea (NRF) (No.2017M3A7B4042024).

References

1. Mo J, Zhang Y, Xu Q et al (2009) Photocatalytic purification of volatile organic compounds in indoor air: a literature review. Atmos Environ 43(14):2229–2246
2. Sidheswaran MA, Destaillats H, Sullivan DP et al (2012) Energy efficient indoor VOC air cleaning with activated carbon fiber (ACF) filters. Building Environ 47(1):357–367
3. Organization WH (2000) Air quality guidelines for Europe. J Toxicol Environ Health Part A 71(91):V

4. Lee CY, Song MJ (2001) Removal characteristic of toluene by titanium dioxide photocatalyst. J. Korea Soc Environ Adm 7:367–374
5. Lee BJ, Heo J, Jung DY, Kim SH, Ryu HS, Shuai JF, Choi MJ, Im SG, Yang WH (2017) Correlation Relationship between personal exposure and biological monitoring for airborne toluene in an industrial complex and general environments. J Environ Health Sci 43(4):324–333
6. Elke K, Jermann E, Begerow J, Dunemann L (1998) Determination of benzene, toluene, ethylbenzene and xylenes in indoor air at environmental levels using diffusive samplers in combination with headspace solid-phase microextraction and high-resolution gas chromatography–flame ionization detection. J Chromatogra A 826(2):191–200
7. Hempel-Jørgensen A, Kjærgaard SK, Mølhave L (1998) Cytological changes and conjunctival hyperemia in relation to sensory eye irritation. Int Arch Occup Environ health 71(4):225–235
8. Kim YH, Yang WH, Son BS (2006) Risk assessment by toluene source emission model in indoor environments of new houses. J Environ Health Sci 32(5):398–403
9. Agency for Toxic Substances and Disease Registry, Public health statement for toluene. Toxic Substances Portal-Toluene (2015)
10. Health Canada, Residential Indoor air quality guidelines-Toluene (2011)
11. Son B, Breysse P, Yang W (2004) Volatile organic compounds concentrations in residential indoor and outdoor and its personal exposure in Korea. Environ Int 29(1):79–85
12. Kim H, Tabuchi S, Tanabe S (2005) Measurement and evaluation of indoor air quality by passive method in Korean house. In: Processings: Indoor Ai, pp 712–717
13. Anthwal A, Chan-Goo P, Kweon J, Min-Young K, Ki-Hyun K (2010) The temporal and spatial distribution of volatile organic compounds (VOCs) in the urban residential atmosphere of Seoul. Korea. Asian J Atmosp Environ 4(1):42–54
14. Oh SM (2003) TiO_2 photocatalust as a VOC remover. J Korea Soc Environ Adm 9(4):407–412
15. Muraswa S, Takaokam Y (1994) Several applications of TiO_2 photocatalyst for water and air purification. In: The first international conference on advanced oxidation technologies for water and air remediation
16. Watanabe T, Kitamura A, Kojima E, Nakayama C, Hashimoto K, Fujishima A (1992) Photocatalytic activity of Ti02 thin film under room light. In: The first international conference on TiO_2 photocatalytic purification and treatment of water and air
17. Zhi-Lin LI, Qiao SJ, Fan YW (2008) Study on treatment of dyeing wastewater with mixed phase nano-titania as photocatalyst. Chem Bioeng
18. Sun F, Wu M, Li W et al (1998) Relationship between optical surface properties and photocatalytic activity of titanium dioxide. Chinese J Catalysis
19. Dibble LA, Raupp GB (1990) Kinetics of the gas-solid heterogeneous photocatalytic oxidation of trichloroethylene by near UV illuminated titanium dioxide. Catalysis Lett 4(4–6):345–354
20. Lin L, Chai YC, Zhao B, Wei W, He DN, He BL, Tang QW (2013) Photocatalytic oxidation for degradation of VOCs. Open J Inor Chem 3:14–25
21. Cortés A, Alarcón-Herrera MT, Villicaña-Méndez M et al (2011) Impact of the kind of ultraviolet light on the photocatalytic degradation kinetics of the TiO_2/UV process. Environ Progress Sustain Energy 30(3):318–325
22. Chen DH, Ye XJ, Li KY (2010) Oxidation of PCE with a UV LED photocatalytic reactor. Chem Eng Technol 28(1):95–97
23. Mo J, Zhang Y, Xu Q et al (2009) Photocatalytic purification of volatile organic compounds in indoor air: A literature review. Atmosp Environ 43(14):2229–2246
24. Boyjoo Y, Sun H, Jian L et al (2017) A review on photocatalysis for air treatment: from catalyst development to reactor design. Chem. Eng. J S138589471630897X
25. Riffat SB, Zhao X (2007) Preliminary study of the performance and operating characteristics of a mop-fan air cleaning system for buildings. Building Environ 42(9):3241–3252
26. Mehrvar M, Anderson WA, Moo-Young M (2002) Preliminary analysis of a tellerette packed-bed photocatalytic reactor. Adv Environ Res 6(4):411–418
27. Pareek VK, Cox SJ, Brungs MP et al (2003) Computational fluid dynamic (CFD) simulation of a pilot-scale annular bubble column photocatalytic reactor. Chem Eng Sci 58(3):859–865

28. Mo J, Zhang Y, Rui Y et al (2008) Influence of fins on formaldehyde removal in annular photocatalytic reactors. Building Environ 43(3):238–245
29. Kumar J, Bansal A (2013) Photocatalytic degradation in annular reactor: modelization and optimization using computational fluid dynamics (CFD) and response surface methodology (RSM). J Environ Chem Eng 1(3):398–405
30. Shin DH (1998) A study on design optimization of stocker type municipal waste incinerator for low emission, Ph.D. dissertation, Department of Mechanical Engineering, Korea Advanced Institute of Science and Technology

Lift and Drag Trend of Exocoetus Volitans Model in the Wind Tunnel

A. F. Osrin, N. A. Razak, Aizat Abbas, and Zarina Itam

Abstract Flying fish is a fascinating animal that has the abilities to swim and glide. These unique abilities have attracted interest amongst scientist and researchers alike. Several studies have been performed to better understand flying fish aerial locomotion and aerodynamics performance. However, understanding biological beings' aerodynamics characteristics such as flying fish have always been more than educated estimation. Researchers have approximated the physics of flying fish based on known aerodynamics principle of other flying animals with similarly aerodynamics parameters. In the present study, we manufactured and tested a model of Exocoetus Volitans flying fish in the wind tunnel. The experiment evaluated the lift and drag coefficients generated by Exocoetus Volitans wings by subtracting the aerodynamics forces contribution from other fish parts. The experiment evaluated the performance at various angles of attacks starting from $-15°$ to $+45°$ and airspeeds ranging from 10 to 15 m/s. The results show that as the angle of attack increases, the lift and drag coefficients also increase even beyond 20°. The maximum lift coefficient is achieved when the angle of attack is at 40°.

Keywords Flying fish · Aerodynamics force coefficients · Wind tunnel experiments

A. F. Osrin · N. A. Razak (✉)
School of Aerospace Engineering, Engineering Campus, Universiti Sains Malaysia, 14300 Nibong Tebal, Penang, Malaysia
e-mail: norizham@usm.my

A. Abbas
School of Mechanical Engineering, Engineering Campus, Universiti Sains Malaysia, 14300 Nibong Tebal, Penang, Malaysia

Z. Itam
Civil Engineering Department, College of Engineering, Universiti Tenaga Nasional, 43000 Kajang, Selangor, Malaysia

P. Rajendran et al. (eds.), *Proceedings of International Conference of Aerospace and Mechanical Engineering 2019*, Lecture Notes in Mechanical Engineering,
https://doi.org/10.1007/978-981-15-4756-0_22

1 Introduction

Exocoetus Volitans or also known as tropical two-wing flying fish can be commonly found in the tropical and subtropical sea [1]. Despite the name, flying fish does not use the same mechanism of flying as bird does. Flying fish does not flap their wing, instead, flying fish utilise gliding mechanism to escape from its predators. This species of fish has highly modified pectoral fins and streamlined bodies which allow them to perform gliding for a considerable distance when they exit the water surface [2].

The gliding process of Exocoetus Volitans starts by building up momentum through swimming towards the surface at high speed. The fish then breaks the surface by angling its body upward and rapidly beating its own tail (up to 50 Hz), which result in a self-propelled leap out of water. As it breaks out through the water, the fish spreads its pectoral fins out to generate lift for gliding flight. According to [2], the leap may be reaching a height up to 4 ft, and the gliding distances may reach as far as 655 ft [3]. References [4] and [5] found that flying fish can achieve flight speed ranging between 10 and 20 m/s.

Exocoetus Volitans (Fig. 1) has the ability to perform multiple gliding flights. This is achieved by taxiing [6]. Taxiing is when the fish dips its lower caudal fin in the sea and beats its tail. This action propels the fish forward gaining forward speed which is then used to glide. The objective of this work is to investigate the aerodynamic lift and drag performance of Exocoetus Volitans at various angles of attacks and airspeeds using the wind tunnel.

2 Materials and Methods

The experimental work performed in this research was conducted at the Aerodynamics Laboratory II, School of Aerospace Engineering, Universiti Sains Malaysia.

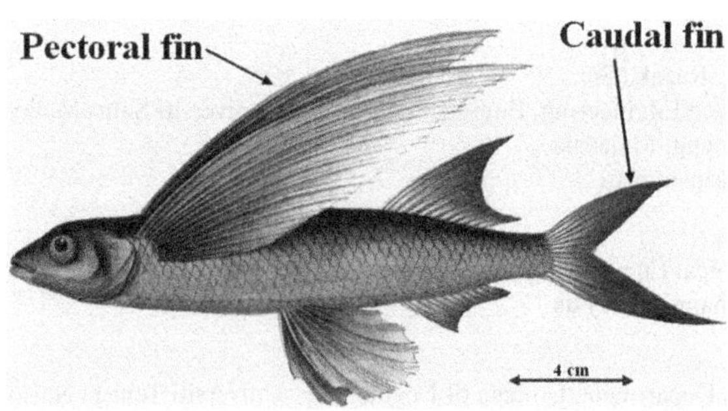

Fig. 1 Exocoetus Volitans [7]

The model was tested in the open-circuit wind tunnel capable of reaching speed up to 25 m/s. The test section measures 35 cm width and 40 cm height. The length of the test section is 40 cm. The wind tunnel is equipped with 3 axes external balance mounted on the side of the test section.

Two different flying fish models were manufactured with one model consist of the fish body, caudal and pectoral fins parts while the second model only consists of the fish body. A 3D drawing of the flying fish model was first designed using Solidworks 2015 software as shown in Fig. 2. The dimension of the model was based on the actual fish size and data [2]. The length of the body measures 20 cm with 3.4 cm width. The span for each wing is 13 cm with a maximum chord of 2.8 cm. Table 1 shows the parameters of the flying fish model tested.

Materials used to fabricate the flying fish model were balsa sheets for making the pectoral and caudal fin and high-density polystyrene foam for making the fish body. Aluminium rods with a diameter of 0.5 cm were used as the support sting. The body and fins parts were formed and sculpted into the desired shape before being sanded

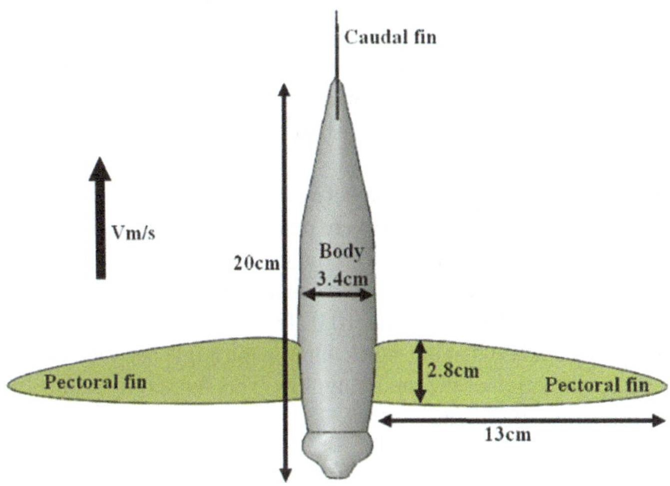

Fig. 2 The dimension of Exocoetus Volitans wind tunnel model

Table 1 Flying fish model parameters

Parameters	Value	Unit
Body length	0.2	m
Pectoral fin area	6.352×10^{-3}	m^2
Body area	5.134×10^{-3}	m^2
Wingspan	0.261	m
Aspect ratio	*10.72*	
Dihedral angle	*7*	degree
Wing thickness	3×10^{-4}	m

and polished to capture the shape of Exocoetus Volitans with high detail. The epoxy coating was applied on the body and fins to make all the surfaces smooth. Balsa sheets were cut according to the design to make the caudal fins. The fins were then stiffened by applying Cyanoacrylate which was allowed to be absorbed into the balsa wood. The mounting stand part was made using two straight aluminium rods beams welded together perpendicularly. The pectoral fins were attached to the fish body by inserting the pectoral fins at 7° dihedral angle on the body. Araldite was utilised to consolidate the joint between the body and pectoral fins. The model was fixed to the mounting stand at the centre of mass of the fish model.

The flying fish body prototype with its support stand was placed in the middle of the wind tunnel test section. The support stand was coupled to the multi-axes external balance. Before conducting the wind tunnel test, calibration was performed on the model to minimize the error and improves the accuracy and traceability of the measurement. After the calibration, the model was tested in the wind tunnel at the predetermined attack angles and airspeeds. A PCD30-A data acquisition system was used to capture the aerodynamics forces and stored in the computer via USB connection. The multi-axes external balance is an in-house built. It consists of four identical and parallel beams installed with strain gauges. One end of the parallel beams is anchored while the other end was attached to the support mount. The external balance can measure vertical and horizontal forces experienced by the model inside the test section. Figure 3 shows a flying fish model installed in the test section of the wind tunnel.

The tested angle of attack of the model was adjusted from −15° to 45° with an increment of 5° manually. The angles were determined using a digital inclinometer placed on the body of the fish model. The forces experienced by the model for each angle attack was recorded for 30 s at 1 kHz sampling frequency. This process was repeated at three different airspeeds of 10.0, 13.5 and 15.0 m/s before the angle is adjusted. It should be noted that the attack angle is the angle between the longitudinal body axis and the direction of the incoming flow. This work only reports the average value of the force measurements.

3 Results

Figure 4 shows the overall aerodynamic lift coefficient generated by the flying fish model plotted against the angle of attack at three airspeeds. The total lift and drag coefficients were obtained by dividing the lift and drag forces with dynamic pressure multiplied by the total area respectively. It is observed that as the angle of attack increases, the lift also generally increases. Significant lift increment is obtained when the angle of attack lies between 0° and 15°. Between 20° and 25°, the lift experienced a slight drop before levelling around 1. The lift trend between the three airspeeds tested is similar. Using numerical simulation, Deng et al. [8] observed a maximum lift coefficient of 1.03 at 35° for Cypselurus hiraii flying fish. Slightly higher lift is measured for the lowest airspeed case. The other two measurement cases show

Fig. 3 Exocoetus Volitans model in the test section of the wind tunnel

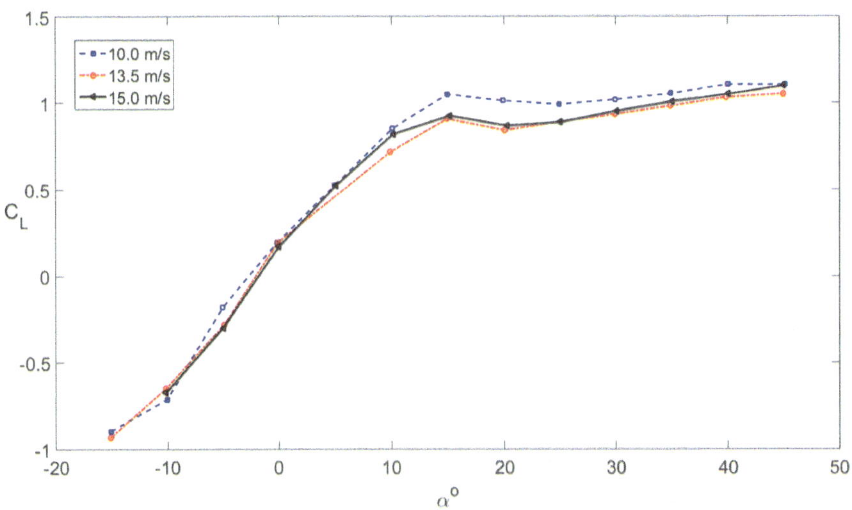

Fig. 4 Total lift coefficient measured on the flying fish model versus angle of attack at three different airspeeds

almost similar lift coefficient values. The lift measured did not follow typical airfoil lift trend as there was no abrupt drop in lift between 10° and 20° attack angles usually associated with stall. The reason is that the pectoral fin thickness is small and the flow regime lies in the lower end of the Reynolds number. Thin wing with the sharp leading edge of operating at low Reynolds number experiences trailing edge stall where the drop in lift is not so abrupt.

The total drag coefficient of the flying fish model is shown in Fig. 5. The drag coefficient decreases when the angle of attack is increased from −15° up to −5°. This is expected since the attack angle starts from negative value. As the angle is increased further, the trend shows the drag coefficients increased almost linearly up to 45°. The peak drag coefficient reached 0.7 when the fish was tested at the airspeed of 10 m/s at the same angle of attack. Higher airspeed tests generated slightly lower drag. The comparison of three plots revealed that at lower airspeed or lower Reynolds number result higher drag coefficient was obtained. This can be explained due to viscous effects that are relatively more significant at lower Reynolds number causing higher drag.

Figure 6 shows the lift coefficients generated by the pectoral fins plotted against the angle of attack. This is obtained by subtracting the total lift generated by the fish model with the force generated by the body only model. It shows that the magnitude of lift force generated by the pectoral fins is substantially lower. The maximum value obtained is 5.9 at 45°. Figure 4 revealed that most of the lift force is indeed generated by the pectoral fin. Furthermore, there is also a lift force generated by the fish body. This is only true when the angle of attack is higher than 10°. There is only a slight

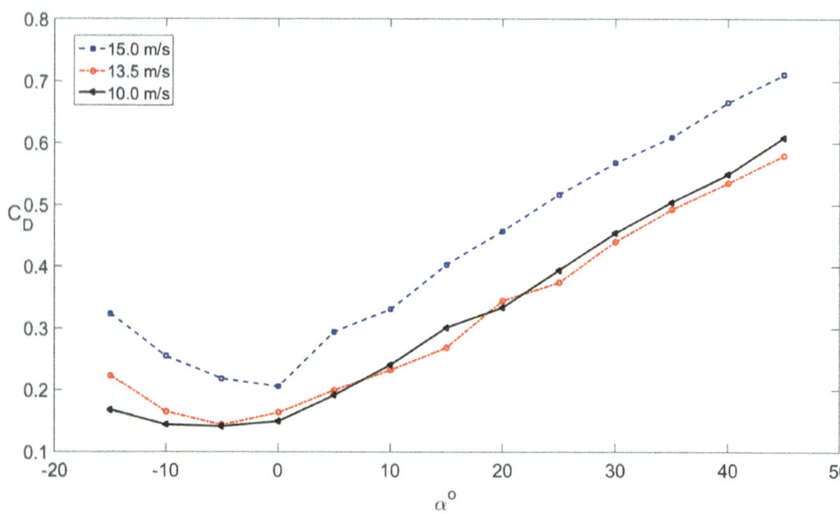

Fig. 5 Total drag coefficient measured on the flying fish model versus angle of attack at three different airspeeds

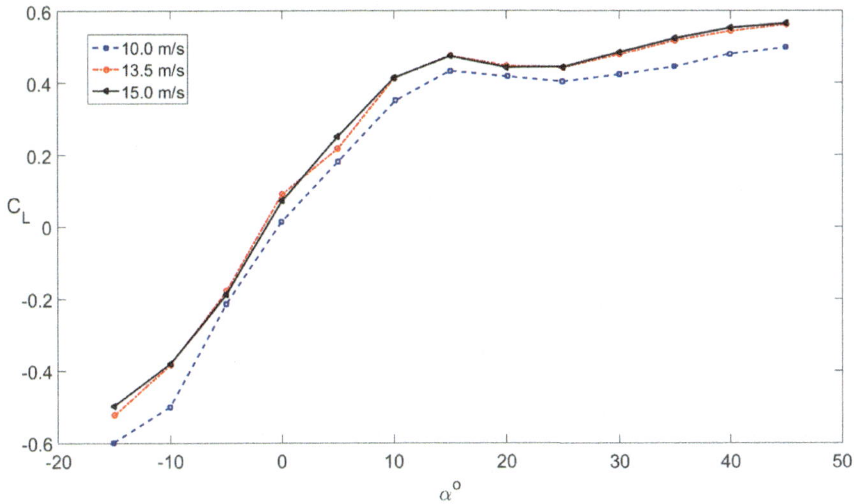

Fig. 6 Lift coefficient generated by the wings of the flying fish model versus angle of attack at three different airspeeds

distinction in the trend and magnitude of the lift force measured between the three airspeeds tested.

4 Conclusion

The work evaluated the lift and drag characteristics of Exocoetus Volitans flying fish in the wind tunnel. The parameters tested were the angle of attack and airspeeds. The lift and drag forces measured revealed some insight into the flying capability of flying fish. Since flying fish glide through the air, it is clear from the results that the fish should operate at low attack angles. This is to minimize the drag even though higher lift can be obtained at high angles of attack. It is important to note that unlike the fish model tested, real flying fish has the ability to modify and morph the pectoral and caudal fin in-flight. These changes almost certainly lead to variation in the aerodynamic forces generated since it depends on the shape, size and orientation of the fins.

Acknowledgements This work is supported by Bridging Grant (6316310) funded by Universiti Sains Malaysia and UNITEN grant J510050834. The authors would also like to acknowledge the assistance provided by H.T.S. Keey, L.H. Wea, M.H.F. Mahizam, M.A. Husny and N.H.N. Azmi for the preparation of the experiment setup.

References

1. Flying fish—Fish Grouping Homepage. https://www.britannica.com/animal/flying-fish. Last accessed 2019/6/23
2. Park H, Choi H (2010) Aerodynamic characteristics of flying fish in gliding flight. J Exp Biol 213:3269–3279
3. About flying fish gliding ability, National Geographic Homepage. https://www.nationalgeographic.com/animals/fish/group/flying-fish/. Last accessed 2019/05/27
4. Fish FE (1990) Wing design and scaling of flying fish with regard to flight performance. J Zool 221:391–403
5. Davenport J (1994) How and why do flying fish fly? Rev Fish Biol and Fisher 4(2):184–214
6. Hertel H (1996) Take-off and flight of the flying fish. In: Katz MS (ed) Structure-form-movement. Reinhold Publishing Company, New York, pp 218–224
7. Alasdair JE, Christopher WG (1987) The fishes of Saint Helena Island, South Atlantic Ocean. II. The pelagic fishes. J Nat Hist 21(6):1367–1394
8. Deng J, Zhang L, Liu Z, Mao X (2019) Numerical prediction of aerodynamics performance for a flying fish during gliding flight. Bioinspir Biomimet 14(4):046009

Flatwise Compression and Buckling Characterizations of Adhesive-Free Additively Manufactured Defected Architected Structures

Khameel B. Mustapha, Mohammad Saad, and Yousif A. Abakr

Abstract In the light of its versatility, the additive manufacturing technology is shaping up to be a boon for the fabrication of advanced structures. As demands soar for the accelerated adoption of this emerging technology, understanding the mechanics of parts derived from the process is crucial to reaping its full potential. In this study, six different unit cells are used to build up different architected structures with artificially introduced missing unit cells. The six structures are screened using the finite element analysis after which focus is restricted to two, which are then produced via the fused filament fabrication method. Assessments of the compressive resistance of the architected structures are conducted. Results indicate a considerable gradual deterioration of the buckling and compressive resistance with an increasing number of missing cells. At higher compressive loads, rupture of unit cells, cracks, minor wrinkling and tearing of the architected structures were observed.

Keywords 3D printing · Buckling response · Compression response · Polymeric architected structures · Imperfections

1 Introduction

Composite structures offer the kind of tremendous performance requirements continuously sought after in aviation, automobiles, spacecraft, and marine industries. Structurally-efficient, these structures provide properties (enhanced strength-to-weight ratio, high thermal insulation, improved stiffness, acoustic damping etc.) that positioned them for wide applications in various domains [1, 2]. Many classifications of composite materials exist, but our focus in this study is on sandwich composite structures, whose mechanical behaviour has been a subject of huge volume of research studies [3]. The basic composition of sandwich structures is well-established as shown in Fig. 1. Essentially, a sandwich structure consists of a

K. B. Mustapha (✉) · M. Saad · Y. A. Abakr
Department of Mechanical, Materials and Manufacturing Engineering, University of Nottingham (Malaysia Campus), 43500 Semenyih, Malaysia
e-mail: khameelb.mustapha@nottingham.edu.my

P. Rajendran et al. (eds.), *Proceedings of International Conference of Aerospace and Mechanical Engineering 2019*, Lecture Notes in Mechanical Engineering,
https://doi.org/10.1007/978-981-15-4756-0_23

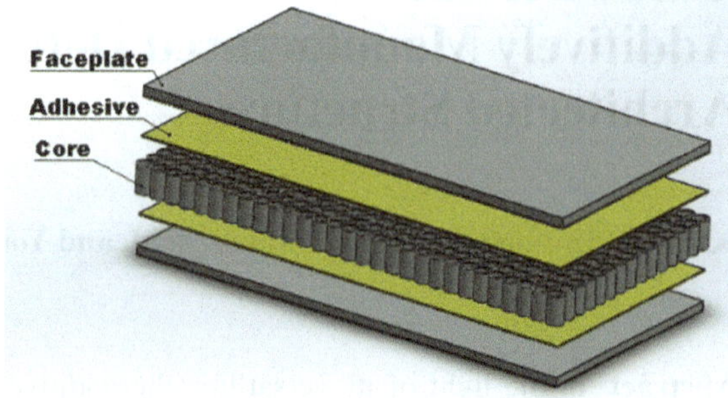

Fig. 1 The composition of a typical sandwich structure

lightweight core bonded, with the aid of an adhesive agent, to faceplates. The core, adhesive agent, and faceplates contribute different functions to the emergent properties of the final structural components. Principally, the faceplates sustain normal stresses and loads in sandwich structures. Metallic materials (e.g. steel, aluminium, etc.), or non-metallic materials (e.g. glass-fibre reinforced composites, wood etc.) are good candidates for the pair of faceplates [4]. In contrast to the faceplates, the core is derived from low-strength materials with specified shear rigidity. Designed to prevent inward and outward buckling of the structure, the core contributes to out-of-plane shear and compressive strengths of the structure [5]. Materials for the core may also be metallic (e.g. aluminium foam) or non-metallic such as balsa, PU, PET and PES foams. In-between the core and the faceplate is the adhesive agent, whose major roles are to bond the core material with the thin faceplates; distribute the shear stress that arises from loadings, and maintains the integration of the structure's constituent parts during loading [4].

Sandwich structures can generally be categorized as periodic or non-periodic. In the current study, our investigations align with the periodic architected sandwich structures. In terms of structural function, the periodic sandwich structures may have a core that experience stretching-dominated deformation [6–8]. This makes them superior to the traditional foam cores in which the deformation is dominated by the bending of the cell walls [6]. Typically, a series of repetitive unit cells form the basis of periodic sandwich structures. Among the periodic sandwich structures, the behaviour and fabrication of those with honeycomb cores (triangular, hexagonal, quasi-triangular and square unit cells) have received tremendous attention [9–11]. These cores have found wide applications in the fabrication of airplane components. Yet, the deployment of honeycomb cores is fraught with challenges. For instance, they are derived from thin sheets (hence lower buckling resistance) [12]. Additionally, they suffer from a multitude of interdependent defects that could be traced to costly complicated manufacturing processes [13]. Wadley [14] reviewed a catalogue of classical methods for producing sandwich structures. Nevertheless, additive manufacturing has emerged as a panacea to some of the challenges associated with the traditional methods [15, 16].

A large body of literature exists on the measurement of the response and theoretical analyses of sandwich structures. A comprehensive account of past studies is contained in Birman and Kardomateas [17]. A major inference from the earlier studies points to the fact that the properties of periodic sandwich structures depend on the material of the network, the topology of the core as well as relative density of the unit cell. However, the understanding of the effects of topology-induced imperfections on the overall structural performance of periodic sandwich structures remains completely unclear.

Many studies exist on the effect of imperfections on the properties of sandwich structures with classical cores [18–20]. However, to the authors' best knowledge, very little work has been done on the effect of topology-induced imperfections on the structural performance of additively manufactured periodic lattice sandwich structures, which is important if these structures are to be put to safe use in various applications [21]. Most structures contain flaws. Oftentimes, the flaw's distribution depend on the processing route employed to fabricate the structure, and the evaluation of the rapid propagation of such flaws or how such flaws worsen under applied loads is important for modern applications [22]. Here, we studied the performance of polymeric, adhesive-free architected structures with a select set of unit cells to offer preliminary insight into their structural response under compressive loads with varying degrees of imperfections. Simulation and physical experimental approaches are applied to facilitate an assessment of the mechanical response.

The presentations of the design details, fabrication and testing are provided in Sects. 2 and 3, while Sect. 4 contains a summary of the conclusions from the study.

2 Materials and Methods

The flow of the investigations follows a four-pronged stage of design, analyses, 3D printing and experimental tests. In what follows, we utilized the properties of polylactide acid plastic (PLA), a biodegradable, thermoplastic material.

2.1 Structural Design and Analysis

The design process involves identification and conceptualization of the geometric features of the architected structures of interest. The architected structures are sandwich-like structures but with periodic cores formed from repeating unit cells. The required dimensions of the unit cells are finalized based on the common convention in the literature. The three-dimensional realizations of the architected structures are done in the PTC Creo. For compatibility with downstream finite element analysis as well

Table 1 A highlight of the designed and analyzed monolithic architected structures

Name	Architected structure	Unit cell
Bilayer hexagon		
Single layer hexagon		
Circular corrugated		
Bi-triangle layer		
Single layer triangle		
Hollow triangle		

as the requirement of 3D printing, we relied predominantly on the solid modelling capability of PTC Creo rather than surface modelling tools. Mostly, the base is first generated using the top plane as the reference plane for the geometry creation. The design of the core then follows. The unit cell is then patterned accordingly. The model is saved as file format IGES (for export to ANSYS) and file format STL format (for CURA, a slicing software for the 3D printing workflow). Table 1 shows a collection of unit cells considered.

After the design process, we carried out initial finite element analyses (FEA) of all structures derived from the unit cells in Table 1 with the aim to investigate how an imperfection due to missing unit cells affect their stiffnesses under flexural, tensile and compression loadings. The FEA acts as a screening tool to determine the structures to focus on for the final experimental tests. Table 2 lists the material properties of PLA used in the analysis as extracted from CES Edupack®. During the FEA, the skins and the core of the structures were modelled with SOLID187 (over 40,000 elements are used after convergence study). A representative meshed structure is shown in Fig. 2. For all buckling analyses, the fixed-free boundary condition is employed (to replicate a close approximation of the experiment). This boundary condition requires

Table 2 Material properties of PLA

Properties	Value	Units
Density	1240.1	kg/m^3
Young's modulus	3302.6	MPa
Poisson's ratio	0.38	
Compressive ultimate strength	65.983	MPa

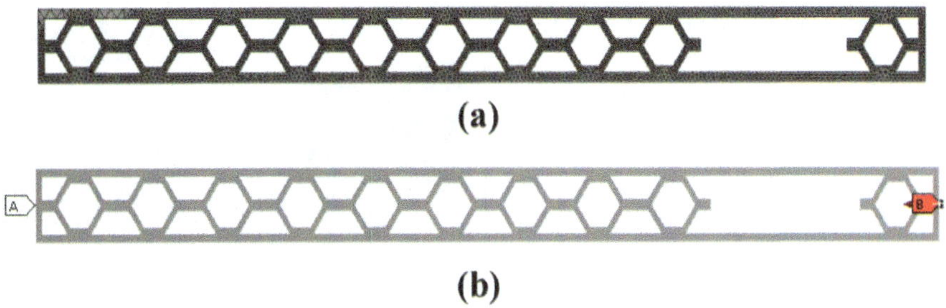

Fig. 2 Representative images from finite element analysis: **a** a meshed structure, **b** compressive load (buckling analysis)

the left end of the structure to be fixed while a compressive remote force is applied on the opposite surface as depicted in Fig. 2. For the flatwise compression test, the bottom surface is fixed while a quasi-static velocity-controlled load is applied to the top surface.

2.2 Additive Manufacturing (AM)

AM is a general term for different layer-by-layer digital fabrication techniques, and the sheer diversity of the processes under the umbrella of AM has led to the development of an ISO standard (ISO 17296-2:2015) [23]. In this study, a 3D printer that works on the concept of fused deposition modelling (FDM) process is employed to fabricate the specimens used for the final physical testing. Group-wise, FDM falls under the material extrusion process [23]. It is a widely used method due to its simplicity. However, one of the major concerns with this method is the poor mechanical properties of the fabricated parts because of dependence on thermoplastics-based filaments. A comprehensive description of the process can be found in [24]. Nevertheless, the FDM process can overcome the limitations in the traditional methods of sandwich fabrication, such as cost, production time, innovation in design complexity, and materials range.

2.3 Mechanical Response Measurements

For the physical tests, both flatwise compressive and in-plane buckling responses of different specimens were assessed. The two tests were carried out using a universal testing machine (LRK50 plus by Lloyd instruments). Attempts were made to follow ASTM C365 and ASTM E2954 standards as closely as possible for the flatwise compression and buckling tests. The dimensions of the tested specimen (L × width × thickness) for both tests are 145 mm × 30 mm × 7 mm. The loading speed of

Fig. 3 Experimental set up: **a** compression test, **b** buckling test

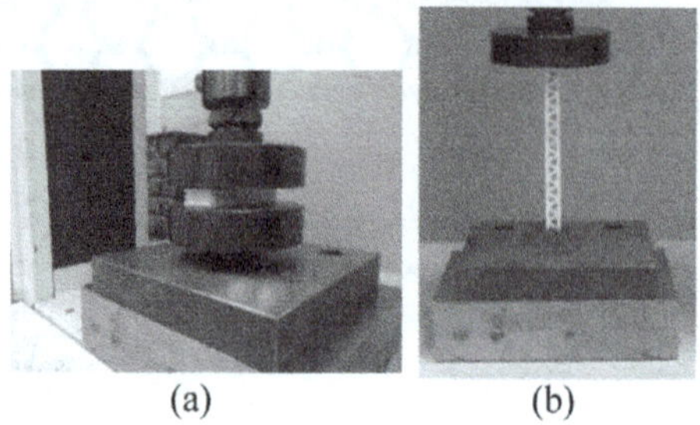

5 mm/min and 10 mm/min was employed for the buckling and compression tests, respectively. The specimen setups are depicted in Fig. 3a, b for the flatwise compression and buckling tests, respectively. For the buckling test, an arbitrary slipping of the base is prevented by using modelling clay to provide additional restraint to the base of the specimen.

3 Results and Discussion

Figure 4 highlights the comparison of the critical buckling loads for the six designs we considered in this study. These results are obtained from the buckling analysis with ANSYS. The results are based on the linear elastic buckling analysis of the structure under a compressive load. In general, different types of compression or buckling instabilities can be observed in sandwich structures [25]. The simulation results provide a virtual assessment of the buckling capability of each design. Each of the designs was subjected to a static compressive load of 15 kN. It is noticed from Fig. 4 that the buckling load, for each of the structure, is less than 15 kN. In Fig. 4

Fig. 4 A comparison of the buckling resistance for the structures with varying number of missing cells

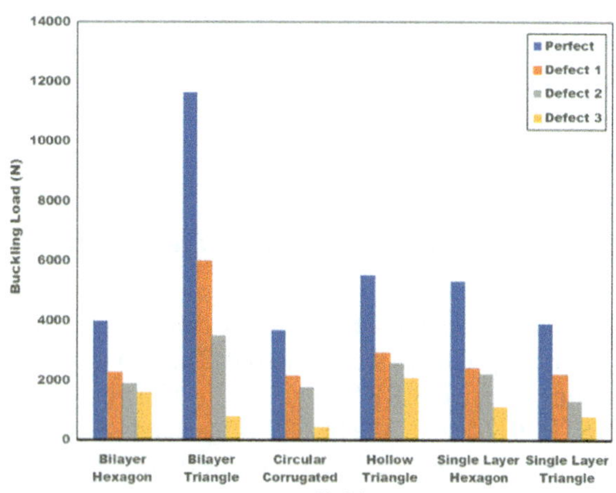

as well as subsequent figures, Defect 1, Defect 2 and Defect 3 denote structures with 1, 2 and 3 missing unit cells, respectively.

A few other deductions can be made from this chart. First, the architected structure with a set of bilayer triangle unit cells is endowed with the biggest buckling load. Second, a progressive decline in the buckling load is observed as the imperfection (induced by missing unit cells) is introduced into all structures. Each of the simulated structures has a length of approximately 145 mm, and the presence of four missing cells results in more than 50% reduction in the buckling capability of a majority of them. Third, among the single-layered structures, the circular corrugated sandwich structure experienced the most dramatic loss in buckling resistance. Further, between the two bilayer structures considered, the design with a bilayer triangle experienced a higher loss of buckling resistance in the presence of missing cells.

The preceding simulation result is complemented by the experimental results from flatwise compression and in-plane buckling tests of the 3D printed specimens, the discussions of which follow. For the tests, only the experimental observations relating to the behavior of the circular corrugated and the bilayer hexagon-based structures are reported. Figures 5 and 6 illustrate the stress-strain curves from the compression

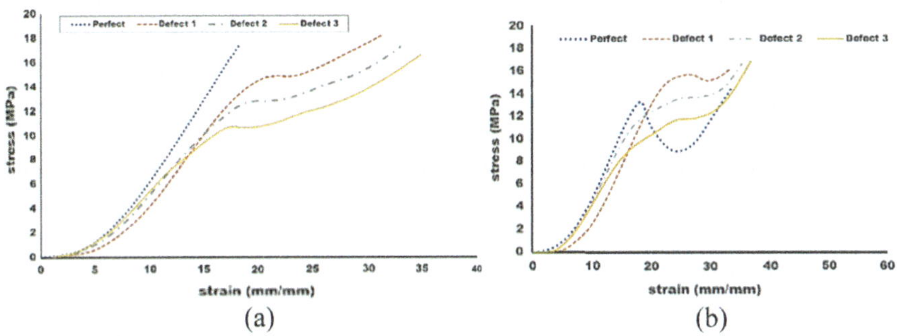

Fig. 5 Stress-strain curves from flatwise compression tests, **a** circular corrugated structure, **b** bilayer hexagon structure

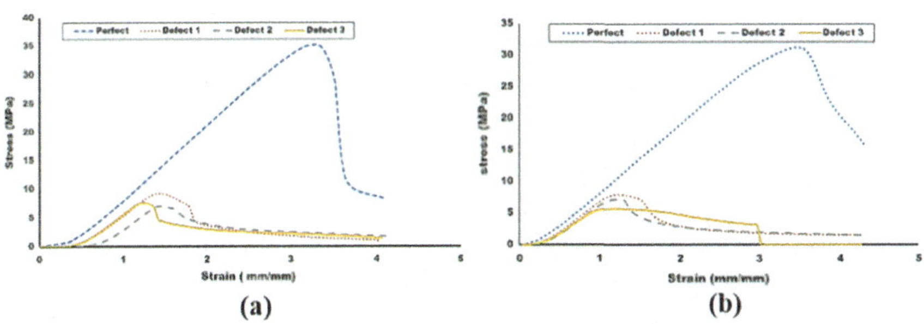

Fig. 6 Stress-strain curves from buckling tests **a** circular corrugated structure, **b** bilayer hexagon structure

and buckling tests, respectively. In the case of the compression test, the load limit was set to 30 kN. The stress values presented here for the compression test is based on the area in contact with the machine platen. Ultimately, as this stress is transferred uniformly to the relative area occupied by the core, it is expected that the stress within the core will be slightly higher than those in Figs. 5 and 6. Nonetheless, for the circular corrugated structure, the perfect sample exhibits a steeply rising stress-strain relationship, while the samples with defects displayed two linear regions separated by a slightly curvy region with a peak transition point. The transition point on the stress-strain curve decreases from around 16 MPa (for 1 missing unit cell) to 10 MPa (for 3 missing unit cells). In contrast, the bilayer hexagon structures show a very pronounced transition from linear to nonlinear behaviour for the perfect sample. The collapse of a unit cell is responsible for the sudden drop in the curve of the perfect structure in Fig. 5b. However, the defected samples display a pattern similar to that obtained for the circular corrugated samples. Further, a brief duration of densification is witnessed in the stress-strain curve where changes in stress become constant over the strain for a small period [26]. As expected, the compressive strength is found to decrease with an increase in the number of missing unit cells. In short, structures with three missing cells experienced the most dramatic changes.

As for the buckling tests, it can be inferred that the bilayer hexagon exhibits a lower failure stress compared to the single-layered circular corrugated structure. Thus, the circular corrugated structure can withstand more compressive load than the bilayer hexagon structure. The failure of a sandwich column under an in-plane compression is caused by a number of competing mechanisms, two of which are global and local buckling instabilities [27]. Several experimental studies have been reported in the past with respect to the buckling response of sandwich structures. For instance, Fleck and Sridhar [28] conducted buckling tests on sandwich columns. Their observations revealed that different failure modes exist depending on the material and geometric properties of the structure. Although several of the earlier studies have employed the classical adhesive-bonded sandwich structures, the sample 3D printed specimens in our studies revealed similar possible failure modes. For instance, we witnessed the fracture of the faceplates, cracks in the architected core and wrinkling, all in agreement with the observations of previous studies [29, 30]. These failure modes are shown in Fig. 7. The failure that typically creates de-bonding in traditionally manufactured architected structure is shown in Fig. 7a, which involves the buckling of the faceplates in an outward direction. Face yielding was seen in the circular corrugated specimen depicted in Fig. 7c. In addition to the above, we further analysed the various failures in the tested specimen at micro level with the help of a Scanning Electron Microscope (SEM). Layer tearing on the surface of the faceplate under buckling is shown in Fig. 8a. Penetrating cracks in the faceplate is displayed in Fig. 8b. An image from a micro-level examination of fracture in the unit cells of the hexagon-based corrugated structure is revealed in Fig. 8c.

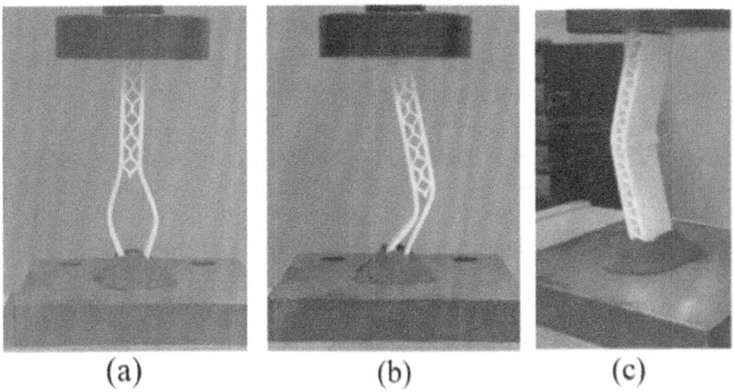

Fig. 7 Selected failure modes observed in buckling experiments: **a** outward face wrinkling failure, **b** inward face wrinkling in bilayer hexagon with four missing cells, **c** dimpling and face yielding in a circular corrugated perfect structure

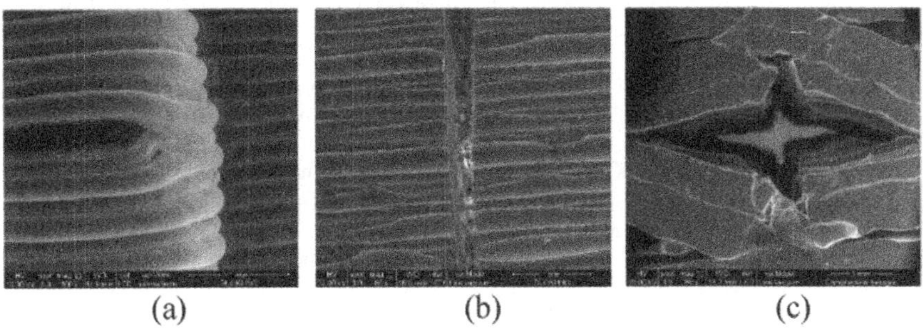

Fig. 8 SEM observation of microscale-level failure in the unit cell during buckling: **a** surface tearing of layer, **b** surface crack, **c** fracture in unit cells of hexagon corrugated structure

4 Conclusion

We studied the flatwise compression and buckling responses of additively manufactured architected structures with missing unit cells. The investigated structures were designed in PTC Creo. Fused deposition modelling, a 3D printing technique, was used to fabricate specimen samples for physical experimental tests. The tests and analyses offered some basic insights into the effect of missing unit cells on the mechanical behaviour of the structures. From the results, the buckling resistance of the bilayer hexagon (in the bilayered group) and the circular corrugated design (among the single-layered group) were found to be more sensitive to the effect of missing unit cells. Physical tests via compression and buckling tests of samples of these two structures lead to the inference that: (i) the compressive strength diminishes linearly with the number of missing unit cells; (ii) the bilayer hexagon structures show a very pronounced transition from linear to nonlinear behaviour in the stress-strain curve for the perfect sample; and (iii) at higher compressive loads, rupture of unit cell, cracks and minor wrinkling and layer tearing occur in both circular corrugated and the bilayer hexagon-based architected structures.

Acknowledgements This work was supported in part by Malaysia's Ministry of Education (MOE) under the Fundamental Research Grant Scheme (FRGS/1/2018/TK03/UNIM/02/1). The authors acknowledge the support of the technical staff of the M3 workshop (especially Mr. Muhammad Nasuha Mazha) and the FOSE 3D Printing Hub team (Mr. Noor Hashimi Mohamad Nor and Mr. Mohd Saiful Bahrin).

References

1. Wu Z, Xiao J, Zeng J, Liu J (2014) Compression performance of integrated 3D composite sandwich structures. J Sandwich Struct Mater 16:5–21
2. Mustapha KB (2018) Manufacturing, applications and mechanical properties of lightweight wood-based sandwich panels. In: Reference module in materials science and materials engineering. Elsevier
3. Allen HG, Neal BG (2013) Analysis and design of structural sandwich panels. The Commonwealth and International Library: Structures and Solid Body Mechanics Division, Elsevier Science
4. Wong YJ, Arumugasamy SK, Mustapha KB (2019) Development of a computational predictive model for the nonlinear in-plane compressive response of sandwich panels with bio-foam. Compos Struct 212:423–433
5. Daniel IM, Abot JL (2000) Fabrication, testing and analysis of composite sandwich beams. Compos Sci Tech 60:2455–2463
6. Deshpande V, Ashby M, Fleck N (2001) Foam topology: bending versus stretching dominated architectures. Acta Mater 49:1035–1040
7. Fleck N, Ashby M, Deshpande V (2002) The topology of cellular structures, in new approaches to structural mechanics, shells and biological structures. Springer
8. Ashby M (2006) The properties of foams and lattices. Philosoph Trans Royal Soc London A: Mathem Phys Eng Sci 364:15–30
9. Gibson L, Ashby M (1997) Cellular solids: structure and properties. Cambridge solid state science series
10. Kim B, Christensen RM (2000) Basic two-dimensional core types for sandwich structures. Int J Mech Sci 42:657–676
11. Aktay L, Johnson AF, Kröplin B-H (2008) Numerical modelling of honeycomb core crush behaviour. Eng Fract Mech 75:2616–2630
12. Wadley HN (2006) Multifunctional periodic cellular metals. Philosoph Trans Royal Soc London A: Mathem Phys Eng Sci 364:31–68
13. Gaydachuk A, Slivinskiy M, Golovanevskiy V (2006) Technological defects classification system for sandwiched honeycomb composite materials structures in materials forum, pp 96–102
14. Wadley HN (2002) Cellular metals manufacturing. Adv Eng Mater 4:726–733
15. Parandoush P, Lin D (2017) A review on additive manufacturing of polymer-fiber composites. Compos Struct 182:36–53
16. Ngo TD, Kashani A, Imbalzano G, Nguyen KT, Hui D (2018) Additive manufacturing (3D printing): a review of materials, methods, applications and challenges. Compos B Eng 143:172–196
17. Birman V, Kardomateas GA (2018) Review of current trends in research and applications of sandwich structures. Compos B Eng 142:221–240

18. Wadee MA (2000) Effects of periodic and localized imperfections on struts on nonlinear foundations and compression sandwich panels. Int J Solids Struct 37:1191–1209
19. Chen M, Zhu X, Lei H, Chen H, Fang D (2015) Effect of defect on the compressive response of sandwich structures with carbon fiber pyramidal truss cores. J Appl Mech 7:1550004
20. Yuan W, Song H, Lu L, Huang C (2016) Effect of local damages on the buckling behaviour of pyramidal truss core sandwich panels. Compos Struct 149:271–278
21. Maskery I, Sturm L, Aremu AO, Panesar A, Williams CB, Tuck CJ, Wildman RD, Ashcroft IA, Hague RJM (2018) Insights into the mechanical properties of several triply periodic minimal surface lattice structures made by polymer additive manufacturing. Polymer 152:62–71
22. Altenbach H, Öchsner A (2010) Cellular and porous materials in structures and processes. Springer, Vienna
23. ISO, ISO 17296-2:2015 (2015) Additive manufacturing—general principles—Part 2: overview of process categories and feedstock
24. Wang X, Jiang M, Zhou Z, Gou J, Hui D (2017) 3D printing of polymer matrix composites: a review and prospective. Compos B Eng 110:442–458
25. D'Ottavio M, Polit O (2015) Linearized global and local buckling analysis of sandwich struts with a refined quasi-3D model. Acta Mech 226:81–101
26. Mamalis AG, Manolakos DE, Ioannidis MB, Papapostolou DP (2005) On the crushing response of composite sandwich panels subjected to edgewise compression: experimental. Compos Struct 71:246–257
27. Ji W, Waas AM (2007) Global and local buckling of a sandwich beam. J Eng Mech 133:230–237
28. Fleck N, Sridhar I (2002) End compression of sandwich columns. Compos A Appl Sci Manuf 33:353–359
29. Fagerberg L (2004) Wrinkling and compression failure transition in sandwich panels. J Sandwich Struct Mater 6:129–144
30. Arun KY, Chandradharappa SM, Basavarajappa S, Krishna M (2010) Failure mechanism in rigid PUF sandwich structures with varied skin materials under bending load. J Reinf Plast Compos 29:1626–1640

Technical and Economic Viability Assessment of Different Photovoltaic Grid-Connected Systems in Jordan

Salah Abdallah and Dana Salameh

Abstract In this work, the technical and economical evaluation of the application of different Photovoltaic (PV) on grid systems was studied based on experimental results and theoretical models. Six types of 20 kWp PV grid-connected systems working at Applied Science Private University, Jordan were involved in study. The Six types of different PV systems studied were: Poly-Crystalline South directed (Poly S), Mono-Crystalline South directed (Mono S), Mono-Crystalline East West directed (Mono EW), Poly-Crystalline East West directed (Poly EW), Thin-Film directed to the south, and a Concentrated PV type with automatic two axes tracking (Con Tracker). For the 20-kWp grid connected systems, the yearly production, the yearly savings, the initial investment costs and the Operating & Maintenance (O&M) costs were estimated, evaluated and compared to get the most beneficial investments by using different economical methods. Con Tracker system presented the most feasible system with higher Net Present Value (NPV) (71733.06 JD), Internal Rate of Return (IRR) (45%), and short Payback Period (PBP) (3 years) than those values of Thin-Film with NPV (42638.15 JD), IRR (37%) and PBP (3 years), Poly S with NPV (44887.23 JD), IRR (34%) and PBP (3 years), Mono S with NPV (48267.89 JD), IRR (33%) and PBP (3 years), Mono EW with NPV (40998.52 JD), IRR (29%) and PBP (4 years), finally Poly EW with NPV (35793.14), IRR (28%) and PBP (4 years).

Keywords Technical assessment · Economic assessment · Photovoltaic systems

---

S. Abdallah (✉) · D. Salameh
Mechanical and Industrial Engineering Department, Applied Science Private University, Amman, Jordan
e-mail: salahabdalah@asu.edu.jo

D. Salameh
e-mail: d_salameh@asu.edu.jo

P. Rajendran et al. (eds.), *Proceedings of International Conference of Aerospace and Mechanical Engineering 2019*, Lecture Notes in Mechanical Engineering,
https://doi.org/10.1007/978-981-15-4756-0_24

1 Introduction

The energy sources in Jordan depend on imported gas and oil for electrical power generation and traffic. Jordan is blessed with enough solar radiation level, where the annual solar radiation is between 5 and 7 kWh/m^2. The importance of investment of renewable energy, especially in PV systems, is due to the need to produce electrical energy with suitable costs for all economic sectors. Also, to reduce environmental pollution associated with the burning of fossil fuels and emission of CO_2 [1].

PV systems witnessed a huge grow in the last four years between 2013 and 2016 at the world level. New added power capacity in 2013 was 38.4 GW. In year 2015, the estimated added power capacity was about 50 GW, and the new added power capacity in 2016 was about 75 GW with more than 50% increase in PV systems installations. Global PV systems installations were expected to grow in the near future in South–East Asia, Latin America, Middle East, India, and North Africa [2–4]. In [5], Denholm and Margolis studied the limits of deployment of PV systems in electric power systems. They founded that under high penetration levels and operations rules, the system would produce extra photovoltaic power generation during some times in the year. The limited flexibility of base load generators gave large values of unusable photovoltaic power generation when PV produced 10–20% of a system's energy. In [6], Abdallah performed an experimental study to evaluate the effect of tracking systems on the electrical power generation at the output of flat plate PV modules. The results indicated that there was an increase of power generation gain up to 43.8, 34.4, and 15.7% for the two axes tracking, EW tracking, vertical and north–south tracking, respectively, as compared with the fixed surface directed to the south with slope angle equals 32° in Amman-Jordan. In [7], two axes sun tracking system based on PLC control system was designed and studied. The experimental study was conducted to evaluate the effect of tracking system on the energy gain. The collected energy on the tracking surfaces was measured and compared with that on a fixed surface directed to the south with tilt angle equals 32°.

Chel et al. studied the costing and sizing of stand-alone photovoltaic systems based on sunshine hours in different locations in the world. This study was conducted according to the load of 3.65 kWh/day. The capital cost and unit cost of generated electrical power by stand-alone PV systems respectively for India [8]. Kaldellis conducted experimental measurements at different locations all over Greece in purpose to determine the optimum configuration of stand-alone PV systems. He used a numerical code 'PHOTOV-lll' to determine the optimum slope angle, the installation cost, the system components and dimensions [9]. Shum and Watanbe made a comparison study about a deployment aspects of photovoltaic systems in the USA and Japan. Japanese deployment strategy of photovoltaic technology was based on a dominant category with mass market potential. In the USA, the cumulative PV installations were divided among many types of systems such as on grid and off grid [10].

In [11], Quaschning performed a detailed technical and economic analysis with computer simulations of non-tracked and two axes tracking, and solar thermal parabolic trough power plants. He found that there were preferred regions for installing of tracking photovoltaic systems in comparison to the fixed systems, and solar thermal trough. For regions with low level of solar radiation, the power generation of solar thermal systems was lower than of photovoltaic systems, and for regions of high solar radiation solar thermal system with parabolic trough provides the best solution. De la Tour et al. [12] studied the cost of photovoltaic technology and the cost of electrical power generated by PV modules out to 2020 using experience curve modules. They founded that from 2011 to 2020, the prices of PV modules would decrease up to 67%. The cost of electrical power that generated by modules would reach that of national grid electricity by 2020 in the countries of solar radiation of 2000 kWh/year or above, such as Spain, Italy and California. Albrecht [13] used portfolio theory to determine the cost competitiveness of electrical energy generated by PV systems. Based on the used assumptions, the cost of electrical energy generated by PV generators could be competitive in the period from 2015 to 2040.

Abdallah et al. [14] built the mathematical models to compare the cost of electricity of three power generation systems in Jordan diesel power generation, national grid, and PV power generations. The results of study showed that the cost of electricity generated from PV generation system could provide alternative power source with lower value than the diesel generation system. Also, national grid still produced the lowest cost as compared with other generation systems. In [15], Badran et al. used the fuzzy sets methodology to evaluate different possible solar technologies for electric power generation in Jordan, namely, photovoltaic and solar ponds. The criterion for evaluation was based on many parameters, i.e., cost, maturity, storage capacity, efficiency, capacity factor, availability, land usage and safety. The results found showed that PV technology was the best choice for electrical power generation in Jordan. Abdallah [16] evaluated the application of two types of sun tracking systems in Jordan based on mathematical model and experimental results. The total solar energy collected for a fixed surface directed to the south with slope equals 32°, a vertical single axis, and two axes sun-tracking surfaces were compared. Theoretical and experimental power gain was 51.53% and 52.37% respectively in case of two axes tracking. Theoretical power gain was 40.72% and experimental power gain was 37.83% according to the one axis vertical tracking system. He founded that the most feasible system was two axes tracking with cost gain of 34%, then vertical single axes tracking with cost gain of 31%, as compared with fixed system directed to the south with tilt angle equals 32°. In [17], Hammad et al. made a technical and economical evaluation of fixed and open loop two axes sun tracking photovoltaic systems connected to the national grid, and installed at the Hashemite University in Zarqa, Jordan. The system power was 7.98 kWp for both, and they were monitored for one year from February 2014 to February 2015. The electrical power gain from using tracking system was 31.29% as compared with fixed photovoltaic system. The feasibility study was performed over 20 years, and the results showed that the fixed photovoltaic system was more feasible in Jordan.

In the last years, Jordan witnessed a huge increase in application of PV systems in real life on grid systems or off grid systems. The aim of this work is to estimate the cost of electricity generated from different types of PV systems based on various economical methods which are: NPV, IRR and PBP. For this purpose, six types of

PV grid-connected systems working at Applied Science Private University, Amman-Jordan, having a name plate capacity of 20 kWp were studied. Five of these PV systems are fixed with tilt angle of 11° which are: Mono-Crystalline S directed, Mono-Crystalline EW directed (which means that half of these PV modules directed to the east and the other half is directed to the west), Poly-Crystalline EW, Poly-Crystalline S directed and Thin-Film S directed. The sixth type is a Concentrated PV type with closed loop automatic two axes tracking. The cost estimation will be provided according to national grid electricity cost and according to the Jordanian market prices of PV modules and tracking equipment. The feasibility of each system is evaluated and compared by using Methodology.

2 Methodology

In the present study, methodology was divided into 2 major parts as shown below.

2.1 Theoretical Model of Energy Collection on Fixed PV Systems and Two Axes Tracking PV System

The major source for the mathematical expressions of the energy collection at the output of six different types of flat plate PV modules is Duffie and Beckman [18]. These six types practically used in Applied Science Private University are shown in Fig. 1.

The technical data characteristics for each system are shown in Table 1.

The mathematical model will be calculated for average days for each month of the year then will be multiplied by the number of days in each month, and the yearly collection of energy will be the total sum of energy collection for 12 months. Applied Science Private University is located in Amman, Jordan where latitude angle $\Phi = 32°$.

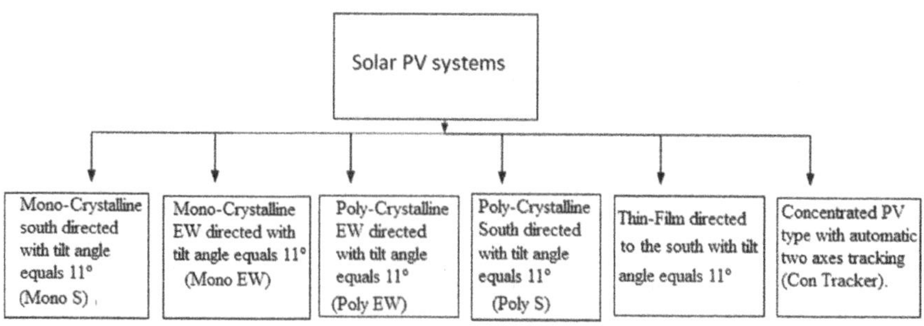

Fig. 1 Six types of 20 kWp PV grid-connected systems

Table 1 The technical data characteristics for each PV system type

PV system type	Rated max. power	Rated voltage	Rated current	Efficiency	Dimension
Units	(W)	(V)	(A)		(m * m)
Mono EW	260	30.8	8.46	16.2%	0.99 * 1.65
Mono S	260	30.8	8.46	16.2%	0.99 * 1.65
Poly EW	250	30.4	8.24	15.6%	0.99 * 1.65
Poly S	250	30.4	8.24	15.6%	0.99 * 1.65
Thin-film	125	59.7	3.37	8.9%	1.001 * 1.402
Con tracker	250	30.4	8.24	15.6%	0.99 * 1.65

The declination angle, δ, is expressed as:

$$\delta = 23.45\sin\left(360\left(\frac{284+n}{365}\right)\right) \tag{1}$$

where n is the number of day in the year. The Zenith angle θ_z is given by

$$\cos\theta_z = \cos\Phi\cos\delta\ \cos\omega + \sin\Phi\ \sin\delta \tag{2}$$

where ω is the hour angle. Sunset hour angle ωs is

$$\omega s = \cos^{-1(-\tan\delta\ \tan\Phi)} \tag{3}$$

The number of daylight hours is

$$N = \frac{2}{5}\left\{\cos^{-1(-\tan\delta\tan\Phi)}\right\} \tag{4}$$

Atmospheric transmittance for beam radiation τ_b is

$$\tau_b = a_0 + a_1 e^{-\left(\frac{k_1}{\cos\theta z}\right)} \tag{5}$$

where standard atmospheric constants a_0, a_1 and k_1 are:

$$a_0 = r_0(0.4237 - 0.00821)(6 - A)^2 \tag{6}$$

$$a_1 = r_1(0.5055 + 0.00595)(6.5 - A)^2 \tag{7}$$

$$k_1 = r_k(0.2711 + 0.1858)(2.5 - A)^2 \tag{8}$$

where A is the altitude of the observer which equals for Amman 0.8 km, r_0, r_1, and r_k are the correction factors for certain climate type [18].

Angle of incidence θ is given by

$$\theta = \cos^{-1}(\sin\delta\sin\Phi\cos\beta - \sin\delta\cos\Phi\sin\beta\cos\gamma + \cos\delta\cos\Phi\cos\beta \\ \cos\omega + \cos\delta\sin\Phi\sin\beta\cos\gamma\cos\omega + \cos\delta\sin\beta\sin\gamma\sin\omega) \quad (9)$$

The ratio of beam radiation on tilted surface to that on a horizontal surface R_b is given by:

$$R_b = \cos\theta / \cos\theta_z \quad (10)$$

Hourly integrated extraterrestrial radiation on a horizontal surface for an hour period in the absence of atmosphere I_o is

$$I_o = \frac{12 * 3600 * Gsc}{\pi}\left(1 + 0.033\cos\left(\frac{360 * n}{365}\right)\right) \\ * \left[\cos\Phi\cos\delta(\sin\omega_2 - \sin\omega_1) + \left(\pi\frac{(\omega_2 - \omega_1)}{180}\right)\sin\delta\sin\Phi\right] \quad (11)$$

where Gsc is the solar constant which equals 1367 W/m^2, ω_2 and ω_1 are the upper and lower limits of certain hour.

Hourly extraterrestrial radiation on a tilted surface I_{0t} is

$$I_{0t} = I_0 * R_b \quad (12)$$

Hourly integrated radiation on a tilted surface and on the earth surface in MJ/m^2 is I_{0tu}

$$I_{0tu} = I_{0t} * \tau_b \quad (13)$$

To transfer the units from MJ to WH we can use the following formula 1 MJ = 277.77 WH. For a plane that is continuously tracking the sun about two axes to minimize the angle of incidence

$$\beta = \theta_z \quad (14)$$

$$\gamma = \gamma_s \quad (15)$$

The yearly theoretical electrical power generation at the output of PV modules for different types of PV systems is shown in Table 2. The reference is Duffie and Beckman [18].

The yearly experimental electrical power generation at the output of six types of 20 kWp PV grid-connected systems working at Applied Science Private University,

Table 2 The yearly theoretical electrical power generation

PV system type	Theoretical yearly power generation (kWh/kWp)
Mono EW	1430.14
Mono S	1770.9
Poly EW	1433.95
Poly S	1772.97
Thin-film	1737.98
Con tracker	2644.44

Table 3 The yearly experimental electrical power generation

PV system type	Experimental yearly power generation (kWh/kWp)
Mono EW	1598.91
Mono S	1737.46
Poly EW	1473.48
Poly S	1646.35
Thin-film	1563.44
Con tracker	2333.85

Amman-Jordan were studied is shown in Table 3. The reference is from the collected experimental data by data logger.

2.2 *Economic Analysis*

PV systems costs and savings. Table 4 shows the yearly cost of the generated electricity (JD) of 20 kW systems which is also considered as the yearly savings or the productivity of 20 kW systems in JD from using PV systems instead of using electricity from National Grid.

Column 1 is the yearly experimental electrical power generation at the output of PV modules for 1 kW system at Applied Science Private University in Amman (Table 5).

$$\text{Column 2} = \text{Column 1} * 20\ \text{kW} \tag{16}$$

$$\text{Column 3} = \text{Column 2} \div 12\ \text{months} \tag{17}$$

$$\text{Column 4} = \text{Column 3} * \text{electricity tariff JD/kWh} \tag{18}$$

$$\text{Column 2} = \text{Column 4} * 12\ \text{months} \tag{19}$$

Table 4 The Productivity of 20 kW systems calculation

	Column 1	Column 2	Column 3	Column 4	Column 5
PV system type	Experimental yearly power generation (kWh/kWp)	Yearly production (kWh of 20 kW system)	Monthly production (kWh of 20 kW system)	Monthly cost of the generated electrical energy (JD)	Yearly cost (savings) of the generated electrical energy (JD)
Mono EW	1598.91	31978.2	2664.85	512.2	6146.4
Mono S	1737.46	34749.2	2895.77	571.32	6855.84
Poly EW	1473.48	29469.6	2455.8	458.48	5501.76
Poly S	1646.35	32927	2743.92	532.44	6389.28
Thin-film	1563.44	31268.8	2605.73	497.07	5964.84
Con tracker	2333.85	46677	3889.75	825.78	9909.36

Table 5 Electricity tariff

Level number	Levels	Fils/kWh	kW consumption
1	First level	42	1–160
2	Second level	92	161–300
3	Third level	109	301–500
4	Fourth level	145	501–600
5	Fifth level	169	601–750
6	Sixth level	190	751–1000
7	Seventh level	265	More than 1000

In case of consumption of 1000 kWh per month, the first 160 kWh will be multiplied by 42 Fils, while the next 139 kWh will be multiplied by 92 Fils, the next 199 kWh will be multiplied by 109 Fils and so on. The cost of installation of 20 kW systems for each system is summarized in Table 6. The PV systems' lifetime is stated for 25 years.

Initial cost of installing 20 kW systems is the actual price for building the 20 kW systems from the Jordanian market nowadays for all systems. National Grid electricity impact study is an initial cost paid to the National Grid electricity to study the effect of connecting these systems to the National Grid electricity before the installation of systems. For Con tracker, the cost of the tracking equipment will be added. Annual O&M cost is the annual expenses paid for operating and maintaining the PV systems at Applied Science Private University. For Con tracker, an extra cost will be added each year due to the added electric elements and control devices that need to be maintained. Since the tracking equipment will be replaced every 10 years because of degradation, the entire cost of tracking equipment replacement will be

Table 6 The cost of installing 20 kW systems

Time (years)	Cash flows (JD)	Mono EW	Mono S	Poly EW	Poly S	Thin-film	Con tracker
t = 0	Initial cost of building 20 kW systems	15,400	15,400	14,000	14,000	11,900	11,900
	National Grid electricity study	1500	1500	1500	1500	1500	1500
	Cost of the tracking equipment	–	–	–	–	–	3358
t = 1–25	Annual O&M cost	300	300	300	300	300	300
	Annual cost due to the added electric elements	–	–	–	–	–	150
t = 10	Tracking equipment replacement	–	–	–	–	–	3358
t = 20		–	–	–	–	–	3358/2

added at year 10 and half of the cost will be added at year 20 for the last 5 years. The savings of installation of 20 kW systems for each system is summarized in Table 7.

The yearly savings or the productivity of 20 KW systems instead of using electricity from National Grid is calculated in Table 3. There is a 13% linear decrease in rated power after beginning after the first year. This will form a geometric gradient

Table 7 The Savings of 20 kW systems

Time years	Cash flows	Mono EW	Mono S	Poly EW	Poly S	Thin-film	Con tracker
t = 1–25	Yearly savings of the generated electrical energy (JD)	6146.4	6855.84	5501.76	6389.28	5964.84	9909.36
t = 2–25	Rate of decreasing per year (%)	5.42	5.42	5.42	5.42	5.42	5.42

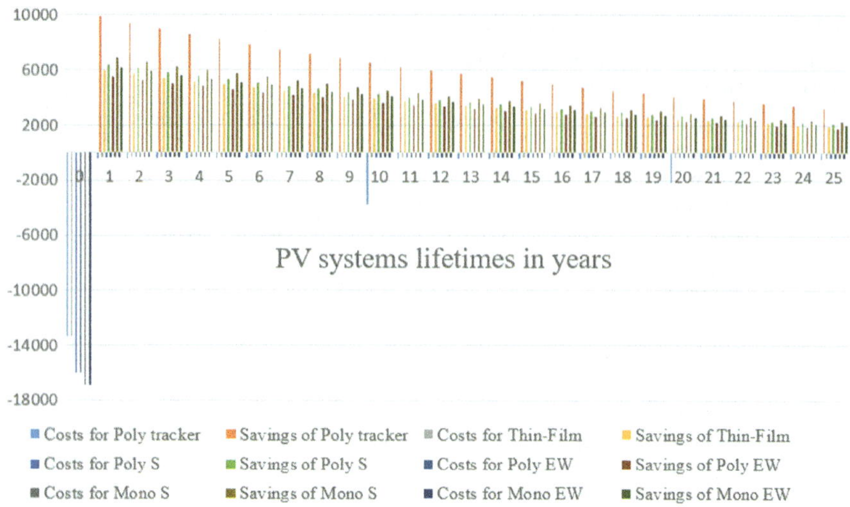

Fig. 2 Cash flow diagram for PV systems running costs and savings

series which is a cash flow series that decreases by a constant percentage each period. This percentage will mostly equal 5.42% each year. This decrease will continue to the end of the life span. Figure 2 shows the cash flow of the PV systems running costs and savings over 25 years. The biggest savings is obtained with the Con Tracker system.

PV systems feasibility. In the economic viability aspect, NPV, IRR, and PBP methods are evaluated in this paper. NPV presents the difference between the present values of the net cash inflow in the future and the initial investment as shown below:

$$NPV = PVP - PVC = \sum_{n=0}^{N} \frac{B_n}{(1+\mathrm{i})^n} - \sum_{n=0}^{N} \frac{C_n}{(1+\mathrm{i})^n} \tag{20}$$

PVB = present value benefit, and PVC = present value cost. NPV of the investment indicates an acceptance criterion if the benefits are greater than the total costs. $\mathrm{B_n}$ = expected benefit at the end of the year n, $\mathrm{C_n}$ = expected cost at the end of year n, i = discount rate, n = project duration in years, N = project period. IRR is calculated by the subtraction of the total present value benefit from total present value cost as shown below:

$$\sum_{\mathrm{n=0}}^{\mathrm{N}} \frac{\mathrm{B_n}}{(1+\mathrm{IRR})^{\mathrm{n}}} - \sum_{\mathrm{n=0}}^{\mathrm{N}} \frac{\mathrm{C_n}}{(1+\mathrm{IRR})^{\mathrm{n}}} = 0 \tag{21}$$

The PV project is acceptable and viable if IRR is greater than the discount rate. PBP is calculated as the following:

$$\sum_{n=1}^{N}(B_n - C_n) = 0 \tag{22}$$

In terms of economics, If the PBP presents a short value (short payback periods), the project will be acceptable. Otherwise, longer PBP shows a worse investment.

3 Results and Discussion

In the parameters evaluation for economic analysis, the currency in all the economic calculations is in JD, where 1 US dollar equals 0.7 JD. The project lifetime of PV systems is stated for 25 years. The interest rate is fixed at 3.25% per year, which a commonly used bank's interest rate paid on deposits linked with the Jordanian Dinar currency. Results of the economic comparison between different PV systems used are shown in Table 8.

According to the NPV, systems are ranked from the most profitable to the least as follows: Con tracker, Mono S, Poly S, Thin-film, Mono E-W and Poly E-W. Based on IRR, all the systems are profitable with percentage of profit per year as follows from the highest to the lowest: Con tracker, Thin-film, Poly S, Mono S, Mono E-W and Poly E-W. The decision made for choosing the best system which is Con tracker system is the same regardless of the method used. As NPV increases, the savings increase. As IRR increases, a wider margin of safety is achieved. The payback period is 3 years for Con tracker, Thin-Film, Poly S and Mono S, which is preferable than 4 years for Mono EW and Poly EW.

The results of this work are different from the results of Hammad et al. [17], that they found that the electrical power gain from using two axes tracking system was 31.29% as compared with fixed PV system, and they concluded that fixed PV systems was more feasible in Jordan. The results of the experimental data and theoretical models indicate that the power gain is about 50% from using two axes tracking as compared with fixed south directed PV system. The source of error in their work, as we think, is related to the inaccuracy of open loop control system or bad operation of the two axes tracking system they used.

From the analysis of the experimental data, theoretical models, and feasibility studies, it can be seen that the electrical power generated by PV on grid systems in Jordan can provide an alternative electrical power source with lower cost than

Table 8 Results of the economic comparison between different PV systems used

Economic parameters	Mono EW	Mono S	Poly EW	Poly S	Thin-film	Con tracker
NPV (JD)	40998.52	48267.89	35793.14	44887.23	42638.15	71733.06
IRR (%)	29	33	28	34	37	45
PBP (years)	4	3	4	3	3	3

national grid power supply. Jordan needs full practice of different types of PV on grid systems, especially PV system with two axes tracking, for mid-size commercial and industrial buildings. If adding to this the ecological effect of using PV systems, as a clean electrical power supply with zero CO_2 emissions, all of these results leads to a conclusion that it is important to increase the investment in the PV systems applications in Jordan.

4 Conclusions

In this work, the technical and economical evaluation of the application of different PV on grid systems in Jordan is studied based on experimental results and theoretical models. Six types of PV grid-connected systems working at Applied Science Private University in Amman, having a power capacity of 20 kWp each were studied.

From the models built, it is seen that using of the different types of PV systems in Jordan can provide alternative electrical power source with lower annual cost than national grid. The PV systems and tracking equipment costs were evaluated based on Jordanian market prices. After using different economical methods, it is found that the application of all six systems is feasible, but Con Tracker system presented the most feasible system with higher NPV (71733.06 JD), IRR (45%), and short PBP (3 years) values those values of Thin-Film with NPV (42638.15 JD), IRR (37%) and PBP (3 years), Poly S with NPV (44887.23 JD), IRR (34%) and PBP (3 years), Mono S with NPV (48267.89 JD), IRR (33%) and PBP (3 years), Mono EW with NPV (40998.52 JD), IRR (29%) and PBP (4 years), finally Poly EW directed with NPV (35793.14), IRR (28%) and PBP (4 years).

Taking into consideration the above mentioned results, it is recommended to start using the PV on grid systems widely, especially PV system with two axes tracking, to power mid-size commercial and industrial buildings in Jordan.

Acknowledgements The authors are grateful to the Applied Science Private University, Amman, Jordan for the full financial support granted to this research work.

References

1. Kabariti M (2005) Identification of national energy policies and energy access in Jordan. National research center report, Beirut, Lebanon. The American University of Beirut
2. Association EEPI (2014) Global market outlook for photovoltaics
3. Programme IEAPPS (2016) A snapshot of global PV (1992–2015). In: Mary Brunisholz IP (ed) Report IEA-PVPS TI-29: International Energy Agency
4. Programme IEAPPS (2017) Snapshot of global photovoltaic markets 2016. In: Mary Brunisholz IP (ed) Report IEA-PVPS TI-31
5. Denholm P, Margolis R (2007) Evaluating the limits of solar photovoltaic (PV) in traditional electric power systems. Energy Pol 35(5):2852–2861

6. Abdallah S (2004) The effect of using sun tracking systems on the voltage-current characteristics. Energy conversion and management, pp 671–1679
7. Abdallah S (2004) Two axes sun tracking system with PLC control. Energy Convers Manag 45:1931–1939
8. Chel A, Tiwari G, Chandra A (2004) Sizing and cost estimation methodology for stand-alone residential PV power system. Agile Syst Manag 45:2745–2760
9. Kaldellis J (2004) Optimum techno economic energy autonomous photovoltaic solution for remote consumers throughout Greece. Energy Convers Manag 45:2745–2760
10. Shum K, Watanbe C (2007) Photovoltaic deployment strategy in Japan and the USA an institutional appraisal. Energy Pol 35:1186–1195
11. Quaschning V (2004) Technical and economic system comparison of photovoltaic and concentrating solar thermal power systems depending on annual global irradiation. Sol Energy 77:171–178
12. De la Tour A, Glashant M, Meniere Y (2013) Predicting the costs of photovoltaic solar modules in 2020 using experience curve models. Energy 62:341–348
13. Albrecht J (2007) The future role of photovoltaic: a learning curve versus portfolio perspective. Energy Pol 37:2296–2304
14. Abdallah S, Muslih I, Abu Husain W (2010) National grid, diesel, and photovoltaic power generation systems in Jordan: an engineering and economical evaluation. Energy Sources Part B 5:370–383
15. Badran O, Abdulhadi E, Mamlook R (2010) Evaluation of solar electric power technologies in Jordan. Jordan J Mech Ind Eng 121–128
16. Abdallah S (2004) Automatic sun-tracking system applications in Jordan: a cost comparative evaluation. Jordan J Appl Sci Univ 40–49
17. Hammad B, Al- Sardeah A, Al-Abed M, Nijmeh S, Al-Gandoor A (2017) Performance and economic comparison of fixed and tracking photovoltaic systems in Jordan. Renew Sustain Energy Rev 80:827–839
18. Duffie J, Beckman W (2006) Solar engineering of thermal processes, 3rd edn. Wiley, Hoboken

Aircraft Arrival Delay Under Uncertainty: G-Queue Modelling Concept

Lay Eng Teoh

Abstract In response to tremendous growth of air traffic, achieving demand-capacity balance by ensuring airport runway's adequacy and maximizing airport efficiency is extremely crucial. It is of utmost important especially to the airport operators in tackling flights delay under uncertainty. As such, this paper aims to propose a modeling framework for aircraft arrival delay, in the form of a G-queue (i.e. a queuing system with negative arrivals). In particular, this paper highlights that inbound flight delay can be modeled as a generalized G-queue network, by presenting a schematic structure to model aircraft arrival delay (with G-queue concept). Besides, a four-step framework is discussed to obtain numerous key data inputs for model development. In addition, a novel approach, namely *K*-GLE, is addressed accordingly in order to capture general distribution of interarrival and service time. Concisely, this paper supports the beneficial application of G-queue in modelling aircraft arrival delay. It is anticipated that the proposed methodology would be useful to assist the airport operators to improve on-time performance when dealing with arrival throughput under uncertainty (including stochastic demand).

Keywords Aircraft arrival delay · G-queue · Negative customers

1 Introduction

Globally, it is expected that 13.5 trillion passenger-kilometer will be flown by the year 2030 and the number of commercial flights will almost double from 26 million to 48.7 million [1]. However, airspaces have a fixed capacity and the existing number of airports is apparently not adequate to accommodate such demand increment. Consequently, airport congestion, which often resulting in substantial flight delays and cancellations, has become an increasingly important phenomenon worldwide [2].

L. E. Teoh (✉)
Universiti Tunku Abdul Rahman, Cheras, 43000 Kajang, Selangor, Malaysia
e-mail: teohle@utar.edu.my

P. Rajendran et al. (eds.), *Proceedings of International Conference of Aerospace and Mechanical Engineering 2019*, Lecture Notes in Mechanical Engineering,
https://doi.org/10.1007/978-981-15-4756-0_25

In practice, flight delay occurs when two or more aircraft seek to use a runway, taxiway, gate or any other airside facility simultaneously [3]. A flight is considered to be delayed if it arrives or departs 15 min after the scheduled time. In general, a flight can be delayed due to several reasons, including aircraft mechanical problems, bad weather, air traffic management initiatives or misaligned crew/aircraft due to previous delayed flight [4]. In the third quarter of the year 2015, it is approximated that around 17.5% of all scheduled flights in the United States (US) suffered from delays while the flight delays for the European market was reported to be almost 21%. This actually denotes an alarming highlight, i.e. more than 2800 flights in the US and 6000 flights in Europe were delayed daily. Besides, flight delay may have a propagation effect, i.e. it would affect passengers' connectivity and consequently delay compensation costs for the airlines [5]. Pyrgiotis et al. [6] highlighted that one hour of flight delay may result in seven hours of delay for the airline's entire fleet. And, the operating costs incurred due to flight delays include crew cost, maintenance cost, fuel burn and emission cost. In the US, air traffic delays reached an all-time peak in the year 2007 and the nationwide impact was estimated to be more than $30 billion, including $8.3 billion for airlines cost and $16.7 billion for passengers cost.

In practice, the formation and propagation of aircraft queues is inherently stochastic in view of the fact that queues do not depend solely on the schedule of flights and the availability of airport capacity, but also driven by the exact sequencing of aircraft arrivals and departures (with the mixture of aircraft, human factors in aircraft operations, etc.). In addition, the operating conditions (e.g. weather and winds) are also stochastic and they would impact airport operations directly. As such, steady state queuing models are inapplicable to analyze delay because airport demand typically varies strongly and hence the dynamic characteristics of the airport queues thus become apparent. Similarly, airport capacity often varies significantly throughout the day especially due to weather conditions. Thus, the use of steady-state expressions often yields very poor approximations. Moreover, steady-state results do not apply when the demand rate of airport exceeds the service rate [7]. As such, deterministic models fail to capture the stochastic nature of airport operations and hence the queuing analysis for airport operations has to be carried out during its transient phase instead of steady-state [8]. However, there is still a lack of use of continuous models with stochastic elements [9].

Correspondingly, this study aims to look into the application of G-queue, i.e. a queuing system with negative arrivals (customers), in modeling aircraft arrival delay. Conceptually, G-queue is capable to capture the occurrence of uncertainty that may impose a great impact, especially on aircraft arrival delay. By applying G-queue concept, the proposed model of aircraft arrival delay is anticipated to ease airport congestion by reducing aircraft queue length with a lower waiting time. Certainly, this would benefit not only air travelers and airlines, but also the airport operators in managing aircraft operations in a better manner.

The remainder of this paper is organized as follows. In Sect. 2, the literatures relating to airport congestion, flight delay, aircraft arrival and queuing models are reviewed and discussed accordingly. Section 3 presents the proposed model structure of applying G-queue to model aircraft arrival delay under uncertainty. This section also discusses the proposed four-step framework for data compilation in order to obtain the relevant key data inputs for further analysis. Note that the derivation of the set of mathematical equations and also the results analysis are not included in this paper in view of the fact the aim of this paper is to propose a modeling approach (with G-queue concept) instead of mathematical derivation and results generation. Lastly, Sect. 4 concludes the papers by addressing several aspects to extend the proposed model.

2 Literature Review

In view of the close relation between airport congestion and flight delay, this section first presents a brief overview on airport congestion and flight delay, followed by the review on aircraft arrival and G-queue (to model inbound flight delay).

2.1 Airport Congestion and Flight Delay

According to the existing literatures, airport congestion and flight delay can be mitigated from four major aspects, namely infrastructure expansion, Air Traffic Flow Management (ATFM) measures, aircraft operational control and demand management measures. Regardless of timely planning, majority of airport congestion problems get resolved tactically on the day of operations through demand management measures and it generally involves a trade-off between airport congestion mitigation and capacity utilization maximization. To do this, a proper scheduling mechanism is required to enhance the existing practices, either administratively (e.g., slot control) or economically (e.g., congestion pricing). In particular, Jacquillat and Odoni [4] highlighted that such mechanisms could lead to a significant delay reduction. However, given the variability of airport operations, airport capacity is not a fixed quantity but depending on several operational factors, including weather conditions, aircraft landings/take-offs proportion and the runway configuration in use [10, 11]. And, it was also found that some of these factors are not determined exogenously in advance.

In order to support strategic planning, macroscopic airport congestion models play a role to aggregate airport operations in providing computationally efficient estimates of the relationships between flight schedules, airport capacity and flight delays. The existing macroscopic congestion models are based on econometric models [12–14], deterministic queuing models [15] or stochastic queuing models [6, 8, 16]. However, these models did not consider endogenous element (e.g., flights schedules and queue length) that might affect the magnitude of delay significantly [17].

Practically, LMINET [18] and LMINET2 [19] are two queuing network models that model airports as classical $M(t)/E_k(t)/1$ queues. In comparison to LMINET,

LMINET2 makes use the information of aircraft itineraries. Besides, The Approximate Network Delays (AND) model is capable to model both deterministic and stochastic effects [20]. The modelling approach in AND and LMINET2 are similar but they possess different calculating strategies for local queuing delays. In addition, DELAYS, a stochastic delay model which is originally developed by MIT, is able to approximate the behavior of $M(t)/E_k(t)/n$ queueing systems.

Estimating flight delay with classical queuing model depends greatly on the rates at which aircraft arrivals and departures are serviced. In the existing models, service rates are generally kept constant over the day [21, 22] or varied using ex-post (actual) operational data (e.g., meteorological conditions, capacity estimates, etc.) [23]. In particular, the queuing model of Pyrgiotis and Odoni [24] included the variations in arrival and departure service rates that occur in response to the changes in daily flight schedules caused by aircraft delays. However, these variations are introduced into the model manually but not systematically [17]. Besides, clustering techniques were also adopted to generate capacity profiles by considering probabilistic weather forecasts and ex-ante (a forecast or estimate) in queuing models [25]. However, these variations in service rates are exogenous, i.e. they did not consider endogenous element.

2.2 *Aircraft Arrival*

With regard to aircraft arrival, there are some related literatures, namely [7, 23, 26]. In particular, Lovell et al. [7] discussed data filtering approach to calibrate aggregate models of internal delays and flights cancellation. They highlighted that internal delays are related to queueing effect at arrival airport (i.e. the relationship between airport capacity and scheduled demand). And, they revealed that aircraft arrival delay is not necessarily due to a shortage of airport capacity. Numerous other factors, such as traffic congestion, weather, late crew arrival and long boarding time may significantly lengthen the previous turn-around and the en-route segment of the flight. Comparatively, en-route congestion and weather are the key factors that cause delays.

Hansen et al. [23] used several queueing models to estimate delay savings from four dimensional trajectory precision. They adopted both deterministic (which comprises micro and macro models) and stochastic queueing model (i.e. DELAYS model) to perform the analysis of delay savings for arrival queue. Their results showed that average delay savings is about 35% and delay predictions from various models are found to be strongly collinear over a wide range of congestion levels.

Focusing on aircraft arrival, Thiagaraj and Seshaiah [26] used M/G/1 model to measure the performance of an airport runway by considering various traffic mixes and operational conditions. They also outlined the limits of using analytical approach and they built a simulation procedure in performing their analysis. Their results highlighted that time-varying Poisson arrivals is a good assumption given heavy utilization and some amount of unscheduled aircraft. It was also found that the transient and steady state delays increase polynomially as the service rate is lowered. In addition, the delays generated by those systems with a low server variance can often be lower than those generated by a higher service rates and variance.

2.3 G-Queue

Under certain circumstances, some queueing systems may encounter failure or breakdown due to various causes [27]. As such, one of the important elements in modeling queuing system is the existence of various external factors (uncertainty) which may pose great impact on the entire queuing network. Such queue is termed as G-queue, i.e. the queue with negative arrivals (customers). In particular, G-queue considers an additional flow of negative customers along with ordinary (positive) customers. When negative customers arrive, the queuing system is affected if and only if customers are present in which a positive customer is removed (according to some killing disciplines) from the system without receiving any service [28]. In other words, negative arrivals have no effect if the system is empty.

Most of the literature on G-queue assumed that the server is reliable on a permanent basis and always available to customers regardless of the arrival of negative customers. Apparently, this assumption seems to be unrealistic in the real systems. For G-queue, there are several removal approaches including individual removal, batch removal, disaster, triggered movement and random work removal [29]. Comparatively, the two simplest forms of killing disciplines are RCH (removal of a positive customer at the head) and RCE (removal of a positive customer at the end) [25]. Remarkably, G-queues were first introduced by Gelenbe [28] in modelling neural networks. After Gelenbe's works were presented [30–32], numerous studies related to negative customers were published. Most of these papers can be found in the bibliography of Do [29]. Do [29] highlighted that choosing an appropriate killing discipline would enrich the modeling capability of G-networks, such as the applicability to model breakdown and packet losses. In term of applications, negative customers can also represent removal signals of works in various systems, e.g. the decision to cancel requests for resources (inhibition signals) in neural networks. Negative customers also play a role to resemble viruses and orders of demand [25, 27, 33].

Notably, the studies on G-queue can be classified according to two characteristics, namely discrete-time and continuous-time. To name a few, Harrison and Pitel [34] adopted Laplace transform method to obtain the sojourn time distribution of an M/M/1 queue in which negative arrivals form a Poisson process. They also analyzed and compared basic performance measures for queueing systems with different queueing disciplines and killing strategies. For M/G/1 queues, Jain and Sigma [35] derived a Pollaczek-Khintchine expression with disasters (DST) while Bayer and Boxma [36] conducted Wiener-Hopf analysis for M/G/1 queue in which positive customers are allowed to be killed only at the end of the service. Besides, Gelenbe's original concept was generalized by Zhu and Zhang [37] where positive and negative customers may cancel out each other.

Recently, Koh et al. [27] analyzed a single-server queue with negative customers by adopting RCH killing strategy. In particular, the interarrival time of positive and negative customers are assumed to have exponential distribution and the numerical method introduced in Koh [38] was applied to determine the stationary queue length distribution. Their results (with exponential and gamma distributed service time) were found to be numerically comparable. However, there is no clear explanation on the application of the proposed approach in solving a practical problem which possesses a complex form with general distribution.

Chin et al. [39] derived a set of equations to find the stationary queue length distribution, by considering Poisson negative arrivals. Their results showed that the inclusion of negative customers is useful in reducing mean queue length and also the congestion level of the queueing system. Subsequently, Chin [40] developed a new approach to find the performance measures of G-queue. In particular, he developed a novel numerical approach, i.e. *K*-layers geometrical linear extrapolation (*K*-GLE) which is capable to capture the general distribution of the interarrival and service time. Yet, the application of the proposed approach in actual practice which possesses a greater complexity still requires further investigation.

In overall, although the afore-mentioned studies are relatively related to flights delay, they still possess some limitations, as stated below:

(a) Aircraft arrival and service rates are in fact endogenous, i.e., they depend on the schedules of flights and the corresponding queue lengths. However, some existing studies did not capture endogenous element explicitly in analyzing flight delay.
(b) Many studies had been analyzed without considering negative customers (which could be either endogenous or exogenous element). Given the stochastic nature and the variability over time of the airport operations, this aspect should be taken into consideration in analyzing flight delay.
(c) Flight delay is actually determined not so much by the average but by the probability that the delay for a given aircraft will be greater than some amount. Yet, there is no clear indication on how to capture aircraft arrival delay (in terms of queuing probability) under uncertainty.
(d) Some studies require the assumptions on the distribution of interarrival and service time. This may not be relevant to the actual practice under uncertainty.

In order to fill the gaps as mentioned above, a G-queue model is proposed to model aircraft arrival delay under uncertainty.

3 G-Queue Model for Aircraft Arrival Delay

3.1 Model Structure

A schematic representation of airport runway system is displayed in Fig. 1. In overall, airport runway system (for aircraft arrival) can be modeled as a single server by considering two major elements, namely *demand* and *supply*. With regard to the demand perspective, arriving aircraft can be interpreted as the arriving customers requesting service to land at the airport runway and this would constitute to *arrival demand rate*. When the arriving aircraft approaches the destination airport for landing, it might be the case that it cannot be landed as the airport runways are usually the limiting resources at the airport. Thus, arriving aircraft may have to wait if the airport runway is not available for landing. As such, the formation of waiting queue (for landing) constitutes to a queuing network for which a longer queue would result in a longer waiting time. In particular, the extent for which the airport runway to accept the arrival of aircraft (i.e. arriving aircraft can be landed) is termed as *arrival service rate*.

In terms of the arrival at the runway system, there are two possible types of arrivals (customers), namely positive and negative customers. Positive customers

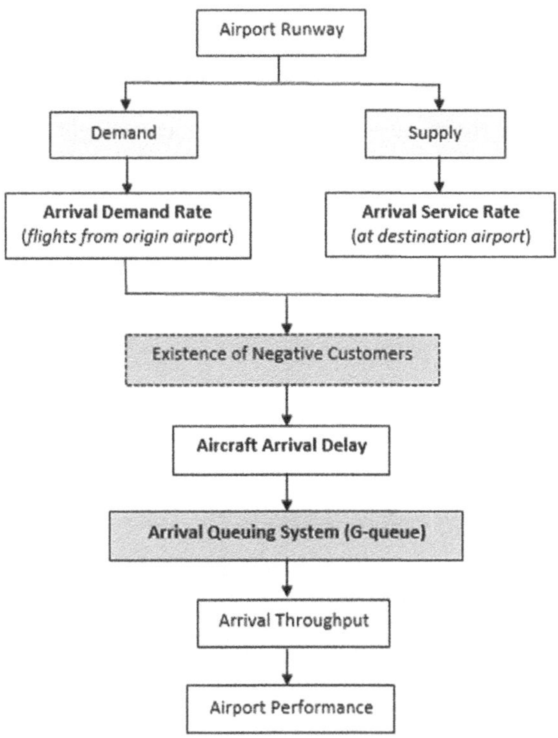

Fig. 1 A schematic representation of airport runway system for aircraft arrival

refer to ordinary landings that happen on time (as per flight schedule) while *negative customers* refer to a particular uncertainty (e.g. bad weather) which will result in not only *aircraft arrival delay*, but also affect the utilization of airport runway in terms of arrival time, queue length, etc. The types of arrival delay include runway access delay (for taxiing process) and landing delay. The existence of negative customers in the *arrival queueing system* at the airport would then constitute to the formation of *G-queue*. In practice, the arrival rate of aircraft and the runway configurations appear to be a function of observed congestion under uncertainty. Besides, a high demand level for aircraft arrival rate and service rate reduction due to the occurrence of uncertainty (e.g. weather disturbances) can also lead to extreme arrival delay.

By considering negative customers (as depicted in Fig. 1), the proposed G-queue model for aircraft arrival delay should be capable to quantify the relation between flight delay and airport congestion under uncertainty. Besides, it can accurately estimate not only the queuing delay for each individual flight, but also the extent of airport congestion, runway queue length and *arrival throughput* which consequently constitute to the overall *airport performance.* Yet, this requires a complete derivation of a set of mathematical equations in order to yield the expected outputs (note that the derivation of equations is not within the scope of this paper).

3.2 Four-Step Framework

As displayed in Fig. 2, four steps are required in order to obtain the key data input in modeling aircraft arrival delay in the form of G-queue. The presented framework in Fig. 2 commences by extracting numerous flight records of the arrival operations of

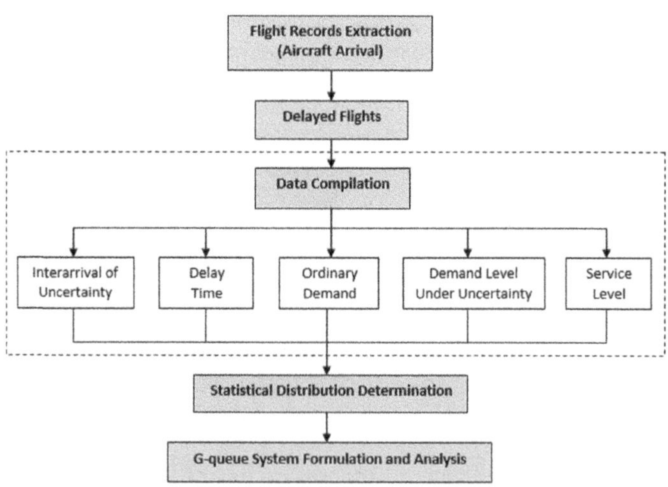

Fig. 2 Four-step framework for data compilation

aircraft (*step* 1). This can be done by accessing the relevant website of a particular airline. For instance, the link of AirAsia can be reached via http://www.klia2.info/tools/check-flight-status in order to access the arrival records of AirAsia that contains the operating records of arrival aircraft. By checking the obtained records, only those flights that have been delayed will be extracted (*step* 2) in order to perform the analysis on aircraft arrival delay. As mentioned in Sect. 1, delayed flights refers to those arrive 15 min after the scheduled time.

For all delayed flights, further data compilation can then be performed accordingly (*step* 3). This includes the compilation on interarrival of uncertainty, delay time, ordinary demand, demand level under uncertainty and service time. In particular, these are the key data inputs to be applied by G-queue model in analyzing inbound flight delay. By having these data in place, statistical distribution determination will be carried out (*step* 4) in order to identify the respective distribution type that to be used by the proposed G-queue model. The determination of the distribution is important as it also plays a role to determine the approach in solving G-queue model. In particular, the *K*-GLE approach developed by Chin [40] can be adopted for the general distribution of interarrival and service time. For brevity purpose, the derivation of the set of mathematical equations with *K*-GLE approach is not covered in the scope of this paper. Yet, the guidelines on how to perform the necessary derivations can be seen in Chin [40].

With the aid of the four-step framework, data compilation was performed by accessing the records of aircraft arrivals of AirAsia [41]. In particular, 30 days of aircraft arrivals (for September, 2018) were extracted for a total of 7961 inbound flights. It was found that about 10% of flights were delayed with an average *delay time* of 29.05 min (per flight). Besides, the average *ordinary demand*, *demand level under uncertainty* and *interarrival of uncertainty* (for each flight) were found to be 5.46, 5.34 and 61.17 min respectively. This shows that the runway is available to serve aircraft arrivals under ordinary condition. However, inbound delay would happen when the uncertainty occurs as the duration of *service level* (about 5.43 min/flight) exceeds the *demand level under uncertainty*.

4 Conclusions

By considering both positive and negative arrivals, this paper deals with the modeling of inbound flight delay under uncertainty by making use the concept of G-queue. To the best of the understanding of the author, this is the first study that applies G-queue network in air transportation sector. The presented methodology should be useful particularly to the airport operators to tackle aircraft arrival delay viably. Besides, this paper also contributes to the theoretical concept of G-queue in practical use. In fact, air transportation system is a highly complicated network and hence the complete analysis is not easy and straightforward as it involves huge data compilation and it requires the consideration of various inter-related factors (e.g. the close-relation of aircraft arrival and departure). In view of its complexity, equations derivation and results analysis are not covered in this paper. They will be left as future works.

Acknowledgements The author would like to extend her gratitude to Mr. Chin Ching Herny for his advices in applying G-queue model. And, the author would like to express her appreciation to Universiti Tunku Abdul Rahman for all kind of support in working out this paper.

References

1. Airbus (2015) Global market forecast flying by numbers 2015–2034. Technical report D14029463, Airbus
2. EUROCONTROL PRC (2015) An assessment of air traffic management in Europe during the calendar year 2014—Performance Review Report
3. De Neufville R (1976) The problem of capacity and delay (Chap. 3). In: Airport systems planning. Macmillan, London
4. Jacquillat A, Odoni AR (2018) A roadmap toward airport demand and capacity management. Transp Res Part A 114:168–185
5. Santos BF, Wormer MMEC, Achola TAO, Curran R (2017) Airline delay management problem with airport capacity constraints and priority decisions. J Air Transp Manag 63:34–44
6. Pyrgiotis N, Malone KM, Odoni A (2013) Modelling delay propagation within an airport network. Transp Res Part C 27:60–75
7. Lovell DJ, Churchill AM, Odoni AR, Mukherjee A, Ball MO (2007) Calibrating aggregate models of flight delays and cancellation probabilities at individual airports. 7th USA/Europe Air Traffic Management R&D Seminar
8. Gupta S (2010) Transient analysis of D(t)/M(t)/1 queuing system with applications to computing airport delays. Massachusetts Institute of Technology, Master of Science in Operations Research
9. Jacquillat A, Odoni A, Webster M (2013) Dynamic control of runway configurations and of arrival and departure service rates of JFK airport under stochastic queue conditions. ESD Working Paper Series
10. Gilbo E (1993) Airport capacity: representation, estimation, optimization. IEEE Trans Control Syst Technol 1(3):144–154
11. de Neufville R, Odoni A (2013) Airport systems: planning, design and management, 2nd edn. McGraw-Hill, New York
12. Kwan I, Hansen M (2000) US flight delay in 2000's: an econometric analysis. 90th Transportation research Board Annual Meeting
13. Morrison S, Winston C (2008) The effect of FAA expenditures on air travel delays. J Urban Econ 63(2):669–678
14. Xu N (2007) Method for deriving multi-factor models for predicting airport delays. Ph.D. dissertation. George Mason University, Fairfax, VA
15. Hansen M (2002) Micro-level analysis of airport delay externalities using deterministic queueing models: A case study. J Air Transp Manag 8(2):73–87
16. Kivestu P (1974) Alternative method of investigating the time dependent M/G/K queue. Master thesis, Massachusetts Institute of Technology
17. Jacquillat A, Odoni AR (2015) An integrated scheduling and operations approach to airport congestion mitigation. Oper Res 63(6):1390–1410
18. Long D, Lee D, Johnson J, Gaier E, Kostiuk P (1999) Modelling air traffic management technologies with a queueing network model of the National Airspace System, Technical report NASA/CR-1999-208988, NASA Langley Research Centrem Hampton, VA

19. Long D, Hansan S (2009) Improved prediction of flight delays using the LMNET2 system-wide simulation model. In: AIAA Aviation Technology, Integration, and Operations Conference
20. Baspinar B, Ure NK, Koyuncu E, Inalhan G (2016) Analysis of delay characteristics of European air traffic through a data-driven airport-centric queuing network model. IFAC-PapersOnLine 49–3:359–364
21. Pyrgiotis N (2011) A stochastic and dynamic model of delay propagation within an airport network for policy analysis. Ph.D. thesis, Massachusetts Institute of Technology
22. Jacquillat A (2012) A queueing model of airport congestion and policy implications at JFK and EWR. Master thesis, Massachusetts Institute of Technology
23. Hansen M, Nikoleris T, Lovell D, Vlachou K (2009) Use of queueing models to estimate delay savings from 4D trajectory precision. In: 8th USA/Europe Air Traffic Management research and Development Seminar
24. Pyrgiotis N, Odoni N (2015) On the impacts of scheduling limits: a case study at newark international airport. Transp Sci. https://doi.org/10.1287/trsc.2014.0564
25. Nikoleris T, Hansen M (2012) Queueing models for trajectory-based aircraft operations. Transp Sci 46(4):501–511
26. Thiagaraj HB, Seshaiah CV (2014) A queueing model for airport capacity and delay analysis. Appl Math Sci 8(72):3561–3575
27. Koh SK, Chin CH, Tan YF, Teoh LE, Pooi AH, Goh YK (2018) Stationary queue length of a single-server queue with negative arrivals and non-exponential service time distributions. MATEC Web Conf 189:02006
28. Gelenbe E (1989) Random neural networks with negative and positive signals and product form solution. Neural Comput 1(4):502–510
29. Do TV (2011) Bibliography on G-networks, negative customers and applications. Math Comp Modell Int J 53(1–2):205–212
30. Gelenbe E (1993) G-networks by triggered customer movement. J Appl Prob 30(3):742–748
31. Gelenbe E (1993) Learning in the recurrent random neural network. Neural Comput 5(1):154–164
32. Gelenbe E (1994) G-networks: a unifying model for neural and queueing networks. Ann Oper Res 48(5):433–461
33. Lee DH, Kim K (2014) Analysis of repairable Geo/G/1 queues with negative customers. J Appl Math
34. Harrison PG, Pitel E (1993) Sojourn times in single-server queues by negative customers. J Appl Prob 30(4):943–963
35. Jain G, Sigman K (1996) A Pollaczek Khintchine formula for M/G/1 queues with disasters. J Appl Prob 33(4):1191–1200
36. Bayer N, Boxma OJ (1996) Wiener-Hopf analysis of an M/G/1 queue with negative customers and of a related class of random walks. Queueing Syst 23(1–4):301–316
37. Zhu Y, Zhang ZG (2004) M/GI/1 queues with services of both positive and negative customers. J Appl Prob 41(4):1157–1170
38. Koh SK (2013) Maintenance of deteriorating non-exponential single server queue. Ph.D. thesis, Universiti of Malaya, Malaysia
39. Chin CH, Koh SK, Tan YF, Pooi AH, Goh YK (2018) Stationary queue length distribution of a continuous-time queueing system with negative arrival. J Phys Conf Ser 1132:012057
40. Chin CH (2018) New approach for finding performance measures of continuous-time server queue with negative customers. Universiti Tunku Abdul Rahman, Malaysia, Master of Science
41. AirAsia (2018) Aircraft arrivals. http://www.klia2.info/tools/check-flight-status

Evaluation of Combustion Characteristics on Simple Cylindrical Combustion Chamber for Different Operating Conditions and Alternative Fuels

Sim Sing Mei and Nurul Musfirah Mazlan

Abstract This study investigates the influence of different operating conditions including fuel temperature, droplet sizes, fuel injector and type of fuel on combustion characteristics of a simple cylindrical combustion chamber. Alternative fuels namely Jatropha Bio-synthetic Paraffinic Kerosene (JSPK), Camelina Bio-synthetic Paraffinic Kerosene (CSPK), mixture of 50% JSPK with 50% Jet-A (50JSPK/50Jet-A) and mixture of 50% CSPK with Jet-A (50CSPK/50Jet-A) are used while Jet-A is used as baseline fuel. At every cases, droplet sizes are varied at 20–40 μm, fuel temperatures are varied from 300 to 500 K and the spray half cone angles are varied from 30° to 50°. Results obtained show that smaller droplet size will lead to high combustion rate and thus provide better efficiency whereas higher fuel temperature produced better evaporation in the chamber. At higher spray half cone angle, shortest penetration length is produced and took shortest time for droplets to reduce its diameter. Among the four tested fuels, 50JSPK/50Jet-A and 50CSPK/50Jet-A have the shortest time to evaporate which indicated that both fuel mixtures have higher evaporation rate due to lower boiling point and density.

Keywords Spray penetration · Alternative fuels · Droplet lifetime · Fuel properties

1 Introduction

Increasing number of air passengers has led to an increase of airplanes needed for transportation purpose but causes increases in gas emissions including pollutants that are harmful for human health and greenhouse. There are few ways in reducing gas emissions, including the usage of alternatives fuels instead of the petroleum-based fuel. Even though the usage of biofuels will be beneficial, however the biofuels have lower energy density, poor high-temperature thermal stability and storage instability [1].

S. S. Mei · N. M. Mazlan (✉)
School of Aerospace Engineering, Universiti Sains Malaysia, Engineering Campus, Nibong Tebal 14300, Pulau Pinang, Malaysia
e-mail: nmusfirah@usm.my

P. Rajendran et al. (eds.), *Proceedings of International Conference of Aerospace and Mechanical Engineering 2019*, Lecture Notes in Mechanical Engineering,
https://doi.org/10.1007/978-981-15-4756-0_26

Fuel properties play an important role in combustion characteristics. Difficulty of the fuel to be atomised depends on the viscosity of the fuel [2]. As the viscosity of a fuel is higher, the velocity of the spray injection will be reduced which leads to creation of larger droplet size [3]. Fuel with lower surface tension will lead to better spray properties because the atomisation process occurs easily [3]. High boiling point indicates less volatility and vice versa [4]. It is stated that increased volatility affects combustion performance by improving stability and the efficiency of the combustion [2]. Fuel with lower boiling point or higher volatility should have shorter droplet lifetime compare to those with higher boiling point [4]. The temperature of the fuel itself will define the efficiency of the vaporisation. By increasing the initial temperature, the droplet particle slows down in high rate and the penetration length is reduced [5]. The penetration is shorter due to the faster rate of complete vaporisation. High initial temperature and velocity are preferable since it increase the vaporisation rate.

Despite fuel properties, operating conditions such as spray half cone angle, droplet size, droplet temperature, and initial combustion pressure and temperature also influence combustion emission. Different fuel injection pressures, injection timings, droplet size distribution and other spray characteristics are used to study on its effect on combustion [6]. Simulation of spray vaporisation and penetration was done to investigate the vaporisation of alternatives fuel under different operating conditions [5]. The influence of fuel spray angles and injection timing on combustion and emission characteristics were observed, in which the narrow entrance from injector which is lower spray half cone angle will increase the air-fuel mixing within the chamber and thus increase the combustion efficiency [7]. At narrow spray angle, the production of NO_x were larger due to fuel rich region, thus early injection strategy is encouraged to reduce the emission [8]. The changing of droplets size along the combustion chamber indicates the evaporation rate of the fuel droplets. As the fuel injected from the injector, the fuel droplet evaporates, and mix with the surrounding air. During the process, the size of the droplet decreases along the path [9].

In this present study, the influence of important parameters namely droplet diameter, spray cone angle, fuel temperature and fuel properties on droplet lifetime and spray penetration are presented using Computational Fluid Dynamic (CFD) to visualise the flow behaviour inside the chamber. The CFD simulation is performed for a simple 2D cylindrical combustion chamber for various spray cone angle ranging from 30° to 50°, droplet diameter ranges from 20 to 40 μm, fuel temperature ranges from 300 to 500 K, and using two different alternative fuels and two mixtures of alternative fuels with Jet-A. The two alternative fuels are Jatropha Bio-synthetic paraffinic kerosene (JSPK) and Camelina Bio-synthetic paraffinic kerosene (CSPK), while the two mixture fuels are containing a blend of 50% JSPK with 50% Jet-A (50JSPK/50Jet-A) and a blend of 50% CSPK with 50% Jet-A (50CSPK/50Jet-A). JSPK and CSPK were tested in civil aircraft as a drop-in fuel and were found to reduce nitrogen oxide (NO_x) emission although an increment in carbon monoxide (CO) was observed. Incomplete combustion due to the poor atomisation process is a main factor leads to increment of CO. Therefore, this present study is beneficial to visualise flow behaviour inside the chamber which leads to the poor atomisation.

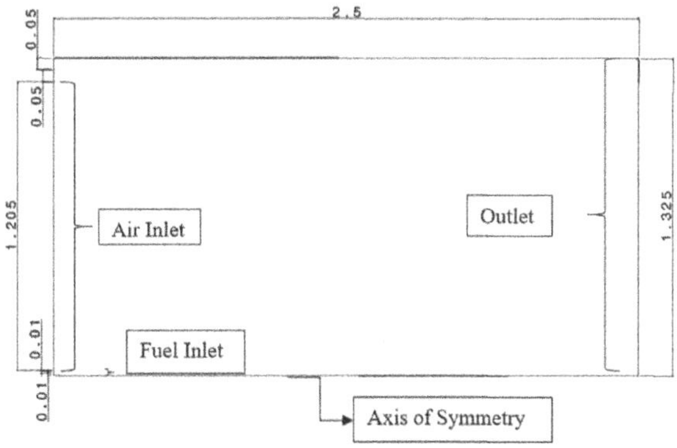

Fig. 1 Combustion chamber configuration (all dimensions in m)

Table 1 Combustion zones length

Zone	Length (m)
Primary zone	0.6310
Secondary zone	0.4205
Dilution zone	1.4485

2 Methodology

2.1 Combustion Chamber Configuration

A simple 2D cylindrical combustion chamber as shown in Fig. 1 was used for the evaluation. The chamber has a fuel inlet, air inlet and outlet. Total length of the chamber is 2.5 m. The chamber is divided into three zones; Primary zone, Secondary zone and Dilution zone. The length for each zones was predicted based on study conducted by Mark and Selwyn [14] who set the primary zone, secondary zone and dilution zone at 25.24, 16.82 and 57.94% from the total combustion length respectively. Based on that percentage, the length of each zone for the combustion chamber in this study was calculated using Eq. (1) and tabulated in Table 1.

2.2 Meshing and Sensitivity Analysis

Five different types of meshes were performed and the comparison between the CFD simulation for each meshes was compared with Mark and Salwyn [10] who obtained the maximum combustion temperature of 2443.4 K. The simulations were run for those 5 different types of meshes from coarsest to finest, and the percentage differences obtained is shown in Table 2.

Table 2 Percentage difference and duration of the simulation for all five meshes

Type of mesh	No. of nodes	No. of elements	Time taken for simulation (min)	Percentage difference (%)
Coarsest	1800	1716	0.13	2.72
Coarse	2378	2280	1.30	3.54
Fine	18,271	18,000	2.95	2.32
Finer	46,431	46,000	7.55	1.91
Finest	78,061	77,500	4.73	4.77

Among all the meshes, finer mesh has the lowest percentage difference with the literature thus were used for other simulations performed in this study. For all the set up and solutions, the iteration numbers were set to be 1500 iterations.

$$Zone\ Length\ Percentage\,(\%) = \frac{Zone\ Length}{Total\ Length} \times 100 \tag{1}$$

2.3 Simulation Set-up and Boundary Condition

The model used in this study is non-premixed. In non-premixed combustion, air and fuel enter the combustion chamber in different stream. The CFD modelling setup for all the cases is shown in Table 3. Velocity at y-axis differs according to the defined spray half cone angle. Table 4 summarises the fuel properties considered in this study.

As this simulation is in 2D, velocity for both X-axis and Y-axis are required. Velocity for X-axis considered in this study was 100 m/s while the velocity at Y-axis was determined by using trigonometry right triangles formula as shown in Eq. 2. In the equation, θ represents spray half cone angle in degree.

$$V_y = V_x \tan\theta \tag{2}$$

Table 3 CFD modelling setup

Set up	Value
Oxidizer temperature (K)	1000
Mass flow rate (kg/s)	0.0002
Particle diameter (μm)	20–40
Fuel temperature	300–500 K
Inlet temperature	1000 K
Outlet temperature	1800 K
Wall temperature	1200 K

Table 4 Fuel properties [4, 10, 11]

Fuel properties	Jet-A	JSPK	CSPK	50JSPK/50Jet-A	50CSPK/50Jet-A
Latent heat (kJ/kg)	226	251	251	238.5	238.5
Specific heat, c_p (J/kgK)	2093	2010	2010	2051.5	2051.5
Density, ρ (km^3)g	780	749	753	764.5	766.5
Boiling point (K)	462	512.25	515.15	487.125	488.58
Vapor temperature (K)	460	510	513.15	485.125	486.58
Vapor pressure (Pa)	90.12	239.03	252.56	150.51	155.18
Viscosity (cST)	8.0	3.663	3.336	5.8315	5.668

3 Results and Discussion

3.1 Influence of Spray Half Cone Angle on Combustion Performance

In this study the spray half cone angle is defined earlier to understand its influence on the combustion performance particularly temperature distribution, spray penetration, and droplet lifetime. Figure 2 shows temperature contour of Jet-A for various spray half cone angles. As spray angle increases, the region of maximum flame temperature is longer and wider due to the interaction between air and spray. At all tested angles, as the fuels are heated up, the temperature increases from the beginning and achieved maximum temperature at the primary zone and consistently decreases downstream along the chamber. However slight increment in the maximum temperature is observed as spray half cone angle increases. The maximum temperature for each of the angle is 2438.57 K, 2459.85 K and 2477.57 K for 30°, 40° and 50° spray half cone angles respectively.

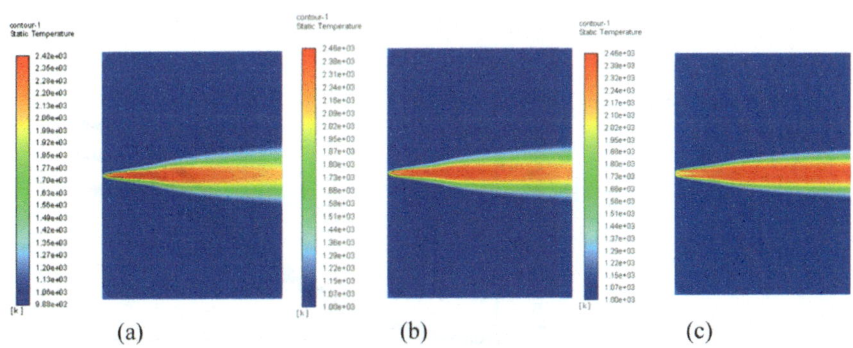

Fig. 2 Temperature contour of Jet-A for **a** 30°, **b** 40° and **c** 50° spray half cone angles

The location of the maximum temperature occurred in the chamber is plotted in Fig. 3. It can be seen that at all spray cone angles, the maximum combustion temperature occurs at primary zone region although the farthest distance is observed for 50° spray half cone angles. Consequently, the combustion did not have enough time to cool down before leaving the combustion chamber. Increasing the spray angle to 50° causes poor ignition performance due to the combustion that occurs near to the secondary zone. For that reason, the longer combustion chamber or more cooling air is required to cool down the temperature before leaving the chamber. Moreover, increasing the spray angle causes poor atomisation thus will produce a large amount of unburned hydrocarbon and smoke.

Spray penetration is defined as the maximum distance of the spray when injected into either stagnation or crossflowing air. The spray penetration is governed by two forces which are kinetic energy of the liquid jet, and the aerodynamic resistance of the surrounding gas [12]. Spray angle may influences the penetration of the spray due to the later forces.

Figure 4 depicts penetration of Jet-A spray at various spray cone angles. Finding revealed that the longer spray penetration is observed for narrow spray angle. As the spray angle becomes wider, the penetration length decreases. Low penetration for the wide cone angle happen due to the incurring of surrounding gas resistance.

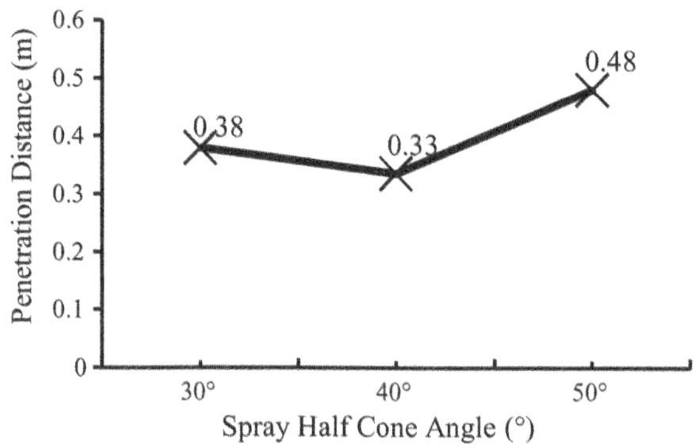

Fig. 3 Location of the maximum temperature for different spray half cone angle

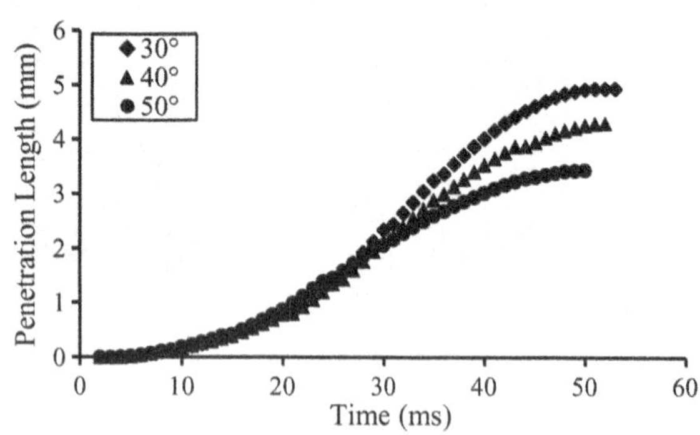

Fig. 4 Penetration of the Jet-A spray for different spray half cone angle

Results obtained are consistent with the finding obtained in [8] who found that the development of spray from injector with narrow spray half cone angle was longer since time taken for the spray from the narrow angle injector to impinge is longer.

Observation on droplet lifetime at various spray cone angle after time start of injection is depicted in Fig. 5. The evaluation of droplet lifetime is important as it determines time required to ensure complete combustion. The lifetime of the droplet is influenced by physical properties of the ambient air, the relative velocity between droplet and the surrounding medium, the properties of the liquid droplet, and initial condition of droplet. In this present study, the air properties, droplet properties and droplet initial condition are kept constant for all tested spray angles. Initial droplet set for this investigation was 20 μm. Velocity of the air is kept as low as 1 m/s to present stagnant flow.

The velocity of liquid droplet at x-axis is assumed to be 100 m/s while the velocity at y-axis is defined from trigonometry formula as shown in Eq. (1). The droplet velocity at the y-axis depends on the spray angle. Increasing the spray angle causing the y-axis velocity increases. As observed, time taken for the fuel droplet to evaporate is shortest at the angle of 50° due to increases of relative velocity between droplet and surrounding air. However insignificant difference observed for the angle of 30° and 40°. After time start of injection, droplet diameter is maintained until a certain point and reducing until it is fully evaporated. The similar trend observed in Boateng [13] who found that as a droplet is heated up during the starting of combustion, constant droplet diameter is observed due to thermal expansion offsetting initial weak evaporation. Between the three spray half cone angles, it is seen that wider spray half cone angle provides faster reduction of particle diameter. Faster reduction of particle diameter for wider spray cone angle promotes air fuel mixing, which is consistent with the study conducted by Zhang et al. [14]. Moreover, the heated-up period for droplet with wider spray angle is shorter compared to 30° and 40° spray angles.

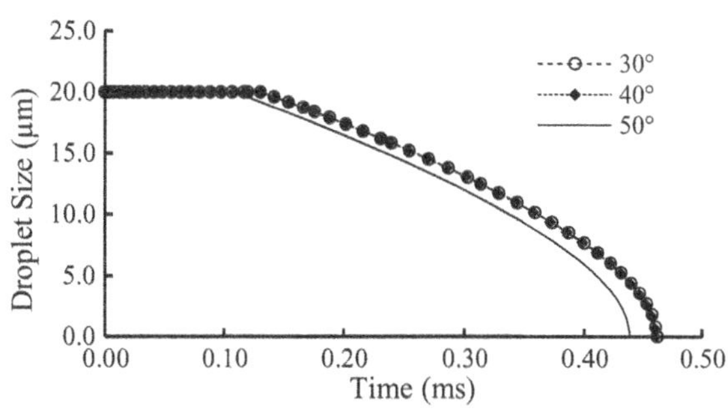

Fig. 5 Jet-A particle diameter reduction against time for various spray cone angles

3.2 Effect of Fuel Temperature on Combustion Characteristics

The effect of fuel temperature on droplet lifetime and spray penetration is shown in Figs. 6 and 7. The evaluation performed for Jet-A fuel in which the spray cone angle is set at 30° and the droplet size is 20 μm.

As shown in Fig. 6, rate of particle diameter reduction is slower as the initial fuel temperature reduces. This indicates that higher initial temperature is required to reach a higher rate of combustion. The particle diameter reduction for initial temperature at 300 K is the slowest, where the line stay further from the other two temperatures. As for initial temperature of 500 K, it took a shorter time for the particle to combust and reduced in diameter. By comparing these three temperatures, low initial temperature is not encouraged in a combustion chamber as it may slows down combustion rate while increases the gas emission.

Figure 7 shows the penetration of the spray for fuel temperature from 300 to 500 K. It is shown that lower temperature produces longer penetration. Evaporation rates should increase with fuel temperature [2]. According to the fuel properties, the boiling point of the Jet A fuel is around 462 K. When the initial temperature is at the

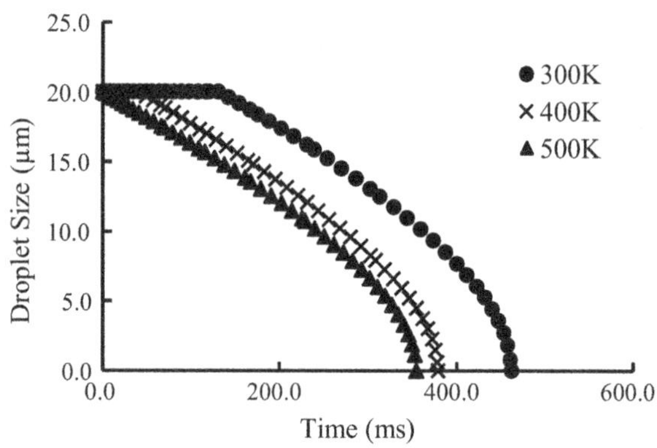

Fig. 6 Jet-A particle diameter reduction against time for various initial fuel temperature

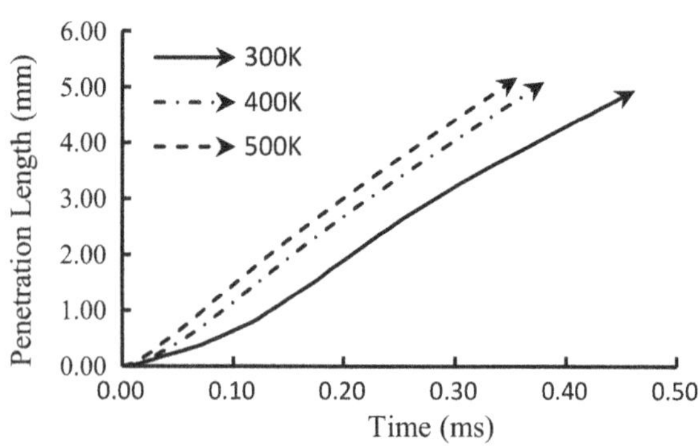

Fig. 7 Graph of penetration for different fuel temperature

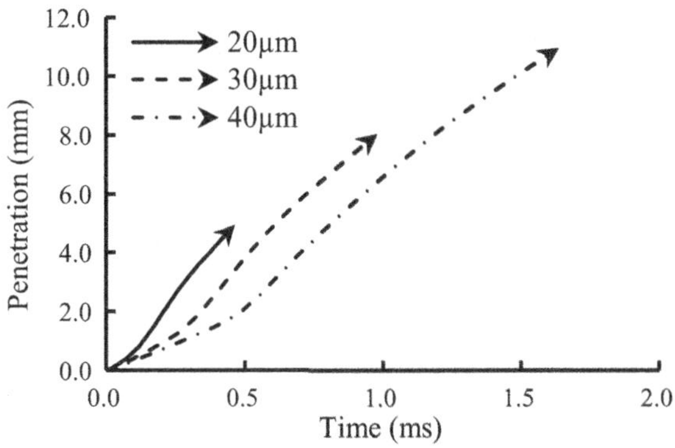

Fig. 8 Penetration length of Jet-A droplet at various droplet diameter

lowest which is at 300 K, the amount of heat is not sufficient for atomisation process. Theoretically, when the temperature is at 500 K which is near to boiling point, the penetration length should be the shortest as the atomisation process occurs in highest rate compared to the other two temperature.

3.3 Influence of Droplet Size on the Combustion Characteristics

The influence of droplet size on the penetration is depicted in Fig. 8. The graph shows penetration of Jet-A at the initial temperature of 300 K and the 30° spray half cone angle. Penetration of the droplet increases as droplet diameter increases. This observation consistent with finding obtained in Csemány and Józsa [15] who found that droplet inertia depends on its size, whereas the larger droplets accelerates slower compared to smaller droplets. It is found that smaller droplets accelerate faster in the initial phase, and also decelerates fast. When the droplet decelerates, the kinetic energy in the particle is small, however the droplets will keep moving forward where it gained enough energy for evaporation to occur. Due to this situation, when the droplet is at 40 μm, it took time for to accelerate to certain level and thus penetrated deeper in the chamber.

3.4 Influence of Alternative Fuels on Combustion Characteristics

Fuel properties also play an important role during the combustion process. Quality of the fuel injection can be affected by both the physical properties and chemical properties of the fuel. The fuel properties such as viscosity, surface tension and

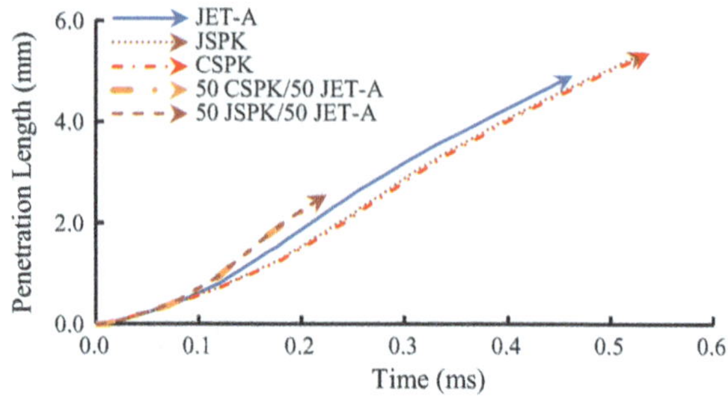

Fig. 9 Penetration length for different type of fuel

density will affect the penetration length of the spray. The influence of alternative fuels on penetration length and droplet lifetime is shown in Fig. 9.

Among all the tested fuels, JSPK has the longest penetration length followed by CSPK, Jet-A, 50CSPK/50Jet-A and 50JSPK/Jet-A. It indicates that 50CSPK/50Jet-A and 50JSPK/Jet-A have higher evaporation rate compared to the other three fuels. Evaporation rate of a fuel depends on its properties which include density, volatility, boiling point and viscosity. The comparison between JSPK and CSPK fuel properties shows both fuels have almost similar density, boiling point and viscosity thus providing almost similar penetration. Jet-A fuel has the highest viscosity among all the fuel, hence provides longer penetration length as the fuel is highly viscous and heavy [4]. Although density and viscosity of Jet-A are higher than JSPK and CSPK, thus should producing longer penetration length, however due to its lower boiling point the penetration of the spray is shorter than those two fuels. Fuel with lower boiling point has higher volatility consequently improves evaporation process. Since Jet-A has lower boiling point than both JSPK and CSPK, the evaporation of the fuel is much faster. Meanwhile as JSPK and CSPK have higher boiling point thus lower volatility, longer time for both fuels to evaporate observed. Mixing JSPK and CSPK with Jet-A improved the properties of the fuels resulted to improvement in evaporation process and spray penetration as shown for 50JSPK/50Jet-A and 50CSPK/50Jet-A.

4 Conclusions

This study evaluates the influence of droplet diameter, fuel temperature, spray cone angles and alternative fuels on droplet lifetime and spray penetration using CFD. A simple 2D cylindrical combustion chamber were used for evaluation. Temperature profiles inside the combustion chamber are presented. Results obtained show that higher spray half cone angle produced shorter penetration length and took shortest time for droplets to reduce its diameter. To make sure the fuel mixed with the air completely, the fuel injector and its spraying angle should be designed properly so that the penetration length is adequate for the mixing to occur. In addition, increasing fuel temperature particularly for high viscous fuel help in producing better evaporation in the chamber. In term of droplet size, smaller droplet size will lead to high combustion rate and thus better burning efficiency. Moreover, evaluation on alternative fuel shows that 50JSPK/50Jet-A and 50CSPK/50Jet-A had the shortest time to evaporate where it indicated that both fuel mixtures have higher evaporation rate due to lower boiling point and density.

Acknowledgements The authors would like to thank Universiti Sains Malaysia Research University grant (1001/PAERO/8014019) for sponsoring this work.

References

1. Zhang C, Hui X, Lin Y, Sung C-J (2016) Recent development in studies of alternative jet fuel combustion: progress, challenges, and opportunities. Renew Sustain Energy Rev 54:120–138
2. Lefebvre AH, Ballal DR (2010) Gas turbine combustion alternative fuels and emissions. Alternative Fuels and Emissions (3rd edn). Taylor and Francis Group, United State
3. Chen P-C, Wang W-C, Roberts WL, Fang T (2013) Spray and atomization of diesel fuel and its alternatives from a single-hole injector using a common rail fuel injection system. Fuel 103:850–861
4. Mazlan NM (2012) Assessing/optimising bio-fuel combustion technologies for reducing civil aircraft emissions. Cranfield University, England
5. Azami MH, Savill M (2016) Modelling of spray evaporation and penetration for alternative fuels. Fuel 80:514–520
6. Agarwal AK, Dhar A, Gupta JG, Kim WI, Lee CS, Park S (2014) Effect of fuel injection pressure and injection timing on spray characteristics and particulate size–number distribution in a biodiesel fuelled common rail direct injection diesel engine. Appl Energy 130:212–221
7. Dimitriou P, Weiji W, Peng Z (2015) A piston geometry and nozzle spray angle investigation in a DI diesel engine by quantifying the air-fuel mixture. Int J Spray Combust Dyn 7:1–24
8. Kim HJ, Park SH, Lee CS (2016) Impact of fuel spray angles and injection timing on the combustion and emission characteristics of a high-speed diesel engine. Energy 107:572–579
9. Jääskeläinen H, Khair MK (2019) Diesel spray formation and mixing. https://www.dieselnet.com/tech/diesel_combustion_spray.php. Last accessed 2019/04/20
10. Mark CP, Selwyn A (2016) Design and analysis of annular combustion chamber of a low bypass turbofan engine in a jet trainer aircraft. Propul Power Res 5:97–107
11. Mazlan NM, Savill M, Kipouros T (2016) Evaluating NO_x and CO emissions of bio-SPK fuel using a simplified engine combustion model: a preliminary study towards sustainable environment. Proc Inst Mech Eng Part G: J Aerosp Eng 231:859–865
12. Lefebvre AH, McDonell VG (2017) Atomization and sprays. CRC Press
13. Boateng AA (2016) Combustion and flame. In: Boateng AA (ed) Rotary kilns, 2nd edn. Butterworth-Heinemann, Boston, pp 107–143
14. Zhang T, Dong B, Zhou X, Guan L, Li W, Zhou S (2018) Experimental study of spray characteristics of kerosene-ethanol blends from a pressure-swirl nozzle. Int J Aerosp Eng
15. Csemány D, Józsa V (2017) Fuel evaporation in an atmospheric premixed burner: sensitivity analysis and spray vaporization. Processes 5:1–16

Design and Development of Renewable Energy Water Pump

Man Djun Lee, Pui San Lee, Jasper Ling, and Heng Jong Ngu

Abstract This study aims to develop a water pump that utilizes natural hydro energy as driving force to deliver water to a higher ground. The conceptual design of using water wheel to extract kinetic energy from water flow, and transfer the energy to power multiple piston pump was created based on the extensive literature review findings. The actual prototype is then built and modified to suit the actual environment considerations. Findings show that single pump is able to produce maximum pressure head of 7.14 m and the maximum volume flowrate achieved is 19.2 l/h (320 ml/min). However, when multiple piston is connected in series (in this research three pistons is used), the maximum water head increased to 13.77 m and the maximum volume flowrate about 19.2 l/h. This result shows that the water pump can be used in remote area or places at higher ground that does not have constant water access. Performance of the whole system can be improved by several factors such as adding more blades to the water wheel, steeper angle and better piston shaft design for water pump, and also proper water sealing of the whole system to prevent head loss and increase the overall performance.

Keywords Natural energy · Water pump · Rural development · Water supply · Piston pump

1 Introduction

Energy is defined as the source of power or the ability of matter to work because of its mass, movement, electric charge, etc. [1]. There are several types of natural energy that had been discovered until today such as electrical, solar, kinetic, potential, nuclear, wind, hydro energy, etc. The law of conservation of energy states that the

M. D. Lee (✉) · P. S. Lee · H. J. Ngu
School of Engineering and Technology, University College of Technology Sarawak, Sibu, Malaysia
e-mail: man.djun@ucts.edu.my

J. Ling
School of Built Environment, University College of Technology Sarawak, Sibu, Malaysia

P. Rajendran et al. (eds.), *Proceedings of International Conference of Aerospace and Mechanical Engineering 2019*, Lecture Notes in Mechanical Engineering,
https://doi.org/10.1007/978-981-15-4756-0_27

sum of the energy of a system is always constant, and it is not possible to destroy nor create energy, it can only be relocated or transform into another form of energy [2]. With this law in mind, scientist and professionals had been trying to develop devices that can utilize the natural energy and the principle of conservation of energy such as turbine and pump [3]. Pump is a device that impart energy to its fluid medium. In the case of water pump, they relocate the energy provided to them to the water. The results are usually the increment of water pressure and change in water velocity. The conversion of energy involved in this case is normally from any form of energy that is provided to the pump, to hydro power [4]. There are many parameters to determine a pump's performance. Mass flow rate of fluid through the pump is one of the most common factors to determine the pump's performance. However, there is another determinant that governs a pump's performance, that is by its net head H or known as Bernoulli head [4]. Net head has its dimension of length. Even when the pump is operating, the net head is considered as the same column height of water.

Generally, water pump is powered by electrical energy. This had put those who lives in remote area where the accessibility to electricity and water is a problem and this translated into a hard situation. In Sarawak state alone, the ratio of urban to rural area is about 52%:48% with 1.2 million of populations living in rural settings. Diving in deeper, the rural area is made up of 6235 villages and that is about 200,000 homes (excluding unexplored region). 1919 (30%) of villages have no access to constant electricity and that is 40,000 homes and 250,000 people [5]. Accessibility to the rural villages is one of the main challenges especially for Malaysia. The use of renewable energy for water pumping applications is attractive in rural areas of many developing countries [6]. However, considering there is no proper road structure to access to most of the rural villages, importing bulky machinery such as electrical generator and solar panel or other resources such as fuel is often as considered dangerous work [7]. Thus, this study aims to design a water pump that can utilize available hydro energy and without the help of electricity to pump water to the rural villages for their development. In this aspect, the research questions are:

1. What is the suitable mechanism to pump water by using energy from river in Malaysia?
2. What are the design considerations to develop such water pump?
3. What is the performance of the water pump?

This study is designed to addressed these issues. In most of the present industry, turbomachinery such as pump and turbine are an essential asset. For example, agriculture industry require pump to deliver water and fertilizers to the crops. Housing industry as well require pump to deliver water to higher floor for residents' usage. Thus, by providing a renewable energy water pump to these industries will be able to help in term of reducing the overall operation cost because it uses renewable and clean energy. On the other hand, Malaysia is a country that acquire plenty of natural resources compared to other countries. Malaysia grows its own agricultural products and is internationally dominant in its production line such as rubber and pepper. To maintain these crops in such humongous scale, water delivery system must be at top notch status and it is costly to do so. Thus, by implementing the renewable energy water pump, it helps the agriculture sector to save cost on energy consumption and the extra capital can be focused on ameliorating the economy even further. As mentioned earlier, implementation of renewable energy water pump succor in Malaysia's economy. Improved economy indicates improved lifestyle to the citizen as well. Not only that, the renewable energy pump does not require any external energy source such as electricity. Thus, by using renewable energy will help in reducing the pollution index in Malaysia and provide Malaysian a healthier lifestyle.

2 Literature Review

Due to the unique geographical location of Malaysia (Malaysia is located near to the equator), Malaysia has an equatorial climate nature that is hot and humid all year round. However, despite that, Malaysia has a relatively high rainfall per year, measured 98 in. (248.92 cm) a year compared to other developed country such as United States has only 32.21 in. (81.81 cm) rainfall per year [8, 9]. Having this amount of rainfall per year indicates that Malaysia has higher cloud coverage and this make Malaysia a unfavorable location for installing solar panel. Other than that, Malaysia's forest type is dominated by a very dense and vast biodiversity tropical rainforest which is known as lowland dipterocarp forest which mainly consist dipterocarp trees from family of Dipterocarpaceae [10]. Dipterocarp trees from Dipterocarpaceae family are mostly large forest emergent species and the maximum height they can grow up to range from 40 to 70 m which allow them to gain the most sunlight compared to other tree species [11]. However, due to this overwhelming height, they can block up to 70% of sunlight reaching the ground which make the approach of using solar energy to operate the water pump another step further from us [12]. In addition, wind energy could not be used as the power source because wind turbine is absurdly huge and required large space to install it. Installing one wind turbine requires 25 m diameter empty field and if we were to install multiple wind turbine, the distance between two adjacent turbine has to be five times their turbine diameter (which is 25 m normally). Not only that, the possible power output by a single normal windmill is only 0.0756 W per m^2 land area which is relatively small compared to the power required by an average water pump which is 150 W [13, 14]. Lastly, as mentioned earlier it is also inconvenient and dangerous to bring in bulky equipment into remote area because of transportation difficulty [7]. In short, review from literature suggest that it is more convenient to use hydro energy compared to other form of renewable energy to power up water pump is remote area.

2.1 Hydro Energy

Water itself can act as a medium to carry energy such as thermal energy, potential energy and kinetic energy. By utilizing the principle of conservation of energy, several methods were invented to extract energy from water. Hydroelectric is a technology that uses water (hydro) to generate electricity (electric) and dam in one of the common hydroelectric. Dam is a barrier/reservoir that traps water in place, and then release the water to a turbine system to generate electricity. That is the rough concept of how dam uses water to generate electricity. When the water is trapped, potential energy carried by the water is at its maximum. As the release valve is opened, water is gushed out from the valve and potential energy is converted to kinetic energy. As the water rushes down with its maximum velocity, it will pass through a series of turbine. At that instant, the shaft is turned by the water and kinetic energy is converted to mechanical energy and then to electricity.

Hydraulic Ram is the only pump that does not rely on other form of power but only kinetic energy in the water to pump water. The key mechanism that allows hydraulic ram to pump water without the help of external energy is the water hammer effect. Water Hammer Effect is a phenomenon that utilizes the incompressible characteristic of water to increase the pressure within itself. Imaging the water as a rigid body and travel at a certain velocity. If the water were to be stopped suddenly, the entire system will experience a sudden pressure propagation shock due to the water is incompressible. This might be bad for the system because if the water were to travel in a high velocity and stopped suddenly, it might damage or even blow up the entire system [14]. However, the water hammer effect is not necessarily bad all the time, not in the case of hydraulic ram. Figure 1 shows the basic components of hydraulic ram.

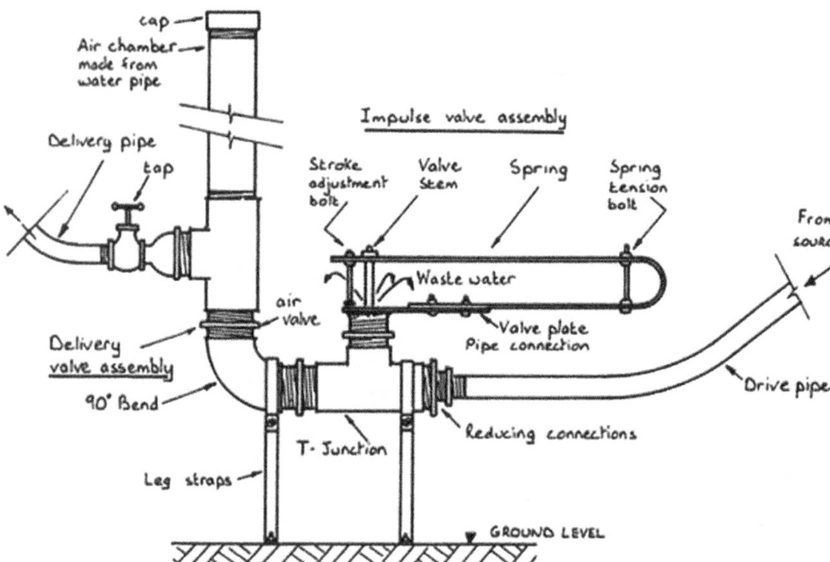

Fig. 1 Schematic diagram of hydraulic ram [15]

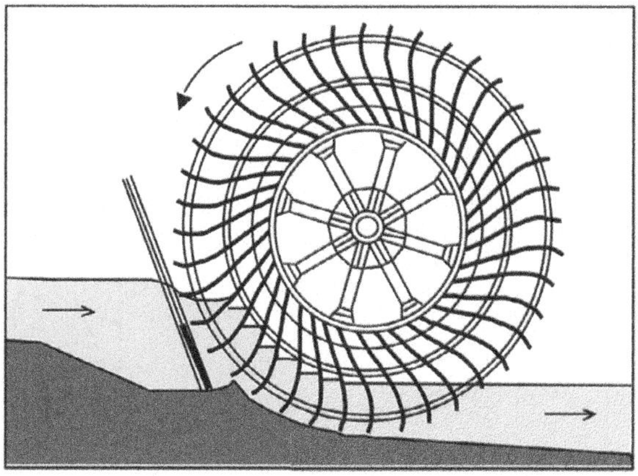

Fig. 2 Schematic diagram of water wheel [16]

The whole working process of hydraulic pump will start from water entering the drive pipe from a certain height and flow out from the impulse valve (because the air valve will be closed). As the water had gained enough velocity, it will shut the impulse valve and the water will experience a sudden pressure spike (water hammer effect). As the pressure of the water increases, it will force the air valve to open and water will flow to the air chamber. As the water in air chamber increases, the volume of the air will decrease and its pressure will increase (obeying Boyle's Law, $P_1 V_1 = P_2 V_2$). As the air chamber gained enough pressure, it will then push the water into the delivery valve and finally to the designated destination. Thus, the entire process involved no external energy such as electricity nor solar energy. However, the pressure head of such mechanism is not high making it its limitation in water delivery.

Water wheel is one of the oldest methods that uses water to generate energy. It was recorded that the first water wheel ever invented was during the first century of BC [16]. The working principle of water wheel is similar to that of dam. It is normally installed on water source. As water passes through its "wheel" that has "fins" on it, the wheel will rotate and convert the kinetic energy to mechanical energy. Figure 2 illustrates the working principal of a water wheel.

2.2 Pumps

There are two main types of pump which are positive-displacement machines where fluid is transferred into a closed volume, and dynamic machines where there is no closed volume. There are many parameters to determine a pump's performance. Mass flow rate of fluid through the pump is one of the most common factors to determine the pump's performance. However, there is another determinant that governs a pump's performance, that is by its net head H or known as Bernoulli head [4]. Net head has its dimension of length. Even when the pump is operating, the net head is considered as the same column height of water. Fluid can be put into two extreme situations

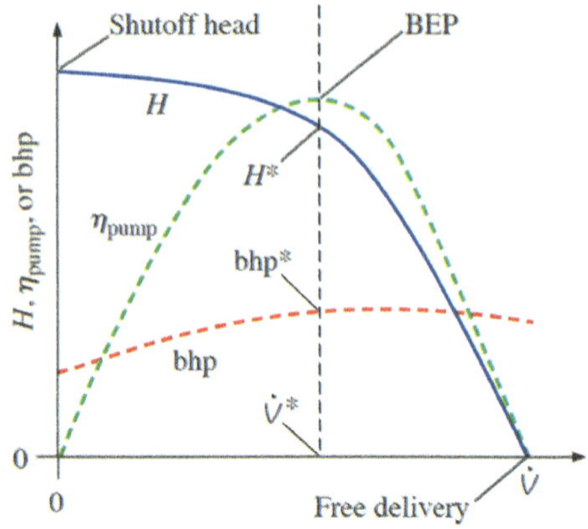

Fig. 3 Pump performance curve [4]

responding to the net head the fluid possesses during that particular condition. For instance, when a fluid has high net head, the mass/volume flow rate of the fluid could be zero ($\dot{V} = 0$). This normally happens when the outlet is block, and the net head in this extreme situation is called the shutoff head. The other extreme situation on the other hand, is when the net head is equal to zero ($H = 0$). At this instant, the mass/volume flow rate will be at its maximum and this flow rate is addressed as free delivery. These two situations is called the extreme situation because this is when the pump does no useful work (extreme situation 1) or there is no load on the pump (extreme situation 2) [4]. Pump performance graph is acquired when these two situations and pump's reaction under different situation is plotted into a graph. The example of pump performance curve is shown in Fig. 3. Thus, pump user has to be aware that a pump can only perform under its own performance curve and should always refer to the pump's performance graph as a guide for the pump to work well under different circumstances [4].

2.3 Gap Analysis

After extensive review in literature on journal articles, books and internet sources in the field of natural energy water pump and its application domain, several gaps are found such as most research were focusing on integrating photo voltaic technology (solar energy technology) with water pump. However, solar energy is not feasible in Malaysia remote area due to majority of Malaysia's tree are Dipterocarpaceae trees which has extraordinary height that will cover the sunlight and high cloud cover frequency. On the other hand, limited study had been conducted on micro-hydroelectric in the past five years (2014–2018). There are demands in Malaysia because Malaysia has quite a number of remote areas compared to other developed countries [17]. Apart from that, most of the recent research focus on pumps running on electricity or electricity generated by other renewable energies, very little study had focused on developing micro-hydro pump. In this aspect, this study is designed to fill these gaps.

3 Methodology

This study started with literature review on the natural energy water pumping concepts and key parameters governing the performance of the pumps. Upon determining the suitable design considerations and operating parameters, a small-scale model is then developed and tested in laboratory. The performance of the system is recorded and optimized in the later stage. This an applied research. To achieve the objective of this study, the methodology plan is shown in Fig. 4. This study is divided into 3 phases which will be explained below:

Phase 1: Conceptual Design and CAD Modeling

In this phase of work, software will be employed to design conceptual model of the device. Initially, the conceptual model would be developed from literature review findings. Published materials related to natural energy water pumps are reviewed

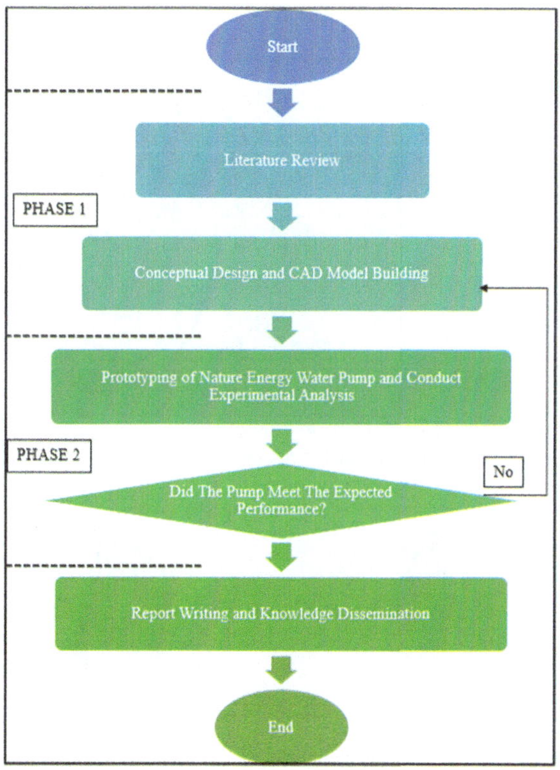

Fig. 4 Research methodology

and from the findings key parameters for the design are generated. The conceptual design is shown in Fig. 5. The concept modelled after water wheels with angle of blade about 45° with width of blade 1 ft long with diameter of the wheel of 4 ft (Fig. 6). The system utilized the flowing river to rotate the wheel too drive a shaft which in turn the shaft will turns 3 piston pumps using cam mechanism to create pressure to deliver water (Fig. 7). The piston pumps are connected to check valves to prevent backflow.

Phase 2: Rapid Prototyping and Functionality Testing

In this phase, a lab scale prototype is developed based on the output from phase 1. The lab scale prototype developed is shown in Fig. 8. Investigation on the prototype is carried out to ensure the functionality of the prototype. The prototype is tested in laboratory by running water through a 1.0 hp pump with average head 25 m and the flowrate is being controlled and set constant to 50 L/min with pipe diameter of 1 in. by ball valve. The prototype is partially submerged in the 1000 L tank covering about 30% of the water wheel. This will roughly create flow velocity of 1.65 m/s. The purpose of this is to simulate actual river flow conditions in Malaysia with average flow velocity of between 1.5 and 4 m/s [18]. The data collected to determine the performance of this prototype are such as pressure head from gauge meter with the working of single or triple pumps and the flowrate that generated from the pumps to plot out pump curve for this system.

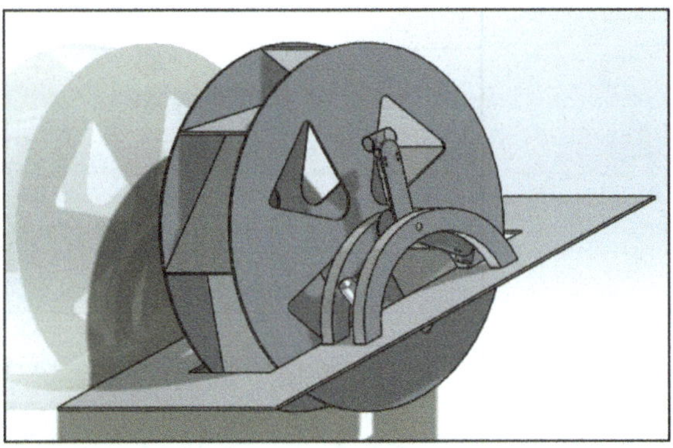

Fig. 5 Conceptual model

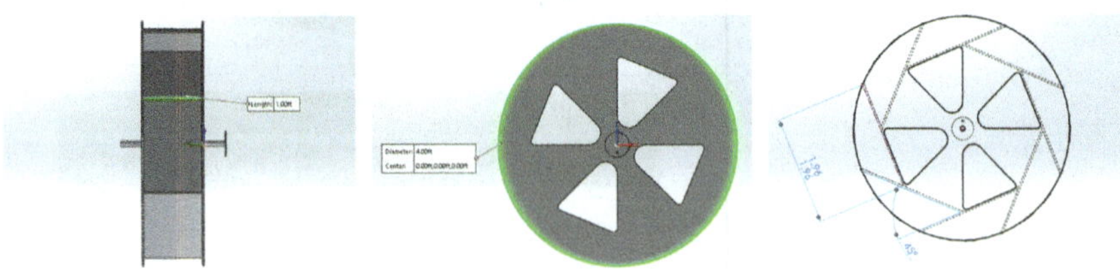

Fig. 6 Details of conceptual model

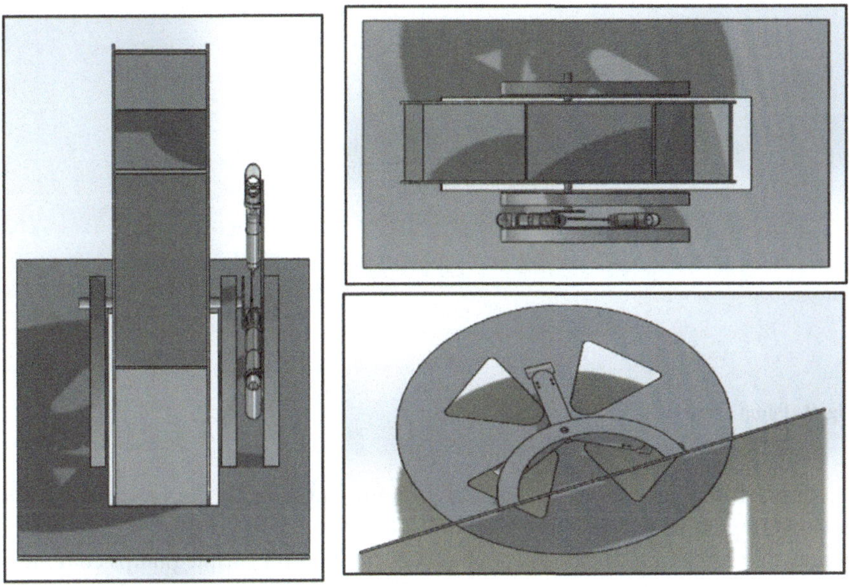

Fig. 7 Driving mechanism for pump

Fig. 8 Lab scale prototype

4 Results and Discussions

Based on the data collected from laboratory testing, the pump performance curves are generated based on single pump and triple pump configurations. The pump curves are shown in Fig. 9.

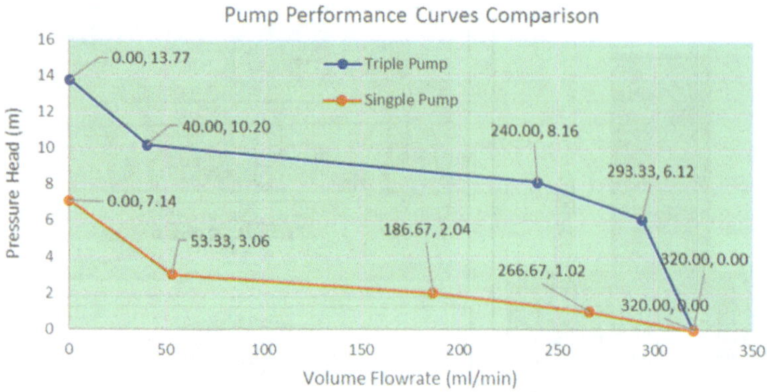

Fig. 9 Pump curves

Results from Fig. 9 show that with driving only single pump, the system is capable of delivering maximum pressure head of 7.14 m and maximum flow of 320 ml/min. Meanwhile, with similar pumps attached together to form a triple pumps configuration, the system is capable of delivering 13.77 m of pressure head and 320 ml/min of water flowrate. The findings indicate that using river flow as a means to pump water is possible. The water wheel mechanism is significant to drive up to three piston pumps with simulated river flow velocity of 1.65 m/s. The findings indicate that the pressure head generated is quite significant and suitable to deliver water to higher ground. The main cause of low water flowrate is that the size of piston used in this study is relatively small. Increasing the size of piston will significantly improve the flowrate for piston pump. Another limitation for this study is that the water wheel has only eight blades installed. After consecutive studies, it is to realized that the efficiency of water wheel can be increased by increasing the number of the blades, or decrease the angle between two blades [19].

The main aim of this study is to design a water pump suitable for the use in remote area that is not connected to national power grid. These areas do not have electricity to power an electrical water pump and also for people who lives in the higher ground because the outcome of this study indicate that for such application is possible with adequate pressure head. Not only that, the pump is beneficial to the agriculture sector as well. Often, there are limited electric access in the planting area due to the challenging geographical conditions and it would be difficult and costly for the owner instill an electric grid system when the area is massive. Thus, the pump can be used to deliver water for agricultural irrigation. In addition, as the water wheel collected the kinetic energy from the water flow, the energy is then converted to rotational energy. Other than delivering water, the rotational energy can be used to run an alternator to produce electricity as well. A simple modification as adding in a gear train could have increase the rotational speed and is decent for generating electrical energy.

5 Conclusion

This study is set out to develop a renewable energy water pump to assist rural development without assess to national power grid. A lab scale prototype was developed to test the concept generated from literature review findings. The primary energy required to drive the pump is flowing energy from the river to drive the water wheel. Upon collecting the kinetic energy from the river, it is converted to rotational energy and mechanical energy to power the pump and thus no external source of unrenewable energy is involved. The core concept throughout the experiment is to impart and extract energy from water to execute other significant task where in this study is to deliver water to secondary location usually at an elevated height. The findings suggested that the it is possible to use river flow as a mean to pump water and the pressure head is significant. The design could be improved further to increase the overall performance such as increase size of piston, increase the number of blades and modify the blade angles.

Acknowledgements Authors would like to express gratitude for the financial support received from University College of Technology Sarawak (UCTS) under internal research grant scheme (UCTS/Research/4/2018/8).

References

1. Hornby AS (2015) Oxford advanced learner's dictionary. Oxford University Press, Oxford
2. Mohamed NS, Sulaiman ZA (2012) Physics college. IPTA Publication Sdn. Bhd, Malaysia
3. Nehrenheim E (2018) Energy and natural resources
4. Cengel YA, Cimbala JM (2014) Fluid mechanics, fundamentals and applications, third edition in SI units. McGraw Hill Education, New York
5. Shiun C (2016) Rural electrification in Sarawak, Malaysia: challenges for Mini-Hydro & Solar Hybrid Solutions, Sarawak
6. Mohamed ES, Papadakis G, Mathioulakis E, Belessiotis V (2005) The effect of hydraulic energy recovery in a small sea water reverse osmosis desalination system; experimental and economical evaluation. Desalination 184:241–246
7. Argaw N (2004) Renewable energy water pumping systems handbook, Denver, Colorado
8. Saw SH (2007) The population of peninsular Malaysia. Institute of Southest Asian Studies Singapore, Malaysia
9. National Centers for Environmental Information, State of the Climate (NOAA) (2017) National climate report for annual 2017, published online January 2018, retrieved on April 15, 2020 from https://www.ncdc.noaa.gov/sotc/national/201713
10. World Wide Fund for Nature (WWF). The Malaysian Rainforest, Malaysia
11. World Wide Fund for Nature (WWF). Low Land Forest
12. Engkik S, Saw LG, Chung RCK (2004) Tree flora of Sabah and Sarawak. Sabah Forestry Department, Forest Research Institute Malaysia (FRIM), Sarawak Forestry Department, Malaysia
13. Hitachi. Hitachi Home Electronics (TM-60L)
14. Thorley ARD (2004) Fluid transients in pipeline systems: a guide to the control and suppression of fluid transients in liquids in closed conduits. Professional Engineering Publishing

15. Watt SB (1975) A manual on the hydraulic ram for pumping water. Intermediate Technology Publications
16. Viollet P-L (2017) From the water wheel to turbines and hydroelectricity. Technological evolution and revolutions. C R Méc 345:570–580
17. Behrouzi F, Nakisa M, Maimun A, Ahmed YM (2016) Renewable energy potential in Malaysia: hydrokinetic river/marine technology. Renew Sustain Energy Rev 62:1270–1281
18. Saupi AFM, Mailah NF, Radzi MAM, Mohamad KB, Ahmad SZ, Soh AC (2018) An illustrated guide to estimation of water velocity in unregulated river for hydrokinetic performance analysis studies in East Malaysia. Water (Switzerland) 10
19. Sritram P, Suntivarakorn R (2017) Comparative study of small hydropower turbine efficiency at low head water. Energy Procedia 138:646–650

CAD-Based 3D Grain Burnback Analysis for Solid Rocket Motors

Ahmed Mahjub, Qummare Azam, M. Z. Abdullah, and Nurul Musfirah Mazlan

Abstract Propellant grain burnback analysis is crucial for solid rocket motor design and performance prediction. Unlike 2D grain configurations, 3D configurations are complex, hence, simulating their burnback inside the rocket combustion chamber is a tedious and time-consuming process. This study proposes a fast and simple approach for modeling, and evaluating the burning area of 3D propellant grains, based on particular features available in a commercial CAD software. A common 3D Finocyl grain configuration available in the literature has been taken as a test case. The results obtained from the proposed approach were compared with the published experimental data and showed good agreement. The proposed approach can handle any arbitrary complex grain geometry and provide fast and reliable analysis for the preliminary design stage of solid propellant rocket motors.

Keywords Solid rocket motor · Burnback analysis · CAD · Finocyl propellant grain

1 Introduction

Propellant grain burnback/regression analysis is critical to decide what grain configuration/shape is suitable for a certain solid rocket motor (SRM) design. The motor performance i.e.; the thrust/pressure-time profile is highly dependent on the variation of burning surface area during the propellant combustion phase [1]. Grain shapes used in SRMs can range from simple 2D geometries (like end-burning and cylindrical) to the complex 3D geometries (like finocyl and wagon wheel) [2]. Frequently,

A. Mahjub · Q. Azam · N. M. Mazlan (✉)
School of Aerospace Engineering, Engineering Campus, Universiti Sains Malaysia, 14300 Nibong Tebal, Penang, Malaysia
e-mail: nmusfirah@usm.my

M. Z. Abdullah
School of Mechanical Engineering, Engineering Campus, Universiti Sains Malaysia, 14300 Nibong Tebal, Penang, Malaysia

P. Rajendran et al. (eds.), *Proceedings of International Conference of Aerospace and Mechanical Engineering 2019*, Lecture Notes in Mechanical Engineering,
https://doi.org/10.1007/978-981-15-4756-0_28

the grain design is a tedious process that includes many iterations in order to reach the final shape with a specific geometry that satisfies specified mission requirements.

Various methods and tools are used to design and analyze different types of grain configurations, including analytical, numerical, and drafting/CAD techniques. Although analytical methods have widely been used since they can provide a fast and exact solution for simple shapes, their use in complex geometries is limited and usually requires an intensive parameter input process [3, 4]. Numerical methods provide an approximate solution for all grain shapes, but typically the solution quality and computation time depend heavily on the mesh preparation which makes it not suitable in conceptual and preliminary design stages [5, 6]. Moreover, both analytical and numerical methods often require that the initial geometry of the grain to be analyzed is prepared in a CAD modeler. On the other hand, drafting methods have been traditionally used as the first technique for predicting the grain regression and it was manually performed in the past [7]. With the increase of computer power and the developing of sophisticated computer-aided design (CAD) software, drafting techniques have become more popular in grain burnback analysis and overcome most of the limitations posed by other methods. PIBAL [8] and ELEA [9] are CAD-based software developed to predict the performance of SRM using 2D and 3D grains. Kamran and Guozhu utilized parametric modeling and offset feature of CAD software to simulate the surface regression and optimize the design of 3D finocyl and radial slot propellant grains [10, 11].

The grain regression in previous studies is achieved by manually offsetting the predefined burning surfaces by equal burn thickness in a normal direction. The burnt surface area is then obtained by calculating the change in grain volume at each burn step and dividing the incremented volume by the burnt thickness. This process is repeated until the grain burns completely. Obviously, there is high human interaction in this procedure which increases both the processing time and possibility of errors. A more automated approach is presented by Püskülcü and Ulas [12] who used the CAD Unigraphics NX software to parametrically model the initial grain geometry, and performing the burnback by only changing theses parameters through a built-in command "part family". Although this approach was relatively easy to apply and gave satisfactory results, there still a considerable amount of user intervention as the parameters need to be changed manually at each burn step. Additionally, the burning area evaluation still relies on the calculation of the change in volume.

The objective of this article is to provide a methodology for 3D grain burnback analysis that is general, affordable, and relatively simple which can be directly used to facilitate the preliminary design of SRMs. The approach adopted in this work is based on CAD Creo Parametric 3D Modeling Software [13]. Creo is a professional 3D CAD product design and development software with powerful analysis tools and parametric/direct modeling capabilities. The 3D solid model of the initial grain geometry is created parametrically, dynamic parameters/dimensions are related to each other using Relation command, the burning surfaces area are selected and analyzed using Analysis tool, the 3D model is regenerated/regressed for each burn step using Regenerate command by equal thickness defined in the model relation.

2 Geometrical Modeling and Burnback Analysis

2.1 3D Grain Modeling

A very common 3D Finocyl (Fin in Cylinder) grain configuration is selected, due to its suitability for a wide range of SRM applications in which high volumetric fraction and dual-thrust profile are desired. The geometrical parameters that define the Finocyl grain are grain length (L1), grain outer diameter (R), fin depth (LS), circular port radius (R1), fin-half width (R2), fin height (L2), and the number of fins (N) as shown in Fig. 1. Using these parameters, the 3D model is created using “Sketch”, “Extrude”, and “Pattern” commands in three steps. First, the cylindrical part is created by extruding two circle sketches representing the parameters R1 and R2 on both sides by half the length value (L1-LS) in each direction. Second, the fin part is created by extruding the same parameters R1 and R2 in one direction by the fin depth value LS, a sketch of one fin is created using parameters L2 and R2 and extruded in the model. Finally, the “pattern” command is used to rotate the fin feature around the axis by the number of fins value N.

Geometrical parameters and the 3D model of the finocyl grain test case are given in Figs. 2 and 3 respectively.

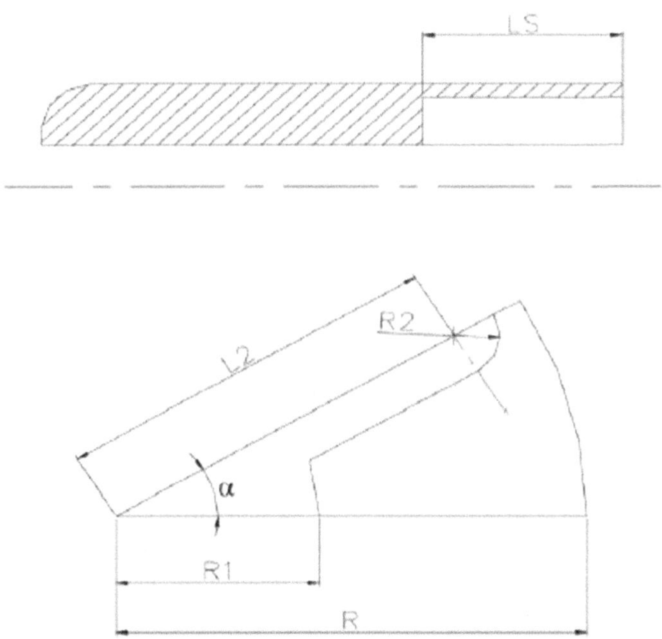

Fig. 1 Parameters of finocyl grain [12]

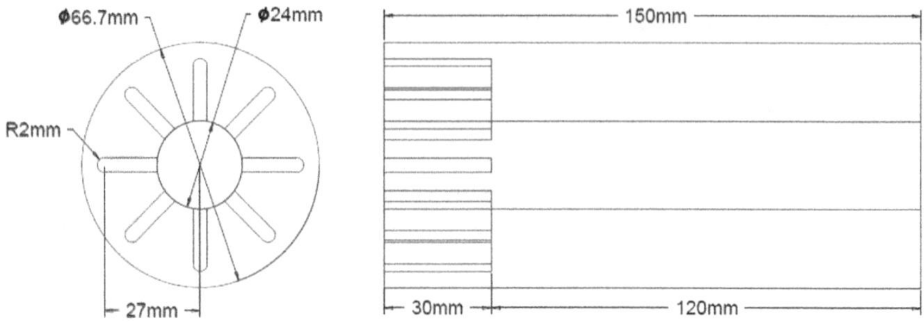

Fig. 2 Finocyl grain test case geometry

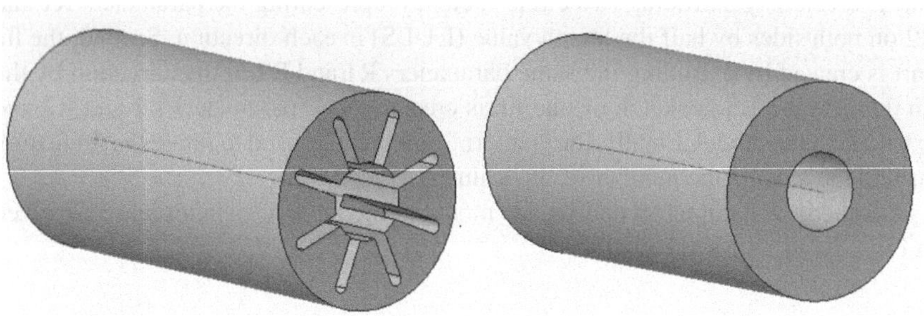

Fig. 3 Finocyl grain test case 3D model (front and rear view)

2.2 Burnback Analysis

After the grain model is prepared, a conditional relation is defined using the "Relation" tool that controls the geometrical parameters and allows corresponding model dimensions to be automatically changed at each burn step. The grain is set to burn internally (radially) and on both ends (axially), which means the only the outer surface is inhibited. Two "IF" statements are written. The first one, IF R1 < R then R1, R2 increase radially and L1, L1-LS decrease axially by an equal burn step (thickness) of 1 mm (burning mode). The second one, IF R1 $\geq$ R, then all parameters return to their initial values and (no-burning mode).

"Analysis" tool is now used to evaluate the burning area of the grain. The burning surfaces (uninhibited) are arbitrarily decided according to the design requirements and must be selected manually only once on the initial geometry of the model. "Measure Area" feature in the "Analysis" tool is used for that, which also provides an option for saving the measured area on the graphic window or as an editable text file for any further analysis.

Finally, a useful command in Creo Model Toolbar "Regenerate" is used to regenerate new models from the parent one according to the dimensional relationship previously set to the model. New models are automatically generated by only clicking "Regenerate" command for each burn step (see Fig. 4), the measured area is

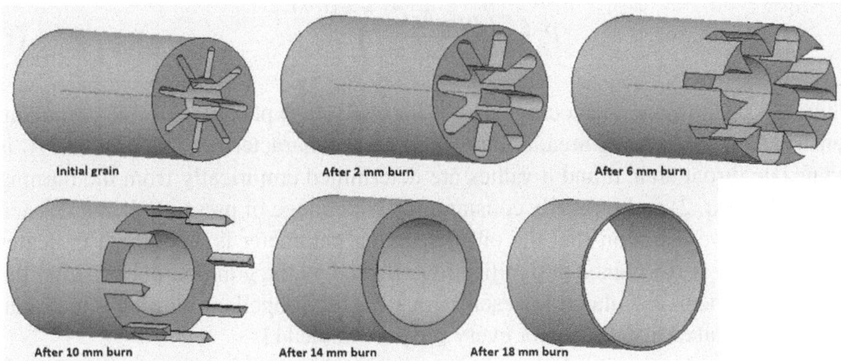

Fig. 4 Grain burnback at different burn steps

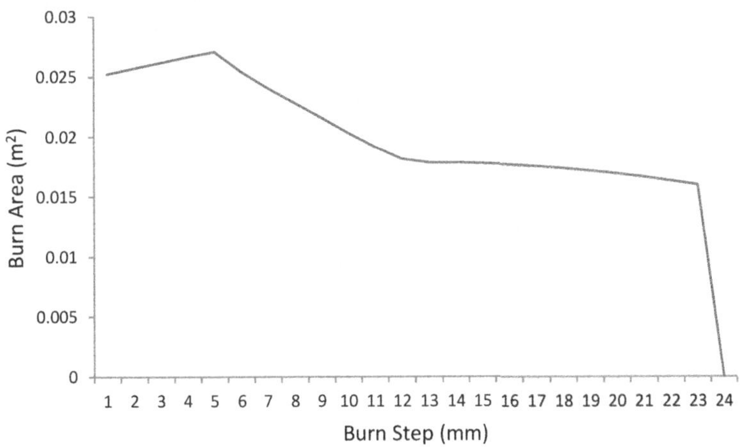

Fig. 5 Burning area at each burn step

changing accordingly and can be copied directly from the graphic window or can be exported at any time from the stored text file. A complete table or plot of burning areas and burn steps can then be easily prepared as shown in Fig. 5.

3 Performance Prediction

To predict an SRM performance, the burning area data obtained from the burnback analysis is used in the well-known steady-state ballistic model which produces results in terms of chamber pressure P_c versus time t [1, 14]. P_c is calculated from Eq. 1, which is typically sufficient for the preliminary design stage.

$$P_c = \left(\frac{a \rho_p A_b C^*}{A_t} \right)^{1/(1-n)} \tag{1}$$

where a is the burning rate coefficient, n is the pressure exponent, ρ_p is the propellant density, A_b is the burning area, C^* is the propellant characteristic velocity, and A_t is the nozzle throat area. a and n values are determined empirically from the burning rate r_b law (Eq. 2) and typically constant in a wide range of pressure. It can be seen from the above equation that the only changing parameter is A_b, which indicates that the motor performance is significantly affected by the grain burnback. After the pressure has been calculated for each burn step, the propellant burning rate r_b can then be calculated from Eq. 2 for every pressure value [1].

$$r_b = a P_c^n \tag{2}$$

It is also known that r_b is essentially the burnt distance (thickness) Δx divided by the elapsed time Δt, i.e.; $r_b = \Delta x / \Delta t$. Where Δx is the arbitrary chosen burn step and r_b has been calculated above, therefore, Δt is easily computed at every Δx and its total value is the total burning time t_b needed for the propellant grain to completely burn.

The obtained burnback data A_b and Δx, along with the propellant data and experimental pressure-time curve from a ballistic motor static test performed by Püskülcü and Ulas [12] are used to validate our approach. The HTPB/AP/Al composite propellant was used for the static test, its a and n values were 7.25×10^{-5} and 0.32 respectively. ρ_p and C^* values were and 1450 m/s and 1740 kg/m^3 respectively. The ballistic motor had a nozzle throat diameter of 11 mm. The comparison of the predicted pressure-time curve with the experimental one in Fig. 6 shows good similarity in the steady-state region of the two curves. The transient region (tail-off) in

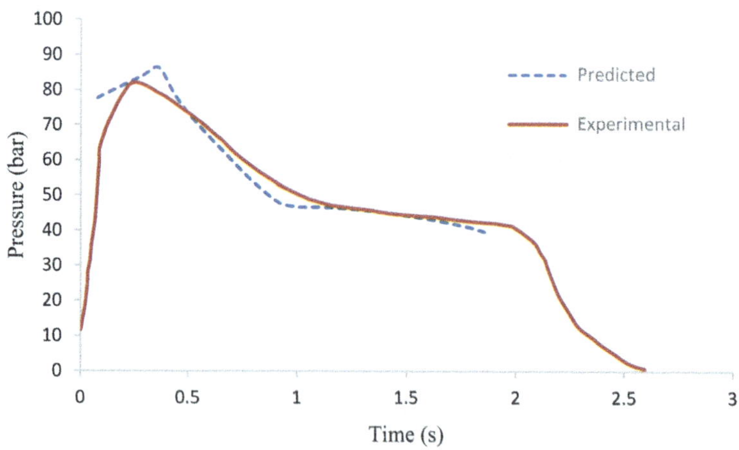

Fig. 6 Comparison of the experimental [16] and predicted pressure-time curves

the curve is not covered by the steady-state model, unsteady-steady state internal ballistic models are used for that to investigate more burning phenomena such as erosive-burning [15], which is out of the scope of this research.

4 Summary and Conclusion

This effort explains in detail a CAD-based approach for the burnback analysis of 3D solid propellant grain shapes. The CAD software utilized is Creo 5.0, which has advanced and powerful geometrical analysis tools. A 3D finocyl grain configuration is taken as a test case to demonstrate the reliability of the suggested approach. The burnback data are obtained in terms of burning area versus burn step, thereafter, used as an input for the internal ballistic model to predict the SRM performance. The output results in terms of the pressure-time profile are obtained and then compared with the published experimental data. The comparison reveals that our approach is good enough and produces acceptable data for the preliminary design stage of SRM. The proposed methodology provides a fast, easy, and affordable tool to perform the burnback analysis for any 3D grain configuration.

Acknowledgements The authors would like to thank Ministry of Higher Education Malaysia's Fundamental Research Grant Scheme (203/PAERO/6071435) and Universiti Sains Malaysia Research University grant (Grant No: 1001/PAERO/8014019) for funding this work.

References

1. Sutton GP, Biblarz O (2016) Rocket propulsion elements. Wiley, Hoboken
2. Brooks WT (1972) Solid propellant grain design and internal ballistics. NASA SP-8076, vol 8076. NASA Special Publication
3. Hartfield R, Jenkins R, Burkhalter J, Foster W (2003) A review of analytical methods for solid rocket motor grain analysis. In: 39th AIAA/ASME/SAE/ASEE joint propulsion conference and exhibit, p 4506
4. Coats D, Nickerson G, Dang A, Dunn S, Kehtarnavaz H (1987) Solid performance program (SPP). In: 23rd Joint propulsion conference, p 1701
5. Mesgari S, Bazazzadeh M, Mostofizadeh A (2019) Finocyl grain design using the genetic algorithm in combination with adaptive basis function construction. Int J Aerosp Eng
6. Ata Y, Kurtulus DF, Arkun OU (2018) Development of a 3D grain burnback simulation tool for solid rocket motors In: Advances in sustainable aviation. Springer, Berlin, pp 65–90
7. Barron J (1968) Generalized coordinate grain design and internal ballistics evaluation program. In: 3rd Solid propulsion conference, p 490
8. Dauch F, Ribereau D (2002) A software for SRM grain design and internal ballistics evaluation, PIBAL. In: 38th AIAA/ASME/SAE/ASEE joint propulsion conference & exhibit, p 4299
9. Saintout E, Le Roux A, Ribereau D, Perrin P (1989) ELEA—A tool for 3D surface regression analysis in propellant grains. In: 25th Joint propulsion conference, p 2782
10. Kamran A, Guozhu L, Rafique AF, Zeeshan Q (2013) ±3-Sigma based design optimization of 3D finocyl grain. Aerosp Sci Technol 26:29–37

11. Kamran A, Guozhu L (2010) Design and optimization of 3D radial slot grain configuration. Chin J Aeronaut 23:409–414
12. Püskülcü G, Ulas A (2008) 3-D grain burnback analysis of solid propellant rocket motors: part 2—modeling and simulations. Aerosp Sci Technol 12:585–591
13. CAD Software Solutions. PTC Inc. https://www.ptc.com/en/products/cad/creo/. Last Accessed 10 Sept 2019
14. Davenas A (2012) Solid rocket propulsion technology. Newnes
15. Willcox MA, Brewster MQ, Tang K-C, Stewart DS, Kuznetsov I (2007) Solid rocket motor internal ballistics simulation using three-dimensional grain burnback. J Propul Power 23:575–584
16. Püskülcü G, Ulas A (2008) 3-D grain burnback analysis of solid propellant rocket motors: part 1—ballistic motor tests. Aerosp Sci Technol 12:579–584

Numerical Simulation of Suddenly Expanded Flow from Converging Nozzle at Sonic Mach Number

Ambareen Khan, Mohd Azmi Ismail, and Nurul Musfirah Mazlan

Abstract Low pressure created at the blunt base results in a substantial increase in the base drag. Controlling the base drag is required by regulating the base pressure. This study presents a turbulent numerical analysis of suddenly expanded flow in a 22 mm suddenly expanded duct with a converging nozzle which operates at sonic Mach number. A rib is placed circumferentially along the inner duct diameter. The rib is located at a different position of diameter duct namely 1D, 2D, 3D, and 4D. The height of the rib is increased from 1 to 4 mm, and the width is maintained at 3 mm throughout the study. The nozzle pressure ratio of the flow is ranging from 1.5 to 5. The numerical analysis carried out is also compared with the duct without rib. The results obtained reveal that the rib changed the flow field and causes variations in base pressure. The rib at the 2D location is found to be the base pressure controller at a moderate level. While 1D reduces the pressure, but the ribs placed at 3D/4D results in a significant increase in the base pressure level for all nozzle pressure ratios of the present study. The pressure variations and velocity field are also studied using simulation results, and they show that these parameters are influenced mainly with the use of rib of a particular location and the height of the rib.

Keywords Nozzle pressure ratio · Base pressure · Passive control · Sudden expansion · Sonic flow

1 Introduction

Low pressure created at the blunt base results in a substantial increase in the base drag. Regulating the base pressure is an option to control the base drag. It is known that flight Mach number is an essential factor affecting the base pressure [1]. Figure 1

A. Khan · N. M. Mazlan (✉)
School of Aerospace Engineering, Engineering Campus, Universiti Sains Malaysia, 14300 Nibong Tebal, Penang, Malaysia
e-mail: nmusfirah@usm.my

M. A. Ismail
School of Mechanical Engineering, Engineering Campus, Universiti Sains Malaysia, 14300 Nibong Tebal, Penang, Malaysia

P. Rajendran et al. (eds.), *Proceedings of International Conference of Aerospace and Mechanical Engineering 2019*, Lecture Notes in Mechanical Engineering,
https://doi.org/10.1007/978-981-15-4756-0_29

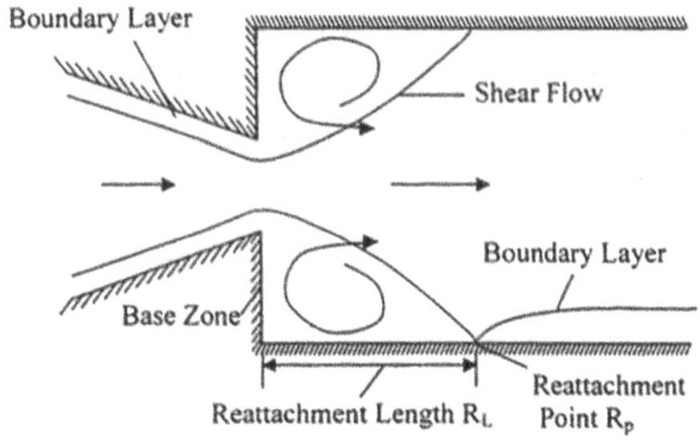

Fig. 1 The sudden expansion flow field

shows the sudden expansion flow field inside a duct of a convergent nozzle. When the flow expands in the duct, the shear layer from the nozzle lip initially gets separated and then attaches itself with the duct wall at certain distance. The length from the nozzle exit to the reattachment point is known as the reattachment length [2, 3].

Experimental work has shown that the base pressure control in a suddenly expanded axisymmetric duct and the wall pressure does not influence by the micro-jets [4]. Chaudhary et al. [5] studied the consequence of the tests to control pressure flow in the recirculation zone. In their work, an orifice diameter microjet served as an active control was located at ninety-degree intervals along a pitch circle diameter of 1.3 times the converging-diverging (CD) nozzle exit diameter in the base region. The Mach numbers of the abruptly expanded flows studied for the base pressure range from 1.1 to 3 and the obtained wall pressure distribution is depicted for Mach number 1.6 and 1.8 respectively. Axi-symmetric round brass tubes were used for jets, and cross-sectional area of ducts was 2.56 times the nozzle exit area. L/D ratio of the broadened pipe varied from 10 to 1 and NPR was shifted from 3 to 11. Anderson and Williams [6] studied the influence of initial flow direction on the intense base pressure in supersonic axisymmetric flow for the flow over boat tails and flares. In the study, they assumed either a constant or rising pressure along the jet mixing region. Pathan et al. [7] investigated the compressible flow experimentally over a backwards-facing step and further studied the flow visualisation for a wide range of NPR and L/D ratios for dimpled and non-dimpled conditions. They concluded that dimple is very useful in controlling the base pressure. They also found that as the length of the duct increases gradually, the control is becoming very useful for all L/D ratios 4, 6, 8 and 10. However, wall pressure is not getting affected by the use of dimple at various locations.

Most of the studies mentioned above performed experimentally. Meanwhile, the use of numerical simulation using turbulence model for the variation of base pressure using a passive control, especially in the high-speed flow and suddenly expanded flows from a nozzle into the duct is limited. For instance, studies on numerical investigations of microjet control effects in sudden expansion in the duct using the pressure-based method is found in [8–10]. The same method also used for wing model [11]. Therefore, this study will use an ANSYS computational model of converging nozzle and duct as a platform to investigate the variation of base pressure and effectiveness of control in the form of annular ribs using k-ε turbulence model.

2 Methodology

The physical model of the converging nozzle and duct with rib is shown in Fig. 2. The ribs width W = 3 mm (fixed) and height H (varied from 1 to 3 mm) are used. The nozzle inlet is of 30 mm diameter and outlet is of 10 mm diameter. Duct length is 132 mm, and the duct outlet is of 22 mm in diameter with the area ratio (ratio of nozzle exit area to duct area) of the model is 4. The flow from the nozzle expands in the duct and passes through the duct where it undergoes several stages of compression and expansion due to expansion fan/shock wave and finally reaches atmospheric pressure at the outlet. The flow through the nozzle is considered to be turbulent in nature; hence k-ε standard model is applied for the compressible flow field.

The k-ε turbulence model is among the famous model which provides economy, robustness, and sufficient accuracy for many kinds of flow situations. The k-ε turbulence model employed in this work is available in the ANSYS Fluent software. The flowing gas is considered to be ideal, and the iterations are repeated until the residuals are within $10e^{-5}$. The unstructured mesh of rectangular type was refined to capture velocity and pressure. The grid sizes are changed several times to give enough accuracy with convergence. Grid size with nodes of about 48,441 and element size of 47,988 was decided for the domain after a grid independence study performed using three different element sizes as shown in Fig. 3.

The simulation is validated with the results presented in Rathakrishnan [12] who performed experimental work using 5 ribs placed at equidistant space in the duct as shown in Fig. 4. The results of base pressure variation with different NPR (ratio of stagnation pressure to atmospheric pressure) and L/D ratio, i.e., the ratio of length to the diameter of the duct, are obtained. This same work is initially repeated to validate the numerical results, having control with ribs and without ribs. The flow in duct without ribs is also called as 'no control' flow.

The base pressure variations obtained experimentally and numerically using k-ε turbulence model are compared in Fig. 5. The nozzle operating at sonic Mach number

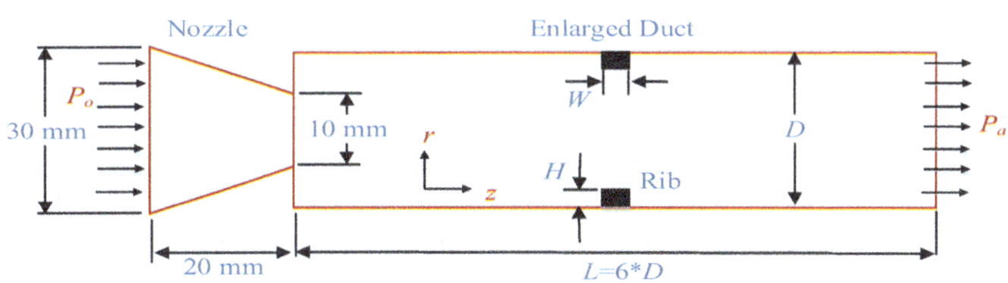

Fig. 2 Nozzle and duct arrangement with rib

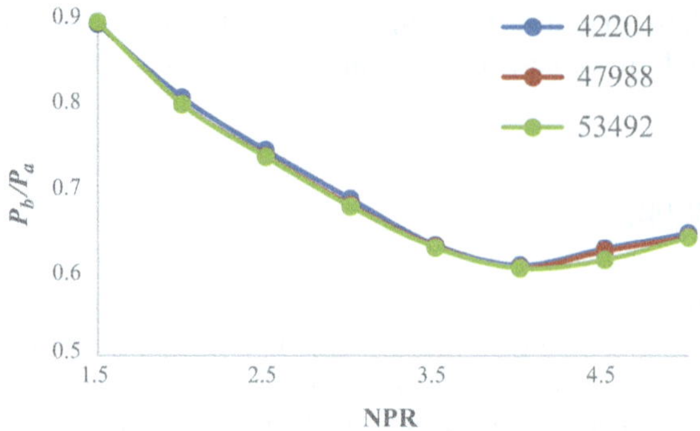

Fig. 3 Three different element size having converged values

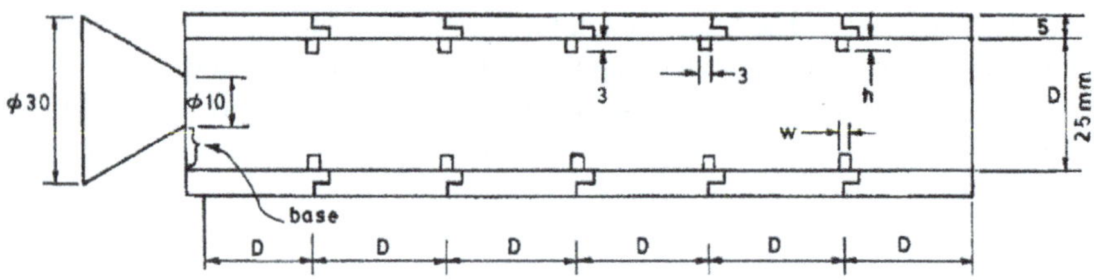

Fig. 4 Duct with 6 ribs used in experimental study [12]

Fig. 5 Comparison of experimental and numerical results

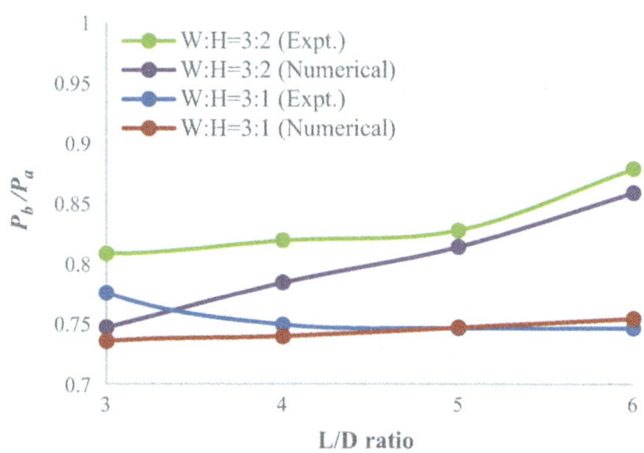

and NPR = 2.458 were employed in the experimental study. The results show that the numerical results from this work and experimental values are in an agreement for duct having six ribs of aspect ratio 3:1 and 3:1. It is well known that flow will achieve sonic conditions at the exit of the nozzle for a primary pressure ratio of 1.89. This is possible with assumptions that the flow is isentropic, where the effects of viscosity and the boundary layer associated with it, are neglected. However, in reality flow is neither isentropic nor inviscid. Hence, minor discrepancy obtained is quite apparent, and it can be considered to be in acceptable limits. Moreover the compressible flow nature of fluid at the sonic Mach number is always tedious to deal with. Marginal variations in base pressure are also attributed to the effect of the backflow as well as the boundary layer effects at these L/Ds.

3 Results and Discussion

The numerical results of suddenly expanded flow through nozzle operating at sonic Mach number are discussed in this section. The diameter of the duct and the L/D ratio is fixed at 22 mm and 6 respectively. The base pressure variations for NPR from 1.5 to 5 and aspect ratio from 3:1 to 3:4 are discussed in detail. The pressure distribution for 1D, 2D, 3D, and 4D location of ribs are investigated in detail.

The base pressure instabilities in duct is depicted in Fig. 6. The rib of aspect ratio is increased from 3:1 to 3:4 in step of 1 mm height in every case. The single rib attached circumferentially to the duct is also shifted from 1D to 4D for each aspect ratio of rib. The NPR is kept constant at 2.5 and simultaneously the rib aspect ratio, i.e. its height is increased from 1 to 3 mm. For each rib location, the height is changed, and the corresponding improvement or reduction in base pressure is observed. It is noted that the base pressure increases with shifting of rib position from 1D to 4D irrespective of its any aspect ratio. At 1D location and for any aspect ratio of rib, i.e. when the rib is placed close to the base corner, the base pressure is almost the same. For other rib locations, the base pressure keeps increasing slightly when the aspect ratio is 3:1 and 3:2 while for 3:3 there is a considerable jump in the base pressure at 3D and 4D. Such variations occur due to significant disturbances caused in the flow regime with the presence of rib where the flow expands.

The increase in base pressure of expanding flow with increase in rib location downstream the nozzle exit involves the dynamics of gas expansion. While the increase/decrease in the base pressure with the rib aspect ratio completely depends upon the reattachment point of shear layer. As briefly mentioned in the introduction

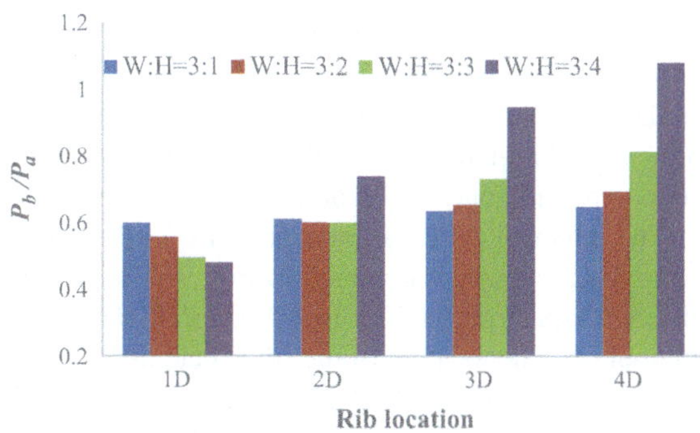

Fig. 6 Increase in base pressure with rib location and its aspect ratio

part about the jet pump action, the corner effects are dominant. When the flow from the nozzle exits at high velocity into the duct, the flow is either under expanded or correctly expanded or over expanded which depends upon the NPR. In this case of sonic Mach nozzle with the described dimensions previously, the NPR required for correct expansion is 1.89. If NPR is lower than 1.89 the flow from the nozzle over expands and for NPR greater than 1.89 the flow under expands. In over expansion, there is an oblique shock in front of the nozzle, and in under expansion, there is an expansion fan. Moreover, the reattachment length gets affected by these flow behaviors. Another level of phenomena coinciding is the flow recirculation occurring at the base corner of the duct. The flow expands, and the shear layer attaches itself with the duct wall in the downstream direction. The fluid at the corner recirculates and dictates the level of low base pressure. Due to low pressure, the fluid from the reattachment point flows back into the corner zone and gets thrown back into the mainstream flow via the shear layer entrainment. This phenomenon of flow reversal and ejection continue in the steady state. The base pressure with this cycle phenomena increases and hence the presence of any barrier attached to the wall will definitely cause either increase or decrease of fluid flow reversal affecting the primary vortex to a greater extent.

The pressure distribution in the duct at various rib heights at fixed location of 1D from the base is shown in Fig. 7. The NPR in these locations is equal to 2.5. When the flow expands from the nozzle, the flows undergoes alternate compression and expansion due to the duct wall. Further downstream of the nozzle, a compression zone can be clearly seen which is similar to the diamond formed axially. Later they diminish as the flow travels longitudinally. As the rib height is increased gradually from 1 to 4 mm, the distance between the successive compressed zones seem to be slightly increased showing the effect of rib. As the rib height increases the strength of these zones increases showing the strong influence of the ribs.

In Fig. 8 the velocity variation in the enlarged duct is shown in contours. The aspect ratio of the rib is 3:3 and its location is varied from 1D to 4D with same NPR of 2.5 in each case. As explained earlier the zone of compression ahead of nozzle lip is at quite some distance. At 2D to 4D a shorter distance of compression zone occurring due to hitting of fluid to the wall. The more velocity variation were observed and the shorter the distance between them shows low velocity and higher corner pressure. The velocity distribution for a duct with rib having aspect ratio 3:3 is much clearer and explains its effect in Fig. 8. It can be seen that the high velocity zones are lesse for 1D position indicating lower base pressure. With increasing rib location from 2D to 4D, the velocity between the nozzle exit and rib corner is higher indicating lower velocity and high pressure in that area. This is also evident by having more number of compressed zones and reduced distance between them.

The effect of NPR on base pressure in the duct where the flow expands from sonic Mach nozzle is depicted in Fig. 9. The NPR is increased from 1.5 to 5 at a different aspect ratio of 3:1 to 3:4 and comparison is made between them and duct without rib. The location of the rib is held at 3D for this analysis. The base pressure decreases consistently for NPR ranging from 1.5 to 5 while for duct having ribs the base pressure increases at different NPR. For rib having a 3:1 aspect ratio although

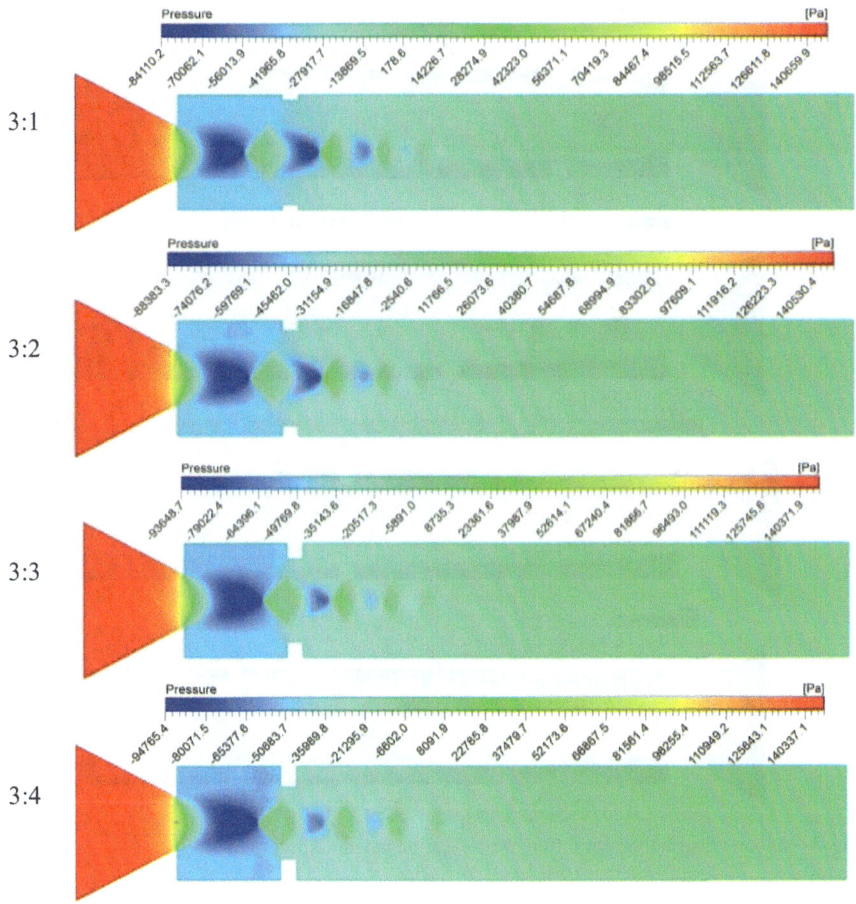

Fig. 7 Pressure distribution for different rib aspect ratio at 1D location

increment in base pressure is unseen but the increment in base pressure is more than that in the smooth duct. For 3:2 and 3:3 aspect ratio the base pressure increases after crossing NPR 2.5. As the NPR increases the flow expands more rapidly with different level of expansion. This cause the shear layer to travel a longer distance before it attaches with the duct wall. Hence, the level of turbulence kinetic energy dissipation and viscous effects of the flow are low. Due to more length traveled by the viscous shear layer, the primary vortex size increases and a lesser amount of fluid reversal due to which base pressure reduces consistently. The rib control effectiveness is highly noticeable at 3:4 due to massive increase at increasing NPR. At all NPRs the use of rib at 3D location tends to change the base pressure irrespective of used rib height. The reversal in base pressure is slower at 3:1 to 3:3 while at the remaining height the reversal is highest.

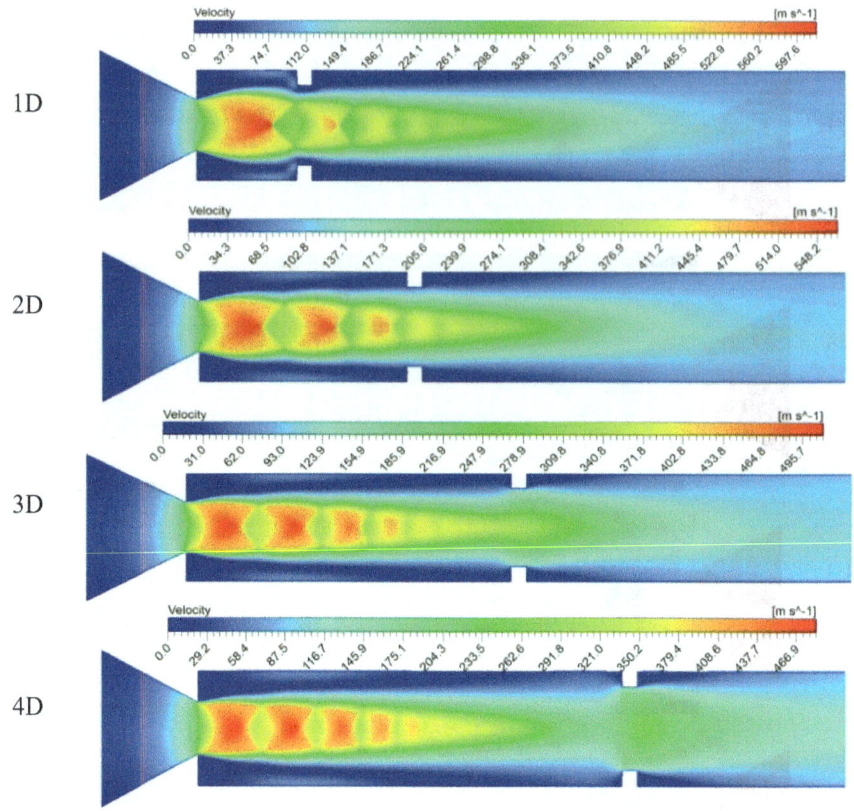

Fig. 8 Velocity distribution for different rib location having aspect ratio 3:3

Fig. 9 Base pressure changes when NPR is increased from 1.5 to 5, and the aspect ratio is changed from 3:1 to 3:3

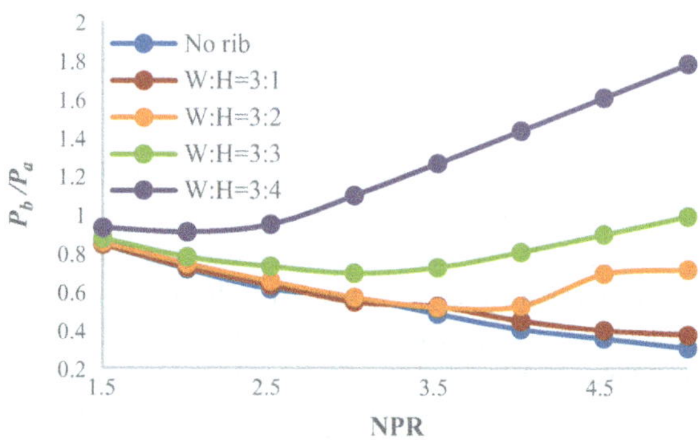

Velocity distribution contour at 3D location of rib in the duct is shown in Fig. 10. The velocity distribution is striking for duct without rib and with rib at different NPRs. The difference in flow field at NPR 2.5 having rib and no rib can be easily perceived. The zones of higher velocity lie ahead of the nozzle exit along the axial tube direction. The diamond shape pattern is affected by the presence of rib and

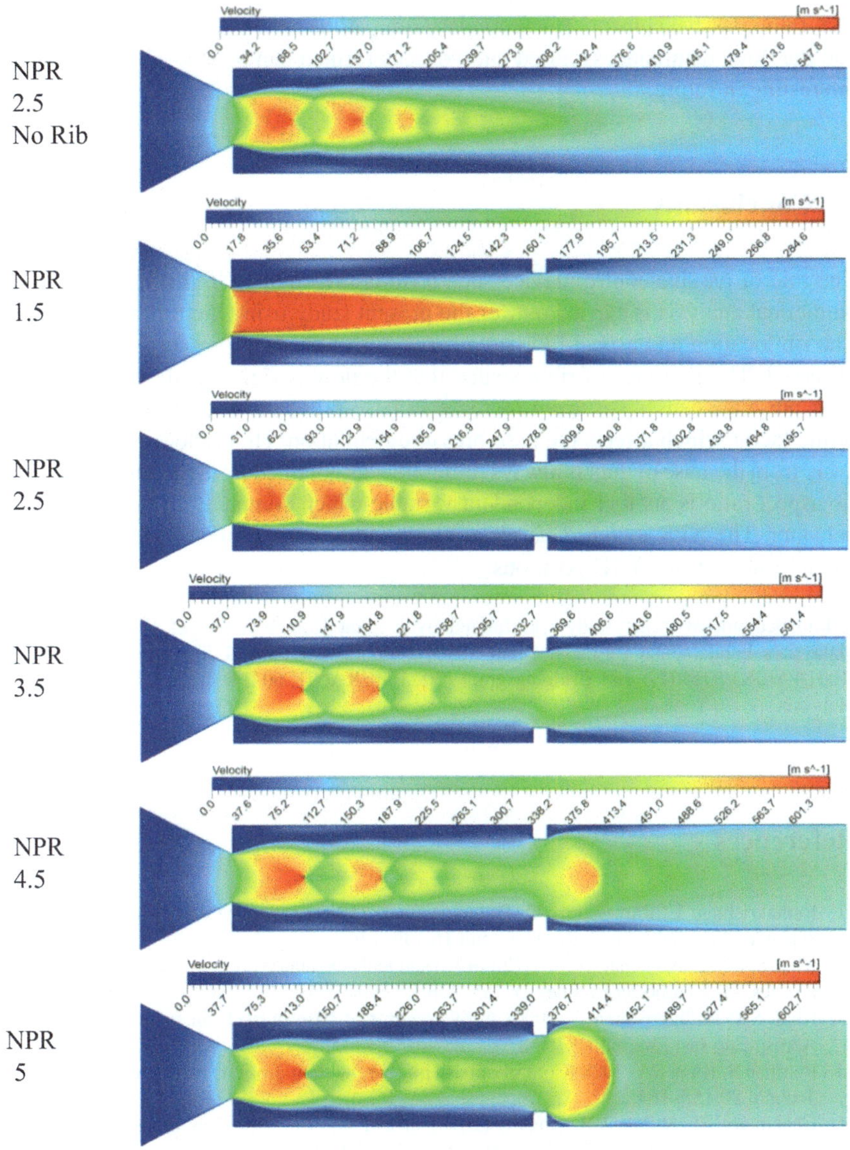

Fig. 10 Velocity in the duct with no rib and with rib having different aspect ratio and NPR

when the NPR is increased the pattern keeps changing. For NPR above 4 the flow exits from the rib location at higher velocity and the boundary is shifted slightly. It can be seen that a zone of secondary vortex at the corner of rib and the exit velocity at the tube end seems to be quite higher due to presence of the passive device. The presence of rib shows one minor character at the downstream of the rib. It can be observed that the flow is disturbed due to sudden re-expansion of flow through the rib throat. This could cause vibrations to be induced in the duct due to this flow disturbing passive control. The secondary drag at the rib exit has to be investigated and the flow disturbances caused and vortices formed can be studied which can be interesting in future.

4 Conclusions

The role of passive control device such a rib is investigated in this research work. Numerical analysis is carried out for the present study using k-e turbulence model. The rib location and aspect ratio are changed, and the base pressure variations are observed. The presence of rib revealed that the flow is significantly disturbed, and their presence causes sufficient increase in base pressure. The location is another prime factor which affects the base pressure regulation. Higher locations of the rib away from the base of duct is always preferable, as obtained from the simulation. The rib aspect ratio is another independent variable which completely changes the flow behavior. The NPR increase causes a massive amount of increase in base pressure when located at 3D or 4D positions.

Acknowledgements This study was funded by a grant from Ministry of Higher Education Malaysia's Fundamental Research Grant Scheme (203/PAERO/6071435), Research University (1001/PAERO/8014019) and Bridging grant (203.PMEKANIK.6071411) from Universiti Sains Malaysia.

References

1. Rathakrishnan E, Ramanaraju O, Padmanaban K (1989) Influence of cavities on suddenly expanded flow field. Mech Res Commun 16:139–146
2. Khan SA, Rathakrishnan E (2004) Control of suddenly expanded flows from correctly expanded nozzles. Int J Turbo Jet Engines 21:255–278
3. Khan SA, Rathakrishnan E (2006) Active control of base pressure in supersonic regime. J Aerosp Eng Inst Eng India 87:1–8
4. Rehman S, Khan SA (2008) Control of base pressure with micro-jets: part I. Aircr Eng Aerosp Technol 80:158–164
5. Chaudhary ZI, Shinde VB, Khan S (2018) Investigation of base flow for an axisymmetric suddenly expanded nozzle with micro JET. Int J Eng Technol 7:236–242
6. Anderson J, Williams T (1968) Base pressure and noise produced by the abrupt expansion of air in a cylindrical duct. J Mech Eng Sci 10:262–268

7. Pathan KA, Khan SA, Dabeer P (2017) CFD analysis of effect of area ratio on suddenly expanded flows. In: 2nd International conference for convergence in technology (I2CT). IEEE, pp 1192–1198
8. Aabid A, Khan A, Mazlan NM, Ismail MA, Akhtar MN, Khan SA (2019) Numerical simulation of suddenly expanded flow at Mach 2.2. Int J Eng Adv Technol 8:457–462
9. Fharukh AG, Alrobaian AA, Aabid A, Khan SA (2018) Numerical analysis of convergent-divergent nozzle using finite element method. Int J Mech Prod Eng Res Dev 8:373–382
10. Khan SA, Aabid A, Baig MAA (2018) CFD Analysis of CD nozzle and effect of nozzle pressure ratio on pressure and velocity for suddenly expanded flows. Int J Mech Prod Eng Res Dev 8:1147–1158
11. Khan SA, Aabid A, Saleel CA (2019) CFD Simulation with analytical and theoretical validation of different flow parameters for the wedge at supersonic Mach number. Int J Mech Mechatron Eng 1
12. Rathakrishnan E (2001) Effect of ribs on suddenly expanded flows. AIAA J 39:1402–1404

Computational Analysis of a Multi-orifice Rotary Injector with Air Core

S. Sahaya Jisha and S. Thanigaiarasu

Abstract The present computation study investigates multi-orifice injection with centre air core. To enhance the mixing process, high pressurized air has been injected at the centre portion of the orifice plate. Numerical simulation has been carried out to study the gaseous jet interaction and liquid jet interaction with air with toluene-air, kerosene-air and water-air as working fluids. Species transport model and Volume of Fluid (VOF) method in Multiphase model are used in the present analysis, which is capable of predicting the interface between these two different fluids each of which is having different mass densities. Initially, dimension of computational domain has been optimized by simulating jet through multi-orifice injector into mixing chamber of different dimensions. Later, grid independent study has been carried out with an optimum computational domain for five different grid sizes. Performance parameters like maximum penetration and volume fraction of liquid profile at different downstream locations has been compared in these five different cases. And one particular grid size is chosen with reasonable computational cost for the rest of analysis. Steady and transient simulations were performed to study the interaction between jet and air. Effect of centre air core on penetration length has been studied and compared. Deformation of cylindrical jet has been captured at different downstream locations and compared in steady analysis. Temporal evolution of the liquid jet forming the spray has been captured at different time-steps and compared in transient analysis.

Keywords Multi-orifice injector · Species transport model · Volume of Fluid (VOF) model · Steady analysis · Transient analysis

---

S. Sahaya Jisha (✉) · S. Thanigaiarasu
Department of Aerospace Engineering, Madras Institute of Technology, Anna University, Tamil Nadu, India
e-mail: sahayajisha@gmail.com

S. Thanigaiarasu
e-mail: sthanigaiarasu@mitindia.edu

P. Rajendran et al. (eds.), *Proceedings of International Conference of Aerospace and Mechanical Engineering 2019*, Lecture Notes in Mechanical Engineering,
https://doi.org/10.1007/978-981-15-4756-0_30

1 Introduction

A liquid injection to the atmosphere is used in many applications such as internal combustion engines, fabric weaving industries, painting companies etc. The geometrical design of the orifice and the number of orifices in the plain-orifice injector defines its injection properties. The spray characteristics formed by a fuel injector is very important for performance of gas turbine combustor and afterburner. It is clear that a more uniformly distributed spray is beneficial to combustion efficiency. Thus, a study on the spray characteristics of an injector is necessary.

Earlier experimental studies to understand the fuel injection properties of plain-orifice injectors are focused on spray behavior as the combustion efficiency is strongly correlated to it. Injection velocity is found to be the most important factor in the atomization of jet formed by plain orifice injector by Zhu et al. [1]. Zhang et al. [2] have furthered the study to understand the influence of three parameters—air velocity, pressure drop across the injector nozzle and injector orifice diameter on the spray characteristics of plain-orifice injector. Based on a very large amount of experimental data, they have arrived at some correlations for mean diameters of drops. Sjoberg [3] explains that a rotary injection leads to greater spray curvature than air swirl for two reasons. The rotating injector creates spray curvature from the sweeping injection itself with radial deceleration caused by drag forces acting on penetrating spray eliminating the need for cross flow. The tangential velocity of the air swirl is weak close to the injector, causing the spray to penetrate straight ahead.

More recent studies on sprays are computational analyses done using commercially available software such as Fluent, to simulate the flow using readily available models [4]. Influence of hole geometry on the downstream flow from the nozzle has been studied using a fully transient 3D-CFD multiphase approach by Brusiani et al. and the references sited therein provide sufficient evidence to conclude that the main spray characteristics (atomization, penetration, and droplet diameter distribution) can be strongly correlated to injector operating conditions and injector geometric layout. With respect to the cylindrical hole shape, the conical holes showed a quite evident reduction of the mass flow rate mainly linked to the reduction of the hole outlet section [5]. Experimental study of multi-orifice injectors done by Motoyuki et al. [6] and Zhong et al. [7] shows the relationship parameters between nozzle geometries, flow rate, and droplet size. They provide a theoretical discussion on fluid velocity in the nozzle orifice.

2 Methodology

Steady and transient simulations are carried out in three dimensional domains using both Species Transport Model and Multiphase model (VOF method). The orifices of the injectors are modelled and a cylindrical domain is chosen for steady simulations. A conical domain is chosen for transient simulation to capture the diffusion of the spray. Computation was conducted in Ansys Fluent and meshing was done in ICEM CFD. The geometry is divided into two sub-domains, the smaller one near the inlet with a higher mesh density. Standard k-ε model is used to simulate the steady flow condition. The spray of toluene in air medium is modelled using species transport model. Second order upwind discretization was used to improve the accuracy. The methodology used is tabulated in Tables 1, 2 and 3. This method was previously used by Malet et al. [4] and Karthikeyan and Sridhar [8] and found to be sufficiently accurate. Initially larger computational domain is considered for the interaction of gaseous jet. It is then reduced in size by observing the species contour along axial locations as well as the mid plane species contour. Performance parameters like species contour of toluene and the species concentration profile are compared for five different cases shown in table with increasing mesh density. When there was no significant change in them with further refinement of the mesh, the grid with lesser elements is chosen, in order to reduce computational time. For transient simulation, sliding mesh model that allows adjacent meshes to slide relative to one another, is used. To implement sliding mesh, two cell zones were created: one for moving injector part and another is the domain. The moving injector part is provided with a constant rotational motion of 1000 rpm and matching mesh interfaces are created (Table 4).

Table 1 Method used in the study

General	Pressure based with absolute velocity formulation	Steady
		Transient
Model	Species transport and	Toluene-air (case 1)
	Multiphase model—VOF method	Kerosene-air (case 2)
		Water-air (case 3)

Table 2 Boundary conditions of non-rotational case

Core inlet	Mass flow inlet
Fuel inlet	Mass flow inlet
Air inlet	Pressure inlet
Outlet	Pressure outlet
Turbulence model	Realizable k-ε model with scalable wall function
Reference pressure	101,325 Pa
Exit pressure	0 Pa
Mass flow rate of fuel	0.003 kg/s
Mass flow rate of core air	0.015 kg/s
Inlet pressure of air inlet	0 Pa

Table 3 Rotational case details

Cases	Geometric details	Mass flow inlet	
	Orifice diameter (mm)	Air (kg/s)	Water (kg/s)
3A	0.5	0.015	0.15
3B	0.5	0.015	0.6

Table 4 Boundary conditions for rotational case

Core inlet	Mass flow inlet
Fuel inlet	Mass flow inlet
Air inlet	Pressure inlet
Outlet	Pressure outlet
Turbulence model	Realizable k-ε model with scalable wall function
Model	Multiphase model (VOF method)
Fuel	Water liquid
Reference pressure	101,325 Pa
Exit pressure	0 Pa
Inlet pressure of air inlet	0 Pa
Rotational velocity	1000 rpm

3 Results and Discussion

3.1 Jet Interaction Without Rotation

Results of Air-Toluene Interaction (Case 1) The species contours along the axis shows the spray diffusion as it occurs because of the interaction of atmospheric air and the air from the core. The aerodynamic interactions between Toluene and air causes surface irregularities on the spray of Toluene and leads to diffusion of the spray. This results in dilution of concentration of Toluene at the surface of the spray along the axis as seen in Fig. 1. As the surface concentration of Toluene decreases along the axis, the core starts to decay and disappears at about 50 mm from the inlet. The diffusion of the spray on the side of the core is higher than the other side due to its interaction with the air core. This causes the spray to attain a distinctive bean-shape. The vector plots at various axial locations as shown in Fig. 1 shows how the interaction between surface of Toluene spray and air causes the air flow around it to circulate and thus improve mixing Fig. 2 shows the diffusion of spray across the orifice at various axial locations. It clearly shows that the species concentration peak gradually decreases and the spray becomes shallower. These trends are in agreement with the species contours in Fig. 1.

Results of Air-Kerosene Interaction (Case 2) Spray diffusion of Kerosene follows the same trend as that of Toluene. Figure 3 shows the spray diffusion from all the

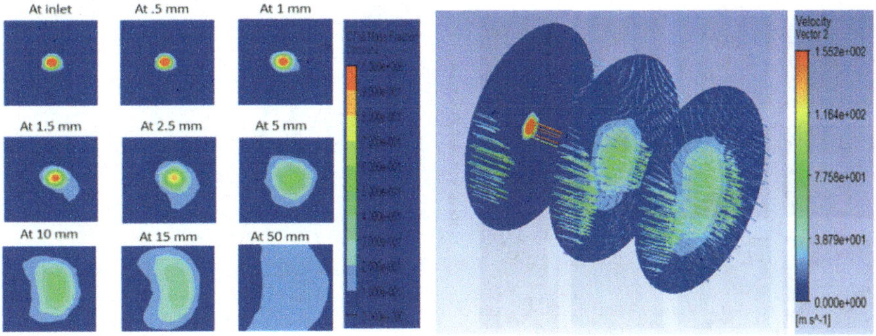

Fig. 1 Species contour and velocity vectors at various axial locations

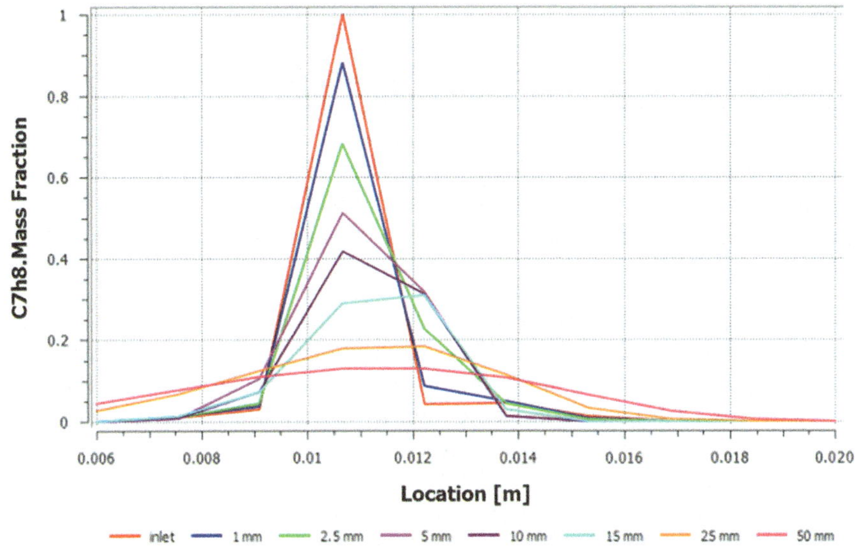

Fig. 2 Species concentration profile at various axial locations

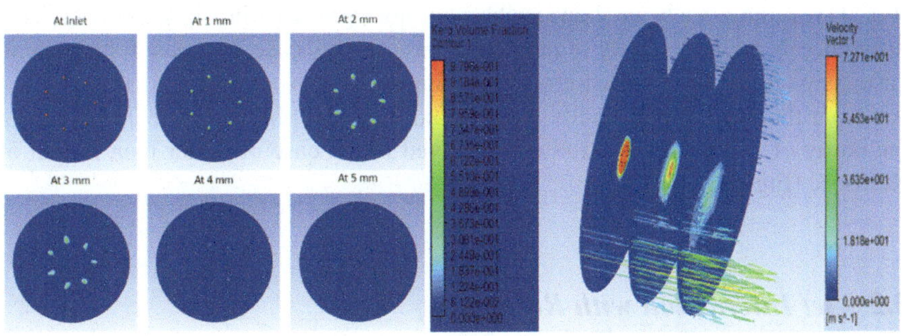

Fig. 3 Volume fraction contour of kerosene and velocity vectors at different axial locations

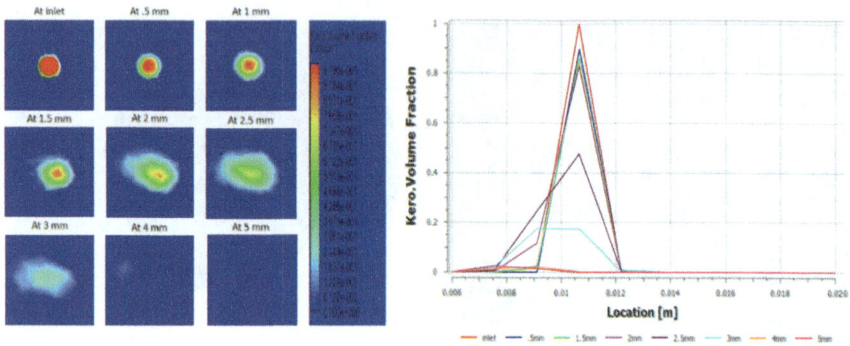

Fig. 4 Volume fraction contour and kerosene volume fraction profile of kerosene at various axial locations

seven orifices. Concentration of spray decreases along the axis as seen in Air-Toluene case. As viscosity for a liquid is higher than that of a gas, the effective mass flow rate decreases, and thus the spray diffuses earlier. The interactions of air core with the spray cause it to diffuse into the core as seen in Fig. 4, the decay of species concentration from the seven orifices is similar. Figure 4 graph shows the decay of species concentration across the orifice as it becomes shallower along the axis and the peak value decreases.

Results of Air-Water Interaction (Case 3) The results from Air-Kerosene case are comparable to that of Air-Water case because of similar densities. Surface tension of Water is higher than that of Air (Surface tension of Water at 20 °C is 0.073 and that of Kerosene is 0.029), hence the spray diffusion occurs later in the case of Water. Kerosene diffuses at 5 mm, Water diffuses at 15 mm. This difference is seen in Figs. 5 and 6 shows it more clearly. The Water liquid spray is more intact for a longer distance than the Kerosene liquid spray as the higher surface tension helps in holding the liquid molecules together.

Figure 6 further shows that the effect of air core as seen in Air-Toluene case (Fig. 1) resulting in a bean-shape is absent or not significant in Air-Water case. This difference is because Water is in liquid phase in this case whereas Toluene is in gaseous phase. A fluid in gaseous phase interacts with another gas (in this case, Air from the core) more readily than a fluid in liquid phase. Figure 7 shows that the decay of species concentration is delayed in the case of Water than in that of Kerosene (Fig. 5).

3.2 *Jet Interaction with Rotation*

The rotational cases 3A and 3B are performed to understand the effect of centrifugal force due to rotation and the interaction of the air core with the fuel.

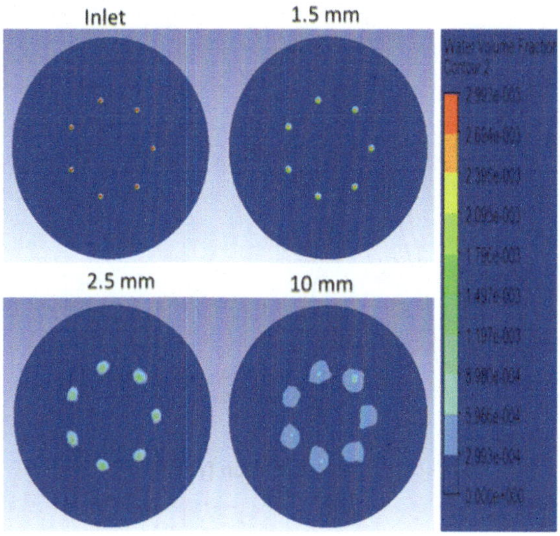

Fig. 5 Volume fraction of water at different axial locations

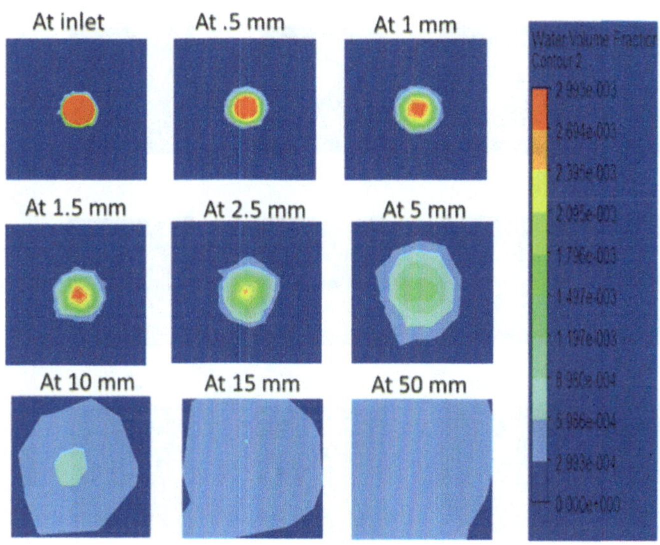

Fig. 6 Volume fraction of water at various axial locations across for a single orifice

Results of Case 3A Figure 8 shows that the flow reaches 0.5 mm after 33 μs and continues to move according to the rotation of the injector. As time increases, the spray diffusion is clearly seen. The image of the spray moves upwards with time as the injector rotates. As the plane examined is very close to the inlet, the peak species concentration is high, close to 1 in the middle of the spray for up to 480 μs. Figure 9

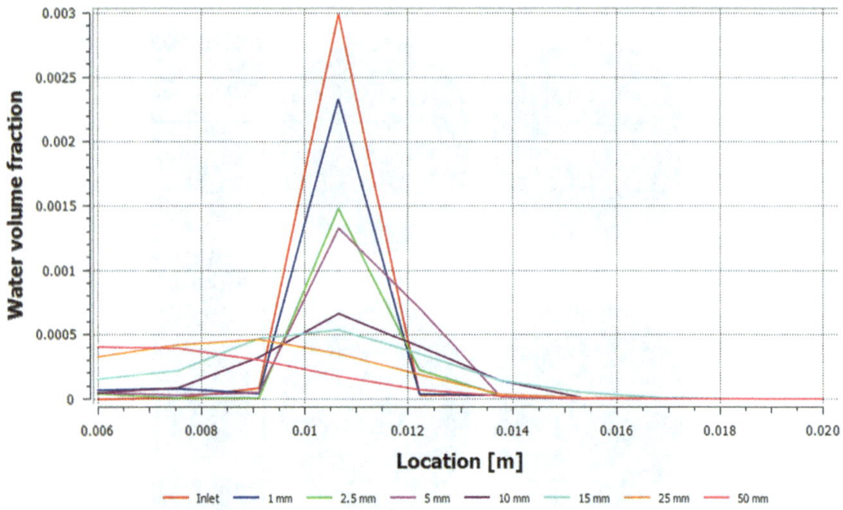

Fig. 7 Volume fraction at various axial locations at one orifice

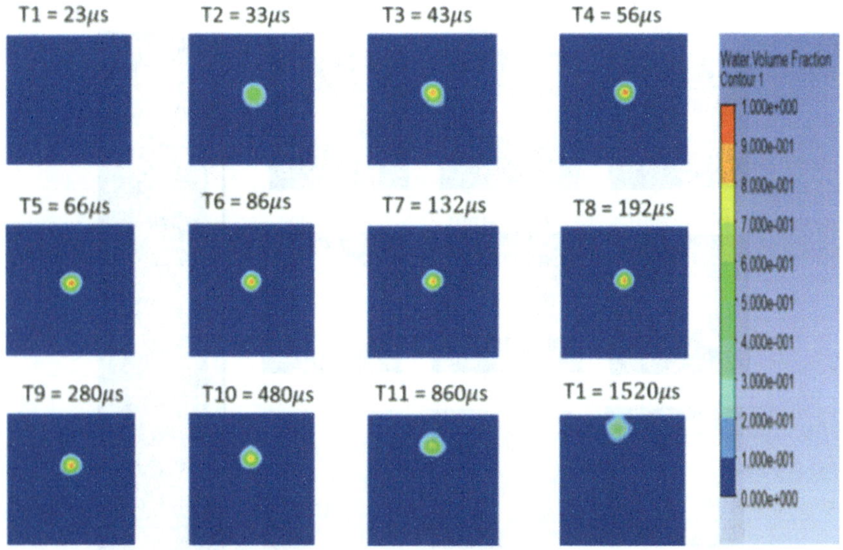

Fig. 8 Temporal evolution of jet at 0.5 mm from the inlet for case 3A

shows that the flow reaches 2 mm only after 56 μs. The centrifugal force acting on the spray can be seen clearly up to 86 μs. As the time step size was increased after 100 μs in order to speed up the simulation and capture the rotation of the orifice plate, the effect of centrifugal force on the spray is not captured by the current setup.

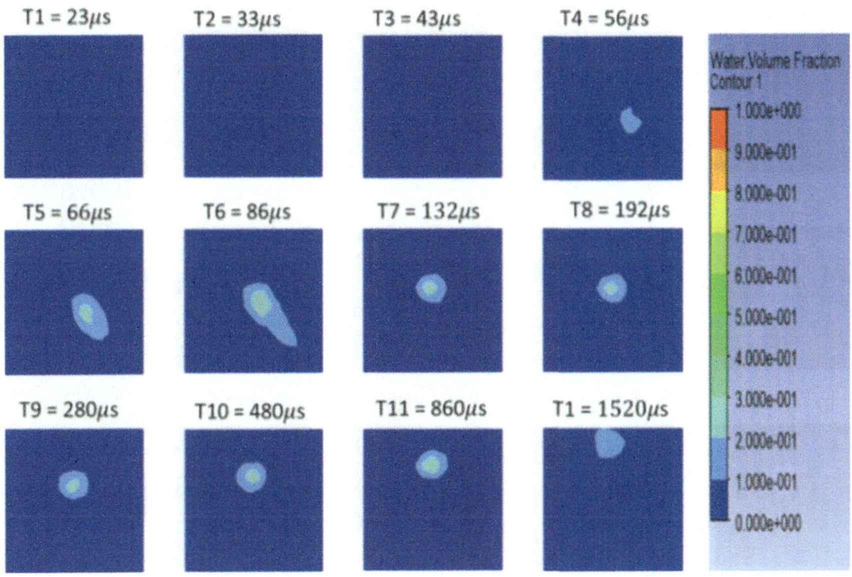

Fig. 9 Temporal evolution of jet at 2 mm from the inlet for case 3A

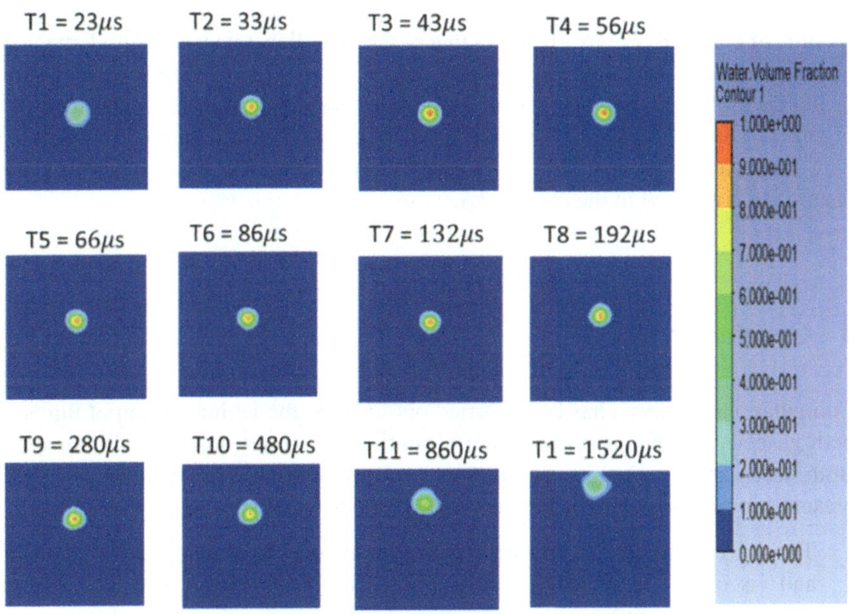

Fig. 10 Temporal evolution of jet at 0.5 mm from the inlet for case 3B

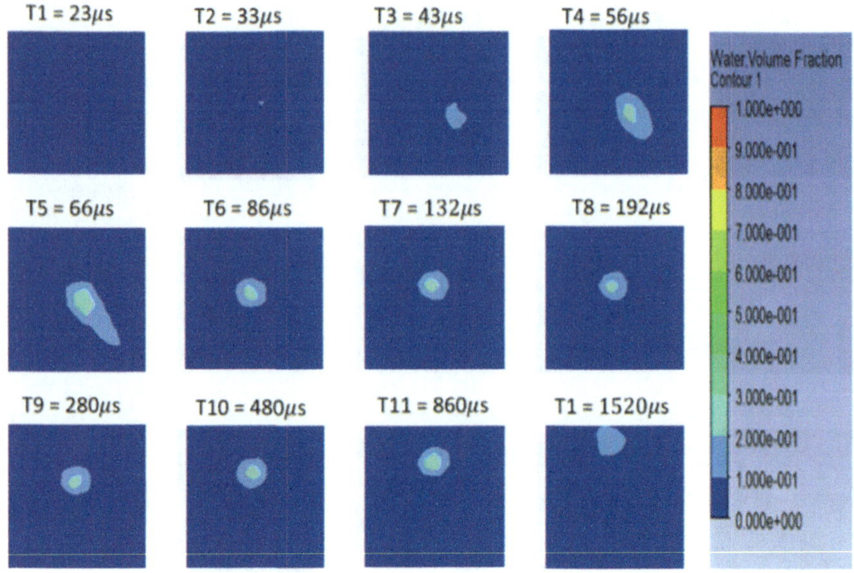

Fig. 11 Temporal evolution of jet at 2 mm from the inlet for case 3B

Results of Case 3B Figure 10 shows that as the mass flow rate is increased, the flow velocity increases and thus the flow reaches 0.5 mm at 23 μs. The spray diffusion is similar to that of Case 3A owing to the fact that same time step sizes are used. Figure 11 shows that the centrifugal force acting on the spray cannot be captured with the current setup at a time step size higher than 2e−6. The flow reaches 2 mm faster than it reached in the case of 3A.

4 Conclusion

Computational analysis has been carried out to study the jet interaction of fluids in both gaseous and liquid phases from a multi-orifice rotary injector with air core. Both steady and transient conditions have been simulated and the main results of the present computational study can be summarized as follows:

1. The spray diffusion is determined by the phase of the jet fluid, the surface tension and viscosity. Lower surface tension results in faster spray diffusion. Higher viscosity results in lower effective mass flow rate and earlier spray diffusion.
2. Fluid in gaseous phase interacts more readily with core air flow than fluid in liquid phase. This is because the surface tension keeps the molecules of the liquid phase intact and reduces the effect of air core.
3. Centrifugal force acting on the spray in rotary injection causes the spray to form a wider cone. This effect can be captured at lower time step size.

Acknowledgements The authors are thankful to Dr. S. Thanigaiarasu, Professor and Head of the Department, Department of Aerospace Engineering, our other staff members and friends for their support and suggestions towards the execution of the study.

References

1. Zhu JY, Hou MY, Chin JS (1986) Experimental study on the atomization of plain orifice injector under uniform cross air flow. ASME
2. Zhang Y, Jun YZ, Wang LX (1987) Experimental study on the atomization of plain orifice injector under uniform and non-uniform cross flowing air stream. ASME
3. Sjoberg M (2001) Rotating injector as a tool for exploring diesel combustion and emission formation processes, TRITA-MMK 2001:13. ISSN 1440-1179 ISRN KTH/MMK/R-01/13-SE
4. Malet J, Foissac A, Vayaboury C, Gélain T, Mimouni S (2012) Numerical study on the influence of simplified spray boundary conditions for the characterization of large industrial safety spray systems used in nuclear reactors. ICLASS
5. Brusiania F, Falfaria S, Pellonia P (2014) Influence of the diesel injector hole geometry on the flow conditions emerging from the nozzle. Energy Procedia 45
6. Motoyuki ABE, Eiji ISHII, Hideharu EHARA (2017) Atomized spray droplet size prediction from multi-hole nozzle for direct injection gasoline engines. JSME
7. Zhong W, He Z, Wang Q, Shao Z, Tao X (2014) Experimental study of flow regime characteristics in diesel multi-hole nozzles with different structures and enlarged scales. Int Commun Heat Mass Transfer
8. Karthikeyan N, Sridhar BTN (2012) Experimental and CFD analysis on coaxial turbulent jets with different velocity ratios. ICMOC

An Experimental Study of Nozzle Geometry Effect on Mixing Characteristics of Two Different Twin Jets

S. Thanigaiarasu, A. Muthuram, S. B. Jabez Richards, T. Vijaya Raj, and Surendra Kumar Yadav

Abstract This experimental work investigates the effect of circular nozzle with non-circular nozzle in a flow-field issuing from circle-ellipse twin jet and the same is compared with circular twin jet for Mach numbers 0.4 and 0.6. For the two twin jet cases, the jets emanated from both nozzles of the twin jets are free jets and had the identical Mach number. In both twin jet cases, for all the Mach numbers, the nozzle spacing (S) is 20 mm. The exit area is 78.5 mm^2 which is same for all the nozzles. The ratio of nozzle spacing to nozzle exit diameter (S/D) for both cases was kept as 2. The results from the radial (Y) direction and shadowgraph pictures indicate that the circle-ellipse twin jet show better mixing enhancement as the entrainment between the ambient fluid and the jets from the nozzles is more. It is also observed that both the nozzle exit Mach number and nozzle geometry strongly influences the mixing of the two jets.

Keywords Jet Mach number · Merging point · Nozzle spacing

Nomenclature

D	Nozzle exit diameter
S	Nozzle spacing
X	Streamwise coordinate along axial direction
Y	Coordinate normal to X and Z axes
Z	Coordinate normal to X and Y axes
M	Mach number

S. Thanigaiarasu · S. B. Jabez Richards
Madras Institute of Technology, Chennai 600044, India

A. Muthuram (✉)
St. Peter's College of Engineering and Technology, Avadi, Chennai 600054, India
e-mail: masrimuthu@gmail.com

T. Vijaya Raj · S. K. Yadav
Prince Shri Venkateshwara Padmavathy Engineering College, Chennai, India

P. Rajendran et al. (eds.), *Proceedings of International Conference of Aerospace and Mechanical Engineering 2019*, Lecture Notes in Mechanical Engineering,
https://doi.org/10.1007/978-981-15-4756-0_31

e Nozzle exit conditions

1 Introduction

The twin jet mixing phenomenon is of essential value in current applications such as exhaust nozzles in aircraft engines for thrust vectoring control, air curtain unit, air conditioning in automobiles, in chemical industries where two chemicals need to be thoroughly mixed, noise reduction, etc., Twin jet is more common in high-speed propulsion systems. Twin jet refers to the jets emanated from the two nozzles to a still atmosphere. The two convergent nozzles when closely spaced merge with each other provides a different flow-field than that of a single jet. The three regions in the flow-field of twin jet such as converging, merging and combining region were experimentally investigated by Moustafa [11] and Kumar and Das [6]. The jets finally converge to form a single jet. Tanaka [13] identified the merge point where the initial merging of two jets occur. This point has approximately zero velocity and it lies on the line of symmetry. The region upstream of this merge point is sub-atmospheric and it is called converging region as observed by Muthuram et al. [8, 9]. The region followed by the converging region i.e., after the merge point is the merging region. The three different flow regions of twin jet are shown in Fig. 1.

Anderson and Spall [1] experimentally and numerically investigated the dual parallel plane turbulent jets and signified that the two jets converge to make a single jet in the far downstream distance. The effect of nozzle spacing in twin jet flow field was experimentally investigated by Rathakrishnan et al. [12] and Yin et al. [14] who revealed that when the nozzle spacing decreases, the merging of the jets occurs faster.

Faghani and Rogan [4] numerically investigated the three flow regions and also the impact of nozzle spacing. Durve et al. [2] carried out a comparison between three jet, two jet and single jet systems and calculated the effects on mixing phenomenon and

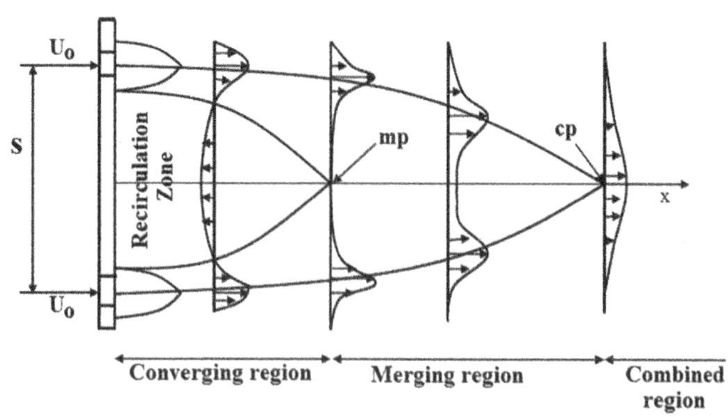

Fig. 1 Schematic diagram of twin jet flow field

turbulent fluctuations numerically by RANS equations. The noncircular or asymmetric nozzle enhances mixing compared to axisymmetric nozzles. Many researchers like Mi et al. [10] and Zaman [15] investigated and proved that the jets from asymmetric nozzle show enhanced mixing performance than the axisymmetric jets. The position of critical mixing regions were used to detect the injection region. The nozzle shape and diameter were also found out by using the location of critical mixing region. It is perceived from the literature survey that previous researchers carried out a very limited research in twin jet with asymmetric nozzles. This present work is to provide the detailed investigation of twin jet with circular and noncircular nozzle combination and the influence of nozzle geometries on growth of twin jets and mixing characteristics.

2 Experimental Work

2.1 Experimental Test Facility and Measurements

The experimental layout of high-speed jet test facility is shown in Fig. 2.

The high speed jet test facility has two 15 HP compressors and storage tank of capacity 20 bar. The moisture separator is positioned after the storage tank to remove the moisture from compressor discharged air. The air filter at the delivery line filters the air before it enters the pressure regulator. The pressure regulator is used to adjust the flow to a desired pressure value. The traversing mechanism has a Pitot probe attached with it so as to measure the total pressure in X, Y and Z directions. The data was acquired by a 16-channel pressure transducer (NETSCANNER™ Model

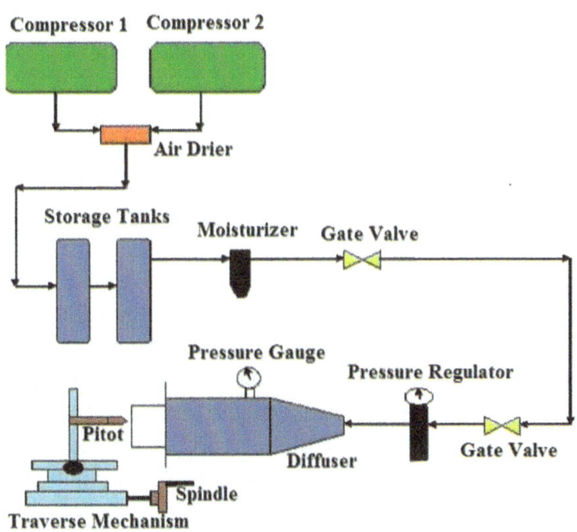

Fig. 2 High-speed jet test facility layout

9116). The nozzles were kept in position in the provision provided after the settling chamber.

The twin jet nozzles were positioned in such a manner that the sharp edges in the asymmetric nozzles face towards the centreline spacing. The distance between the longitudinal axes of individual nozzles in both the circle-circle and circle-ellipse nozzle configuration (S) is 20 mm and the ratio of nozzle spacing to nozzle exit diameter (S/D) is 2. The nozzle exit area for the both circle and ellipse nozzles is kept as 78.5 mm^2 so that the mass flow rate is equal. This experimental study have been carried out for Mach numbers 0.4, and 0.6. The total pressure was measured at numerous downstream locations from the nozzle exit. The readings in the radial direction were obtained for each 2 mm increase in flow direction up to a downstream distance of 20D.

3 Results and Discussion

This paper investigates the mixing characteristics of the twin jet (circle-circle and circle-ellipse) along the radial and transverse direction. The graph is plotted with non-dimensionalized Mach number (M/Me) versus X/D ratio in order to understand the rate of decay and mixing characteristics of the twin jet. From this experimental study, it is obvious that the mixing occurs rapidly in the merging region and there is a significant increase in velocity in the same region.

3.1 Twin Jet Flow Characteristics

Measurements were taken in the radial direction at various axial locations such as X/D = 0, 2, 3, 4, 5, 6, 10, 12, 16 and 20 up to the outer shear layer of the twin jet after which it reaches the atmospheric value. For two different Mach numbers, measurements in the flow field were obtained for S = 20 mm and also for same jet conditions. In the circular twin jet, the flow characteristics and the radial profile are similar on either side of the midpoint of the twin jet. The spreading increases slightly when moves downstream from the nozzle exit which can be identified from the radial profiles. From the plots it is evident that both the twin jets tend to resemble a single jet but with the different location of maximum velocity in the far-field. The jet momentum is conserved in the circular nozzle side whereas in the elliptical nozzle side, the jet decays rapidly. In both cases of the twin jet, the merging of jets takes place at a point known as merge point or free stagnation point. Up to this merge point, the inner shear layer of the jets and the interaction between them leads to the confluence which causes the contra rotating vortices to occur. The region from the nozzle exit up to which these vortices occur is called recirculation zone or the converging region. In the merging region, the strong interaction among the jets and recirculation region was observed and detected by Lin and Sheu [7] with the help of flow visualization and split film probe. After the converging region, there is an added energy in the flow-field due to the combination of two jets so that the velocity increases to a maximum value at a point which is called combined point. Zero velocity and maximum velocity is attained at the point of merging and combined point respectively as the change of pressure energy is converted to kinetic energy as reported by Elbanna and Gahin [3]. After the merging region, the circular twin jet has achieved the single peak value like a single jet at the midpoint between the two nozzles while in case of circle-ellipse twin jet, the peak moves towards the circular nozzle side. The merging region is the region between the merging point and combined point. Here, the mixing and interaction of jets occur in a very rapid manner. The merge point and combined point for circular twin jet occurs in the centerline of the twin jet. However, in circle-ellipse twin jet, the point of merging of the jets and combined point occurs at a different location. In this experimental study, the centerline refers to the line extending from the center of the nozzle spacing along the flow direction. This is considered as the line of reference for the two twin jet nozzle configuration.

3.2 Measurements Along Radial (Y) Direction

In circular twin jet, the radial profiles are similar on the either side of the centerline which is observed from Fig. 3. The location of merging of jets lies between the X/D ratio of 4 and 5 which is shown by zero velocity. The velocity of jets after merging increases as the resultant of both the jets occurs in flow field direction. The inner shear layers of the two jets merge so that there is a greater transfer of momentum and energy between the two jets.

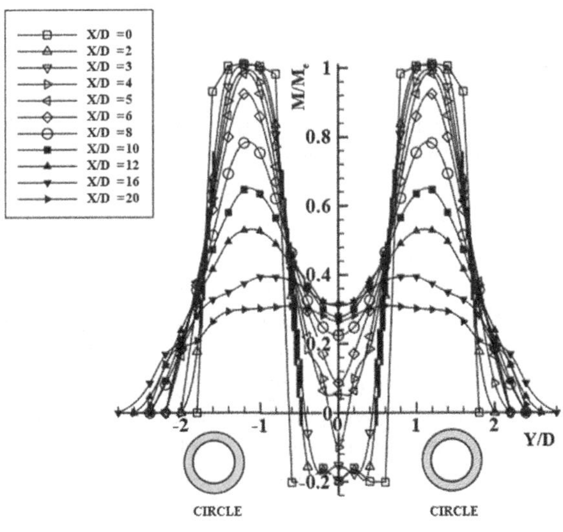

Fig. 3 Radial Mach number decay rate of circular twin jet for M = 0.4

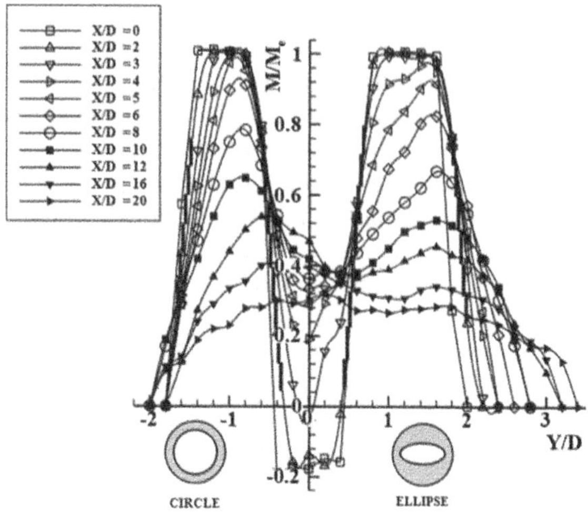

Fig. 4 Radial Mach number decay rate of circle-ellipse twin jet for M = 0.4

Because of the greater momentum transfer, the entrainment rate is high within the jets and also with the surrounding fluid. The maximum velocity was attained by the jet at a point in the centerline or plane of symmetry which is the combined point. From this plot, it is evident that the location of combined point is at X/D ratio of 16 after which the velocity starts decreasing.

But in case of circle-ellipse twin jet, from Fig. 4, the merging point occurs earlier at X/D = 3 which ensures that the mixing occurs at a faster rate. In circle-ellipse twin jet, the combined point refers to the point where the trajectories of maximum velocity points on each jet intersect as experimentally studied by Anderson and Spall [1] The profiles are not similar on either side of the centerline.

There is no plane of symmetry until the self-preserving state as the noncircular elliptical jet causes a greater fluctuation and the center of the jet moves towards the elliptical nozzle side. However, the peak moves towards the circular nozzle side as the decay is less compared to the elliptical jet side.

It is evident from Fig. 5 that the point of merging in circular twin jet moves upstream which lies between X/D = 3 and 4. The combined point exists at X/D = 16. The radial profiles are similar to those obtained by Elbanna and Gahin [3] and Tanaka [13]. It reveals that the jet after the combined point and single jet have similar velocity profiles in the far downstream distance i.e., after the combined point. The velocity then decreases as shown by the velocity magnitude. From Fig. 6, for circle-ellipse twin jet, even though the merge point lies between X/D = 2 and 3, combined point exists at X/D = 12 which ensures that mixing takes place at a faster rate.

Ghanshyam et al. [5] indicated that the asymmetric nozzles lead to enhancement in mixing. The unsymmetrical vortices created in the circle-ellipse twin jet because the entrainment to takes place at a faster rate. Hence, it is also identified that the nozzle geometry has greater influence over the merging region of twin jet which

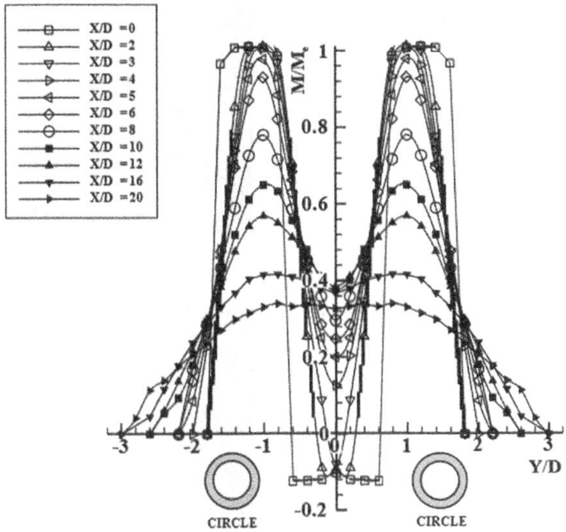

Fig. 5 Radial Mach number decay rate of circular twin jet for = 0.6

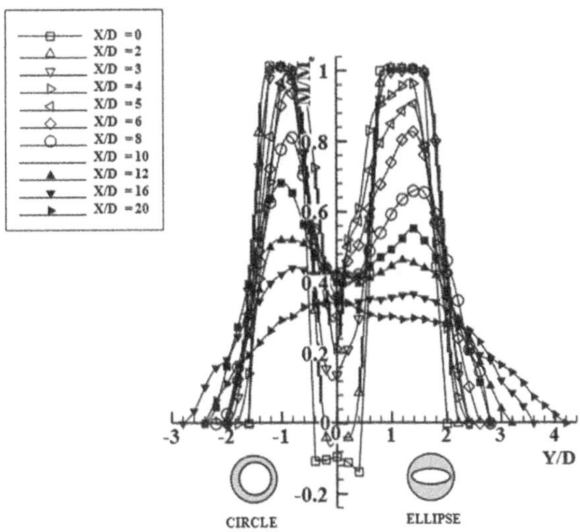

Fig. 6 Radial Mach number decay rate of circle-ellipse twin jet for = 0.6

is clear from the radial profiles. It is also clear that for all the Mach numbers, the spreading is more in the elliptical nozzle side compared to circular nozzle side.

3.3 Optical Flow Visualization for Underexpanded Jets

Figures 7 and 8 show the shadowgraph pictures for the visualization of flow field for nozzle pressure ratio NPR 2. The expansion is mild and weak waves prevail in

Fig. 7 Shadowgraph picture of circle-circle twin jet at NPR 2

Fig. 8 Shadowgraph picture of circle-ellipse twin jet at NPR 2

the jet flow. The upper nozzle is circular and lower is ellipse. The development of expansion is same in both the nozzles for circular twinjet but in case of circle-ellipse twinjet the decay is more in elliptical nozzle side. The shock cells is also less in the elliptical nozzle. Hence, it is evident from the shadowgraph pictures that the mixing is enhanced in circle-ellipse twin jet.

4 Conclusions

From this investigation, it is evident from the merge point and combined point of the radial profiles that the circle-ellipse twin jet shows better mixing enhancement than circle-circle twin jet. It is noted that the recirculation zone which is indicated by the negative value and also the location of merge point decreases as the Mach number increases. This indicates that the mixing is faster between the jets in the merging region. Also, the radial Mach number profiles of combined jet have a propensity to resemble a single jet after a downstream distance of 20D. The results also indicated that the mixing characteristics were strongly influenced by the nozzle geometry. The spreading increases downstream which could be identified from the transverse profiles. It is also evident from the shadowgraph pictures that the decay is faster in the elliptical nozzle side. Hence, it is clear that the mixing is strongly affected by jet Mach number and nozzle geometry.

References

1. Anderson EA, Spall RE (2001) Experimental and numerical investigation of two-dimensional parallel jets. J Fluids Eng Trans ASME 123(2):401–406
2. Durve A, Patwardhan AW, Banarjee I, Padmakumar G, Vaidyanathan G (2012) Numerical investigation of mixing in parallel jets. Nucl Eng Des 242(2012):78–90
3. Elbanna H, Gahin S (1983) Investigation of two plane parallel jets. AIAA J 21(7):986–991
4. Faghani E, Rogan SN (2012) Application of CFD and phenomenological models in studying interaction of two turbulent plane jets. Int J Mech Eng Mechatron 1(1)
5. Ghanshyam S, Sundararajan T, Bhaskaran KA (2003) Mixing and entrainment characteristics of circular and non-circular confined jets. J Fluids Eng 125:835–842
6. Kumar A, Das MK (2011) Study of a turbulent dual jet consisting of a wall jet and an offset jet. J Fluids Eng 133:101201-1
7. Lin YF, Sheu MJ (1990) Investigation of two plane parallel unventilated jets. Exp Fluids 10:17–22
8. Muthuram A, Thanigaiarasu S, Divvela R, Rathakrishnan E (2020) Nozzle geometry effect on twin jet flow characteristics. Int J Turbo Jet Engines. https://doi.org/10.1515/tjj-2019-0049
9. Muthuram A, Thanigaiarasu S, Ezhilan V (2017) Experimental study of symmetrical and asymmetrical twin jet for low Mach numbers. Int J Eng Technol Sci Res 4(11):1025–1030
10. Mi J, Nathan GJ, Luxton RE (2000) Centreline mixing characteristics of jets from nine differently shaped nozzles. Exp Fluids 28(2000):93–94
11. Moustafa GH (1994) Experimental investigation of high-speed twin jets. AIAA J 32(11):2320–2322
12. Rathakrishnan E, Vijayabhaskar Reddy P, Padmanaban K (1989) Some studies on twin jet propagation. Mech Res Commun 16(5):279–287
13. Tanaka E (1974) The interference of two-dimensional parallel jets 2nd report. Experiments on the combined flow of dual jet. Bull JSME 17(109):920–927
14. Yin Z, Zhang H, Lin J (2007) Experimental study on the flow field characteristics in the mixing region of twin jets. J Hydrodyn Ser B 19(3):309–313
15. Zaman KBMQ (1999) Spreading characteristics of compressible jets from nozzles of various geometries. J Fluid Mech 1999(383):197–228

Effect of Alumina Nanofiller on the Mechanical Properties of 2D Woven Biotex Flax/PLA Fiber-Reinforced Nanocomposite

Adnan Amjad, Habib Awais, Muhammad Zeshan Ali, M. Shukur Zainol Abidin, and A. Anjang

Abstract Fiber-reinforced polymer (FRP) composite has become one of the most important applications in the aerospace industry. More than 50% of the streamlined fuselage is made of composite material, and this will continue further due to technological advancement in the field of fiber-reinforced polymer composite material. Natural fiber (Bast, Stem, Leaf) are the most focusing in today's era due to their environment-friendly nature. The development of 2D woven Biotex flax/PLA fiber-reinforced polymer composite integrated with alumina nano-filler is the main objective of the current study. This study mainly discussed the effect of alumina nanofiller using four concentrations from 0 to 3 wt% on the tensile and impact properties of sodium hydroxide treated and untreated biotex flax/PLA fiber-reinforced composite. The results exhibit that the increase in filler loading has improved tensile and impact properties. At 2 wt% filler loading, the tensile strength and impact strength was maximum. Further, increase the filler contents have adverse effects on mechanical properties. The results have witnessed better mechanical properties of treated compared to untreated reinforcement.

Keywords Woven fabric · Nanofillers · Flax fiber · Mechanical properties

1 Introduction

2D woven fabric structures have been used as reinforcement in the fiber-reinforced laminated composite for their many industrial applications such as protection, transportation, construction, and aerospace industry [1, 2]. 2D woven fabric made by the perpendicular interlacement of two sets of yarn called the warp yarn (ends) and

A. Amjad · H. Awais · M. S. Zainol Abidin · A. Anjang (✉)
School of Aerospace Engineering, Universiti Sains Malaysia,
14300 Penang, Malaysia
e-mail: aeaslina@usm.my

M. Z. Ali
School of Materials and Mineral Resources Engineering,
Universiti Sains Malaysia, 14300 Penang, Malaysia

P. Rajendran et al. (eds.), *Proceedings of International Conference of Aerospace and Mechanical Engineering 2019*, Lecture Notes in Mechanical Engineering,
https://doi.org/10.1007/978-981-15-4756-0_32

the filling yarn (picks). As compared to filling yarn, the warp yarn has more twists, stronger and with a uniform appearance. The main weaving construction parameters for 2D woven fabric structures are ends per inch picks and the thickness of warp and filling yarn. The density of the woven fabric is defined as ends per inch. Plain, Twill, and Satin are the basic 2D woven structures [3]. 2D woven fabric is important for its lightweight, low cost, and impact damage resistance. However, during interlacement, the float and crimp of both warp and filling yarn affect the mechanical performance of 2D woven structures [4].

Many types of fibers have been used in 2D woven structures such as human-made or high-performance fibers (Carbon, Glass, Kevlar, Nomex, and UHMWPE) and natural or biodegradable fibers (Jute, Hemp, Flax, Sisal, Banana and Kenaf) [5, 6]. As the world is trying to keep the environment safe and secure, the scientists and researchers are focusing on shifting the use of human-made fibers to natural fibers. Because the human-made fibers are not biodegradable, reproducible and reusable due to which it is not considered as an environment-friendly [7, 8]. That is why the year 2009 was considered the International Year of the Natural Fiber (IYNF) to promote natural fiber and materials [9].

Along with most of the advantages, natural fibers have few drawbacks. One of them is their poor interfacial adhesion. To improve the interfacial properties, natural fibers are pretreated to remove oil, wax, dust, lint, and chemically modified surfaces. All-natural fibers have a hydroxyl (OH) group on its surface, which is involved in hydrogen bonding that reduces the activity of natural fibers towards the matrix. There are several treatments to improve mechanical properties and fiber-matrix adhesion [10]. In 2D laminate composite, the strength of the matrix, fiber-matrix interface, and inter-ply interface plays a vital role in load-bearing and load transferring stages [11, 12]. The 2D laminate composite shows excellent properties when the load is applied in the fiber direction. However, when the load is applied in off-axis to the fiber direction, then the debonding and microcracking will start, which leads toward the inter-laminar delamination. Delamination is the major issue in the 2D laminate composite, which occurs when the bonds that hold the stacked layers together start to a breakdown, which leads to failure of the composite material [13, 14].

Considering this, the addition of nanofillers/nanoparticles in a matrix as a reinforcing agent is a valuable approach to strengthen the matrix and its interface. The addition of nanoparticles/fillers had enhanced the in-plane mechanical properties and inter-laminar shear properties of the laminate composite [15, 16]. A vast amount of work has been done with manmade or synthetic fiber, but still, there is a gap to try different metal oxide nanofillers with different natural fibers and by different techniques. So, the primary objective is to study the influence of alumina nanofillers on the mechanical properties of biotex flax/PLA fiber-reinforced epoxy composite.

2 Experimental

2.1 Materials

Aluminum oxide nanofillers having particle size 50 nm with 99% purity and density of about 3.95 g/cm^3 were supplied from Richest Group Limited, Shanghai, China. The biotex flax/PLA having 2/2 twill weave design and 400 g/m^2 fabric weight with a density of 1.33 g/cm^3 was obtained from Composites Evolution Ltd., Chesterfield, United Kingdom. Pro-Set INF-114/INF-213 infusion epoxy was purchased from China, and Caustic soda NaOH was obtained from a local supplier.

2.2 Methodology

The following factors and their levels that will be considered in this study are given in Table 1. The DOE of this study is the total possible combination of these factors and their levels. So, a total of eight combinations formed four of each treated and untreated reinforcement, respectively. Table 2 showed the detail of each sample concerning nanofiller concentration.

Table 1 Factors and their levels

Factors	Level of factors			
Fiber (Reinforcement)	Treated biotex flax		Untreated biotex flax	
Weave	2/2 Twill			
Nanofiller	Al_2O_3			
Nanofiller concentration (wt%)	0	1	2	3

Table 2 Summary of composite samples

Sample no.	Reinforcement type	Nanofiller concentration (wt%)
S1	Treated biotex flax	0
S2	Treated biotex flax	1
S3	Treated biotex flax	2
S4	Treated biotex flax	3
S5	Untreated biotex flax	0
S6	Untreated biotex flax	1
S7	Untreated biotex flax	2
S8	Untreated biotex flax	3

Pretreatment of 2D Woven Biotex Flax/PLA Fabric

First, the biotex flax fabric was pretreated with a 2% solution of sodium hydroxide (NaOH) having sodium dodecyl sulfate (SDS) as a wetting agent and ethylenediaminetetraacetic acid (EDTA) as a sequestering agent. The wetting and sequestering agent was added in ppm. The fabric was treated for 60 min at 100 °C temperature. After 60 min, the fabric was hot rinse and then a cold rinse. After washing, the fabric was dried at room temperature for 24 h and oven-dried at 80 °C temperature for 60 min. After that, treated and untreated biotex flax fabric was cut into a 300 × 300 mm dimension so they can be used as reinforcement.

Composite Fabrication

The three layers samples having dimensions 300 × 300 mm and the stacking sequence [0/90/0] s was manufactured by vacuum bagging technique using infusion epoxy as a matrix. In all composite samples, all three reinforcement layers are of the same type. The fiber mass to matrix mass ratio was 1:2.5. Before applying the epoxy, the nanofillers was added to the epoxy by weight percentage (0–3 wt%) according to DOE followed by the dispersion of nanofillers by mechanical stirring at a speed of 2500 rpm for 60 min. After mixing of nanofiller, the mixture underwent 10 min degassing process. Then the hardener is added to the epoxy with a ratio of 100:27.4. After the fabrication of composite plates, the samples were removed from vacuum bagging after 24 h then was placed in the oven for the post-curing of composite plates at a temperature of 50 °C for 60 min.

Characterization

After curing, all samples were cut according to ASTM D3039 and ISO 179/180 standard for tensile and Charpy impact testing, respectively. The band saw cutter was used to cutting the testing specimen. The dimensions for tensile testing and Charpy impact testing specimens were 200 × 25 mm and 80 × 10 mm, respectively. The Instron universal testing machine (UTM, 3367, United States) was used for tensile testing, and Zwick/Roell® pendulum impact tester was used for Charpy impact testing. The gauge length for tensile testing was 120 mm with a speed of 2 mm/min while the weight of the hammer for Charpy impact testing was 7 J. Each sample has three repetitions for both measurements.

3 Result and Discussion

3.1 Tensile Properties

The out-turn of alumina nanofillers on the tensile behavior of the laminated composite was determined by Instron 3367, and the results were compared with respect to nanofiller concentration and reinforced fabric treatment. As shown in Figs. 1 and 2, by keeping constant the nanofiller concentration, both the tensile strength and

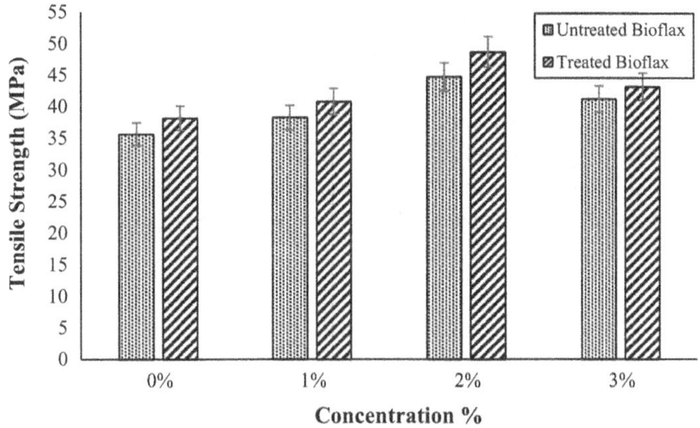

Fig. 1 Tensile strength of composite laminates

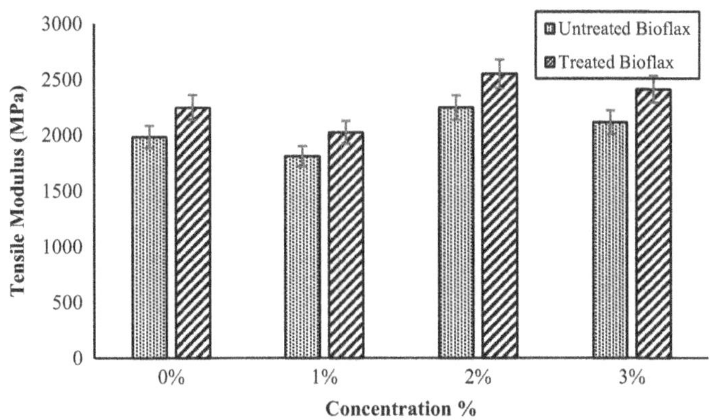

Fig. 2 Tensile modulus of composite laminates

modulus of laminates having treated reinforcement were found to be higher than the untreated reinforcement. This is because of improvement in interfacial adhesion property of reinforcement. The mercerization of reinforcement removes oil, wax, and dust from the surface and opens the pores, which lead to improving the fiber-matrix adhesion.

On the other hand, when the results were compared with respect to the concentration of nanofiller, it was disclosed that both the tensile strength and modulus of the laminated composite were increased with the percentage of nanofillers up to 2%. The addition of 3% shows a negative effect on these properties due to the tendency of nanofillers to attract each other, which causes agglomeration [17].

As compared to 0%, the tensile strength and modulus of treated laminates having 2% alumina nanofillers were improved by 27.35% and 13.35%, respectively, while the tensile strength and modulus of untreated laminates having 2%, nanofillers were improved by 25.33% and 13.37% respectively. The tensile stress-strain curves of the treated and untreated laminate composite are illustrated in Fig. 3 and Fig. 4, respectively. According to stress-strain curves at 1% nanofillers concentration, the percentage of strain is maximum, and it had increased up to (17–19%). This increment may be due to the debonding of nanofillers. Further, the increase in nanofillers concentration at 2 and 3% of the strain % decreases. This decrement may occur due to the brittleness of the laminates by alumina nanofillers agglomeration. Figure 5 shows the tested samples for tensile properties.

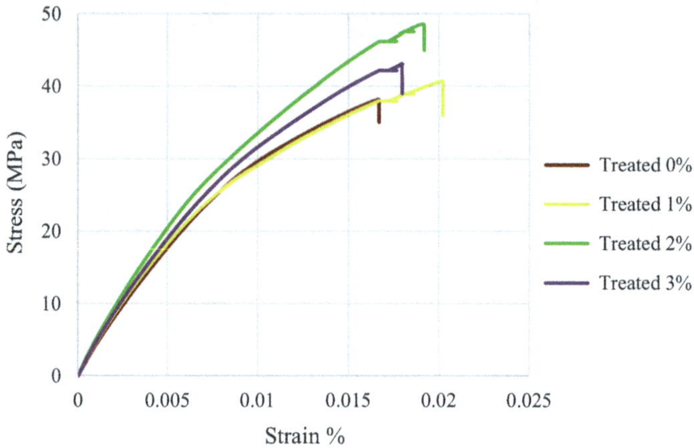

Fig. 3 Tensile stress-strain curves of treated laminates

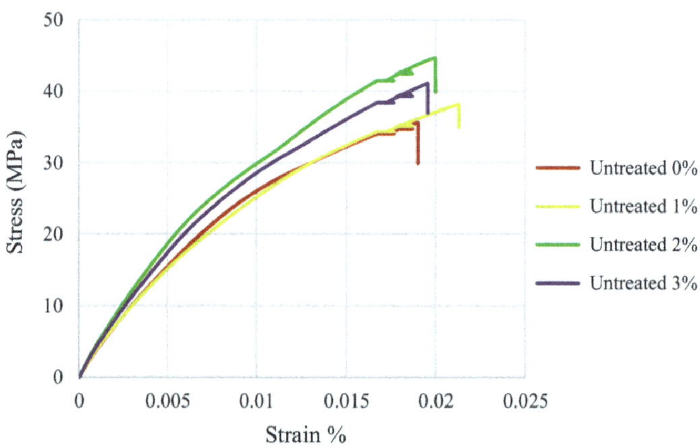

Fig. 4 Tensile stress-strain curves of untreated laminates

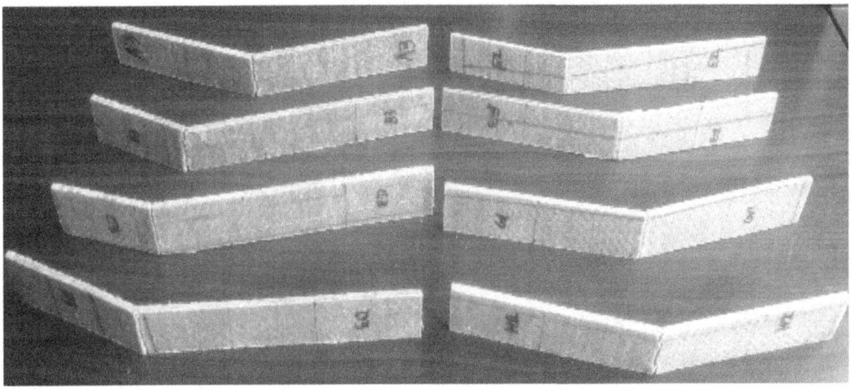

Fig. 5 Tensile properties tested samples of treated and untreated biotex flax fiber-reinforced composite

3.2 Impact Properties

The impact strength of un-notched biotex flax/PLA composite laminates was determined using Zwick/Roell® Charpy impact tester. The results were compared with respect to the concentration of alumina nanofillers and reinforced fabric treatment. Figure 6 shows the impact strength of treated and untreated biotex flax/PLA composite laminates according to the nanofiller concentration %.

The results have shown the improvement in the impact strength of laminates as the concentration of nanofillers increases by up to 2%. Further increase the concentration to 3% the impact strength starts declining may be due to non-uniform mixing of nanofillers in matrix caused agglomeration, which leads to the brittleness of the laminates. At 2%, the impact strength of treated and untreated reinforcement has

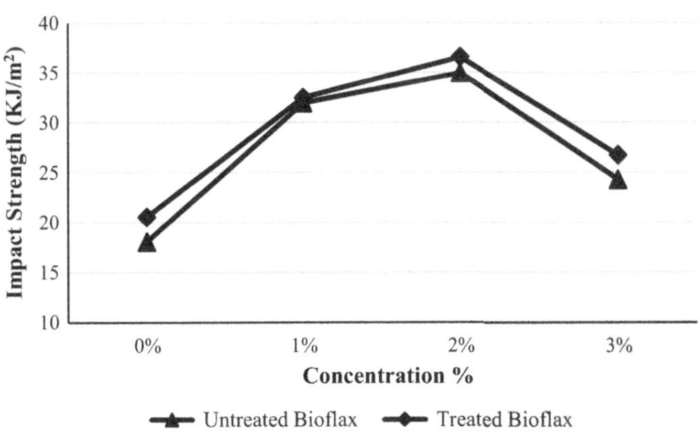

Fig. 6 Charpy impact strength of laminates

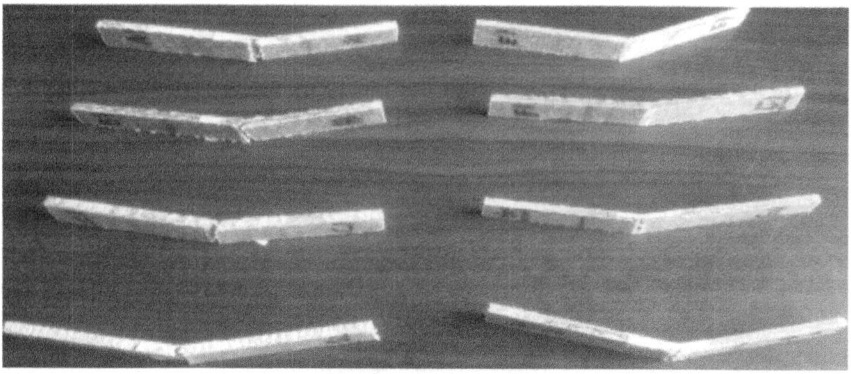

Fig. 7 Impact properties tested samples of treated and untreated biotex flax fiber-reinforced composite

increased up to 73.14% and 84.62%, respectively. Figure 7 shows the tested samples for impact properties.

4 Conclusion

The experimental study on the effect of alumina nanofiller loading on the tensile and impact behavior of biotex flax/PLA reinforced epoxy composite leads to the following conclusions:

1. Treated reinforcement had shown better properties than the untreated reinforcement due to their better fiber-matrix adhesion properties.
2. The inclusion of alumina nanofiller had affected the tensile strength and modulus of the laminates. As the percentage by weight of the nanofiller increases, the tensile strength and modulus increased by up to 2%. The further increase has shown a negative effect.
3. The tensile strength and modulus are maximum at 2%. As compared to the control sample, at 2% of nanofillers, the tensile strength and modulus of treated reinforced laminates improved by 27.35%, and 13.35% and untreated reinforced laminates improved by 25.33% and 13.37% respectively.
4. The addition of nanofillers also affects the impact strength of laminates. At 2% nanofiller concentration, the impact strength is high. Further increase the concentration to 3% the impact strength starts declining may be due to ununiform mixing, agglomeration, and brittleness of the laminates.
5. Therefore, it can be concluded that the addition of nanofiller up to 2% by weight of the matrix is the optimum level of mechanical performance.

Acknowledgements The authors would like to acknowledge the Higher Education Commission (HEC) Pakistan, and Universiti Sains Malaysia FRGS Grant (203/PAERO/6071409) from the Malaysian Government for their assistance and financial support in this study.

References

1. Chen X, Taylor LW, Tsai LJ (2011) An overview on fabrication of three-dimensional woven textile preforms for composites. Text Res J 81:932–944
2. Cicco D, Taheri F (2018) Use of a simple non-destructive technique for evaluation of the elastic and vibration properties of fiber-reinforced and 3D fiber-metal laminate composites. Fibers 6:14
3. Aiman DPC, Yahya MF, Salleh J (2016) Impact properties of 2D and 3D woven composites: a review. In: AIP conference proceedings. American Institute of Physics Inc.
4. Rouf K, Denton NL, French RM (2017) Effect of fabric weaves on the dynamic response of two-dimensional woven fabric composites. J Mater Sci 52:10581–10591
5. Karger-Kocsis J, Mahmood H, Pegoretti A (2015) Recent advances in fiber/matrix interphase engineering for polymer composites. Prog Mater Sci 73:1–43
6. Baghaei B, Skrifvars M, Berglin L (2015) Characterization of thermoplastic natural fiber composites made from woven hybrid yarn prepregs with a different weave pattern. Compos Part A Appl Sci Manufact 76:154–161
7. Sanjay MR, Madhu P, Jawaid M et al (2018) Characterization and properties of natural fiber polymer composites: a comprehensive review. J Clean Prod 172:566–581
8. Saba N, Jawaid M, Alothman OY et al (2016) Recent advances in epoxy resin, natural fiber-reinforced epoxy composites, and their applications. J Reinf Plast Compos 35:447–470
9. Zakriya M, Ramakrishnan G, Gobi N et al (2017) Jute-reinforced non-woven composites as a thermal insulator and sound absorber—a review. J Reinf Plast Compos 36:206–213
10. Mohd Edeerozey AM, Md Akil H, Azhar AB, Zainal Ariffin MI (2007) Chemical modification of kenaf fibers. Mater Lett 61:2023–2035
11. Tessema A, Kidane A (2017) Cross-property interaction between stiffness, damage, and thermal conductivity in the particulate nanocomposite. Polym Test 64:127–135
12. Tessema A, Zhao D, Kidane A, Kumar SK (2016) Effect of micro-cracks on the thermal conductivity of particulate nanocomposite. Springer, Cham, pp 89–94
13. Tessema A, Ravindran S, Wohlford A, Kidane A (2018) In-situ observation of damage evolution in quasi-isotropic CFRP laminates. In: Conference proceedings of the society for experimental mechanics series. Springer, New York, LLC, pp 67–72
14. Tessema A, Ravindran S, Kidane A (2018) Gradual damage evolution and propagation in quasi-isotropic CFRC under quasi-static loading. Compos Struct 185:186–192
15. Tessema A, Zhao D, Moll J et al (2017) Effect of filler loading, geometry, dispersion, and temperature on thermal conductivity of polymer nanocomposites. Polym Test 57:101–106
16. Tessema A, Mitchell W, Koohbor B et al (2018) The effect of nano-fillers on the in-plane and interlaminar shear properties of carbon fiber-reinforced composite. J Dyn Behav Mater 4:296–307
17. Zhang H, Zhang H, Tang L et al (2010) The effects of alumina nanofillers on mechanical properties of high-performance epoxy resin. J Nanosci Nanotechnol 10:7526–7532

Investigating Dielectric Properties in Hybrid PLA-PHA Composites with Sodium Hydroxide Treated Flax Fibers

Elammaran Jayamani, Hari Prashanth Palani Velayuda Shannmugasundram, and KokHeng Soon

Abstract Dielectric materials with high permittivity and low dielectric loss are preferred for increasing insulation. Compared with metals or ceramics, polymer composites have proven to be good alternatives due to their exemplary material properties such as flexibility, machinability and adjustable dielectric properties. Besides, it is important to develop a novel material that has greater dielectric properties and one that uses sustainable materials in order to satisfy the growing field of green engineering. In this research paper, the dielectric constants, and dielectric losses are investigated to find the right composition of PLA-PHA blended, surface treated flax fiber infused biopolymer composites. Also, the factors that affect these properties such as interfacial polarization, dipole polarization have been explained according to the results obtained. The results obtained showed an increased dielectric constants and decreased dielectric loss at certain frequencies due to the inclusion of conductive copper fillers, surface treated flax fibers and a biopolymer mixture.

Keywords Dielectric · Natural fiber · Composites

1 Introduction

The theory of dielectrics has become more prominent ever since the sizes of electrical devices turned smaller, faster and offer higher electrical performance. Dielectric materials have been in the limelight for manufacturing modern integrated circuits which contains enormous amount of transistor that relates the properties of dielectrics [1]. These dielectric properties can be altered in many ways to invent new products that could be used in many venues of evolving, new technologies. For instance, dielectric materials can be used in long distance electricity transmission, applied high energy density storage systems as well as in printed circuit boards [2]. In the past few decades, materials sciences have emerged a new sector of study with biological polymers that can be found in natural circumstances and can be synthesized without

E. Jayamani (✉) · H. P. P. V. Shannmugasundram · K. Soon
Swinburne University of Technology, Sarawak Campus, 93350 Kuching, Sarawak, Malaysia
e-mail: ejayamani@swinburne.edu.my

P. Rajendran et al. (eds.), *Proceedings of International Conference of Aerospace and Mechanical Engineering 2019*, Lecture Notes in Mechanical Engineering,
https://doi.org/10.1007/978-981-15-4756-0_33

having concerns of negative environmental impacts. In this new venture, researchers have been focusing on biopolymers to fabricate tailored materials that can optimize the performance of many above-mentioned electrical devices [3]. Theories and past studies have suggested that the use of conductors inside the sample can assist in conductivity between the molecules, easing the polarization process between them. As a result, the addition of a conductive filler, there are two types of polarization that can be achieved: Which is the electronic and dipole polarization, one of the main factors that affects the dielectric permittivity of the composites. Due to the existence two biopolymers and a conductive filler, it is expected to find new effects of polarization which would not be found in past research.

2 Literature Review

The use of polymers or hybrid polymer composites have recently come into light, due to the various merits they possess concerning dielectric and mechanical properties. The processing capabilities of polymer composites compared to ceramics are considerably simpler. Their availability and affordability is higher and the amount of percentage loading within the composite is much lower than that of ceramics, which does not affect the percolation threshold, dielectric properties and the mechanical strength of it. Hence, it is understood that polymer composites can be a good replacement for ceramic composites. Polymer composites in applications of insulation allow the use of metals such as silver (Ag) and copper (Cu) to be added in their matrices to act as electricity conducting fillers. Dielectric constants (k) can be increased when all-organic polymer composite films are used. As results obtained from Dang and Zheng's (2018) experiments, which involved Polypropylene (PP) and Polyvinylidene fluoride (PVDF), showed approximately 3 times higher permittivity than in pure PP [4]. In addition, organic polymers can be blended with carbon nanotubes (CNTs) and graphene to make polymer nanocomposites fulfil dielectric requirements of the newly engineered material. A study showed that a polymer nanocomposite material that comprises of PVDF and CNTs, which were tested for dielectric properties, exhibited a high dielectric constant, low dielectric loss and improved flexibility in the material. This was due to the nanocomposite's ability to allow homogeneous dispersion of CNTs to suppress the dielectric loss [5].

Polymer composites can be blended with natural fibers, which subsume a high dielectric constant and can be treated with chemicals to modify their structures. This will improve both dielectric and mechanical properties of the blend [6]. A study concerning functionalized *jute* fiber (natural fiber) that was blended with polypropylene (PP), to strengthen its polymer matrix, reportedly found that the tensile strength and Young's modulus had almost a 20% increment due to the bolstered up interfacial interactions between the fibers and the polymer matrices [7]. As a result, polymer composites have been recognized as one of the viable compounds to produce novel materials with high standards of dielectric and mechanical properties.

2.1 Polylactic Acid (PLA) and Polyhydroxyalkanoates (PHA)

PLA has superior mechanical properties with a tensile strength of 53 MPa, elongation at break of 10–100% and a flexural modulus of 350–450 MPa. Making it an optimum candidate for applications in energy storage and insulation [8]. PLA can be blended using the process of solution blending in laboratory conditions whereas at an industrial level, the preferred method would be melt blending. As discussed above, crystalline PLA can be dissolved in chlorinated solvents such as chloroform and dichloromethane. Thereafter, mechanical mixing can be exercised and eventually the solvent can be evaporated using an oven. On the contrary, the melt blending method follows the mixing of components above their melting temperature within an internal blender or a screw blender that forms miscible (homogenous mixture) blends of PLA [9]. PHA and their main variants have been a great field of interest in improving their mechanical strength for applications in biomedical, packaging and in agriculture. PHA can be blended with other polymers that, after investigation, have proved to improve the thermal stability and biocompatibility. PHA and one of its variants polyhydroxybutyrate (PHB) homopolymer has a melting temperature of 180 °C and can be less processable and more brittle. In addition, PHA and its variant are prone to degradation in saline, aerobic and anaerobic conditions. To remedy this, PHA and PHB compounds must either be biosynthesized or blended with other polymers [10]. PHA matrices can be produced by the twin-screw extrusion method, which is at first pelletized and compression molded at 190 °C under 150 bar of pressure and followed through with the melt-blending process [11].

2.2 Flax Fibers

Flax fibers have typically similar stiffness to glass fibers and have a density just half that of glass fibers. Due to this factor, these fiber reinforcements are used in the cores of skiing and surfing equipment, manufacturing of ribbing in cars and for internal shaping of spaceships [12]. It is evident from the tensile testing that the flax fiber-infused polymer composite showed increase in tensile strength after curing [13]. Maya Jacob John (2012) had studied the effect of coupling agents on dielectric permittivity of flax-polypropylene composites. An ethanol and water mixture was used to immerse non-woven flax fibers and drained out. Following this process, they were dried in an air oven at 110 °C. Subsequently, polypropylene sheets with the flax fibers were formed into molds using compression molding and finally cured. The results of the dielectric permittivity was obtained and appeared higher than usual due to the presence of polar groups present in the polymer matrix [14].

2.3 Copper Particles (Microparticles)

Copper in the form of powder has been used in an effort to produce conducting polymer composites. Copper being the filler material that helps conduct electricity with the polymer-matrix being the insulating dielectric material. They have been exploited in the polymer-based composite due to the ease of manufacture, corrosion resistance, low cost, weight and the fact that electrical conductivity was introduced to the composite. Metallic copper powder has a very high conductivity and is one of the best choices for filler materials that can be used for fabrication. However, it is noteworthy to understand that the filler volume should not exceed the percolation threshold (>30% wt%) as exceeding this limit would completely remove the dielectric property of the composite, making it a full time conductor [15]. Hence, it is understood that when the conductive copper particle content remains less than 30 wt%, the composite mimics the properties of semi-conductor, however, better than conventional ones.

3 Methodology

3.1 Natural Fiber Preparation

Flax fibers were used in this investigation which was supplied by Go Green Products Pvt. Ltd (India). The fibers that were procured are non-woven and were immersed in 5 wt% NaOH solution for 24 h. The fibers were then washed with distilled water until all the impurities were removed. Subsequently, the fibers were oven dried at 95 °C for 6 h. It is noteworthy to know that the flax fibers were 'crispy' dry and separate strands of them were obtained. Lastly, these strands were trimmed into extremely short, 5 mm fibers. The fiber lengths were standardized to 5 mm as to a past study showing sophisticated strength models through fiber lengths less than 1 cm [16].

3.2 Composite Fabrication

In this research project there are different ways the novel material has to be fabricated in accordance with the objective of this research. Firstly, to investigate the dielectric properties of the composite material, ASTM-D150 standard will be used to standardize the measurement of dielectric constant and loss tangent. For this investigation, the sample preparation method was compression molding. Compression molding was found to provide fairly good mechanical properties compared to other methods such as injection molding or through extrusion [17]. The molding was performed by GOTECH Hot Press and LS-22071 Hydraulic Hot Press machines. Firstly, pure samples of PLA and PHA were fabricated and after a few trial and error methods, the compression molding process was adapted to produce the best samples of PLA-PHA blended composites. The samples were 50 mm in diameter and 5 mm in thickness in accordance to dielectric testing standards of ASTM D150 [18]. The composite masses with constant PLA-PHA vol% (30 and 70%), alkali-treated flax fibers at 2 vol%, and copper filler vol% were altered from 2 to 8 vol% with an increment of 2% was calculated. The dielectric samples were produced at 150 °C (both bottom and top plate) and exerted a constant 10 tons of pressure for 45 min. The samples were kept under cool down in natural air convection for 4–6 h and were removed from the presses. Secondly, the samples with varying flax fiber content were produced to investigate the changes in dielectric constants and losses when fiber volume percentages were altered from 2 to 8 vol% with an increment of 2 vol%. Hence, 4 sets of dielectric samples were fabricated with varying compositions of NaOH treated fibers, which had undergone the same process of hot pressing. (Temperature = 150 °C; Pressure = 10 tons).

3.3 Dielectric Testing

Dielectric Constant. For investigating the dielectric properties of PLA-PHA composite blend, the samples with varying conductive filler (Copper) and flax fibers, the Agilent 16451 B Dielectric Test Fixture was used supplied by Swinburne University of Technology Sarawak. Every sample that is to be tested were 'sandwiched' between two contacting electrodes and their thickness will be adjusted according to the sample thickness (5 mm) using a micrometer. Also, these samples will be tested in different frequency ranging from 20 Hz to 2 MHz respectively. The dielectric constant was calculated using the formula;

$$\varepsilon' = \frac{C_p}{C_0} \tag{1}$$

C_p is the capacitance of the sample and C_0 is capacitance of the air that is occupied between the electrodes.

C_0 can be calculated using the equation show below. C_0 can be calculated using the equation show below.

$$C_0 = \varepsilon_0 \frac{A}{t} \tag{2}$$

$$\varepsilon' = \frac{C_p t}{\varepsilon_0 A} \tag{3}$$

The dielectric constant can be found when the equation for C_0 is manipulated in the first equation. A is the area of contact with the specimen, t is the thickness of the sample between the electrodes and ε_0 is the permittivity of free space (i.e.

8.854×10^{-12} F) [19]. It is noteworthy to know that since the area of contact and thickness is standardized by ASTM D150, it is safe to assume that C_0 value would be approximately 3.41 pF as the dielectric permittivity of air within the 5 mm height would be roughly the above-mentioned value according to the calculations using Eq. 2.

Loss Factor. The loss factor or dielectric loss is found to assure how much electrical energy is lost from the sample to the circuit. The dielectric loss is calculated using the equation shown below;

$$\varepsilon'' = \varepsilon' \tan \delta \tag{4}$$

$$\tan \delta = D \tag{5}$$

Agilent dielectric fixture will provide two important values, one, the capacitance of the sample, C_p and two, the dissipation factor, D (as displayed in the machine) or otherwise, $\tan \delta$.

4 Results and Discussion

In this investigation, the dielectric constant and dissipation factor was obtained from the samples that were tested in the dielectric fixture. The frequencies at which the samples tested were at 20 Hz, 500 Hz, 1 kHz, 10 kHz, 100 kHz, 500 kHz, 1 MHz and 2 MHz respectively.

4.1 Effect on Dielectric Constants of PLA-PHA Composite at Varying Compositions of Copper Filler

The dielectric constant of PLA-PHA composite samples with varying copper filler percentages (2, 4, 6 and 8 vol%) were subjected to different frequencies and as the volume percentage of copper increased, the dielectric constant values increased gradually from as low as 1.2–3.47. This phenomenon is clearly reflected in Fig. 1. The reason behind this drastic change is when the conductive filler content increases, the conductivity of the sample must have increased. The fact that when an electric field is applied, 'electronic polarization' occurs in a molecular level, where the non-polar PLA-PHA blend gets polarized and when conductivity is increased, the area of contact between the conductor and the polymer blend increases. This eventually increases the polarizability of the sample itself. Hence, due to this increased movement of electrons, the interfacial polarization increases the dielectric constants of the samples. From the graph (Fig. 1) it can be observed that the dielectric constant of samples tested at 20 Hz, is always higher than samples tested at other frequencies.

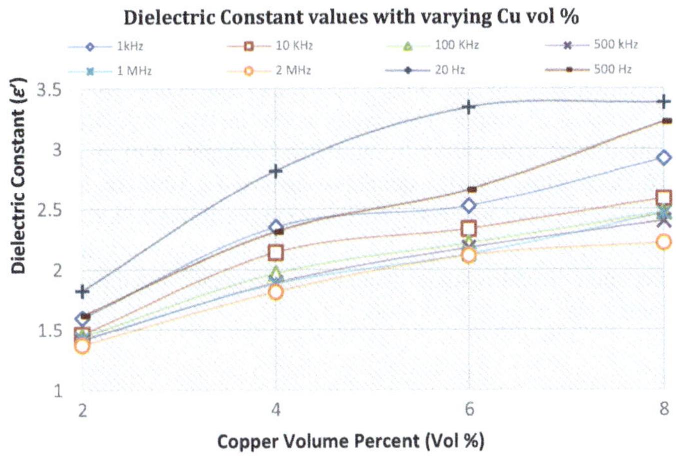

Fig. 1 Dielectric constant comparison with varying copper content

Typically, there is a 25.13% difference between the dielectric constant of the sample when that was tested at 20 Hz and 2 MHz; Also, the difference in dielectric constant of the sample tested at 20 Hz and 2 MHz with 8 vol% copper is a significant value of 1.63.

The primary reason behind this substantial increase is due to the time interval between electronic and dipole polarization. As to when electricity is passed through the sample, electronic polarization happens almost immediately (under 10^{-10} s), however, the dipole polarization takes longer to set in. Non-polar polymers like PLA-PHA blends will have their molecules to orient towards the direction of the electric field and on further increase in the frequency, these molecules will be subjected to more chaotic oscillations, causing the degree of orientation to diminish [20]. Hence at dielectric constant is higher at low frequencies and vice versa.

4.2 *Effect on Dielectric Losses of PLA-PHA Composite at Varying Compositions of Copper Filler*

Most of the fabricated samples that were tested for dielectric loss had shown clear coherence with the theory, which states that when frequency increases the dissipation factor of the dielectric sample decreases as it can be seen from the graph. The highest losses were experienced by samples that were tested at 20 Hz, 1 kHz and 10 kHz. The increase in this loss factor at 20 Hz could have been the result of dipole relaxations happening after the relaxation time that disorients the molecules and throw them away from each other causing them to lose charge at a faster rate. On the other hand, the contribution of higher losses at 1 and 10 kHz could have been the erratic electric fields that supplied electrical and heat energy which would have possibly aggravated the charge dissipating capacity of the samples [21].

The 'relaxation time' is the single most important factor that would affect the dissipation factor of a sample. This factor is the inverse of 'cut-off frequency' at which the interfacial polarization is cut off and dissipation of charge commences. The lowest dielectric loss from the samples were tested at 500 kHz, 1 MHz and 2 MHz at all variations of copper percentages. The samples at these particular frequencies had not enough time to go through dipole polarization to absorb enough amount of charge, hence their dielectric losses were low, as to preserving the miniscule amount of charge which was already been stored [3].

4.3 Effect on Dielectric Constants PLA-PHA Composite at Varying Compositions of Surface-Treated Flax Fibers

The PLA-PHA samples with varying volume percentages of flax fibers were tested from 2, 4, 6 and 8% and their dielectric constant were found which is plotted on Fig. 2. The conductive copper filler percentage was standardized at 6 vol% to conserve the validity of the experiment. From the results it was found that the dielectric constants reduced from 2% fiber content until and further continued to reduce until 8% at all frequencies. It was also observed that the dielectric constant was the highest when the fiber content was kept at 2 vol% at 20 Hz. In this case, the fiber treatment played a big role in the polarization process. The treatment of 5 wt% of NaOH had removed the interaction of polar groups that must have promoted better adhesion with the

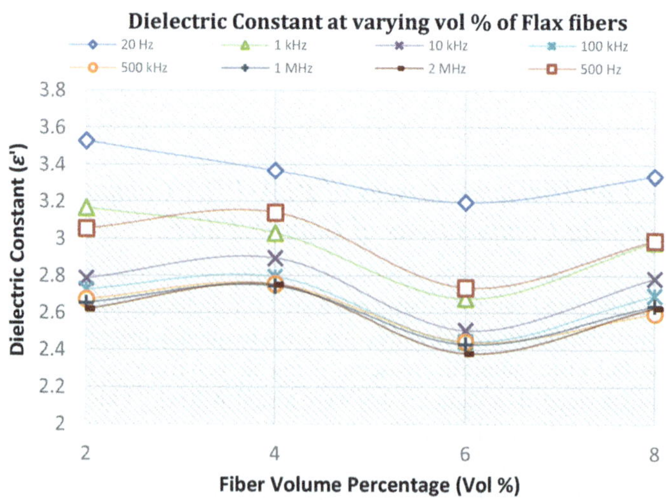

Fig. 2 Dielectric constant comparison with varying flax fiber content

polymers that were employed in the composite sample which had a profound effect on the dielectric constant, increasing it to an astounding value of 3.55 [18].

When the samples were tested at higher frequencies, the permittivity values had decreased and further decreased significantly when the fiber loading was at 6 vol%. The curve took a slight ascend when fiber loading was at 8 vol% but the increase was not of big significance to the results. The absurd drop in dielectric constant values were due to the increased loading of the fibers which were not properly oriented in the PLA-PHA matrix. It was also found that some of the samples had patches of short fibers clumped together which could again, reduce the integrity of the results. A similarity between this particular result can be found from another study which states that the disorientation of fibers would lower the conductivity and polarization of the molecules in the matrix, which is evidently observed in Fig. 2 [22].

4.4 Effect on Dielectric Losses PLA-PHA Composite at Varying Compositions of Surface-Treated Flax Fibers

The dielectric loss values that had been obtained showed high losses of samples from an average of 0.1873–0.07454 that were tested at 20 Hz, 500 Hz and at 1 kHz where the dielectric constants in the previous graphs were the highest at the same frequencies. This is again, due to the fact of high charge gain by the polymer composite, which tends to lose the charge after the cut-off frequency. On the other hand, when the fiber loading increases, the loss factor was reduced due to higher chances of interaction between the polar groups of the natural fibers [19].

5 Conclusion

The results that were obtained after testing appear to validate the theory of dielectrics and past research made on similar biopolymer materials. From the acquired results, the appropriate compositions of PLA-PHA blended hybrid polymer can be identified, depending on the desirable properties that a researcher is looking for to implement in an application. The results vividly showed that this composite material can be put to use in 'low-k' dielectric applications under low to mid (500 Hz–500 kHz) frequency operations. In a technical aspect, the dielectric constants and losses observed in this investigation, indicates the various polarization and fiber orientation that happens when the material is subjected to an electric field; Dipole Polarization, Electronic Polarization, Ionic conductance and Fiber Dipolar Reorientation. Hence, there must be further research in this particular matter that focusses more on molecular motions, charge carriers for energy storage and distribution. Lastly, it can be concluded that, amorphous polymers such as Polylactic acid and Polyhydroxyalkanoates can offer a humungous edge to the dielectric properties at lowered frequencies.

References

1. Martin SJ, Godschalx JP, Mills ME, Shaffer EO, Townsend PH (2000) Development of a low-dielectric-constant polymer for the fabrication of integrated circuit interconnect. Adv Mater 12(23):1769–1778
2. Dang Z-M (2018) Introduction. Dielectric polymer materials for high-density energy storage, pp 1–9
3. Ward AA (2016) Dielectric materials for advanced applications, p 51
4. Dang Z-M, Zheng M-S (2018) Multiphase/multicomponent dielectric polymer materials with high permittivity and high breakdown strength. Elsevier Inc.
5. Wang Y, Xing C, Guan J, Li Y (2017) Towards flexible Dielectric materials with high dielectric constant and low loss: PVDF nanocomposites with both homogenously dispersed CNTs and ionic liquids Nanodomains. Polymers (Basel) 9(11)
6. Elanchezhian C, Ramnath BV, Ramakrishnan G, Rajendrakumar M, Naveenkumar V, Saravanakumar MK (2018) Review on mechanical properties of natural fiber composites. Mater Today Proc 5(1):1785–1790
7. Varghese AM, Mittal V (2018) Polymer composites with functionalized natural fibers. Elsevier Ltd.
8. Madhavan Nampoothiri K, Nair NR, John RP (2010) An overview of the recent developments in polylactide (PLA) research. Bioresour Technol 101(22):8493–8501
9. Hamad K, Kaseem M, Ayyoob M, Joo J, Deri F (2018) Polylactic acid blends: the future of green, light and tough. Prog Polym Sci 85:83–127
10. Ten E, Jiang L, Zhang J, Wolcott MP (2015) Mechanical performance of polyhydroxyalkanoate (PHA)-based biocomposites, Fourteenth. Elsevier Ltd.
11. Crétois R, Delbreilh L, Dargent E, Follain N, Lebrun L, Saiter JM (2013) Dielectric relaxations in polyhydroxyalkanoates/organoclay nanocomposites. Eur Polym J 49(11):3434–3444
12. Mathijsen D (2018) The renaissance of flax fibers. Reinf Plast 62(3):138–147
13. Rahman R, Zhafer Firdaus Syed Putra S (2019) Tensile properties of natural and synthetic fiber-reinforced polymer composites. Elsevier Ltd.
14. John MJ, Tlili R, Anandjiwala RD, Boudenne A, Ibos L (2012) Effect of amphiphilic coupling agent on heat flow and dielectric properties of flax-polypropylene composites. Compos Part B Eng 43(2):526–532
15. Barkoula NM, Alcock B, Cabrera NO, Peijs T (2008) Fatigue properties of highly oriented polypropylene tapes and all-polypropylene composites. Polym Compos 16(2):101–113
16. Sparnins E, Mechanical properties of flax fibers and their composites. http://www.campuscorner.Fibre2Fashion.com. Last accessed 2006
17. Garkhail SK, Meurs E, De Beld V, Peijs T (1999) Thermoplastic composites based on biopolymers and natural fibres. In: 12th international conference on composite materials, pp 1175–1184
18. Jayamani E, Hamdan S, Rahman MR, Bin Bakri MK (2014) Comparative study of dielectric properties of hybrid natural fiber composites. Procedia Eng 97:536–544
19. Triki A, Karray M, Poilâne C, Picart P, Gargouri M (2015) Dielectric analysis of the interfacial polarization of alkali treated woven flax fibers reinforced epoxy composites. J Electrostat 76:67–72
20. Member VSY, Sahu DK, Member YS, Dhubkarya DC (2010) The effect of frequency and temperature on dielectric properties of pure poly vinylidene fluoride (PVDF) thin films, vol 3, pp 17–20
21. Burubai W (2013) Effects of moisture and frequency variations on the dielectric properties of African Nutmeg (Monodora myristica) seeds. 9(1):80–85
22. Huo S, Thapa A, Ulven CA (2013) Effect of surface treatments on interfacial properties of flax fiber-reinforced composites. Adv Compos Mater 22(2):109–121

Fabrication and Characterization of Lightweight Engineered Polypropylene Composites Using Silica Particles and Flax Woven Comingled Structure

Habib Awais, Yasir Nawab, Adnan Amjad, Aslina Anjang, and M. Shukur Zainol Abidin

Abstract The focus of the current study was the fabrication of woven comingled fabrics using flax and polypropylene yarns to improve the impregnation of flax reinforcing yarn in the polypropylene matrix. Hollow glass microspheres (HGM) as fillers were incorporated to tailor the mechanical properties of the resultant comingled composites. The dip and dry method was used to deposit the HGM on these comingled fabrics to achieve uniform distribution of hollow silica particles. Cross-ply composites were fabricated by these comingled fabrics using compression moulding technique. The mechanical properties were measured in terms of tensile, flexural and impact strength. Furthermore, the effect of HGM loading (1.5 and 3%) on mechanical properties was studied and compared with control composites prepared without HGM. It was observed that 1.5% inclusion of HGM had imparted an increase of 13.8 and 5.6% in the tensile strength and modulus respectively by enhancing the brittleness of the composites. However, a further increase in the HGM content from 1.5 to 3% had exhibited a decline of 10.2 and 8.8% in the tensile strength and modulus owing to the agglomeration of silica particles. A similar trend was observed in the flexural properties of the laminates. While, an increment of 10.4% and 18.2% was observed in the impact strength of the fabricated composite laminates by the loading of 1.5% and 3% HGM respectively.

Keywords Flax fibre · Hollow glass microspheres · Comingled composites · Mechanical properties

---

H. Awais · A. Amjad · A. Anjang · M. S. Zainol Abidin (✉)
School of Aerospace Engineering, Universiti Sains Malaysia, 14300 Penang, Malaysia
e-mail: aeshukur@usm.my

Y. Nawab
Faculty of Engineering and Technology, National Textile University,
37610 Faisalabad, Pakistan

P. Rajendran et al. (eds.), *Proceedings of International Conference of Aerospace and Mechanical Engineering 2019*, Lecture Notes in Mechanical Engineering,
https://doi.org/10.1007/978-981-15-4756-0_34

1 Introduction

Thermoplastic composites using continuous fibres are gaining considerable attraction in recent years. Thermoplastic composites exhibit a wide range of advantages over thermoset, particularly short processing time, ease of processability, infinite shelf life, better toughness, excellent impact and corrosion resistance [1]. Such extraordinary characteristics of thermoplastics are expanding new and more challenging application areas, especially in the transport industry. It is well established that the melt viscosities of thermoplastic polymers are much higher compared to thermoset polymers which make it challenging to wet the fibres fully and hinders the product moulding in a single step process [2]. Comingling techniques are commonly deployed at fibre, yarn or fabric level to improve the impregnation as comingling is the most economical way to blend polymers and fibres [3]. Nonwoven comingled structures [4] have been used earlier to fabricated the jute reinforced polypropylene composites and witnessed the improvement in the tensile, flexural and impact properties. Similarly, micro braided comingled yarn [5] and woven comingled fabrics [6] using handloom were developed to improve the mechanical properties of the composite materials. The aforementioned studies have witnessed the improvement of mechanical properties by respective comingling techniques. Hence this research work employs the woven comingling technique.

Flax fibre is among one of the oldest fibre crops in the world. Flax is a bast fibre and belongs to the family "*Linum usitatissimum*". Low density, less energy consumption during production, high specific properties and biodegradability of flax fibres make them an eminent choice for fibre reinforced composites [7]. Several researchers [8–10] have explored the use of flax fibre for the development of bio-composites in order to mitigate the environmental effect of conventional synthetic fibres. The mechanical strength of FRP composites still needs to improve further in order to meet the use function of such materials in engineering applications. Therefore, various organic and inorganic fillers are being utilized by research communities to overcome the strength and cost issues of FRP materials [11–14]. Hollow glass microspheres are another useful inorganic filler having interesting characteristics, particularly good thermal properties and low density. Besides, in various investigations, the incorporation of HGM with phenolic resin [15], epoxy resin [16], polyethylene [17], polyurethane [18] and polypropylene [19] had provided a positive impact on the mechanical, thermal and rheological properties of the ensuing materials.

Seeking the fabrication of eco-friendly lightweight composite materials with desirable properties, the incorporation of HGM on the flax fibre may be an interesting strategy as there has been no comprehensive study in the literature about using HGM with comingled fabrics. Therefore, in the current study, flax woven comingled composites were fabricated using HGM. The synergetic effect of comingling and HGM on flax fibre reinforced composites was assessed in terms of tensile, flexural and impact properties.

2 Experimental

2.1 Materials

Hollow glass microspheres (Sigma Aldrich) with 9–13 μm particle size (density 1.05–1.15 g/ml) were purchased from DJ Scientific, Karachi, Pakistan. The flax 100% yarn (spun) was acquired from Shanghai Vico International Co. Ltd. China, with an average yarn linear density of 2.45 ± 0.15 expressed in terms of English cotton count. The Duralene® polypropylene yarn (filament) was procured from Synthetic fibres, Karachi, Pakistan with the average yarn count of 2400 D and density of 0.91 g/cm^3.

2.2 Methodology

Development of woven comingled fabric. Rapier loom (BeatMax ISL888-II, Ishikawa Seisakusho. Ltd., Japan) was used to manufacture the comingled fabric. The 4/1 satin weave pattern was selected to develop the woven comingled fabric by inserting flax yarn as warp yarn and polypropylene (PP) yarn as weft yarn (Fig. 1). The balanced quality with the same ends and picks (20×20) of comingled fabric was developed for this research work.

Deposition of hollow silica particles. Silica particles (HGM) were deposited on woven comingled fabric using the dip and dry method. First, two solutions were prepared by mixing 25 and 50 g of HGM in one liter of water each, while continuous stirring for 15–20 min with a mechanical stirrer until a white homogeneous solution is obtained. The woven comingled fabric was dipped in this solution with the help of a metal wire mesh to keep the comingled fabric smooth and straight and to avoid

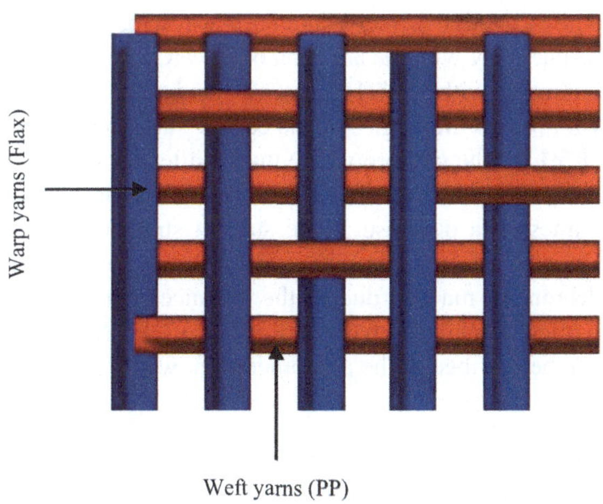

Fig. 1 Schematic diagram of the woven comingled fabric

the accumulation of hollow silica particles at specific points. The woven comingled fabric was then dried in an oven at 100 °C to evaporate the water.

Fabrication of composite laminates. The woven comingled fabrics with HGM and without HGM (control specimen) were utilized to fabricate the cross-ply composite laminates in stacking sequence of $[0/90]_3$ using compression moulding. A temperature of 185 °C and pressure of 3.4 MPa was used during the compression moulding to convert the comingled fabric into composite laminates.

Characterization. The tensile and flexural performance of the fabricated composite laminates was determined using the Zwick/Roell® Universal testing machine (UTM, Z100, Germany) while impact behavior was measured using Zwick/Roell® Pendulum impact tester (HIT50P, Germany). ASTM D3039, ASTM D7264 and ISO 179 were adopted to obtain the tensile, flexural and impact properties of the laminates respectively. The gauge length for tensile testing was 120 mm with a crosshead speed of 2 mm/min. 80 mm gauge length and 1 mm/min crosshead speed was employed for the flexural testing.

3 Results

3.1 Tensile Properties

The effects of HGM loading on tensile behavior of the fabricated composite laminates were determined by the universal testing machine and compared to the control composites. The results have revealed that both properties (strength and modulus) were increased with 1.5% HGM but the addition of 3% HGM has a negative impact on these properties which can be ascribed to the tendency of HGM to attract each other and cause the agglomeration. A growth of 13.8% in tensile strength and 5.6% in modulus was observed with 1.5% HGM addition compared to the control laminates while a decline of 10.2% and 8.8% was witnessed in the tensile strength and modulus respectively with the addition of 3% HGM. The tensile strength and modulus of the laminates with 1.5% HGM was found to be significantly higher than the laminates with 3% HGM (Fig. 2). This can be attributed to the uniform distribution of HGM (1.5%) in the polypropylene matrix after composite fabrication.

The tensile stress as a function of strain (%) was illustrated in Fig. 3 which witnessed a decrease at the start in strain (%) with the addition of 1.5% HGM. On average, 5.5% drop in strain (%) was observed with 1.5% HGM addition. This decrement may be due to the enhancement of laminates brittleness by the silica particles. While, with the addition of 3% HGM, strain (%) increased up to 44%. This can be ascribed to the agglomeration, which leads to the debonding of particles.

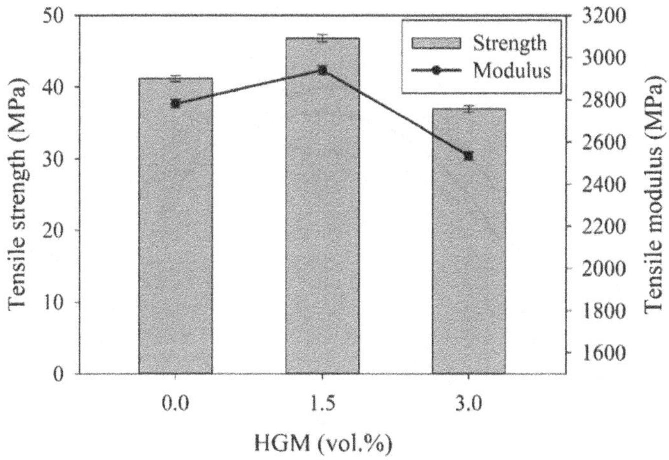

Fig. 2 Tensile properties of composite laminates

Fig. 3 Tensile stress-strain curves of laminates

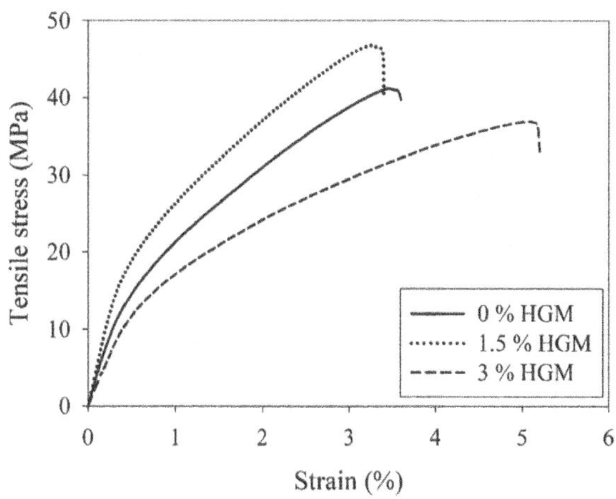

3.2 *Flexural Properties*

3-point bending test was used to elucidate the effect of HGM content on flexural properties of the laminates. Hollow glass microspheres depict the congruent trend of tensile properties during the 3-point bending test (Fig. 4). An increment of 10.3% and 16.8% in flexural strength and modulus respectively was observed with 1.5% HGM incorporation. Similarly, a diminution of 14% and 9.4% was witnessed in the flexural strength and modulus respectively by the incorporation of 3% HGM. The justification of results is similar to the tensile results.

Figure 5 had portrayed the flexural stress and strain trajectories of the fabricated composite laminates. Similar behavior of the tensile stress-strain curves was found

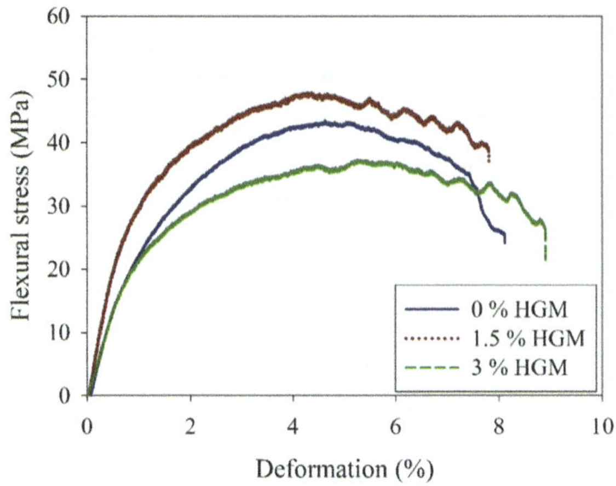

Fig. 4 Flexural properties of the composite laminates

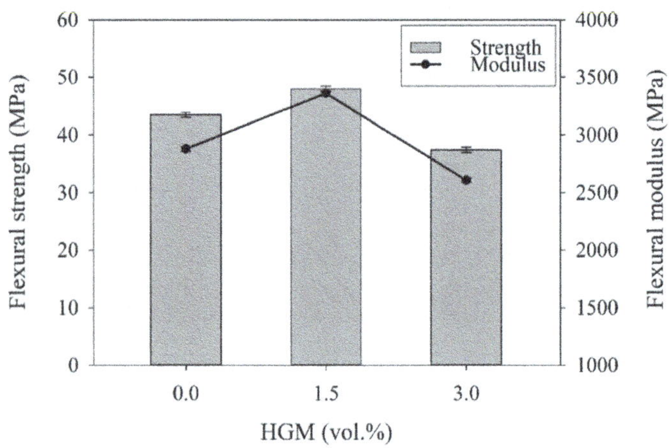

Fig. 5 Flexural stress-strain trajectories of the laminates

in that of the flexural trajectories. An explicit zig-zag crack path was observed in these trajectories (Fig. 5), which is evident that the HGM delays or hinders the crack initiation or its propagation.

3.3 Impact Properties

The impact strength of the un-notched comingled composite laminates was determined using Charpy impact tester and compared with HGM embedded laminates.

Table 1 Charpy impact strength of comingled composites

Material	HGM loading (vol%)	Impact strength (kJ/m^2)
Flax	0	41.5
	1.5	45.83
	3	49.06

The results have shown continuous improvement in the impact strength of laminates with HGM loading (Table 1). The loading of 1.5% HGM had enhanced the impact strength up to 10.4% while 18.2% increment had been imparted by the loading of 3% HGM.

4 Conclusion

Light-weight and high strength polymer materials can be fabricated using hollow glass microspheres (HGM) as fillers. The HGM as fillers had improved the mechanical strength to the resultant composite laminates only when the volume fraction of HGM was lower (1.5%). When the concentration of HGM was increased further (3%), the mechanical properties (tensile and flexural) tend to decrease while continuous improvement in the impact strength was observed by the addition of HGM. Therefore, it can be concluded that incorporation of the HGM does not always consider as a key to improving the properties and the optimum level needs to be defined before the incorporation of HGM in polymer materials.

Acknowledgements The authors are thankful to National Textile University, Pakistan (faculty development program) and Universiti Sains Malaysia (FRGS 203/PAERO/6071410) for providing financial support and resources for this study.

References

1. Takkalkar P, Nizamuddin S, Griffin G, Kao N (2019) Thermal properties of sustainable thermoplastics nanocomposites containing nanofillers and its recycling perspective. In: Inamuddin AA, Thomas S, Kumar Mishra R (eds) Sustainable polymer composites and nanocomposites. Springer, Cham, pp 915–933
2. Ning H, Lu N, Hassen AA, Chawla K, Selim M, Pillay S (2019) A review of long fibre thermoplastic (LFT) composites. Int Mater Rev 6608:1–25
3. Awais H, Nawab Y, Amjad A, Anjang A, Md Akil H, Zainol Abidin MS (2019) Effect of comingling techniques on mechanical properties of natural fibre reinforced cross-ply thermoplastic composites. Compos Part B Eng, vol 177 (June):107279
4. Ameer MH, Nawab Y, Ali Z, Imad A, Ahmad S (2019) Development and characterization of jute/polypropylene composite by using comingled nonwoven structures. J Text Inst 110(11):1–8
5. Kobayashi S, Takada K (2013) Processing of unidirectional hemp fiber reinforced composites with micro-braiding technique. Compos Part A Appl Sci Manuf 46(1):173–179

6. Righetti Souza B, Mello Di Benedetto R, Hirayama D, de Andrade Raponi O, Cristina Miranda Barbosa L, Carlos Ancelotti Junior A (2017) Manufacturing and characterization of Jute/PP thermoplastic commingled composite. Mater Res 20:458–465
7. Woigk W et al (2019) Interface properties and their effect on the mechanical performance of flax fibre thermoplastic composites. Compos Part A Appl Sci Manuf 122:8–17
8. Woigk W et al (2016) Mechanical properties of tough plasma treated flax fibre thermoplastic composites. In: ECCM 2016—Proceedings of 17th European conference on composite materials, June
9. Bourmaud A, Le Duigou A, Gourier C, Baley C (2016) Influence of processing temperature on mechanical performance of unidirectional polyamide 11-flax fibre composites. Ind Crops Prod 84:151–165
10. Tak Lau K, Yan Hung P, Zhu MH, Hui D (2018) Properties of natural fibre composites for structural engineering applications. Compos Part B Eng 136 (September 2017):222–233
11. Abdul Khalil HPS et al (2019) Enhancement of basic properties of polysaccharide-based composites with organic and inorganic fillers: a review. J Appl Polym Sci 136(12)
12. Peng Y, Nair SS, Chen H, Yan N, Cao J (2018) Effects of Lignin content on mechanical and thermal properties of polypropylene composites reinforced with micro particles of spray dried cellulose nanofibrils. ACS Sustain Chem Eng 6(8):11078–11086
13. Anjum F et al (2018) Mechanical, thermal, and microscopic profile of novel glass fiber-reinforced polyester composites as a function of barium sulfate loading. Adv Polym Technol 37(3):929–936
14. Bilalova EA, Prut EV, Kuznetsova OP (2019) Polypropylene composite material and its rheological and mechanical properties depending on the size of the filler $CaCO_3$. IOP Conf Ser Mater Sci Eng 525:012006
15. Yang H et al (2018) Mechanical, thermal and fire performance of an inorganic-organic insulation material composed of hollow glass microspheres and phenolic resin. J Colloid Interface Sci 530:163–170
16. Imran M, Rahaman A, Pal S (2019) Effect of low concentration hollow glass microspheres on mechanical and thermomechanical properties of epoxy composites. Polym Compos
17. Zhu BL, Wang J, Zheng H, Ma J, Wu J, Wu R (2015) Investigation of thermal conductivity and dielectric properties of LDPE-matrix composites filled with hybrid filler of hollow glass microspheres and nitride particles. Compos Part B Eng 69:496–506
18. Jiao C, Wang H, Li S, Chen X (2017) Fire hazard reduction of hollow glass microspheres in thermoplastic polyurethane composites. J Hazard Mater 332:176–184
19. Jung BN, Kang DH, Cheon S, Shim JK, Hwang SW (2019) The addition effect of hollow glass microsphere on the dispersion behavior and physical properties of polypropylene/clay nanocomposites. J Appl Polym Sci 136(14):1–12

Numerical Study on the Effect of Innovative Vortex Generators in the Mixing Enhancement of Subsonic Jets

S. Thanigaiarasu, S. B. Jabez Richards, V. Karthikeyan, A. Muthuram, Surendra Kumar Yadav, and T. Vijaya Raj

Abstract Numerical analysis of subsonic jets with and without tabs (vortex generators with exit area blockage of 4.84%) has been done in the present research paper. Two tabs has been fixed at nozzle exit diametrically opposite. ANSYS CFX software was used for analysis purpose and to understand the flow characteristics at four different Mach numbers 0.4, 0.6, 0.8 and 1.0. The main objective of the present study is to analyze the effect of innovative vortex generators in the form of solid tabs with triangular fins on either side offset to each other on the performance characteristics. Due to counter rotating streamwise vortices, there is large engulfment of masses from ambient air into the core jet, resulting in the reduction of potential core length.

Keywords Subsonic jets · Tabs · Asymmetric · Innovative vortex generators · Mixing enhancement · Stream-wise vortices · Mass entrainment

S. Thanigaiarasu · S. B. Jabez Richards (✉) · V. Karthikeyan
Madras Institute of Technology Campus, Anna University, Chennai, India
e-mail: jabez3011@gmail.com

S. Thanigaiarasu
e-mail: sthanigaiarasu@mitindia.edu

V. Karthikeyan
e-mail: aerokarthi21@gmail.com

A. Muthuram
St. Peter's College of Engineering and Technology, Chennai, India
e-mail: masrimuthu@gmail.com

S. K. Yadav · T. Vijaya Raj
Prince Shri Venkateshwara Padmavathi Engineering College, Chennai, India
e-mail: aero.surendra@gmail.com

T. Vijaya Raj
e-mail: jaywinvijay@gmail.com

P. Rajendran et al. (eds.), *Proceedings of International Conference of Aerospace and Mechanical Engineering 2019*, Lecture Notes in Mechanical Engineering,
https://doi.org/10.1007/978-981-15-4756-0_35

Nomenclature

D	Diameter of nozzle exit
M	Mach number
P_0	Gauge inlet total pressure
P_{0e}	Total pressure at nozzle exit
x	Axial distance along the jet
RT	Right-angled triangular projection

1 Introduction

Jet mixing enhancements of both low and high-speed flows are widely used in numerous applications such as noise suppression in airplanes, improvement of efficiency in combustors, pollutant dispersal to the environment and in reducing infrared radiance of plumes in the case of military jets. The jet noise is minimum when the jet stream mixes with the ambient fluid at the earliest. In the case of gas turbine engines, for better combustion characteristics, the mixing of fuel should be in the proper proportion with the air. The main combustion problem which is generally encountered in the gas turbine engines is the minimal time available for the fuel to mix with the air and ignite to produce the required thrust due to the faster moving air into the core. Therefore, jet exit at the fuel injector should reach the optimal pressure required for better combustion at the earliest. Enhancing the mixing in that regime helps in designing a combustor of shorter length, which is an important aspect in the design gas turbine engines. Another application of jet control is reduction of plume length in the nozzle exit, which serves for military purposes making aircraft to minimal exposure to the enemy radars by reducing its IR signature. For such applications, jet control using both passive and active techniques are in practice. Active control techniques are moderately used due to its complex phenomenon. As far as passive control technique is concerned, it is quite simple and as much effective as active control techniques. Thus, passive control methods are widely preferred for jet control applications, which uses geometrical modifications such as tabs, grooves, vanes, notches, cross wires etc. for controlling the jet stream.

Many researchers have carried out extensive researches on various passive methods of jet controls. An early study by Bradbury and Khadem [1] found that two tabs placed 180° apart at the exit of an axisymmetric jet caused it to bifurcate, causing significant increment in the overall mass entrainment of the fluids. With the obtained results as reference several studies were made by various researchers and the results were reported taking into account the variations in flow field conditions, tab shape, size, number and angle. Behrouzi and McGuirk [2] observed the jet decay as function of tab projected area, tab width, number of tabs, tab orientation angle and tab shape. From the observations, it was found that jet centreline decay weakly dependent on tab orientation angle and tab shape but the effects are quite significant with increase in the number of tabs. Zaman [3] studied experimentally the effect of vortex generators, in the form of small tabs at the nozzle exit over various Mach number ranges. The effects of increasing the number of tabs has been studied in comparison with the corresponding case without a tab. Bifurcations were observed when two tabs were placed diametrically opposite at the nozzle exit.

Venkatramanan and Thanigaiarasu [4] made an experimental and computational study for different jet spread regimes in the nozzle exit. From their investigation they found that the jet decay and the jet spread characteristics have increased substantially using the rectangular tabs. The controlled jet has about 79.04% of potential core length reduction for Mach 0.8 and 80.74% for Mach 1.0. Berrueta and Rathakrishnan [5] studied experimentally the mixing caused by the limiting tab with and without corrugations at the nozzle exit at various downstream distances. It was observed that the presence of tab at the nozzle exit gives positive results than at downstream locations.

Lovaraju and Rathakrishnan [6] investigated the cross-wire effectiveness for subsonic and sonic jet control for axisymmetric jets for various Mach numbers. The cross-wires were found to be effective in promoting jet mixing right from the nozzle exit, at all Mach numbers. Dharmahinder et al. [7] studied the effect of perforated arc tabs in the mixing enhancement of axi-symmetric subsonic and sonic jets. Centerline velocity decay profiles and radial velocity decay profiles of the controlled jet gave a clear insight on the effect of perforated arc tabs.

Asad Ahmed et al. [8] studied the effect of perforated tabs and solid tabs in the jet flow at Mach numbers 0.4, 0.6, 0.8 and 1.0. From their investigation, they found that the effect of solid tabs dominates in far downstream locations compared to perforated tabs. Rathakrishnan [9] studied the effect of rectangular tabs of different aspect ratios, with and without corrugation, in the mixing of subsonic and sonic free jets. For the sonic jet, a maximum core length reduction of 32.3% was achieved with the plain tabs, whereas with the corrugated tabs the core length reduction obtained was 17.63%.

Zaman [10] experimentally investigated the jet spreading characteristics of several nozzles with circular, rectangular and elliptic cross-sections. Jets from the rectangular nozzle fitted with two tabs on the narrow edges and the circular nozzle fitted with four equally spaced tabs were also included in the comparative study over various ranges of Mach numbers. In comparison with the axisymmetric jet, the asymmetric jets spread only slightly faster at subsonic condition than at the supersonic regime. Zaman [11] investigated the effect of triangular tabs, placed at the nozzle exit, on the evolution of free jets. The effect, a large distortion of the jet cross section and a resultant increase in mixing downstream, has been inferred before as a result of a pair of stream wise vortices originating from each tab.

In the present study, tabs are used to create instabilities, which modifies the jet flow structure in the proximity of the nozzle exit. Instead of normal tab configuration used in previous researches for jet control, this paper presents innovative tabs with sharp cornered right-angled triangular fin like projections on solid rectangular tabs with certain offset distance between them on either side.

2 Nozzle Geometry and Tab Placement

In this paper, plain nozzle and nozzle with two different tab investigations are carried out. Plain nozzle is simply a convergent circular nozzle with 40 mm inlet diameter and 20 mm exit diameter with nozzle lip thickness of 5 mm throughout the length of the nozzle. Jet is controlled by placing the tabs (Vortex Generators) at the exit plane of the circular nozzle on both sides of the jet centreline. Tabs used here have a rectangular shaped stem (4 mm × 1 mm × 1 mm) with triangular projections placed at both sides offset to each other asymmetric to the jet centreline. Asymmetry mentioned in this paper is not with respect to the placement of tabs, but due to the extended triangular projections about the jet centreline. Area blockage by the tabs accounts for 15.2 mm^2 which are 4.84% of exit area of plain circular nozzle. Nozzle is designed using SOLIDWORKS 2015 software (Figs. 1 and 2).

Tabs are categorized into two types based on the distance between triangular projections and named as mentioned below:

- RT_0 mm: Tabs with Right angled triangular projection where distance between projections is 0 mm.
- RT_1 mm: Tabs with Right angled triangular projection where distance between projections is 1 mm.

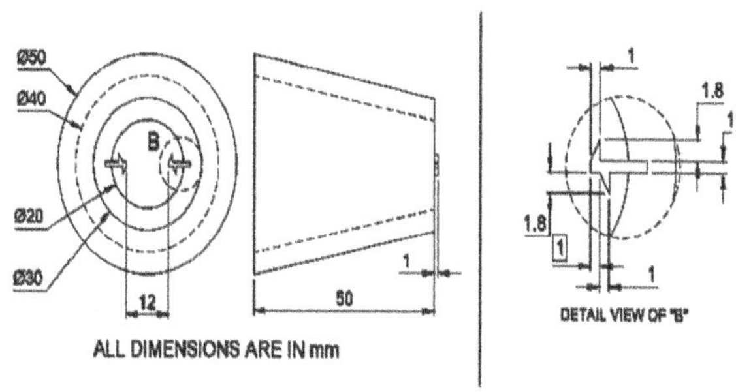

Fig. 1 Nozzle with RT_0 mm tab

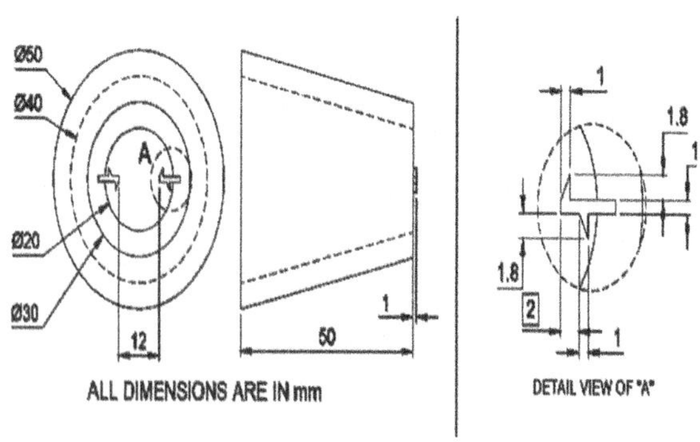

Fig. 2 Nozzle with RT_1 mm tab

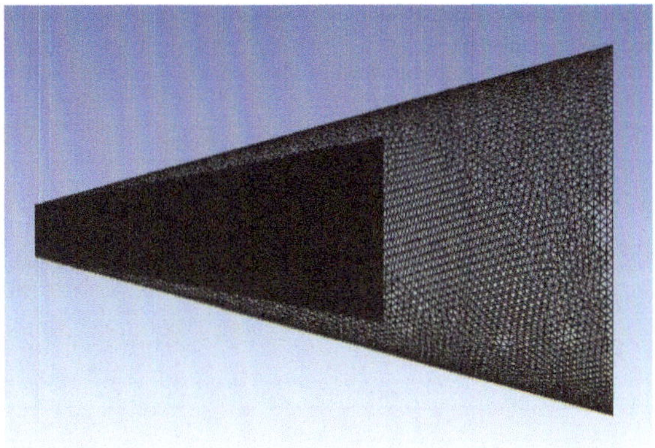

Fig. 3 Unstructured mesh of nozzle with domain

3 Computational Details

3.1 Geometry and Domain

Convergent Nozzle with inlet diameter 40 mm, exit diameter 20 mm and length 50 mm with a lip thickness of 5 mm. Domain for capturing the jet mixing characteristics is designed up to 35 times the exit diameter. Two domains are created such that first domain is up to 20 x/D and next domain up to 35 x/D with a domain interface.

Grid Generation Grids are generated using the ANSYS 16.0 meshing software. Generated grids are unstructured tetrahedrons, where finer meshing with minimum element size of 2 mm is achieved in first domain. In second domain, element size increased by a factor of 1.2. Mesh adaption is done until there are finite number of nodes at required points and thereby grid independency is studied (Fig. 3).

3.2 Boundary Conditions

Defining boundary conditions involves identifying the location of the boundaries and applying information at the boundaries. The data required at a boundary depends upon the boundary condition type and the physical model employed. Poorly defined boundary conditions can have a significant impact on the solution. Inlet of the nozzle is given as pressure inlet, Exit of domain as pressure outlet; sides of the domain are pressure farfield (Opening), nozzle is chosen as wall as shown in Fig. 4.

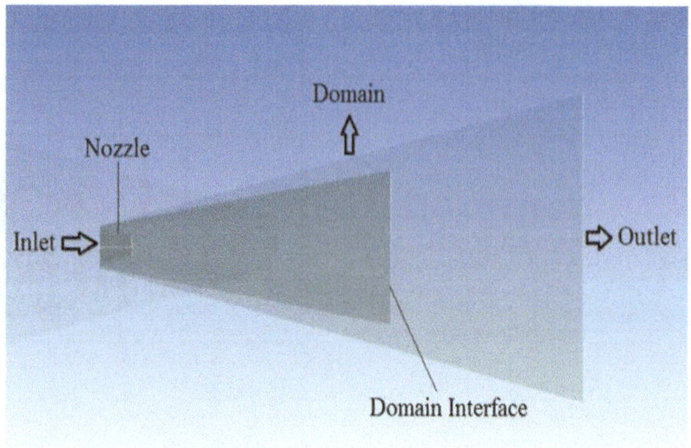

Fig. 4 Boundary details of the model analyzed

3.3 *Solver and Turbulence Model*

CFX 16.0 is used as solver to perform the numerical simulation, in which the finite difference method is used for solving the governing equations. Selecting the turbulence model is most important criteria in performing the numerical simulation. That defines how close the numerical data are matching with the experimental results. In most cases, K-ε turbulence model is used for defining the turbulence characteristics in the jet flow due to its accuracy. But in current study, there is limitation of using K-ε turbulence model (i.e.) tabs which are placed at the exit plane of the nozzle creates a strong adverse pressure gradient which makes the solver unstable using that turbulence model. So, for current study SST K-ω turbulence model is used, since the model has an advantage being stable at the strong adverse pressure gradient regimes. The SST model accounts for cross-diffusion with the K-epsilon and K-omega models. In other words, SST K-ω turbulence model works like K-epsilon in the far field and K-omega near the target geometry. This criterion fulfills flow requirement of the current study and thus SST K-ω is preferred over the standard K-ε turbulence model used in previous research papers.

4 Results and Discussions

4.1 *Centreline Pressure Decay*

Centreline pressure decay profile is a significant measure of jet mixing. The faster the decay, the faster is the jet mixing rate with the entrained ambient fluid mass and so on, which clearly indicates that the core of the jet gets better mixing due to the entrainment of the ambient fluid from the surrounding. The centreline pressure decay clearly shows the extent of the potential core of the jet, which is defined as the axial distance up to which the nozzle-exit velocity remains uniform for subsonic jets. P_{oe} is the gauge total pressure along the jet axis which is non-dimensionalized with the gauge inlet total pressure P_o supplied to the nozzle inlet. Axial distance x is non-dimensionalized with nozzle exit diameter D. Thus, the graph is plotted between x/D in x-axis and P_{oe}/P_o in y-axis to observe the jet decay characteristics in axial direction.

Figures 5, 6, 7 and 8 shows the centreline pressure decay for the jet with Mach numbers of 0.4, 0.6, 0.8 and 1.0 respectively. Base represents the centreline decay of uncontrolled jet whereas RT_0 mm and RT_1 mm represent centreline decay of controlled jet with corresponding tab configurations. Potential core region for uncontrolled jet increases slightly as Mach number increases which may be due to the high kinetic energy possessed by the jet with the supplied inlet pressure to achieve the desired Mach number. Percentage reduction in potential core length for tabbed jet is increasing with increase in Mach number. This is because the mixing efficiency of the vortices generated due to the tabs depends on its vortices strength and also the residual time for the interaction with the ambient fluid.

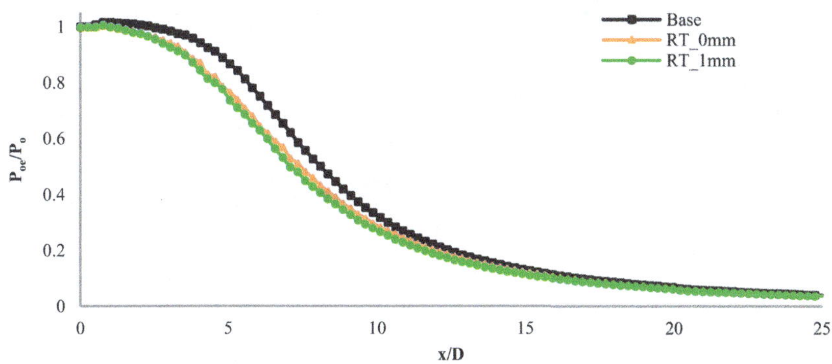

Fig. 5 Centreline pressure decay for M = 0.4

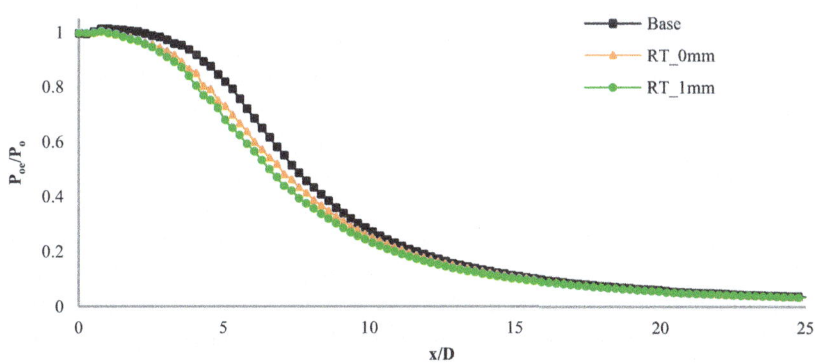

Fig. 6 Centreline pressure decay for M = 0.6

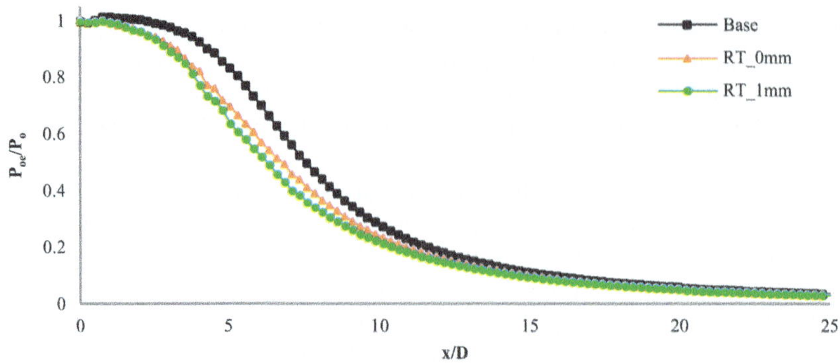

Fig. 7 Centreline pressure decay for M = 0.8

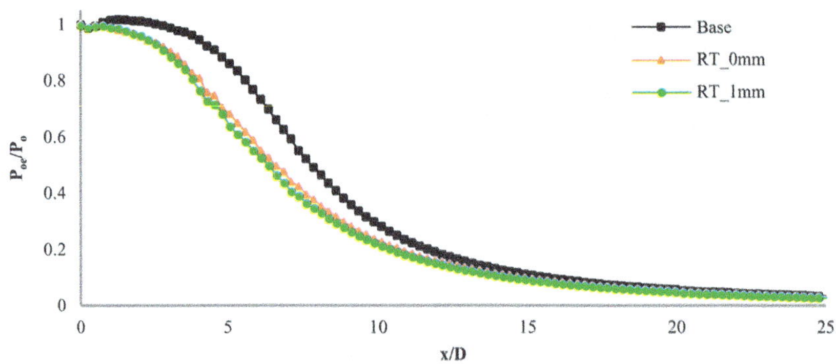

Fig. 8 Centreline pressure decay for M = 1.0

For instance, when comparing Mach numbers 0.4 and 0.6 for present study, the interaction time available for vortices at Mach 0.6 is less than that at Mach 0.4, which depicts that the vortices are transported away faster, since the faster transportation causes lesser vorticity loss, these vortices are helpful in engulfing the ambient fluid into the core region. So, the performance of the tabs in reducing the potential core length increases proportionally with the Mach number. The other factors include, the mass flow accumulation which is more at higher Mach numbers making more pressure gradients and causing the vortices' convective Mach number to be large causing strong eddy formation in the shear layer of the jet stream.

The concept of using the tabs with projections instead of a plain rectangular tab is to promote generation of more vortices from the sharp corners which in turn results in the engulfment of more ambient fluid into the jet core. When the tab configurations RT_0 mm and RT_1 mm are separately compared for centreline pressure decay, nozzle with RT_1 mm shows more characteristic decay. This is because the vortices which are produced at the sharp corners of the projections are counter-rotating type which becomes streamwise vortices due to high kinetic energy of the jet. So, when the projections are placed at a distance of 1 mm, the interaction between the streamwise vortices produced by those sharp corners gets delayed, thereby ensuing minimal loss in vorticity content when compared to RT_0 mm tab configuration.

Thus, observed centreline pressure decay is more for RT_1 mm tab configuration in potential core regime as well as transition regime of the jet. It is also observed that compressibility effect has a role to play in the jet decay characteristics due to the presence of the tabs (i.e.) when Mach number increases, reduction percentage increases. From the above plots the potential core length is maximum for Mach 1.0 which extends to about 4.54 D for uncontrolled jet, whereas potential core length is reduced to 1.37 D for same Mach number of controlled jets with tab RT_1 mm configuration which is a significant reduction.

5 Conclusions

In this paper numerical analysis of base nozzle and nozzle with tabs (2 configurations) has been presented for four Mach numbers. Due to the effect of tabs, the reduction in potential core length is observed because of the enhanced mixing produced by the counter rotating vortices from the sharp corners of tabs. From the numerical results, it is found that the core length was reduced significantly with increase in Mach number along the jet axis centreline. Even the vortex generators enhance the slope of the centreline pressure decay in the transition region when compared to base nozzle's centreline pressure decay. Nozzle with tab named RT_1 mm for Mach number 1.0 shows maximum percentage reduction of potential core about 69.82% when compared with free jet.

References

1. Bradbury LJS, Khadem AH (1975) The distortion of a jet by tabs. J Fluid Mech 70:801–813
2. Behrouzi P, McGuirk JJ (2006) Effect of tab parameters on near-field jet plume development. J Propul Power 22(3):576–585
3. Zaman KBMQ (1991) Effect of tabs on the evolution of an axisymmetric jet. NASA Technical Memorandum 104472
4. Venkatramanan S, Thanigaiarasu S (2017) Experimental and computational investigation of effect of jet control using rectangular tabs. Int J Eng Technol Sci Res 4(11):1080–1087
5. Berrueta T, Rathakrishnan E (2016) Control of subsonic and sonic jets with limiting tabs. Int J Turbo Jet- Engines 34(1):103–113
6. Lovaraju P, Rathakrishnan E (2006) Subsonic and transonic jet control with cross-wire. AIAA J 44(11):2700–2705
7. Dharmahinder SC, Thanigaiarasu S, Elangovan S, Rathakrishnan E (2011) Perforated arc-tabs for jet control. Int J Turbo Jet- Engines 28:133–138
8. Asad Ahmed R, Thanigaiarasu S, Santhosh J, Elangovan S, Rathakrishnan E (2013) Study of slanted perforated jets. Int J Turbo Jet-Engines, 1–12
9. Rathakrishnan E (2013) Corrugated tabs for subsonic and sonic jet control. J Aeronaut Aeros Eng 2(5)

10. Zaman KBMQ (1996) Spreading characteristics and thrust of jets from asymmetric nozzles. NASA Technical Memorandum 107132. AIAA J, 96-0200
11. Zaman KBMQ (1993) Streamwise vorticity generation and mixing enhancements in free jets by delta-tabs. NASA Technical Memorandum 106235, AIAA J, 93–3253

Nonlinear Follower Force Analysis with Ground Static Test Validation of High Aspect Ratio Wing

Norzaima Nordin, Noor Shazwani Muhamad Rafi, and Mohammad Yazdi Harmin

Abstract This paper presents a technique proposed to analyzed the geometrical nonlinearity of High Aspect Ratio (HAR) wing at aspect ratio, AR 16 with the inclusion of follower force. In this study, nonlinear static analysis of HAR Wing model under non-follower and follower force was performed through the Finite Element Analysis (FEA) using MSC NASTRAN software. Two important parameters (force identification, 'FORCE2' and large displacement, 'LGDISP') are considered to demonstrate the FEA nonlinear static analysis of follower force. For validation, a ground static load test was performed using small-scale load application structure through a different load direction applied for non-follower and follower force. It was found that, the selection of LGDISP = 1 in FEA nonlinear static analysis of follower force shows closest results compared to experimental data. This exhibits that combination of large displacement and follower force stiffness in LGDISP = 1 contribute to the accuracy of nonlinear follower force results. Another finding shows that, techniques of determining the load direction in ground static load test is adequately to predict the tip deflection of HAR wing with a minimal percentage error obtained. The comparison for the load case under non-follower and follower force are also demonstrated through FEA and experimental. The results show that the follower force effect indicates a higher deflection than non-follower force for both considered cases; simulation and experimental.

Keywords Finite element analysis · Follower force · Ground static test · High aspect ratio wing

---

N. Nordin (✉) · N. S. M. Rafi
Faculty of Engineering, Department of Aeronautic Engineering and Aviation, Universiti Pertahanan Nasional Malaysia, Kem Perdana Sungai Besi, 57000 Kuala Lumpur, Malaysia
e-mail: norzaima@upnm.edu.my

N. Nordin · M. Y. Harmin
Faculty of Engineering, Department of Aerospace Engineering, Universiti Putra Malaysia, 43400 Serdang, Selangor, Malaysia

P. Rajendran et al. (eds.), *Proceedings of International Conference of Aerospace and Mechanical Engineering 2019*, Lecture Notes in Mechanical Engineering,
https://doi.org/10.1007/978-981-15-4756-0_36

1 Introduction

High Altitude Long Endurance (HALE) type aircraft has become one of the unique class aircraft within the Unmanned Aerial Vehicle (UAV) application which practically used for civilian and military purposes such as for weather forecast, border patrol, as well as Intelligence, Surveillance and Reconnaissance (ISR) mission. Hence, to build a successful HALE aircraft that could perform at high altitude for long endurance and long-range operation requires some important considerations to be made for it to become more practical. One of the considerations includes the need of High Aspect Ratio (HAR) wing design for the aircraft to operate efficiently within that environment. Generally, HAR wings are capable to produce more lift thus to provide the aircraft with a higher lift-to-drag ratio [1] which proven in Eq. (1) in which the total drag coefficient can be defined as:

$$C_D = C_{D0} + C_{Di} = C_{D0} + \frac{(C_L)^2}{\pi.e'.AR} \tag{1}$$

where C_{D0}, C_{Di}, C_L, e' and AR are the profile drag coefficient, induced drag coefficient, lift coefficient, span efficiency factor and aspect ratio of the wing, respectively. As C_D is inversely proportional to AR, thus, a higher AR wing results in a lower overall drag coefficient. Despite all the improvements in HAR wing performance, there are also some drawbacks on its structural design, which led to a geometrical nonlinear issues. This is due to the flexibility issue that had exhibited larger deformation under the same flight condition [2]. This reflects the changes in dynamic behavior and in aeroelastic response, resulting in instabilities [3, 4]. Therefore, it is important to take into account geometric nonlinearities in the design of HAR wings. Geometrical nonlinearity on HAR wing is caused by the effect of continuous acting aerodynamic forces and the large displacement.

To better understand on geometrical nonlinearity behavior, one must also to clarify the consequence of considering conservative forces of Non-Follower Force (NFF) and non-conservative forces of Follower Force (FF) for the accuracy on the aerodynamic-structural analysis. In reality, the direction of the force will change with deformation of the wing due to the nature of the actual aerodynamic which referred to FF. Most of the research efforts dealing with geometric nonlinearities have focused on the NFF; less emphasize on the impact of FF on the HAR wing structure [5]. Kim et al. had conducted an experimental study to compare static response data of qualification test article collected for the cases of FF and NFF [6]. A FF setup was built to resemble realistic loads as in the actual aircraft wings. The results demonstrate that larger deformation had occurred in the FF case than NFF. In analytical study performed by Castellani et al., an aerodynamic force has been considered as FF [7]. A greater tip displacement under FF load was observed compared to NFF load in nonlinear FEM case. A similar study was demonstrated by Xie et al. using Reduced Order Model (ROM) to replace the structure modelling method for both linear and nonlinear analysis [8] with the inclusion of aerodynamic FF effect. There were significant variations between linear and nonlinear approaches due primarily to the impacts of the FF.

The focus of the present study is to validate the geometrical nonlinear analysis in term of wing tip deflection using FEA and experimental study. Force identification and large displacement approach were particularly defined in FEA analysis in order to differentiate NFF and FF applied load. A small-scale load application structure was specifically designed to elucidate a comparison between NFF and FF of HAR wing static response subjected to ground static load test. To this end, the accuracy of the FEA force identification and large displacement approach of NFF and FF analysis will be demonstrated through the validation with experimental results.

2 Finite Element Method and Simulation Setup

In this study, a MATLAB programming was coded with the routine that can establish an input file (.DAT or .bdf) format. This programming is made fully integrated with finite element (FE) solver of MSC NASTRAN. The input file composes of structural information including geometry, meshing, material properties, boundary conditions, force applied and type of solution that need to be performed. Linear and nonlinear static analysis with a solution control code (SOL 101 and SOL 106) and the input file was sent automatically to the solver to carry out FE analysis. The output file (.f06) was post-processed using MATLAB programming for result analysis. This section is also highlighted the consideration of force identification specifically for NFF and FF with the combination of large displacement interface (LGDISP) in the analysis.

2.1 HAR Wing Simulation Model

The HAR wing simulation model is constructed based on Tang and Dowel model [9] as illustrated in Fig. 1. A total number of 17 ribs were equally divided along the leading and trailing spars. The total of wing span length is denoted as L_S, chord length is denoted as, L_C with a constant value 0.05 m while the distance between the rib is denoted as L_R.

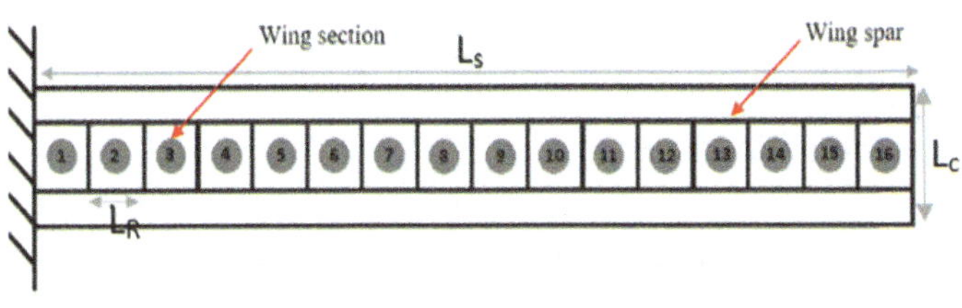

Fig. 1 The layout of HAR wing model

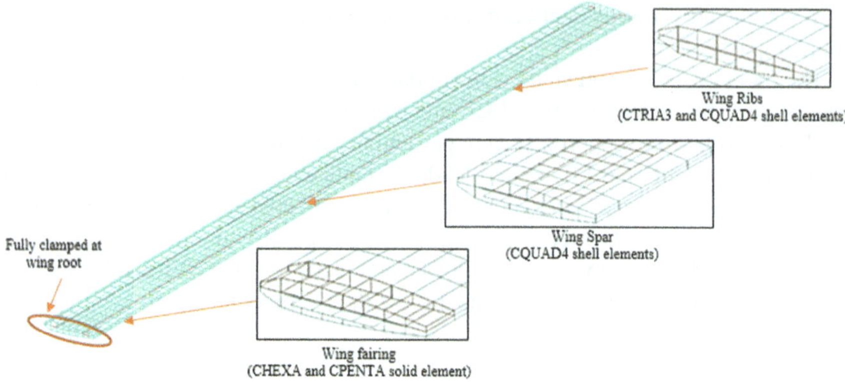

Fig. 2 HAR wing meshing

Table 1 Material properties HAR wing components

Wing component	Material	Properties	Value
Spar	Spring steel	Young modulus, E_s	207 GPa
		Density, ρ_s	7833.41 kg m^{-3}
		Poisson's ratio, υ_s	0.295
Fairings	Styrofoam	Density, ρ_f	127.6 kg m^{-3}
Ribs	ABS	Density, ρ_r	1264.83 kg m^{-3}

2.2 *Meshing, Material Properties and Boundary Condition*

The HAR wing spar was configured as a simple flat plate using CQUAD4 shell element with a fully clamped boundary condition applied at the wing root to resemble a cantilever-like condition. The wing fairing is modelled with CPENTA and CHEXA solid elements while ribs are modelled with CTRIA3 and CQUAD4 shell elements as shown in Fig. 2. The spar elements are rigidly connected to the ribs and fairings elements by multi-point constraint (MPC) elements of rigid bar connections known as RBAR. The material properties entry for spar, fairing and ribs were provided in Table 1 which replicate with the same material used in HAR wing fabrication for experiment.

2.3 *Force Identification*

There are two cases of static concentrated forces applied on nonlinear static analysis of HAR wing model which are NFF and FF cases. For NFF case, the static forces maintain in its initial direction. It is denoted as 'FORCE' in bulk entries which

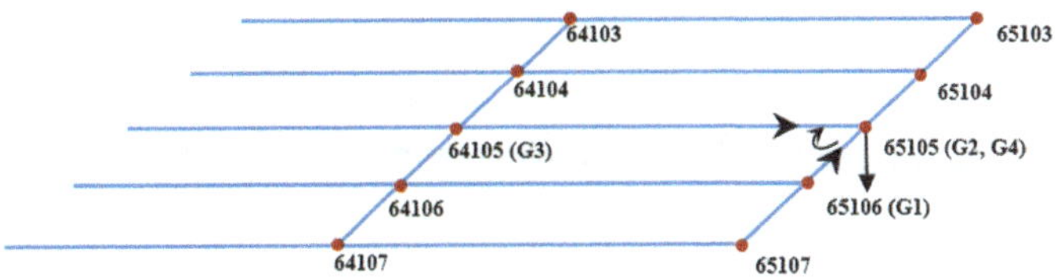

Fig. 3 The direction of vectors from G1 to G2, G3 to G4

Table 2 Example of follower force grid point identification

1	2	3	4	5	6	7	8
FORCE2	SID	G	F	G1	G2	G3	G4
FORCE2	65,105	65,105	−1.0	65,106	65,105	64,105	65,105

represents that the static concentrated force applied a grid point by specifying a vector and the components of a vector are measured in a coordinate system in MSC NASTRAN [10].

Meanwhile for FF case, the direction of the forces moves with the deformation. As a HAR wing deforms, follower force will adjust the direction of the forces to ensure the forces remain normal to the surface. To implement this, two alternative force identification can be used, either using 'FORCE1' or 'FORCE2' in bulk entries. In this study, 'FORCE2' was used to determine the direction of force using four grid points where the direction of the force is parallel to the cross product of the vector from G1 to G2 and G3 to G4. Figure 3 illustrates the direction of the vector and concentrated force applied at the center node of the wing plate.

Table 2 shows a sample format to determine the FF based on grid identification. The table consists of eight columns; FORCE2, load set identification number (SID), grid point identification number (G), the magnitude of the force (F) and identification number of four grid points (G1, G2, G3, G4). The vertical force was applied at the center tip of the wing model, starts from 0.5 to 3.5 N with incremental load of 0.5 N.

2.4 Large Displacement Consideration

The nonlinear effects become prominent in a structure when it encounters a large deformation, which elevates the significant use of geometric nonlinearity approaches. Therefore, implementation of large displacement interface for differential stiffness (LGDISP) is required.

Another consideration need to be highlighted is the nonlinear analysis is follower forces stiffening effect, denoted as FOLLOWK. The interface of FOLLOWK where the follower force stiffness need to be set as YES to include the follower force analysis. For LGDISP = 1, the FOLLOWK parameter is set to YES by default meanwhile, for LGDISP = −1 and LGDISP = 2, the parameter FOLLOWK is ignored. Figure 4

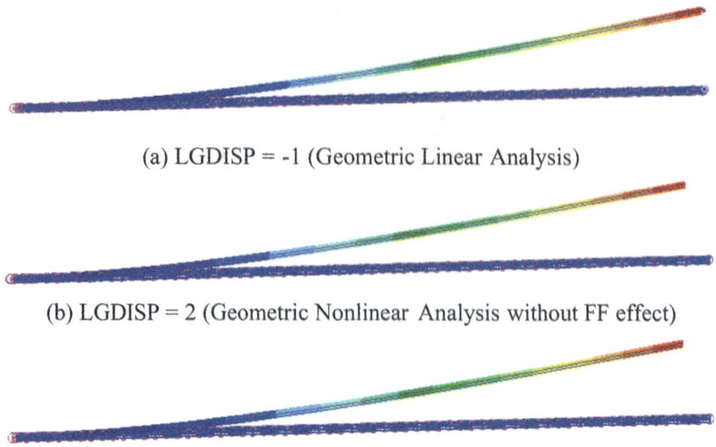

(a) LGDISP = -1 (Geometric Linear Analysis)

(b) LGDISP = 2 (Geometric Nonlinear Analysis without FF effect)

(c) LGDISP = 1 (Geometric Nonlinear Analysis)

Fig. 4 Deformation spectrum due to concentrated forces applied

Table 3 Differential and follower force stiffness

LGDISP	FOLLOWK	Remarks	
−1	No	Large displacement	No
		Follower force	No
1	Yes	Large displacement	Yes
		Follower force	Yes
2	No	Large displacement	Yes
		Follower force	No

shows the sample of deformation spectrum based due to concentrated force applied. Table 3 summarize the options that available for LGDISP and FOLLOWK.

3 Ground Static Test Setup

A ground static test was designed for validating the finite element analysis established in this paper. The setup starts with the fabrication of the wing and the small-scale load application structure followed by determination of load direction and data collections on HAR wing with AR 16.

3.1 *Fabrication of HAR Wing*

The fabrication of the HAR wing covers the construction of spar, ribs and fairing. The spar section of a flat plate is fabricated from spring steel sheet with a dimension of 0.8 m × 0.025 m × 0.00125 m (length × width × thickness). AR 16 was selected based on the maximum wing span length attributes to the limitation of load application structure. Spring steel was chosen due to its high yield strength and its ability to withstand any distortion despite significant deformation. The NACA 0012 profile ribs were fabricated using the 3D printer with a thickness of 0.00125 m and a slot with cross-sectional of spar dimensions. Finally, an aerodynamic profile of the wing was finely shaped with the Styrofoam fairings that are produced using an electrically heated hot wire cutter device, cut from a block of Styrofoam material. This material was chosen due to its lightweight and flexibility to take the deformation shapes without causing the surface of the wing being wrinkles as the wing deflected. All ribs are placed into the spar and distributed equally along the span with a gap of 50 mm to each other. The gap between the ribs is covered by fairings, which complete the overall aerodynamic contour of the wing model. All ribs and fairing are attached permanently with epoxy.

3.2 *Load Application Structure, Determination of Load Direction and Data Collections*

The load application structure for this experiment was made of the mild steel bar as shown in Fig. 5. The top mounting structure (a) was used to attach the HAR wing

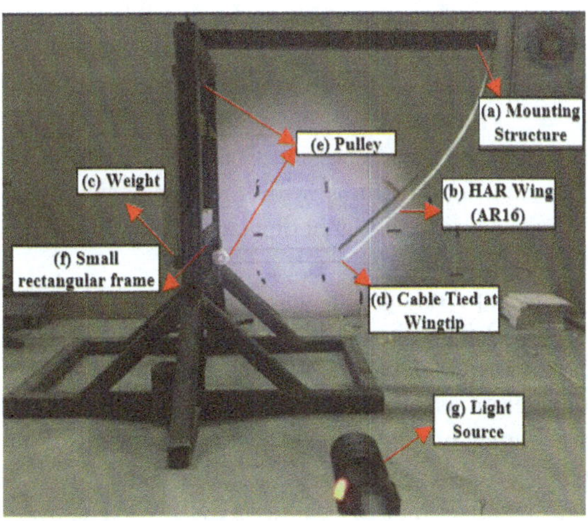

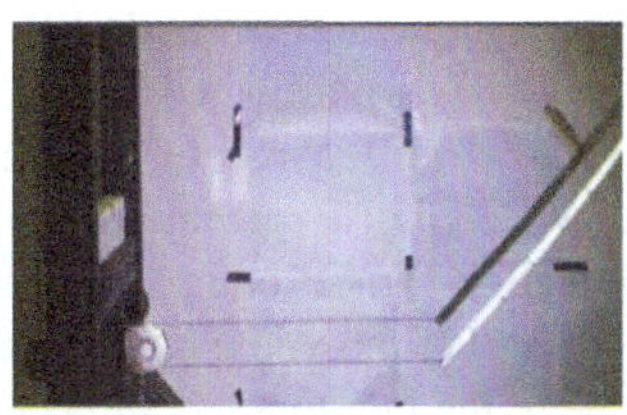

(a) NFF direction

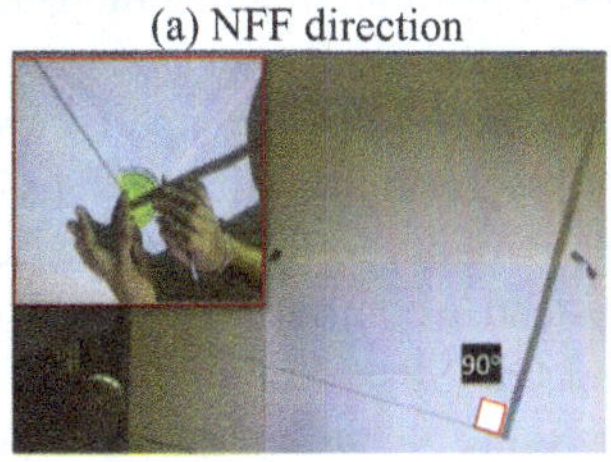

(b) FF direction

Fig. 5 Experimental setup to determine load direction

model (b) and functions as a fixed boundary condition. A cable was attached to the free end of the wing. A measured weight (c) ranging from 50 to 350 g was suspended on the other end of the cable (d) through the pulley (e). The load application structure was designed with the small rectangular frame (f) specifically to allow load direction to translate upward and downward. The experiment was carried out in two cases; NFF and FF cases.

In NFF case, the load direction was defined as constant so that the initial orientation is maintained even after the wing has deformed as shown in Fig. 5a. In FF case, the load direction maintained at 90° between the wing tip and the cable as a wing deform. A light source (g) was added to improve accuracy. A shadow generated on the graph paper, paste on the wall was used for a reference to determine the exact angle. Once the right angle was obtained using a protractor and the reference line was marked, the force direction will be adjusted by translating the pulley attached to the small rectangular frame upwards or downwards until the shadow of the cable overlaps with the marked reference line. The experimental setup for FF shows in Fig. 5b.

In order to measure the deflection of the HAR wing, a Deko Laser Distance Meter was used. An L-shaped steel base plate was set to be as a fixed reference position for placing the measuring device. The value displayed on the screen of the measuring device indicates the distance from the back end of the device to the wing tip.

4 Results and Discussions

4.1 Large Displacement (LGDISP) for Follower Force Analysis

The comparison of LGDISP for follower force case using FEA nonlinear static analysis was performed at different applied load. This particularly focuses on follower force, identified as 'FORCE2' alternately combined with LGDISP options; LGDISP = −1, LGDISP = 1 and LGDISP = 2. Figure 6 shows the outcomes of the LGDISP selection for FF compared to experimental data. The selection of LGDISP = −1 shows that the value of tip displacement found similar with linear analysis. This is due to the fact that this interface is not considering large displacement and follower force effect in the analysis.

Meanwhile, in comparison with the experimental results, LGDISP = 1 shows that closest results compared to LGDISP = 2. This shown that consideration in LGDISP = 2 without follower force effect is inadequate to nonlinear follower force analysis. Instead, a combination of both considerations; large displacement and follower force in LGDISP = 1 contribute to the accuracy of nonlinear follower force results.

Table 4 summarize the percentage difference between FEA using different LGDISP option with experimental results. Analysis using LGDISP = 1 shows a minimal margin error which less than 6.5% compared to analysis using LGDISP = 1. Since the accuracy of the FEA nonlinear static analysis of follower force is

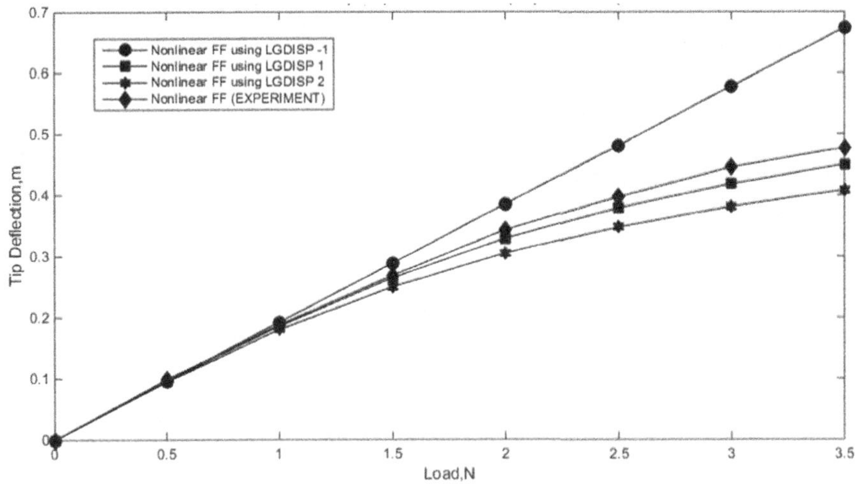

Fig. 6 Comparison between FEA linear and nonlinear static on NFF and FF at AR 13–16

Table 4 Percentage difference in FEA and experimental of FF at different LGDISP condition

Load (N)	Wing tip displacements (m)		
	FF (% difference with experiment)		
	LGDISP = –1	LGDISP = 2	LGDISP = 1
0.5	2.73	4.34	3.03
1.0	2.45	4.15	1.06
1.5	7.40	6.91	1.49
2.0	12.30	11.05	4.08
2.5	21.26	12.47	4.53
3.0	29.55	14.53	6.28
3.5	41.03	14.54	5.65

less than 6.5%, it can be concluded that the result demonstrated are fairly accurate. The presences of the error were assume attributed to some imperfection during the experimental procedure, particularly due to the friction of rope between pulley and small rectangular frame area. This result highlighted the necessity to apply interface LGDISP = 1 with FORCE2 in FEA nonlinear static analysis of follower force.

4.2 *Ground Static Test Validation with FEA Nonlinear Static Analysis*

The experimental data from this paper were collected using the method outlined in Sect. 3 and are used for comparisons against follower and non-follower force from

FEA results. From all of the cases, it can be observed that the FEA of nonlinear static for both cases (NFF and FF using LGDISP = 1) are able to estimate the true tip deflection of the structure with the maximum percentage difference less than 6.5% as presented in Fig. 7 and Table 5. This indicates a good correlation between experimental and simulation data. The inaccuracy of FEA linear analysis thus proved that linear analysis cannot be employed for highly flexible wing structure.

However, there were significant differences result for both cases of FEA and experimental primarily due to the impacts of the follower force as shown in Fig. 7. The follower force are predicted greater deflection than non-follower force cases. This is due to the follower force effects which act to stiffen the structure with a much lower order of hardening effect compare to the non-follower force. This statement

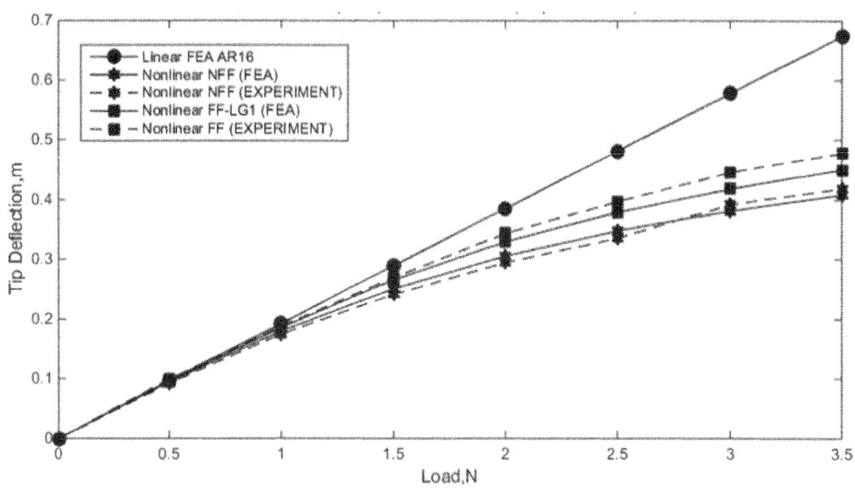

Fig. 7 Plots of nonlinear FEA and experimental results for NFF and FF

Table 5 Percentage difference in FEA and experimental result of NFF and FF

Load (N)	Wing tip displacements (m)					
	NFF		Difference	FF		Difference
	(FEA)	(EXP)	(%)	(FEA)	(EXP)	(%)
0.5	0.095	0.092	3.26	0.096	0.099	3.03
1.0	0.180	0.175	2.86	0.186	0.188	1.06
1.5	0.250	0.241	3.73	0.265	0.269	1.49
2.0	0.305	0.294	3.74	0.329	0.343	4.08
2.5	0.348	0.335	3.88	0.379	0.397	4.53
3.0	0.381	0.392	2.81	0.418	0.446	6.28
3.5	0.409	0.419	2.39	0.451	0.478	5.65

concurred to the definition of follower force where the force remains normal to the surface deflection, thus the concentration of the force to the corresponding structural grid will always be at the highest magnitude.

5 Conclusion

The paper aims to investigate the geometrically nonlinear effects with the inclusion of a follower force effect on the HAR wing model. FEA analysis was used as a computational tool to analyse HAR wing model focusing on nonlinearity aspects. Two main parameters (force identification, 'FORCE2' and large displacement, 'LGDISP') were performed in FEA nonlinear static analysis of follower force. Selection of LGDISP was made based on the closest value of tip displacement compared with experimental data. Ground static test was used as a method to validate with a simulation result at the AR 16. This study highlighted main findings were as followed: (a) for FEA nonlinear static analysis of follower force, LGDISP = 1 was chosen due to this parameter considered both condition; large displacement and follower force stiffness effect (b) the technique used to determine the load direction in ground static load test is effectively can be used at different AR due to minimal margin error between experiment and simulation and (c) the inclusion of follower force loads predicted greater deflection than non-follower force cases due the follower force effects act to stiffen the structure with a much lower order of hardening effect compare to the nonlinear follower force. Minimal error of less than 6.5% for both FF and NFF cases prove that FEA force identification approach is feasible for FF study.

References

1. Abbott IH, Doenhoff AEV (1959) Theory of wing sections including a summary of airfoil data. Dover Publications, New York
2. Afonso F, Goncalo L, Vale J, Oliveira EL, Lau F, Suleman A (2015) Linear vs non-linear aeroelastic analysis of high aspect-ratio wings. In: Congresso de Metodos Numericos em Engenharia 2015. APMTAC, Lisbon, pp 1–9
3. Smith M, Patil M, Hodges D (2001) CFD-based analysis of nonlinear aeroelastic behavior of high-aspect ratio wings. In: 19th AIAA applied aerodynamics conference, fluid dynamics. AIAA, Anaheim, pp 1–10
4. Harmin MY, Cooper JE (2011) Aeroelastic behaviour of a wing including geometric nonlinearities. Aeronaut J 115(1174):767–777
5. Chae S (2004) Effect of follower forces on aeroelastic stability of flexible structures. Aerosp Eng
6. Kim T, Swenson E, Kunz D, Lindsley N, Blair M (2011) Follower-force experiments with geometric nonlinear coupling for analytical validation. In: 52nd AIAA/ASME/ASCE/AHS/ASC structures, structural dynamics and materials conference. AIAA, Denver, pp 1–17
7. Castellani M, Cooper JE, Lemmens Y (2017) Nonlinear static aeroelasticity of high aspect ratio wing aircraft by finite element and multibody methods. J Airc 54(2):548–560

8. Changchuan X, Chao A, Yi L, Chao Y (2017) Static aeroelastic analysis including geometric nonlinearities based on reduced order model. Chin J Aeronaut 30(2):638–650
9. Tang D, Dowell EH (2001) Experimental and theoretical study on aeroelastic response of high-aspect-ratio wings. AIAA J 39(8):1430–1441
10. MSC Software (2007) Bulk Data Entry Descriptions. In: MSC Nastran Quick Reference Guide. MSC

Effects of Circular Tabs on Enhancement of Jet Mixing

S. Thanigaiarasu, Surendra Kumar Yadav, S. B. Jabez Richards, A. Muthuram, and T. Vijay Raj

Abstract The Flow measurements were made in the circular jets using two tabs at nozzle exit, placed diametrically opposite locations. The total pressure was measured along the jet axis (X-Axis) and normal to the jet axis (Y-Axis) at Mach number 0.2. Also, total pressure was measured for free jet to compare the results. The effectiveness of circular solid tabs and circular tabs with hole to enhance the mixing of axi-symmetric subsonic jets has been presented in this paper. It is found that both the circular solid tabs and circular tabs with hole significantly enhance the mixing of jets. The potential core length were reduced by 37.5% and 62.5% for jet of circular tab with hole and solid tabbed jet compared to free jet respectively.

Keywords Potential core · Mixing enhancement · Vortices

Nomenclature

M	Mach number
P_e	Exit Total Pressure
P_o	Settling Chamber Pressure
D	Exit Diameter of Nozzle
X	Distance along the jet axis
Y & Z	Distance normal to the jet axis

S. Thanigaiarasu · S. B. Jabez Richards
Madras Institute of Technology, Anna University, Chennai, India

S. K. Yadav (✉) · T. Vijay Raj
Prince Shri Venkateshwara Padmavathy Engineering College, Chennai, India
e-mail: aero.surendra@gmail.com

A. Muthuram
St. Peter's College of Engineering and Technology, Chennai, India

P. Rajendran et al. (eds.), *Proceedings of International Conference of Aerospace and Mechanical Engineering 2019*, Lecture Notes in Mechanical Engineering,
https://doi.org/10.1007/978-981-15-4756-0_37

1 Introduction

An attempt was made to increase the jet mixing in a flow, based on passive control method by using mechanical tabs at the exit of the nozzle. Bradbury and Khadem [1] made significant contribution in understanding the effect of mechanical tabs on the mixing of jets, and reported that for the low speed jet the insertion of small tabs into the jet can enhance the mixing and decrease the potential core. Bohl and Foss [2] performed experiment to enhance the mixing characteristics of subsonic jet with the use of tabs, and resolute that the tab produces a pair of transverse vortices rotating opposite to each other which gets converted into stream wise vortices as the flow moves downstream of the nozzle exit. It is proved that the small vortices has better mixing efficiency since they have more stability and can travel long distance. Reeder and Samimy [3] did research on low speed flow in water flow channel and used two diametrically opposite tabs which produced stream wise vortices at opposite ends; they found that two vortices rotating in opposite stream wise direction shed from tabs was the reason for the deforming of the jet. Rathakrishnan et al. [4] conducted experiments on a convergent nozzle by placing a cross wire at the nozzle exit along its diameter. The result was an improvement seen in the mixing of the jet at all Mach numbers as the flow moved away from the nozzle exit. Thanigaiarasu and Rathakrishnan [5] found out from their research that in case of an "arc-tab facing in" configuration, the characteristic decay rates becomes faster in comparison with "arc tab facing out" and "rectangular tab" configurations at various levels of blockages. There was a reduction of 80% in the core length for an "arc tab facing in" configuration and correspondingly only 40% reduction was seen for an "arc tab facing out" and "rectangular tab" configurations. Zaman and Reeder [6] extended their study to find the effectiveness of the tabs in supersonic flow and arrived at the conclusion that the mechanical tabs promote better mixing of high speed jets as well as increase the jet mixing for low speed jets. Ahuja and Brown [7] found the effects of tabs on heated subsonic and supersonic jets by using two tabs placed diametrically opposite at nozzle exit. The effect seen was an increase in mixing performance at high temperature jet flow. Zaman and Reeder [6] proved that the tabs also have the capability to weak the strength of the shocks in the supersonic flow which resulted in the reduction of screech noise. Singh and Rathakrishnan [8] investigated that for effective enhancement of mixing of the jet; the tab's length is more effective compared to its width. They also concluded that the length of the tab can be at most equal to the radius of the nozzle exit. This limit of the tab is known by the term "RATHAKRISHNAN LIMIT". Knowles and Saddington [9] reviewed techniques for reducing the infrared (IR) signature for high-speed jets. Clement et al. [10] conducted experiments for a sonic nozzle operating at nozzle pressure ratios from 2 to 7 and examined the effects on the jets using tabs. Their conclusion was that the maximum potential core reduced by 75% at nozzle pressure ratio 7 in comparisons with that of a free jet. More recently, it was found that the tabbed nozzle have been used in fuel injector in supersonic combustion chamber to enhance the mixing of fuel with supersonic air.

Although many important results have been achieved by previous researchers, much work still need in order to understand the mechanism of enhancement of jet mixing caused by tabs. The present research involves studying the effects of circular tabs with and without hole on the mixing of the jets.

2 Experimental Facility

The experiment was carried out in high speed jet test facility in Madras Institute of Technology, India. The jet lab consists of two compressors named as A and B. The compressed air is stored in the external reservoir tanks. The experimental setup is shown in Fig. 1. Compressed air is sent to the settling chamber, where the flow is made to reach equilibrium condition. The constant stagnation pressure 2.865 kPa in the chamber is controlled and maintained with the help of pressure regulating valve. The convergent nozzle is used to expand the stagnant air from the settling chamber to achieve the desired Mach number 0.2 at the nozzle exit.

A Brass made convergent nozzle of 20 mm exit diameter is used for the experiment. "O" ring sealing is used to fix the nozzle at the end of the settling chamber in order to avoid the leakages. A 0.4 mm internal diameter and 0.6 mm external diameter pitot probe is used to measure the pressure. The pitot probe is fixed at the centre of the nozzle with the help of "transverse mechanism" has been used to move the pitot with a uniform pitch of 1 mm in all the three (X, Y & Z) directions. The blockage due to the pitot protrusion is 0.09% which is negligible. Two tabs are fixed diametrically opposite at the nozzle exit, made of brass as shown in Fig. 2.

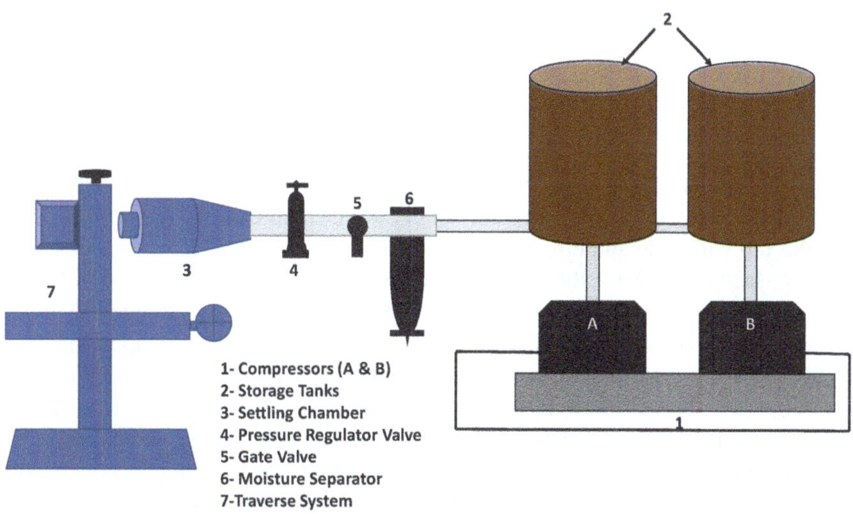

Fig. Schematic diagram of high-speed jet laboratory

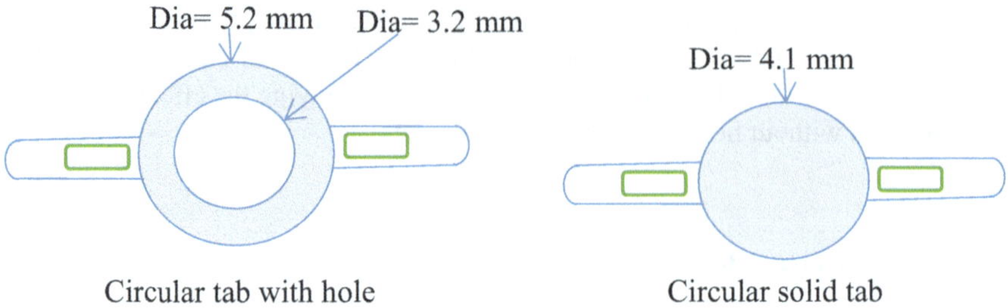

Fig. 2 Schematic diagram of tabs

DSA9116 (Digital Sensor Array) was used to measure the pitot pressure at various locations of the jet. Scanivalve Corporation NETSCAN is the software that helps to communicate the DSA9116 with the Computer.

3 Results and Discussion

It is commonly known that a small projection of tab into the flow with appropriate geometry enhance the mixing of jets, Bhol and Foss [2] recognized that tabs generated a pair of oppositely rotating stream wise vertices that can change the flow structure downstream and works as good mixing promoters. The distance along the jet axis and normal to the jet axis are non-dimensionalized with the exit diameter of the nozzle and represented by X/D and Y/D respectively, while nozzle exit pressure is non- dimensionalized with the settling chamber pressure and represented by P_e/P_o.

3.1 Centerline Pitot Pressure

Centerline pressure decay is a reliable feature to measure the mixing of jets [11]. Figure 3 shows the center line pressure decay at Mach number 0.2 for free jet, jet of circular tab with hole and circular solid tabbed jet. From Fig. 2, it can be observed that for Mach number 0.2, the potential core length reduced from $X/D = 4$ to 2.5 and $X/D = 4$ to 1.5 for jet of circular tab with hole and solid tabbed jet compared to free jet respectively. This amount of potential core length reduction were approximately 37.5% and 62.5%, for jet of circular tab with hole and solid tabbed jet respectively.

The jet coming out through the hole of the circular tab with lesser velocity than the main jet distort the shear layer due to the formation of vortices, which result the engulfment of ambient air into the main jet, thereby reducing the potential core length. It is found that the potential core length reduction obtained in circular solid tabbed configuration was more significant than the one with hole and free jet. Therefore it

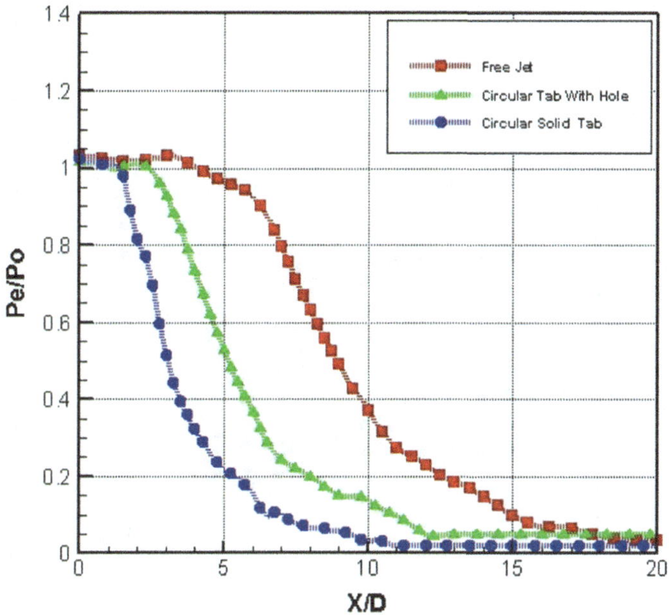

Fig. 3 Center line pressure decay for M = 0.2 jet

can be concluded that the circular solid tabs are good for enhancing the mixing of the jets in low subsonic flow compared to the circular tabs with hole.

3.2 Pressure Profile in Radial Y-Direction

The pressure profiles parallel to the direction of the tabs are observed in the Figs. 4, 5 and 6 for various X/D locations at Mach number 0.2, and it's compared with the profile of uncontrolled jets. Experiments were conducted up to X/D = 10, since beyond X/D = 10 the effect of the tab is negligible and flow attained the zone of established flow.

As observed from Figs. 4, 5 and 6 the circular solid tabs are better when compared to circular tabs with hole and that of the free jets. This is because circular solid tabs have vortices formed from the periphery along with low pressure zone created behind the tab resulting in low momentum fluid. Therefore high momentum fluid of the main jet moves towards the low momentum fluid. At the same time, vortices from the periphery during moving downstream from the nozzle exit becoming large which further improves the mixing. That is why at X/D > 3, the decay of Pe/Po is faster compared to circular tab with hole and free jets.

From the Fig. 6, it can be noticed that the solid tabbed jet is fully established at X/D = 4. However, the jet of the circular tab with hole (Fig. 5) and uncontrolled

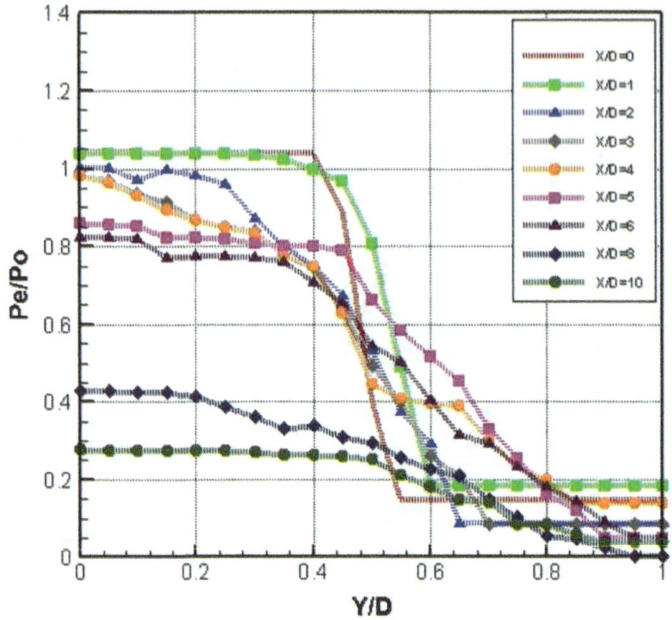

Fig. 4 Pressure profile in Y-direction for free jet at M = 0.2

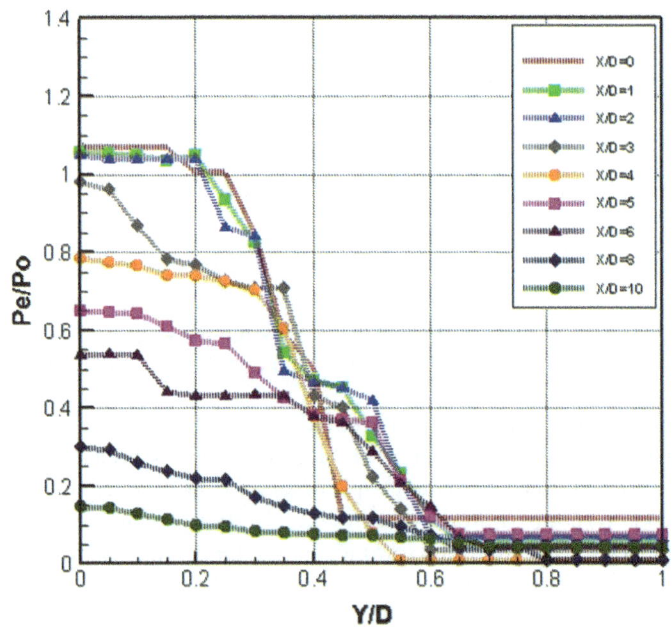

Fig. 5 Pressure profile in Y-direction for circular tab with hole jet at M = 0.2

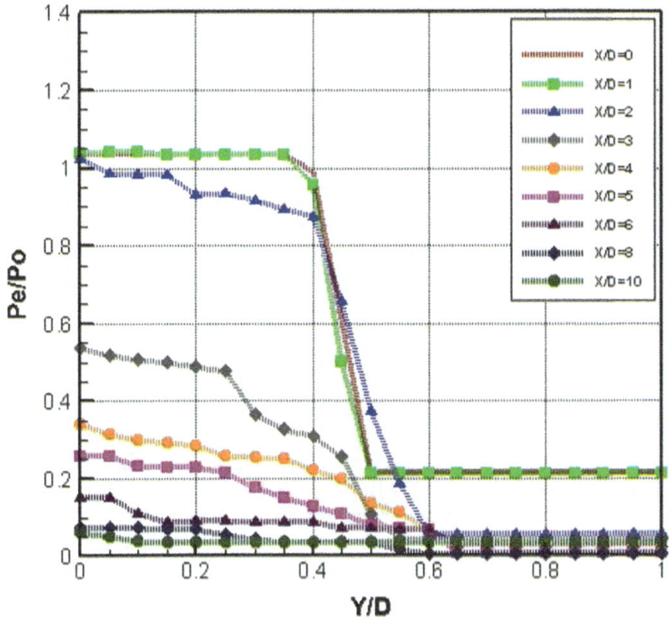

Fig. 6 Pressure profile in Y-direction for circular solid tabbed jet at M = 0.2

jet (Fig. 2) flow is fully established at approximately X/D = 8. This indicates the effectiveness of circular solid tabs to promote subsonic jets mixing, at the same time it can be observed, from Figs. 4, 5 and 6 that the spread rate is faster for solid tabbed jet compared to circular tab with hole, it shows that both the tabs enhances mixing but circular solid tab have better mixing characteristics at low Mach number compared to circular tabs with hole and free jets.

4 Conclusion

This paper studied jet control with circular tabs fixed at the nozzle exit which showed drastic potential core reduction and faster decay for circular solid tabbed jets compared to free jet and circular tabbed jet with hole. The vortices of mixed size shed from the periphery of circular tabs with hole enhance the mixing but in the addition low pressure zone created by circular solid tabs behind the tabs make it more effective compared to the circular tab with hole.

References

1. Bradbury LJS, Khadem AH (1975) The distortion of jet by tabs. J Fluid Mech 70:801–813
2. Bhol D, Foss JF (1996) Enhancement of Passive Mixing Tabs by the Addition of Secondary Tabs. AIAA, pp. 96–054
3. Reeder MF, Samimy M (1993) The evolution of axi-symetric jet with vortex generating tabs. ASME FED, Turbulent Mixing, vol 174. New Orleance, L.A. USA, pp. 1–6
4. Lovaraju P, Rathakrishnan E (2006) Subsonic and transonic control with cross wire. AIAA 44
5. Thanigaiarasu S, Jayaprakash S, Elangovan S, Rathakrishnan E (2008) Influence of tab geometry and its orientation in under expanded sonic jets. Proceeding I Mech. E. Part G J: Aerospace Engineering 222
6. Zaman Reeder MF, Samimy M (1993) Effect of tabs on the flow and noise field of an axi-symmetric jet. AIAA 31:609–619
7. Ahuja KK, Brown WH (1989) Shear flow control by mechanical tabs. AIAA, pp.89–0994
8. Singh NK, Rathakrishnan E (2002) Sonic jet control with tabs. Int J Turbo Jet engines 19:107–118
9. Knowles K, Saddington AJ (2005) A review of jet mixing enhancement of aircraft propulsion applications. J Aerosp Eng Part G 220:103–123
10. Clement S, Rathakrishnan E (2006) Characteristics of sonic jet with tabs. Shock Waves, vol 15. pp 211–214
11. Rathakrishnan E (2009) Experimental studies on the limiting tab. AIAA 47:2475–2485

Analysis of the Impact of Degradation on Gas Turbine Performance Using the Support Vector Machine (SVM) Method

Khairul Fata B. Ahmad Asnawi and Tamiru Alemu Lemma

Abstract Degradation is an important aspect in the operation and maintenance of gas turbines since it affects maintenance costs substantially. Hence, the study of degradation in terms of recoverable and non-recoverable degradation is crucial to formulate a correct maintenance strategy and, as a result, achieve optimum maintenance cost. In this paper, the impact of recoverable and non-recoverable degradation towards compressor discharge pressure, fuel gas flow, and exhaust gas temperature are measured during the start of run period that reflects the time period from the new gas turbine condition to the first scheduled offline crank wash, which normally approximates to 8000 running hours. For the study, a three-unit single speed light industrial gas turbine that drives an electrical generator to power up an offshore platform located in a tropical climate is considered. The measurement of the parameters has been conducted using the support vector machine (SVM) method.

Keywords Gas turbine · Recoverable degradation · Non-recoverable degradation · Support vector machine

1 Introduction

Gas turbine has been a popular choice of prime mover for power generation and mechanical drive in various industries. At present, the available output power of gas turbines in the market can range from micro-turbines at 50 kW to heavy-duty industrial turbines at a capacity higher than 200 MW [1, 2]. Operating a gas turbine, however, requires significant capital expenditure and high maintenance and operating costs. Wan et al. [3] remarked that more than 50 million Euro was required to operate a gas turbine for a period of 40 years, which represents a higher cost when compared to other machines. It is also common that gas turbine degrades with time that results in an increase in cost. Hence, understanding the degradation mechanism of gas turbine is crucial for operators to avert excessive cost.

K. F. B. A. Asnawi (✉) · T. A. Lemma
Mechanical Department, Universiti Teknologi PETRONAS, Seri Iskandar, Malaysia
e-mail: khairul_1732@utp.edu.my

P. Rajendran et al. (eds.), *Proceedings of International Conference of Aerospace and Mechanical Engineering 2019*, Lecture Notes in Mechanical Engineering,
https://doi.org/10.1007/978-981-15-4756-0_38

In recent years, many studies have been conducted to investigate gas turbine degradation with the aim to find best maintenance strategy to reduce cost. As an example, Botros et al. [4] investigated gas turbine recoverable and non-recoverable degradations for aero-derivative gas turbine compressor fleet. Meanwhile, Igie et al. [5] focused only on recoverable degradation impact on heavy duty gas turbine with different offline and online turbine wash intervals. Washing effectiveness by varying detergent ratio was also covered in their study. Very recently, Cruz Manzo et al. [6] established a model using Simulink to predict degradation for gas turbine compressor and generator. A numbers of other authors have as well looked at gas turbine degradation, however, to the authors' knowledge, the fixed speed gas turbine and gas turbine with part load operation are the least explored aspects.

Hence, this paper will elaborate further on gas turbine degradations in forms of recoverable and non-recoverable degradations and how it impacts critical gas turbine performance parameters, namely compressor discharge pressure (CDP), fuel gas flow (FGF), and exhaust gas temperature (EGT). The focus of this study will be three units of gas turbine generators operated at part load between 50 and 60% and located in a tropical offshore environment. Support Vector Machine (SVM) technique will be used to develop performance model for the gas turbine generators. The timeframe of study is during the start of run period which refers to new engine condition to the first offline crank wash which scheduled approximately at 8000 running hours.

2 Gas Turbine Degradation

Gas turbine degrades due to various reasons such as fouling, erosion, and corrosion. Generally, there are two types of degradation known as recoverable and non-recoverable. Recoverable degradation is normally caused by fouling which mostly affects axial compressor blades due to impurities and dust passing through the air inlet system. Recoverable degradation can be removed by an offline crank wash while the non-recoverable degradation remains at the gas turbine and it can only be removed by component replacement [7–11].

2.1 Recoverable Degradation

Compressor fouling is the most discussed recoverable degradation type compared to turbine fouling. Compressor fouling always contributes to 70–85% performance degradation to gas turbines compared to other degradation type [7, 12]. Even though there is an air filtration system, some amount of impurities and dirt pass the filtration system and stick to the compressor blades. The impurities will change the profile of the compressor blades air-foil shape, leading-angle, surface roughness, and air flow area [7, 13] which subsequently reduce efficiency.

Compared to compressor fouling, turbine fouling is less affecting gas turbine performance. Zwebek and Pilidis [14] reported the impact of turbine fouling in combine cycle power plant only 0 to (−5%) for non-dimensional flow and 0 to (−2.5%) for turbine efficiency.

2.2 *Non-recoverable Degradations*

Erosion. Erosion in compressor refers to abrasive loss of material by hard materials suspended in the gas flow stream. Diakunchak [12] reported the hard materials will cause significant effect to the compressor blade which lead to change in surface roughness, blade geometry, and surface area [9]. Major impact of compressor erosion is on the reduction of the flow and compressor efficiency [10, 13, 15]. Zwebek [14] reported that compressor erosion may reduce the compressor efficiency by up to 2.5% when compared to baseline condition.

Erosion reduces gas turbine efficiency by reducing the turbine's ability to convert the flue gas to useful power output. Based on the work of Diakunchak [12], a 2.5% efficiency drop due to erosion may cause a 9.25% power loss and 6.75% increase in the heat rate. This loss of power due to erosion is significant, but its likelihood of happening is rare. Another effect observed by Amare et al. [16] is that when the erosion gets severe, the turbine inlet temperature increases but the exhaust temperature reduces.

Tip Clearance. Rotor tip clearance increases over time and will exceed its limit. It causes inefficient sealing across compressor or turbine blade and introduces leakages in between stages inside the gas turbine. For compressor, immediate indication of tip clearance issue is reduction in the compressor discharge pressure [3, 17]. Frith and Graf et al. [17, 18] studied the impact of compressor discharge pressure, flow capacity, and compressor efficiency when the compressor tip clearance increases. The results show that when the tip clearances increases in range of 1.4−3%, the compressor discharge pressure reduces by about 3%, flow capacity reduces at 4.5%, and compressor efficiency reduces by 2.5%. Meanwhile on turbine tip clearance, apart from reduction in turbine efficiency which is seen in most of the cases, Melcher et al. [19] mentioned that the increase in tip clearance also causes an increase in exhaust gas temperature.

3 Support Vector Machine (SVM)

Support vector machine (SVM) is part of the supervised learning techniques in Machine Learning. Fundamental principle of SVM is to generate hyperplane to divide two types of training samples into two classes by creating maximum margin between two support points called 'support vector' for optimal decision hyperplane [20]. It falls in classification algorithm which is able to classify the future input based on

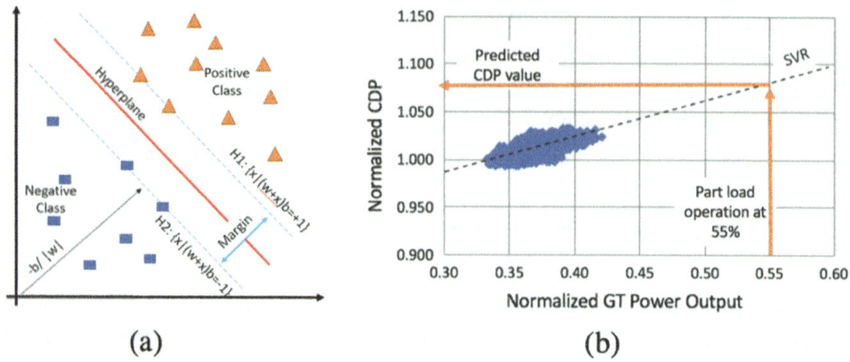

Fig. 1 Representation of linear SVM classification (**a**) and prediction using SVR (**b**)

learned input and output data [20–22]. SVM can also be used as a regression tool [23–25] and has been applied in the field rotating machinery problems which include diagnostic and fault findings [18, 26].

Figure 1a shows linear classification to separate two classes of data represented by square and triangle. Those data set are separated by linear hyperplane which act as boundary in the middle of two classes that separates positive class represented by triangle and negative class represented by square. There are imaginary separation margins in between those classes where the hyperplane is considered midpoint. Those points are called support vectors and contain all the necessary information used to design classifier algorithm [20, 22]. This hyperplane can be modified to create a regression line normally known as support vector regression (SVR).

According to Yin et al. [20] and Widodo et al. [22] both sets of data, for example, those in Fig. 1a, can be mathematically expressed as;

$$(x_1, y_1), \ldots, (x_1, y_1), \quad x \in M^n, \quad y \in \{+1, -1\} \tag{1}$$

Data input is x_i ($i = 1, 2, \ldots, M$), where M is the number of data. Two classes are assigned to the data as positive class $y_i = 1$ and $y_i = -1$ for negative class. These data can be divided by hyperplane using the following function;

$$f(x) = W^T x + b = \sum_{j=0}^{M} wjxj + b = 0 \tag{2}$$

W is considered as M-dimensional vector and b is scalar to use to find hyperplane position. To create optimal hyperplane the maximum margin between imaginary line shall be maximum. Hence, the following equation is applicable.

$$y_i[(w \cdot x_i + b)] \geq 1 \quad i = 1, 2, \ldots, l \tag{3}$$

Alternatively, the following equation also can be considered.

$$(w \cdot x_i) + b \geq 0 \quad if \; y_i = 1 \tag{4}$$

$$(w \cdot x_i) + b \leq 0 \quad if \; y_i = -1 \tag{5}$$

Furthermore, constraint in Eq. (6) is considered to simplified previous function.

$$\Phi(w)\frac{1}{2}\|w\|^2 = \frac{1}{2}(w \cdot w) \tag{6}$$

To solve this problem, the function can be converted to Lagrange function as in Eq. (7).

$$L(w, b, \alpha) = \frac{1}{2}w^T \cdot w - \sum_{i=1}^{n}\alpha_i\left[y_i\left(w^T \cdot x_i - b\right) - 1\right] \tag{7}$$

α_i is Lagrange multiplayer that satisfy $\alpha_i \geq 0$ and $i = 1, \ldots, l$. When the constrain at Eq. (6) become maximum, the corresponding point shall meet condition at Eqs. (8) and (9).

$$\frac{\delta L}{\delta b} = 0 \quad \rightarrow \quad \sum_{i=1}^{n}\alpha_i y_i = 0 \tag{8}$$

$$\frac{\delta L}{\delta w} = 0 \quad \rightarrow \quad \sum_{i=1}^{n}\alpha_i y_i x_i = w \tag{9}$$

This will derive the following equation when condition at Eqs. (8) and (9) is considered in Lagrange function in Eq. (7).

$$w(\alpha) = \sum_{i=1}^{l}\alpha_i - \frac{1}{2}\sum_{i,j=0}^{M}\alpha_i\alpha_j y_i y_j x_i \cdot x_j \tag{10}$$

Equation (10) is valid when meet the following condition,

$$\alpha_i \geq 0, \quad i = 1, \ldots, M, \quad \sum_{i=1}^{M}\alpha_i y_i \tag{11}$$

At this stage x_i can be solve by this function,

$$f(x) = sign\left(\sum_{x_i \in X_{SVM}} \alpha_i y_i\right) \tag{12}$$

x_i with condition

$$w = \sum_{x_i \in x_{SVM}} \alpha_i x_i \tag{14}$$

are recognized as support vectors when

$$x_i \cdot s \quad \text{with} \quad \alpha_i \geq 0 \tag{15}$$

4 Degradation Impact to GT Performance

4.1 Methodology

To investigate the impact of degradation on gas turbine performance at part load operation, a three-unit gas turbine located in a tropical offshore environment has been considered. The gas turbine runs at load between 50 and 60%. In this study, the time frame is focused during the start of run period, which is defined as the time period from zero RH to first offline crank wash. The methodology for the study follows the steps below;

1. Collect performance data for SN868, SN869 and SN890 GT. Each data collected from the gas turbines contains information during new, dirty, and clean conditions as per Table 1.
2. Normalize and filter the data to remove outliers.
3. Generate load based SVR model for each data set (Fig. 1b)
4. Predict performance data for specific part load to ensure consistency of the predicted data. In this case 55% load is considered which represents the average load of the gas turbines.
5. Measure impact recoverable and non-recoverable degradation to compressor discharge pressure, fuel gas flow and exhaust gas temperature.

The performance impact towards recoverable and non-recoverable degradation is computed using the following two Equations;

Table 1 GT running hours for all condition

GT	Running hours		
	New	Dirty	Clean
SN868	29	8245	8294
SN869	84	8095	8151
SN870	63	8494	8629

$$Recoverable\ degradation(\%) = (dirty\ condition - clean\ condition) / Total\ degradation \quad (16)$$

$$Non\text{-}recoverable\ degradation(\%) = (new\ condition - clean\ condition) / Total\ degradation \quad (17)$$

4.2 Results and Discussion

New condition refers to the factory conditions of the gas turbines with the least fouling and best tip clearances. While dirty condition refers to degraded gas turbine due to both recoverable and non-recoverable degradations. Clean condition is the GT condition with only non-recoverable degradations. All of the gas turbines under study have undergone offline crank wash between dirty and clean condition. During the offline crank wash, the gas turbine rotated while stream of detergent and water is injected to the compressor suction to remove impurities at the gas path. Figure 2 shows the trend of the performance parameters that consist of compressor discharge pressure, fuesl gas flow and exhaust gas temperature. All of the gas turbines show similar pattern of CDP, FGF and EGT during the transition from new condition, dirty, and clean conditions.

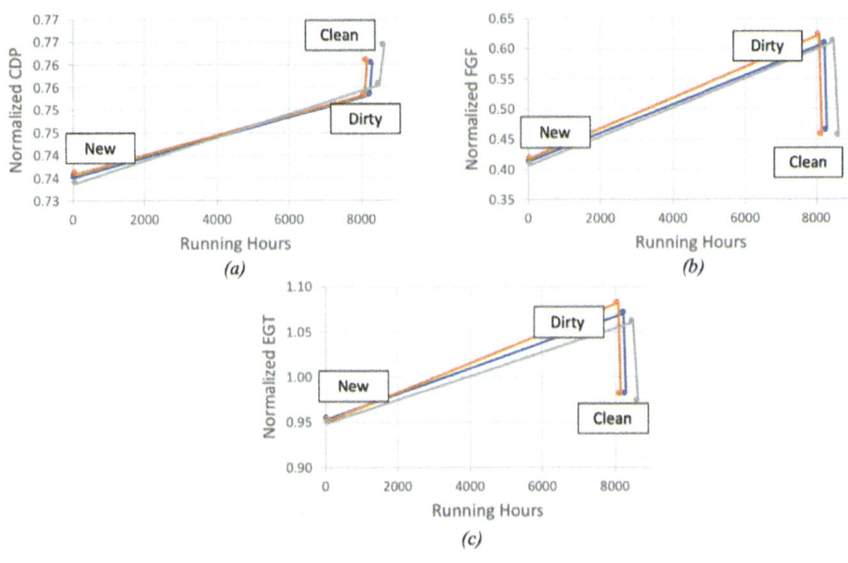

Fig. 2 Transition of the CDP (**a**), FGF (**b**) and EGT (**c**) for all GTs in study from new to dirty and clean condition for GT operating at 55% part load

Table 2 CDP ranking related to FGF and EGT increment rate ranking

	CDP	FGF	EGT
Most	SN869	SN869	SN869
Least	SN870	SN870	SN870

Compressor Discharge Pressure (CDP). Over time, the axial compressor of the compressor fouls and CDP is expected to reduce due to the reduction in its efficiency [7, 13]. After the crank wash is done, some of the degradation will be removed, hence it is expected that the CDP will be increase back close to CDP at clean condition. However, as seen in Fig. 2a, all the three gas turbines do not show the expected pattern when transitioning from new to dirty condition. Instead, the actual compressor discharge pressure increases from dirty to clean condition. One possible explanation for this that because of the increase in backpressure in the turbine section caused by the non-recoverable degradation. When the non-recoverable degradation increases, a higher pressure is required by the air to overcome the backpressure as also explained by Kurz et al. [27]. Hence, it is reflected in the increase in the compressor discharge pressure. To support this argument, compressor discharge pressure is ranked between all three GTs to see the most and the least compressor discharge pressure. Then FGF and EGT also ranked in the same manner where the most and least increment rate is identified as in Table 2. From the ranking, it can conclude that the SN 869 has the highest compressor discharge pressure, which is also has highest FGF and EGT increment rate that represent backpressure from turbine side. The opposite effect is observed on the SN870 engine. Hence, CDP increases can be said due to increase in backpressure at the turbine back end cause by non-recoverable degradation. However, CDP improves for all GTs from dirty to clean condition. During the crank wash, substantial amount of fouling is removed from the axial compressor hence, more efficient, which contribute to higher CDP.

Fuel Gas Flow (FGF) and Exhaust Gas Temperature (EGT). FGF and EGT show same pattern for transition from new to dirty and clean condition. It proofs that relationship between FGF and EGT is closely related. Figure 2a, b shows both FGF and EGT at the lowest point during new condition. This is the point where the GT efficiency is the best due to no fouling at axial compressor and the rotor tip clearances are very minimal at axial compressor and GT. When GT continue operating from new to dirty condition FGF and EGT increases from initial value to the highest level due to combined effect of recoverable and non-recoverable degradation. At the dirtiest condition, axial compressor efficiency drop which cause reduction in air flow delivery. To compensate the air flow reduction, more fuel is required to deliver same amount of power. Increase in fuel subsequently increase the firing temperature which reflect in EGT reading.

Offline crank wash conducted at approximately 8000 running hours to removes the fouling on the axial compressor blades. The effect seen during the transition from dirty to clean condition where the FGF and EGT reduces. However, those values will not go beyond new condition. At this phase, almost all the foulant at axial compressor

Table 3 Impact of recoverable and non-recoverable degradations

	SN868 (%)		SN869 (%)	SN870 (%)	Average (%)
FGF	Recoverable	74.0246	81.2285	76.5916	77.3183
	Non-recoverable	25.9754	18.6615	23.4084	22.6817
EGT	Recoverable	75.7524	77.8291	78.8999	77.4938
	Non-recoverable	24.2476	22.1709	21.1001	22.5062

are gone and left only non-recoverable degradation in forms of turbine tip clearance and blade erosion. Hence, the value of FGF and EGT at this phase is represent only non-recoverable degradation.

Degradation Impact to FGF and EGT. To determine impact of recoverable and non-recoverable degradation, FGF and EGT values during new, dirty and clean condition are used for calculation. Only FGF and EGT is considered since CDP data is not meeting the desired behavior. The calculation shows that at the dirtiest condition, 77.32% of FGF increase due to recoverable degradation while 22.68% is due to non-recoverable degradations. The similar impact also happened to EGT where 77.50% increase is due to recoverable degradation while remaining 22.50% is due to non-recoverable degradation. That value almost consistent between all three units of the GT as per Table 3. It indicates dominant contributor to degradation is recoverable degradation which can be recovers through schedule offline crank wash. Even though the non-recoverable percentage is lower, it keeps growing and expected become dominant when approaching *end of run* period.

5 Conclusions

This study to determine the impact of critical GT parameters namely FGF and EGT towards recoverable and non-recoverable degradations during its start of run period. Three units of GT driving electrical generator operating at part load installed at tropical offshore location is studied. The following are the findings;

1. SVR used to predict the CDP, FGF and EGT at specific power. In this study, all the condition is predicted using SVR at gas turbine load 55%.
2. From the observation, CDP increases from new condition to dirty condition which contradict with expected behavior. Based on literature review and data analyzed, this phenomenon is due to back pressure in turbine section cause CDP react when the degradation happens. Hence, CDP impact to degradations is not determine.
3. Both FGF and EGT shows similar pattern when transit from new to dirty condition and finally to clean condition after undergoing offline crank wash. It shows close relationship of amount of fuel gas required and firing temperature.

4. The amount of degradation impact to FGF and EGT is consistent among all GTs studied. 77.32% of FGF increase due to recoverable degradation while 22.68% is due to non-recoverable degradations. While EGT experience 77.50% increase due to recoverable degradation while remaining 22.50% is due to non-recoverable degradation.
5. Recoverable degradation is dominant during start of run period. However, it will recover after offline crank wash. Unlike recoverable, non-recoverable degradation stays. In the future, non-recoverable degradation will grow and become dominant.

References

1. Soares C (2015) In: Gas turbines: a handbook of air, land and sea applications, 2nd Edn. Elsevier
2. Sabri HAR, Rahim ARA, Majid H (2013) Concepts of turbomachinery selection for upstream oil and gas facilities project. In: 2nd International conference on mechanical, automotive and aerospace engineering, Kuala Lumpur
3. Wan A, Chen J, Zheng L, Hall P, Ji Y, Gu X (2018) Prognostic of gas turbine: a condition-based maintenance approach based on multi-environmental time similarity. Mech Syst Signal Process 109:150–165
4. Botros KK, Hartloper C, Golshan H, Rogers D (2016) Recoverable versus unrecoverable degradations of gas turbines employed in five natural gas compressor stations. J Eng Gas Turbine Power 138(2)
5. Igie U, Pilidis P, Fouflias D, Ramsden K, Laskaridis P (2014) Industrial gas turbine performance: compressor fouling and on-line washing. J Turbomach 136(10):101001
6. Cruz Manzo S, Panov V, Zhang Y (2018) Gas path fault and degradation modelling in twin-shaft gas turbines. Machines 6(4):43
7. Tahan M, Tsoutsanis E, Muhammad M, Abdul Karim ZA (2017) Performance-based health monitoring, diagnostic and prognostics for condition-based maintenance of gas turbine: a review. Appl Energy 198:122–144
8. Petek J, Hamilton P (2005) Performance monitoring for gas turbines. Orbit 25
9. Boyce MP (2012) Gas turbine engineering handbook, 4th edn. Elsevier
10. Kurz R, Meher-Homji C, Brun K (2014) Gas turbine degradation. In: Turbomachinery and pump symposia (Pump & Turbo 2014). Houston
11. Hepperle N, Therkorn D, Schneider E, Staudacher S (2011) Assessment of gas turbine and combined cycle power plant performance degradation. In: ASME turbo expo. Vancouver
12. Diakunchak SI (1992) Performance deterioration in industrial gas turbines. J Eng Gas Turbines Power 114(2):161–168
13. Brun K, Kurz R, Simmons HR (2006) Aerodynamic instability and life-limiting effects of inlet and interstage water injection into gas turbines. ASME J Gas Turbine Power 128(3):617–625
14. Zwebek AI, Pilidis P (2004) Degradation effects on combined cycle power plant performance-Part iii: gas and steam turbine component degradation effects. ASME J Gas Turbine and Power 126(2):306–315
15. Kurz R, Brun K (2007) Gas turbine tutorial—maintenance and operating practices effect on degradation life. In: Proceedings of the thirty-six turbomachinery symposium. Turbomachinery Laboratory, Houston
16. Amare DF, Aklilu TB, Gilani SI (2016) Effects of performance deterioration on gas path measurement in and industrial turbines. ARPN J Eng Appl Sci 11(24):14202–14207
17. Frith PC (1994) The effect of compressor rotor tip crops on turboshaft engine performance. ASME J Gas Turbine Power 2

18. Graf M, Wong T, Greitzer E, Marble F, Tan C, Shin HW (1997) Effects on non-asymmetric tip clearance on axial compressor performance. In: ASME international gas turbine and aero-engine congress and exposition
19. Melcher KJ, Kypuros J (2003) Towards a fast response active turbine tip clearance control. In: Symposium on air breathing engines. Ohio
20. Yin Z, Hou J (2015) Recent advances on svm based on faults diagnosis and process monitoring in complicated and industrial processes. Neurocomputing 174(Part B):643–650
21. Hamed A, Tabakoff W, Singh D (1998) Modelling of compressor performance deterioration due to erosion. Int J Rotating Mach 4(4):243–248
22. Widodo A, Yang BS (2007) Support vector machine in machine condition monitoring and fault diagnostic. Mech Signal Proc J 21(6):2560–2574
23. Wu CH, Ho JM, Lee DT (2003) Travel-time prediction with support vector regression. IEEE Trans Intell Transp Syst 2:1438–1442
24. Clarke SM, Griebsch JH, Simpson TW (2005) Analysis of support vector regression for approximation of complex engineering analyses. J Mech Des 127(6):1077–1087
25. Drucker H, Burges CJC, Kaufman L, Smola A, Vapnik V (1996) Support vector regression machines. Neural Information Processing Systems (NIPS), pp 155–161
26. Vapnik VN (1999) An overview of statistical learning theory. IEEE Trans Neural Networks 10(5):988–999
27. Kurz R, Brun K (2000) Gas turbine degradation—what makes the map? In: Turbomachinery symposium, 29th Edn. Houston
28. Allen CW, Holcomb CM, Olievera MD (2017) Gas turbine machinery diagnostic: a brief review and sample application. In: Turbomachinery conference and exposition, GT2017. ASME, North Carolina

Fast and Efficient Collision Avoidance Algorithm for Autonomous Mobile Robot

A. F. Hawary and A. N. Azizan

Abstract An autonomous mobile robot or Unmanned Ground Vehicle (UGV) is a vehicle that operates remotely, without human on-board. UGVs are widely used for mission-based applications that are often hazardous or/and inconvenient for humankind regardless of its sizes. As to enable its autonomous capability, UGVs often equipped with artificial eyes e.g. sensors and on-board camera to assist navigation during mission. In this experiment, we utilized a sweeping technique using ultrasonic sensor (HC-SR04) as the 'eye' of the mobile robot to map the obstacle around it using distance measurement. Meanwhile, an Arduino based controller is used to execute the avoidance algorithm. The ultrasonic sensor is programmed to sweep at six different angles to cover about 150° field of views. Path planning algorithm in this project provides the decision to avoid the obstacle and create temporary new waypoints within split seconds using decision table. We observed that the algorithm that uses decimal method is easy to implement and provides faster decision as compared to the binary method.

Keywords Unmanned ground vehicle · Path planning system · Collision avoidance system · Pulse width modulation

1 Introduction

1.1 Background and Problem Statement

UGV is a vehicle that operates without human on-board. It is commonly used for application that is hazardous and inconvenient for humans. Generally, UGVs are equipped with sensors and camera to perceive surrounding and make decisions about its localization and heading [1]. For the UGV to operate autonomously, a sensor is required to sense and avoid obstacles. Usually, it is a non-contact sensor that capable

---

A. F. Hawary (✉) · A. N. Azizan
School of Aerospace Engineering, Engineering Campus, Universiti Sains Malaysia, Penang, Malaysia
e-mail: aefaizul@usm.my

P. Rajendran et al. (eds.), *Proceedings of International Conference of Aerospace and Mechanical Engineering 2019*, Lecture Notes in Mechanical Engineering,
https://doi.org/10.1007/978-981-15-4756-0_39

of sensing an object at distance through various methods depending of the application [2]. The ultrasound sensors uses echo adapted by is bio-inspired bats, which use ultrasonic waves to navigate in the dark [3]. Having the vehicle to move autonomously, obstacle avoidance is a vital feature for UGVs. Most of modern robots require the element of sensing, computing and communication capabilities, which make them able to execute complex and coordinated operations faster and efficient [4]. The accuracy of obstacle sensing is significantly important to estimate the geometry of the obstacle [5]. In [6], the fundamental in evaluating a collision free-path for self-driving vehicles is its ability to sense the environment and proposed a low-cost ultrasonic sensors in evaluating the proper maneuver to avoid collisions. This provides the information of the size and location about the surrounding objects. A proper algorithm is needed to acquire geometry of the approaching obstacle and the response appropriately [7, 8]. Dynamic steering algorithm which ensures faster execution as the UGV does not have to stop in front of the obstacle which allows the UGV to navigate by avoiding collision [9]. There are many avoidance algorithms proposed in the literature but most of them are computationally expensive and unable to provide real-time decision making. For example, before a robot is deployed in an environment, it is tested using a crowd simulator, as it can support the evaluation of different navigation algorithms under comparable crowd conditions [10]. Nevertheless, little work has been carried out the detail of how to develop them at least on the low-level programming for simple application. Therefore, in this paper we propose an algorithm for collision avoidance for autonomous vehicle platform that would maneuver efficiently using a fast and efficient algorithm for collision avoidance decision making.

2 Methodology

2.1 Collision Detection and Avoidance

As to enable the autonomous capability of UGV, avoiding collision is a must feature. A simple method for avoiding collision can be built using ultrasonic sensor [11]. In this study we utilized HC-SR04 ultrasonic sensor. At fundamental level, an ultrasonic sensor consists of transmitter and receiver units in one module. The transmitter (T) transmits ultrasound and the receiver (R) measures the echo signal based on its time of flight (ToF) as shown in Fig. 1.

The transmitter will only transmit the signal if a trigger signal is activated. The trigger indicates that the signal start to travel at the specific direction at the speed of sound waves which is 343 m/s. Then a return signal sense by the receiver is a signal bounced by the obstacle (if any) and thus the distance between the sensor and the object can be calculated using Eq (1).

$$d_t = at/2 \tag{1}$$

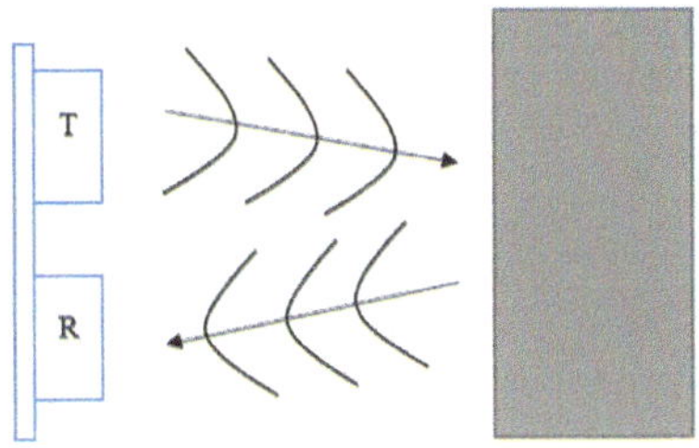

Fig. 1 Working principle of ultrasonic sensor

d_t = Distance travelled
a = Speed of sound or Time of High (Echo) Pulse
t = Time taken

From our previous study, the beam pattern for HC-SR04 ultrasonic sensor is 30° and the sensing distance is 390 cm. Having a static sensor position, the detection angle is just enough for robot maneuvering. However, as to increase the angle of detection, the sensor can be swept using a servo motor.

In this experiment, the servomotor was programmed to sweep at 30° interval. The servo motor will sweep from left to right with 6 different sensing angles. The sensing angle begins with 15° with the increment of 30° and sweep until 165° as shown in Fig. 2. Since the servo motor angle has limitations, 0°–15° and 165°– 180° are the blind spot regions. This method allows the ultrasonic sensor to scan possible object distance at six different angle (15°, 45°, 75°, 105°, 135°, 165°) at one sweep.

The system makes one complete sweep (from left to right) and then returns back to original position to start the next sweep. This system is considered as the sensor reached 165° (on the right most), the sensing angle of 15° (left most) is already left unattended for a while. Therefore, it is better for the sensor to start again at the first position.

Detecting obstacles from a distance is one part and with the addition of maneuverability in the picture, CAS then only can be complete and successful system. In this project, the UGV is equipped with Ackerman steering system using a servo motor attached to the front wheels steering system. When initiated, the servo will rotate to the right or left and at the same time moving the wheels. The servo has the range of 180° rotation; from left to right. However, the wheels have limitation and could

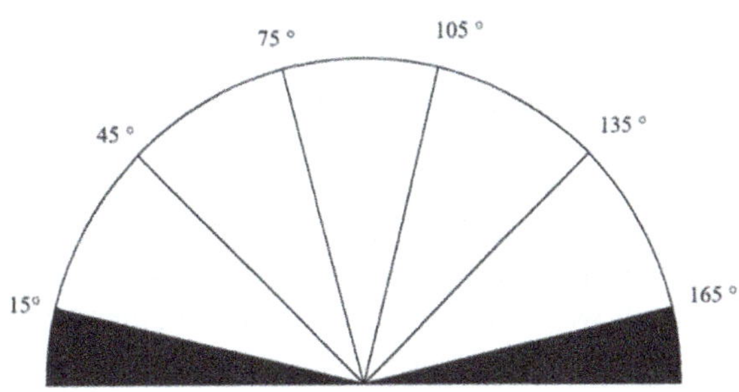

Fig. 2 Ultrasonic sensor sensing angles

Fig. 3 Permanent magnet placement on the wheel

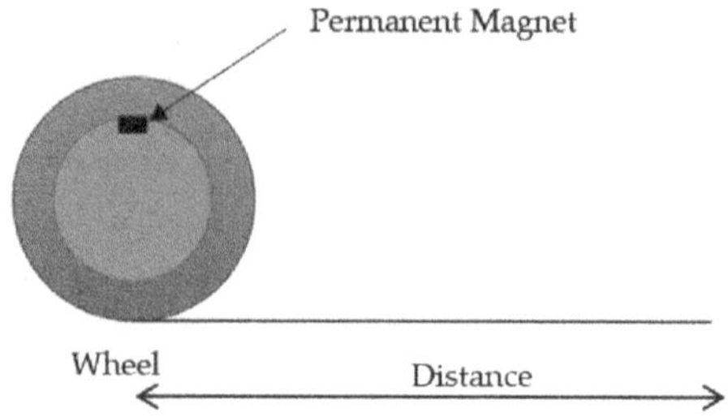

not perform a sharp 0° or 180°, this limitation is typical in vehicle with wheels. Both servos are driven by a PWM signal, ranging from 0 to 255. One of the features of standard UGV is, the vehicle is able to vary its speed from low to high and high to low. By using PWM signal input, the UGV will move at different speeds and hence the speed has become part of the equation.

Two tests are carried out to get the actual speed of the UGV at every PWM signal input. The first test is performed by letting the UGV to run at a PWM signal input at a time until it reaches the maximum speed. A GPS application; *Speedometer GPS Pro* speed tracker is mounted on the UGV to record its actual speed. To support the result by using the GPS, another test is conducted to measure the actual speed by using a wheel encoder using a hall-effect sensor mounted on the wheel. The sensor is placed at the rear wheels, directed towards inside the wheel as a magnet is placed on the inner side of the wheels. The sensor calculate the speed manually by using Eq. (2) below.

$$Speed = \frac{Distance}{Time} \tag{2}$$

From this equation, the variable time can be detected by the control unit and the distance is calculated from the circumference of the wheels by taking into consideration the radius given by Eq. (3) and illustrated in Fig. 3 below.

$$Distance = 2\pi R \tag{3}$$

Which is the location of the magnet on the wheels as shown in Fig. 3.

The minimum operational speed of the UGV is at 1.84 m/s or at PWM signal of 70 as shown in Fig. 4. Within the orange region, the UGV did not have enough power to move constantly and steadily due to its weight.

Another important mapping is that, the controller has to compute the turning angle versus the speed as it may take less time to turn the same angle at faster speed. This mapping is important since the controller will turn according to the position of the incoming obstacle. For instance, if the UGV's wheels steer at maximum angle to the right with the objective of turning 90° and the steering angle is hold for a long time, the UGV will continue to turn right and eventually move in a circle or in U-turn. On the other hand, if the steering angle is only hold for a short time, 90° turn to the

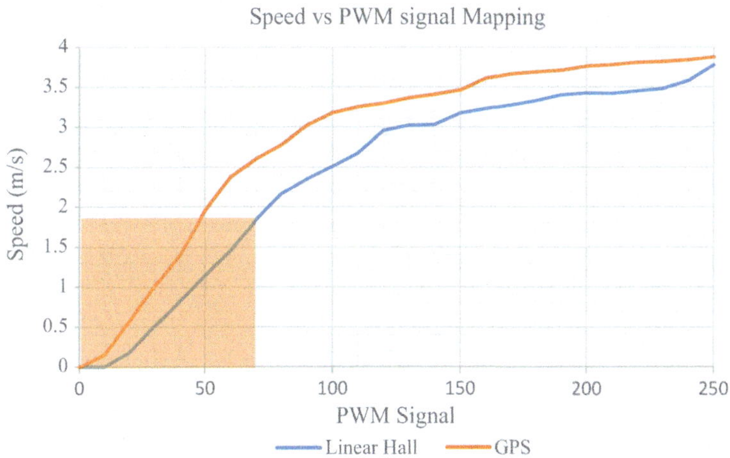

Fig. 4 A comparison of speed profile versus PWM signal between hall-effect sensor and GPS

right could not be achieved. In order to obtain the perfect timing and the best speed needed to perform 90° turn to the right/left, a test is performed. The test for 90° right turn is done by letting the UGV drive forward until it achieves its constant speed and the time start recorded at the first deflection of wheels and stop once the UGV completed the 90° turn. The same data will be applied on the left turn. The result is plotted in Fig. 5.

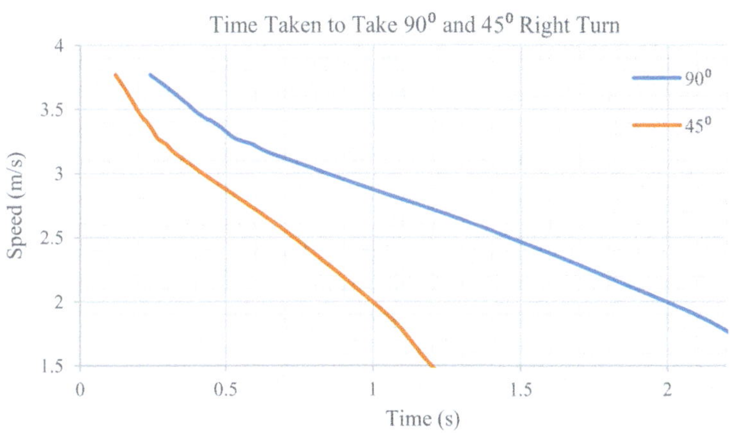

Fig. 5 Time taken for the UGV to turn at certain speed

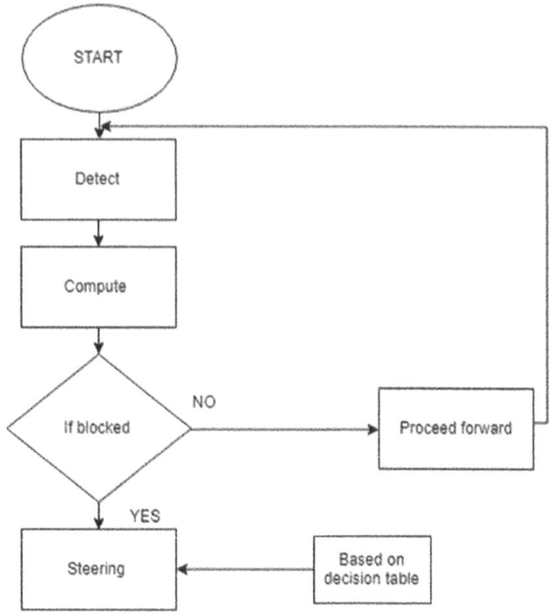

Fig. 6 Collision avoidance flowchart

2.2 *Path Planning*

Path planning is the key element for this UGV in finding ways to create a new path immediately after the original path is found blocked to ensure smooth maneuver [12]. Meaning that the process of decision making by the controller must be fast and efficient enough. A simple strategy is that, when no obstacle is detected, the UGV may proceed to move forward. When there is an obstacle detected, the algorithm will create a new temporary waypoint to the left or right or reverse direction depending on the position of the obstacles. The flow of the algorithm is shown in Fig. 6.

The UGV travels freely and does not controlled by any remote controller or software. The ideal approach is when there are no obstacles detected along the path as the UGV moves forward as displayed in Fig. 7a. However, if in any case the CAS detects obstacles, PPS will set a new pathway by creating a new temporary waypoint. The UGV will use the temporary waypoint as a guidance to the new pathway and then proceed forward as displayed in Fig. 7b.

The algorithm takes inputs and determine the direction according to the obstacle position and turning angle mapping data.

The obstacle input is taken from an array of six different sensing angles (15°, 45°, 75°, 105°, 135°, 165°). A single sweep of six inputs may give 64 possible combination. These inputs are then computed into a decision table that uses a binary system. If the sensor detects an obstacle within 1 m, it is recorded as '1' while if there is no obstacle within the distance, it is recorded as '0'. After the sensor makes on complete sweep, the algorithm will convert the binary value to decimal value.

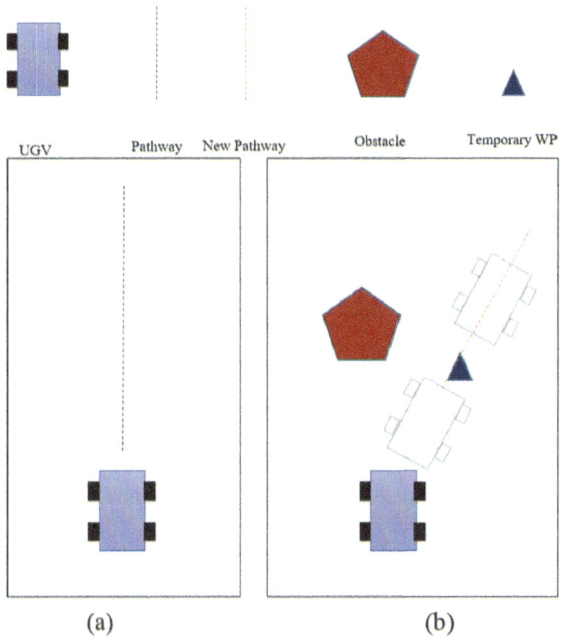

Fig. 7 **a** Ideal pathway **b** pathway with obstacle

The sensing angles are programmed in array; therefore, the binary values that will be received are also in array form. The six input from six sensing angles are then split into two as shown in Fig. 8.

The decimal value serves as the decision table for the maneuver of the UGV by going through a series of conditions as shown in Table 1.

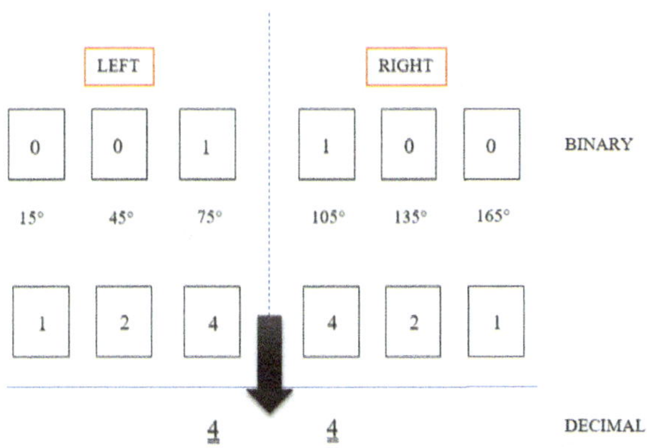

Fig. 8 Array of binary inputs converted into decimal value

Table 1 Decision table based on decimal method

<table>
<tr><th>Condition</th><th>Steering response</th></tr>
<tr><td>If L <= 3 && R <= 3</td><td>Forward</td></tr>
<tr><td>If L => 4 && R <= 3</td><td>Right 45°</td></tr>
<tr><td>If L =>1 && (R == 4|| R == 5|| R == 6)</td><td>Right 90°</td></tr>
<tr><td>If L <=3 && R =>4</td><td>Left 45°</td></tr>
<tr><td>If (L == 4|| L == 5|| L == 6 && (R == 2|| R == 3 || R => 6)</td><td>Left 90°</td></tr>
</table>

3 Result

A simple test was carried out to test the detection by putting an obstacle at a specific along the pathway as shown in Fig. 9.

It is noticed that there are five different scenarios along the pathway where the position of the obstacles is varied. These scenarios aim to evaluate how fast the algorithm makes the decision to avoid the obstacles. The algorithm that uses binary method and decimal method is compared in terms of turning accuracy and the decision speed.

Based on the result (Table 2), it is observed that the decimal method is faster than that the binary method in term decision speed but has slightly less accuracy in

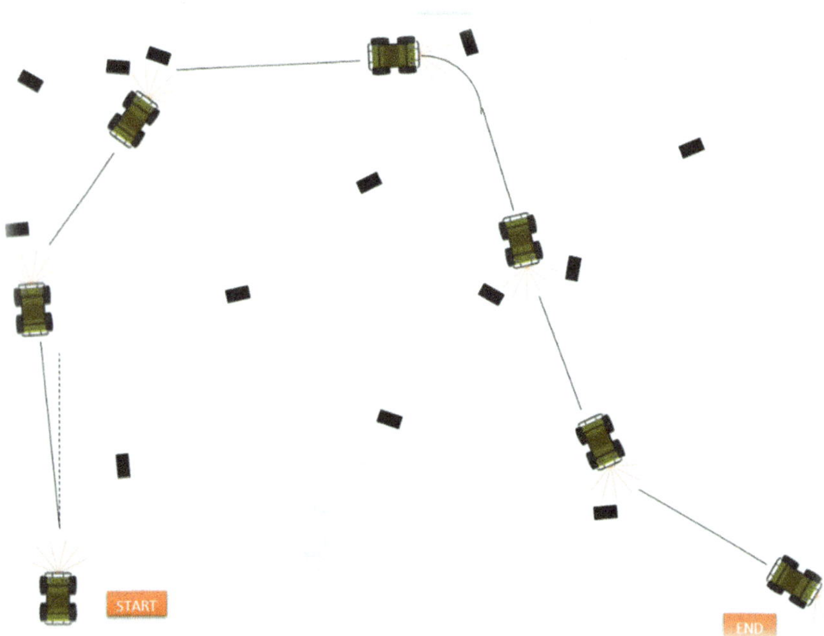

Fig. 9 Path planning and algorithm test scenario

Table 2 Decision speed for different scenarios

	Binary method		Decimal method	
Scenario	Decision speed (s)	Turning accuracy (%)	Decision speed (s)	Turning accuracy (%)
1	2.3	95	0.56	89
2	3.15	93	1.02	85
3	2.4	96	0.8	82
4	3.0	91	1.65	88
5	2.1	90	0.72	87

turning. However, the turning above that 80% is considered acceptable since it did not collide with any of the obstacles along the pathway.

4 Conclusion

Based on the result, it is observed that the algorithm that uses decimal method provides faster decision as compared to the binary one. This is because binary method takes many steps to decide because there are 64 different combination of decision. However, in terms of the turning accuracy, binary method has better accuracy since it moves slowly and take some time to steer the steering. However, with that trade-off, decimal method seems to offer faster and efficient avoidance since the robot did not collide with any of the obstacles.

References

1. Frank LL, Shuzhi SG (2018) Autonomous mobile robots: sensing, control, decision making and applications. CRC Press
2. Huang L, Huang Q, Zhou W (2018) Research on cooperative target detection and tracking of unmanned ground vehicles. In: International conference on artificial intelligence and Big Data(ICAIBD), IEEE, Chengdu, pp 128–133
3. Boguslaw S, Michal L (2014) A Surrounding World Knowledge Acquiring by Using a Low-cost Ultrasound Sensors. Procedia Comput Sci 35:93–100
4. Behzad B, Joel LC, Tullio F et al (2016) Requirements for building an ontology for autonomous robots. Industrial robot. In: Ind robot int j 43
5. Helen O, Dominik H, Marc P (2015) Reactive avoidance using embedded stereo vision for MAV flight. In: IEEE international conference on robotics and automation (ICRA), pp 50–56
6. Marco CD, Zandra BR, Domenico GD (2018) Obstacle avoidance system for unmanned ground vehicles by using ultrasonic sensors. Machines 6(2):18
7. Amir M, Likun X, Tong BT (2015) Vehicle detection techniques for collision avoidance systems: A review. IEEE Trans Intell Transp Syst 16(5):2318–2338
8. Houshyar A, Delpak RZ, Saeid N (2015) Optimisation of nonlinear motion cueing algorithm based on genetic algorithm. Veh Syst Dyn 53(4):526–545

9. Gines B, Francisco B, Jose ES et al (2002) Using infrared sensors for distance measurement in mobile robots. Robot Autono Syst 40(4):255–266
10. Anoop A, Susan LE, Raj K (2018) MengeROS: a crowd simulation tool for autonomous robot navigation. In: AAAI Fall symposium on AI for HRI
11. Kirti B, Sayalee D, Shraddha D et al (2016) Obstacle avoidance robot. Int J Sci Eng Technol Res (IJSETR) 5(2):439–442
12. Shinpei K, Eijiro T, Yoshio I et al (2015) An open approach to autonomous vehicles. IEEE Micro 35(6):60–68

Experimental Investigation of Mixing Enhancement Through Nozzle Exit Modification in Subsonic Flow

T. Vijayaraj, S. Thanigaiarasu, S. B. Jabez Richards, Surendra Kumar Yadav, A. Muthuram, S. Jaichandar, and S. Jagadeshwaran

Abstract There is a need of increase in efficiency and reduction of sound in jet engine where the researchers found that if mixing enhancement is made possible at the area prior to the primary zone of the combustion chamber the efficiency could be increased. As the free jet out of the nozzle taken the time to decay, there is a possibility for the fuel to exit at high speed in an unburnt state which primarily tends to reduce the efficiency of the jet engine. To overcome this problem, a square shaped fin tabs were designed and fabricated and the same was fixed diametrically opposite to the nozzle exit to create formation of vortices, thus the decay is made to be smaller and mixing of air and fuel will be enhanced. The experimental study was made along with data acquisition system to witness the effect of Mach number variation on and pressure ratio with respect to X/D as well as Mach number ratio by having a supply of air at M = 0.4, 0.6, and 0.8 in both the cases.

Keywords Jet control · Potential core & mixing enhancement

1 Introduction

The aircraft propulsion system consists of many parts in which combustion chamber is the main part in producing energy and so the working of jet engine combustion chamber should be effective in having greater efficiency of engine. The efficiency of an engine can be determined by the ratio of output power to that of input power. The output of the engine is based on the best utilization of the fuel which is then combusted to produce power. But due to the design and construction of the engine at the combustion chamber the air from the compressor to the combustion chamber is forced due to the entry having smaller cross sectional area. Because of this action the fuel entering into the combustion chamber may be carried at a greater velocity

T. Vijayaraj (✉) · S. Thanigaiarasu · S. B. Jabez Richards · S. K. Yadav · A. Muthuram
Madras Institute of Technology, Chennai, India
e-mail: jaywinvijay@gmail.com

T. Vijayaraj · S. Jaichandar · S. Jagadeshwaran
Prince Shri Venkateshwara Padmavathy Engineering College, Chennai, India

P. Rajendran et al. (eds.), *Proceedings of International Conference of Aerospace and Mechanical Engineering 2019*, Lecture Notes in Mechanical Engineering,
https://doi.org/10.1007/978-981-15-4756-0_40

without combustion and thus remain as unburnt. So obviously the efficiency will be reduced due to the reduced combustion of fuel and expected power output cannot be met. This can be rectified by lowering the streamline velocity possessed by the air to the combustion chamber. The alternate to the streamline is turbulence will itself be helpful for the reduction of unburnt. The turbulent flow in the air can be produced by forming vortices, a whirling mass of air by making any disturbances at the entry section. So these disturbances in turn form the streamline vortices. When the vortices get formed on the compressed air, the fuel is sprayed and mixing of fuel occurs reducing the travel of fuel has in the simpler nozzle exit. There by vortices formed can be used in increasing efficiency of the Engine as fuel gets effectively used. So many researches try to find a best solution for the vortex enhanced mixing. Because not only in reducing the unburnt, it can also be imparted in reducing the jet noise. Here, the designed square fin tabs can be used to produce the disturbances in case of forming vortices in the combustion chamber, prior to primary zone enhances the turbulence which eventually results in good combustion rate. The technique of designing and fabrication of the square fin tabs to a minimum dimensions to support the formation of vortex is simple and costless mechanical modification of the boundary geometry of existing design. The vortices formation can be analyzed by the data acquisition system in which graph for the experiment by considering various mach numbers of subsonic speed i.e., 0.2, 0.4, 0.6, 0.8. The Graph variations were checked by considering L/D ratio in x-axis and P_E/P_O and m/mc in y-axis. So this is used to analyse the best efficiency providing component.

Paterson [1] measured the characteristics of velocity and turbulence of mixed air in the flow from a multi-lobed nozzle/mixer which could cause large scale stream wise vortices responsible for enhancement of mixing. Cormick and Bennett [2] revealed the interaction of Kelvin-Helmholtz (span wise) vortices with the stream wise vortices where the normal vortices when travel lengthwise, at the end gets squeezed off into small structures. These structures will be intended in formation of turbulence flow which is used in mixing enhancement, thereby producing high level mixing of air. Asad et al. [3] had given the effect of tab geometry where tabs was used in the place of producing disturbances which is of different shapes like rectangle, semicircle at the exit of the Nozzle and they are made as such to withstand the load opposed by the flow on the tabs. The results were compared between the nozzle without any disturbances and with tabs and found that jet core length is getting reduced by 80% by having the tabs at the exit. Thanigaiarasu et al. [4] presented the study on slanted perforated tabs where the perforated tabs were used to reduce the sound and also the forward force of air from the Nozzle exit. This study will be helpful to find the increase in efficiency of the engine by having the comparison between free jet and exit with tabs and finding which produces mixing better and also at what Mach number. The results were marked by plotting centreline velocity and radial velocity on solid tab, free jet and perforated tabs. Dharmahinder et al. [5] in tested the perforated tabs for its effectiveness at various Mach numbers of subsonic and sonic jets at Nozzle pressure Ratio 2 and 3. The Enhancement of mixing is observed due to the reduction of potential core of air towards its length and also velocity reduction from the Nozzle exit due to the disturbances forming vortices. Zaman [6] in his study on stream wise vorticity generation and mixing enhancement in the free-jets by delta-tabs proposed the vortex generation through the delta tabs where he found that jet flow is controlled by the pair of tabs placed at the Nozzle exit. So that it results in the mixing enhancement due to the spread of jet through the vortex formed. The Delta tabs were fitted to the axisymmetric nozzle named because of the nozzle being symmetrical with reference to an axis and nozzles of rectangular shape with the ratio of 3:1, 4:1, 8:1 to observe the mixing variations. Rathakrishnan [7] did the detailed study on rectangular tabs and tabs with corrugations that is series of parallel ridges were made on the tabs. The corrugated tabs are made for adding rigidity and strength. However the result based on reduction of length of the Jets potential core is larger in the plain tabs. There are several various tabs designed by the researches to reduce the potential core of the jets length and to enhance the mixing but the consideration should be given for also the blockage of area of the nozzle exit because the high speed jet is controlled by the lesser blockage.

2 Experimental Facility

The experimental setup given in Fig. 1 is available in High speed jet laboratory at Anna university MIT (Madras Institute of Technology) campus, Chennai, Tamilnadu, India. This setup associated with Data acquisition system through which we can

Fig. 1 High speed jet laboratory

obtain the data about the compressed air pressure and the distances of the travel of air from the exit of the nozzle.

Here the convergent Nozzle is being used which is fixed to the setting chamber having inlet diameter of 40 mm and exit diameter of 20 mm. The settling chamber is of uniform cylindrical shape where the constant pressure is maintained which otherwise called as stagnant pressure. The stagnant pressure is maintained as constant by adjusting the pressure gauge (or) pressure regulator. As the pressure is maintained as constant, we can control the Mach number at the Nozzle exit. The nozzle built diameter is 20 mm taken as 'D, while considering X/D ration. The M/M_E and P_E/P_O values were compared with X/D where X/D varies from 0 to 20. So that the jet core length is calculated for about 400 mm. Before calculating the length of the jet has to be travelled the Mach number is adjusted by maintaining the pressure to a certain value to obtain subsonic Mach number. The subsonic Mach number of 0.4, 0.6, and 0.8 were maintained by having the exit pressure as 11.7 kPa, 28.7 kPa, 53.7 kPa respectively. The jet core length of 400 is divided into 5 mm and so 80 divisions were obtained. For the each and every 5 mm distance, the value of M/M_E and P_E/P_O.

2.1 Free Jet

Figure 2 shows a convergent nozzle of exit diameter 20 mm made of stainless steel is taken for free jet condition. The air is made to flow freely without any disturbances. The Nozzle is fixed to the settling chamber very tightly to avoid the leakage of air. As

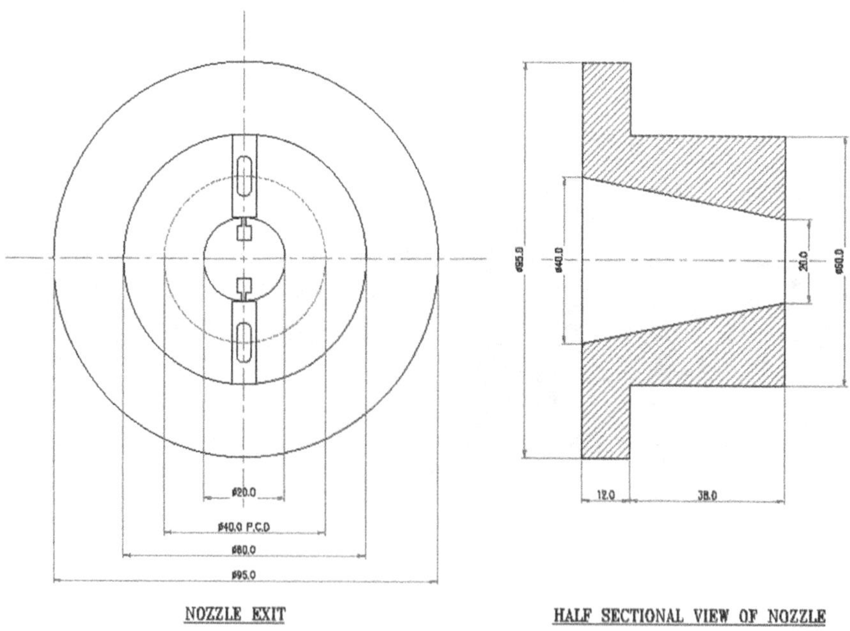

Fig. 2 Free jet

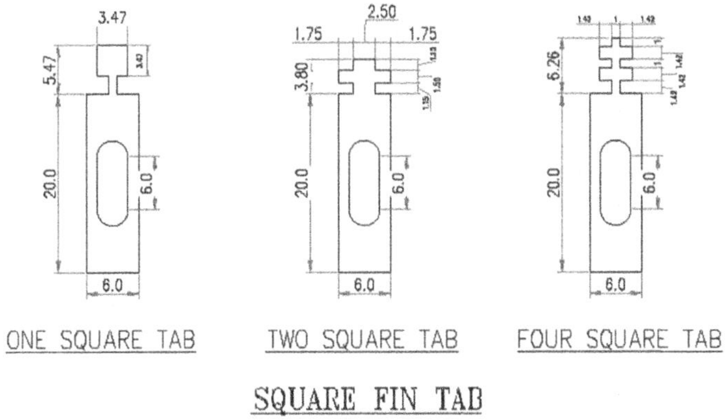

Fig. 3 Squared Fin Tabs

said before the pressure regulator, the pressure along the flow of jets can be measured by a pitot tube of stainless steel of inside diameter 0.4 mm and outside diameter of 0.6 mm.

2.2 *Squared Fin Tab Jet*

Squared Fin Tabs (Fig. 3) are set at the nozzle exit to create distortion in the air flow and they are not in motion, just made to stand still on the diametrically opposite positions. So it is called as passive control method. The Squared Fin Tabs are made of stainless steel and it is of length 6 mm, width 1 mm. The Area of the blockage of Nozzle exit due to tab arrangement is 8.91.

3 Results and Discussion

Thus the mixing enhancement is made possible by forming vortices with the squared tabs. The squared tabs were made of stainless steel which is to be placed at the nozzle outlet which is diametrically opposite to the nozzle exit prior to the primary zone of the combustion chamber. This square tabs are of 1 mm of thickness. The various square tabs were made based on numbers. They are one square, two square and four squares within them. The measured pitot pressure is denoted as PO, the stagnation pressure is the pressure was made to be measured by having a small protrusion outside the nozzle to which the pitot tube is connected and there the air is made to pass through the nozzle and pressure was noted.

3.1 Centreline Pressure Decay

The air from the nozzle without any tabs will have the less decay and thus travel more distances carry out the fuel which is sprayed inside the combustion chamber. For the Mach numbers 0.4, 0.6, 0.8, the exit pressure will be nearly equal to the stagnation pressure for certain value of aspect ratio X/D. This case makes us to come to the conclusion that the pressure decay is zero initially at the nozzle without any disturbances and so we can say that the carrying out of fuel at unburnt will surely occurs.

3.2 Experimental Study for All Three Tabs at Varying Subsonic Mach Numbers

Figures 4, 5 and 6 depict the variation of pressure ratio along the jet with four squared tabs, tow squared tabs, single squared tabs and free jet at various near stream locations is given. It is seen in the Fig. 4 which Experimental study is given for M = 0.4 where there is slight increase in the dip of P_E/P_O up to X/D = 3 due to the unique characteristic of the jet coming out straight from the nozzle.

When there is any disturbances due to the tabs of square shape found in the nozzle exit may result in the formation of vortices. This formation of vortices used in the better decaying of air. This dip is seen in smaller strength as approaching X/D = 0.5 and the dip keeps on lowering as the x/d keeps on increasing. The major differences can be found between the free jet and squared fin tabs at nozzle exit where P_E/P_O decrease is slow in free jet but P_E/P_O decrease is high in exit of nozzle with tabs as

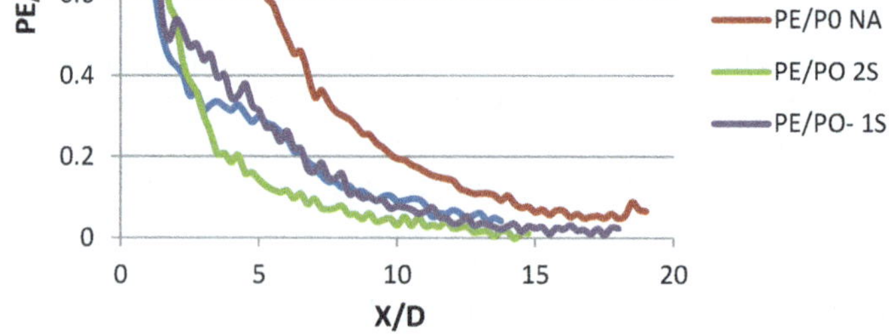

Fig. 4 The variations of P_E/P_O with X/D at M = 0.4

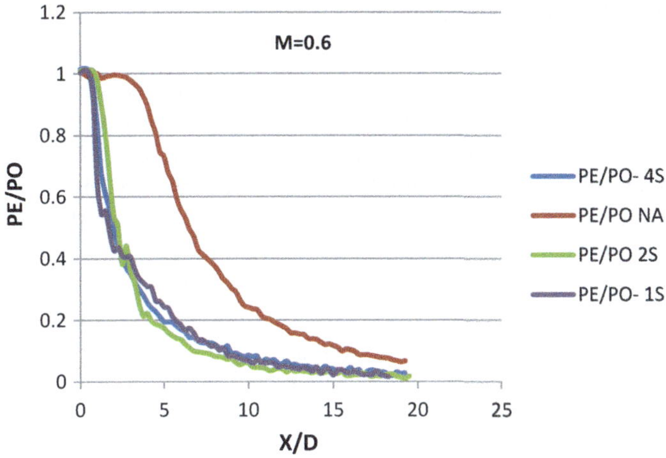

Fig. 5 The variations of P_E/P_O with X/D at M = 0.6

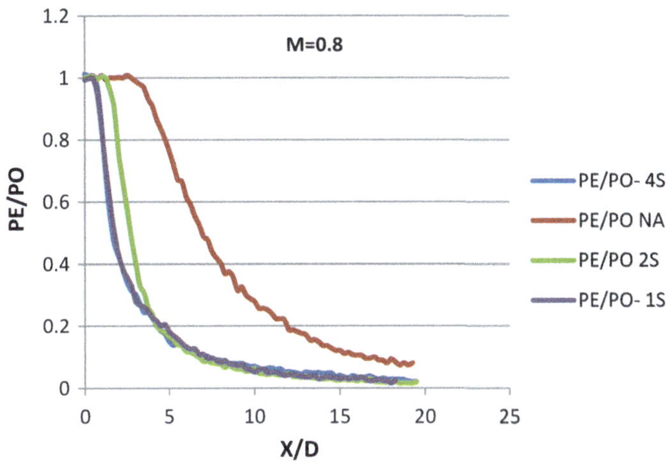

Fig. 6 The variations of P_E/P_O with X/D at M = 0.8

this denotes the decay is larger when there is squared fin tabs than in free jet without disturbances. The Fig. 5 gives the experimental study on variation of P_E/P_O and X/D for M = 0.6 and Fig. 6 gives the numerical study on variation of P_E/P_O and X/D for M = 0.8 where both are merely equal to the diagrammatic representation given in Fig. 4. By observing this one can conclude that the pressure decay is well and good in tabbed suction than in free jet. So P_E/P_O ratio from the high value starts to reduce in a faster rate as the X/D ratio keeps on increasing than that of the nozzle without any square tabs.

3.3 Centreline Velocity Decay

In the three plots, Figs. 7, 8 and 9, the variation of Mach number ratio (M/ME) along the jet with four squared tabs, two squared tabs, single squared tab and free jet at various near streams are given. This is considered as velocity decay because the Mach number is related with velocity and there is a due decrease in Mach number ratio which gives the explanation to the velocity decay due to the disturbance provoking tabs present at the nozzle exit and also the free jet.

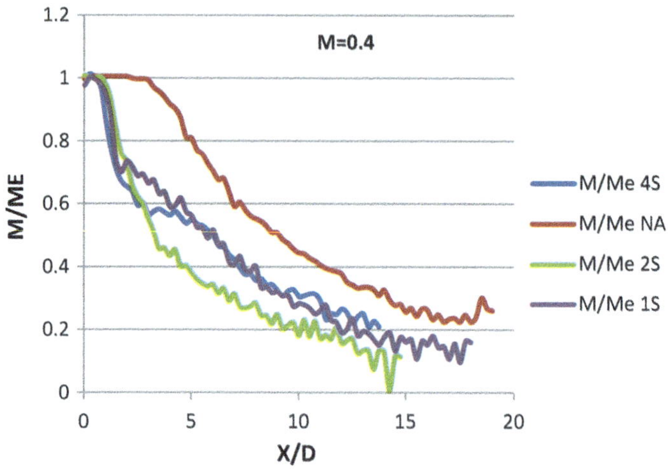

Fig. 7 The variations of M/M_E with X/D at M = 0.4

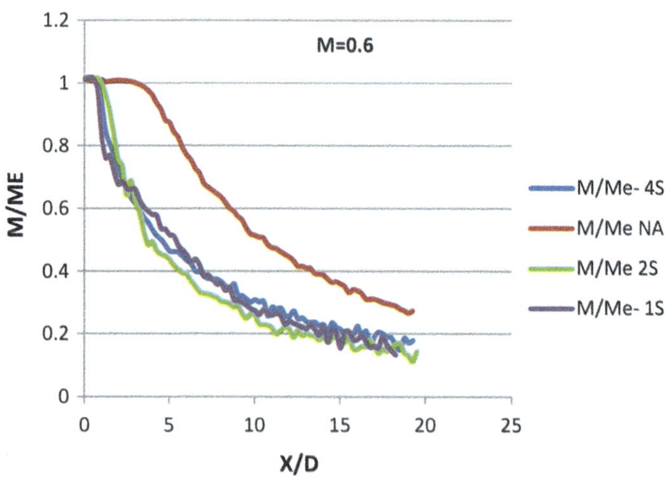

Fig. 8 The variations of M/M_E with X/D at M = 0.6

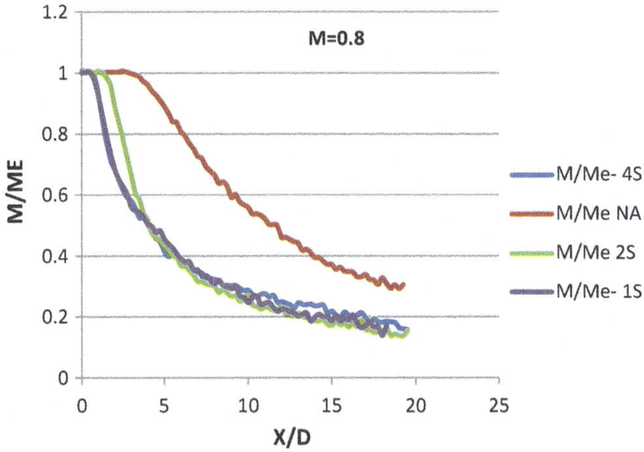

Fig. 9 The variations of M/M_E with X/D at M = 0.8

It is seen in the figures that M/ME = 1 for a certain X/D value that is up to 3 which gives the clear view on equating Mach number M = ME showing the velocity does not getting decayed over the distance.

The dip is continuously increasing as the X/D increase over x-axis. The value of M/ME in various squared tabs keeps on lowering. The decreasing curves are varying more similarly but the variation is large between free jet and nozzle exit with squared tabs. The Fig. 7 gives the numerical study of on variation of M/ME and X/D for M = 0.4 of air passed into the experimented setup. The Fig. 8 gives the numerical study on variation of M/ME and X/D for M = 0.6 and Fig. 9 respectively for M = 0.8. By observing these graphs, we can conclude that the velocity decay is slower in free jet when compared to nozzle exit with square tabs where decay is faster due to the formation of vortices at the exit.

4 Conclusion

Through the results of experimental study on variation of P_E/P_O and M/M_E with respect to X/D for air proved with the Mach Numbers M = 0.4, 0.6, 0.8, it is given that the decay of air flow can be observed to be less in free jet and the decay of air flow is more in the nozzle possessing square fin tabs. So the airflow decay can be more helpful in the enhancement of mixing of air, and also the sound travel gets reduced due to no more travel of air to greater distances. Thus the sound travel is very less and so reduction in sound can also be achieved with increasing the mixing enhancement.

References

1. Paterson RW (1982) Turbofan mixer Nozzle flow field—a benchmark experimental study. J Eng Gas Turbine Power 106:692–698
2. Mc Cormick DC, Bennett JC (1994) Vortical and turbulent structure of a lobed mixer free shear layer. AIAA Paper 32:1852–1859
3. Asad Ahmed R, Thanigaiarasu S, Santhosh J, Elangovan S, Rathakrishnan E (2013) Study of slanted perforated jets. Int J Turbojet Engines 30:347–357
4. Thanigaiarasu S, Jayaprakash S, Elangovan S, Rathakrishnan E (2008) Influence of tab geometry and its orientation on under expanded sonic jets. J Aerosp Eng 222:331–339
5. Dharmahinder SC, Thanigaiarasu S, Elangovan S, Rathakrishnan E (2011) Perforated arc tabs for jet control. Int J Turbo Jet Engines 28:133–138
6. Zaman KMBQ (1993) Stream wise vorticity generation and mixing enhancement in free jet by delta-tabs. AIAA J 93–3253
7. Rathakrishnan E (2013) Corrugated Tabs for subsonic and sonic jet control. J Aeronaut Aerosp Eng (2)

Experimental Modal Analysis of a Simple Rectangular Wing with Varying Rib's Orientation

M. S. Othman, L. Teh, and M. Y. Harmin

Abstract Three oriented rib profiles of a simple rectangular wing model were considered for the experimental modal analysis (EMA). They were namely the best individual, best parallel and best overall oriented rib profiles. The wing was setup as a cantilever-like condition with a roving hammer technique was employed for the testing. Modal properties with respect to natural frequency and mode shape were measured and the results were then compared against the baseline configuration. From the results, the impact towards the bending modes were found very minimal and can be negligible for all the considered oriented rib profiles. Nevertheless, significant increases in torsional frequencies was found for the best overall oriented rib profile when compared with the baseline configuration, showing that the wing becomes much stiffer in twisting direction without any increase in weight. Hence, this offer a possibility in flutter improvement or in weight reduction if the flutter performance to be kept similarly as the baseline configuration.

Keywords Structural dynamics · Wing design · Experimental modal analysis

1 Introduction

Aviation has becoming one of the major catalyst in driving the global economic prosperity and social growth. With an increasing demand in air travel, most of the major airlines have continuously increase their fleets not only for meeting the demand but as well as to gain more profit in order for them to stay competitive in the aviation industry. On one hand, this progression provides positive impact for both economic and social benefits, but on the other hand it turns into the fastest growing sources of greenhouse gas emissions. Due to this reason, a number of initiatives such as ACARE 2020 [1], FLIGHTPATH 2050 [2] and CORSIA [3] has been introduced which bring together all the key players in aerospace and aviation industries in order to seek solutions for reducing the harmful emission from an aircraft.

M. S. Othman · L. Teh · M. Y. Harmin (✉)
Universiti Putra Malaysia, 43400 Serdang, Selangor, Malaysia
e-mail: myazdi@upm.edu.my

P. Rajendran et al. (eds.), *Proceedings of International Conference of Aerospace and Mechanical Engineering 2019*, Lecture Notes in Mechanical Engineering,
https://doi.org/10.1007/978-981-15-4756-0_41

Over the past decades, the design of the aircraft has undergoadvancement in structural, material, aerodynamic, control, propulsion and continuous manufacturing technology. Nevertheless, the external and internal configurations of the aircraft are still more or less similar since the past 50 years. Numerous attempts have been conducted on changing the spars and ribs orientations in the wing interior configuration so that it behaves as an anisotropic manner. One of the earlier attempts was recognized by Harmin et al. [4] who changed all the ribs orientation and the direction of crenellations over a range of ±45° in a metallic wing structure. A study related to aeroelastic tailoring on a more realistic wing structure has also been conducted in simplified NASA Common Research Model (CRM) by using fiber tow steered composite, functionally graded materials (FGM) and curvilinear ribs, spars and stringers [5]. Francois et al. [6] has conducted the first experimental based work on 3D printed wing and vary the whole ribs angle from −15° to +45° and a parametric study on the influence of wing sweep angle and rib orientation on an un-tapered wing structure was also been conducted by the same author [7]. In the latest work by Francois et al. [8] the crenelated skins has been incorporated via modelling and experiment. Besides that, Othman et al. [9] has investigated the bending and torsional modes through altering only a single rib orientation on a simple rectangular wing-box model. In 2017, Chan et al. [10] used NASA CRM to illustrate the idea of varying the rib's orientation in a more realistic aircraft wing structure. The most recent work is by Harmin et al. [11] where the parametric study is conducted for all the possible oriented rib combination within 0° and ± 45°.

All the above mentioned works have shown the tailored configuration provides an impressive enhancement especially in terms of its dynamic characteristics when compared with its corresponding conventional wing configuration. Therefore in this paper, the effect on varying the rib's orientation on modal properties is assessed via the experimental modal analysis

2 Methodology

2.1 *Simulation & Experimental Models*

Figure 1 shows the general representation of the wing model for the baseline configuration including its dimensioning. The wing model is consist of two spars which embody the leading and trailing spars, with a numbers of 10 ribs that were distributed along the span. Note that, an assembly of spars and ribs parts formed the wing-box section showing as a shaded region in Fig. 1. In the case of baseline configuration, the ribs were distributed equally, hence having an equal spacing between them. Both spars and ribs are made from the spring steel sheet metal, whereas the Styrofoam material is employed to complete an aerodynamic planform of the wing model. The spring steel is selected due to its ability to withstand high deformation without suffering any distortion to its mechanical properties, while the Styrofoam is selected due

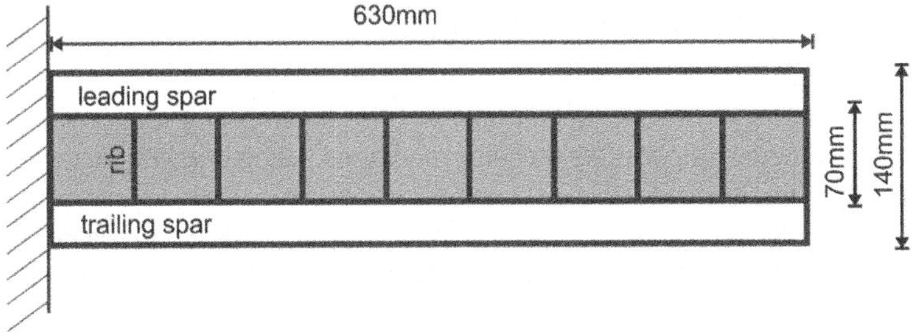

Fig. 1 Baseline wing model

to its lightweight and flexibility to take the deformation shape of the wing. The ribs and spars are connected via bolt and nut and the skins were attached permanently to the wing-box with the use of epoxy glue.

In this work, focus is given to the three wing-box configurations of different oriented rib profiles which were based upon the previous study made by the authors [reference]. These configurations were selected within the three considered parametric category of individual oriented rib, parallel oriented ribs and overall oriented ribs with an optimum flutter speed was taken as a cost function. In this paper, these configurations will be known as the best individual oriented rib, best parallel oriented ribs and best overall oriented ribs.

Figure 2 shows the experimental models of the considered oriented rib profiles. Note that, the baseline configuration was also taken into account as the assessment for the oriented rib configurations will be made against it. It should be noted that all the considered wing configurations are having a similar total mass, which is more or less about 0.15 kg. The weight is maintained such that the thickness of the oriented rib is lesser than its corresponding baseline rib in order to counterbalance due to an extension in its length. This to ensure that, any possible improvement either in static or dynamic characteristics of the proposed oriented ribs configuration should be mainly due to the changes in its effective stiffness.

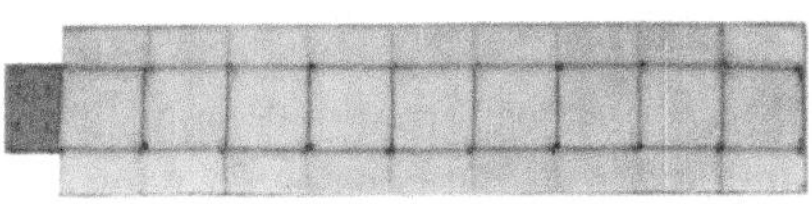

a) Baseline

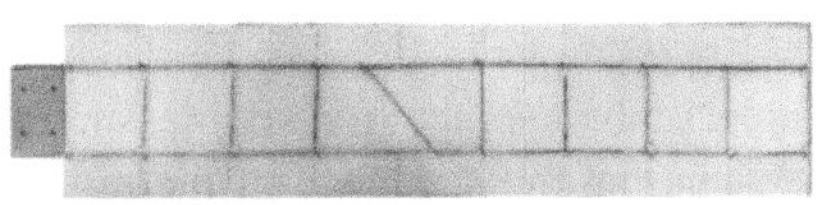

b) Best individual oriented rib profile

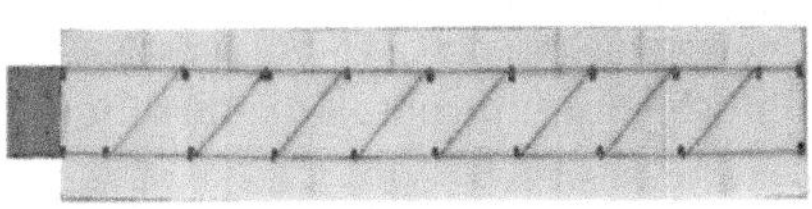

c) Best parallel oriented rib profile

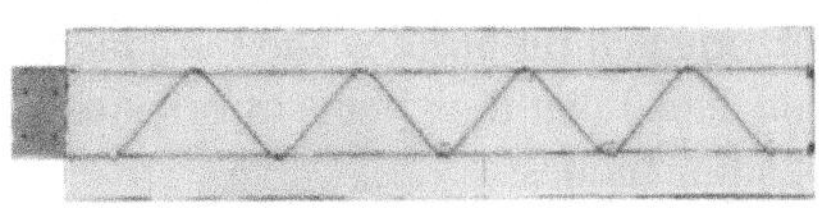

d) Best overall oriented rib profile

Fig. 2 Experimental wing model

2.2 *Experimental Modal Analysis*

In this work, the experimental modal analysis (EMA) of impact testing was employed to identify the modal properties of the wing models namely natural frequencies and its corresponding mode shapes. The system consists of three major components which are the LMS TestLab package, impact hammer and uniaxial accelerometers which represent respectively the data analyzer system, excitation mechanism and transducer. Since the wing models were of a simple configuration with a cantilever-like condition, hence a roving hammer test was in preference than the roving accelerometer technique. This is mainly due to the location of the transducers that can be remain stationary which enable consistency in mass distribution throughout the EMA testing. Nevertheless, care must be taken when positioning the transducers so that they are not at the or close to the nodal point of the modes of interest. Therefore, a finite element analysis of normal mode solution (SOL 103) has to be assessed accordingly in order to identify the suitable positions of the transducers as well as the position of the impact points.

The EMA testing was carried out with the input of hammer force is applied three times and roved on each of the defined impact points. This operation enables the average results to be obtained with the purpose to improve the quality of the measurement. It should be noted that the structure must returned to its rest before the next average of data is taken. As the impact is given, the data analyzer will measure the time domain responses from both load cell and accelerometers. These responses are then being manipulated and transformed into the frequency response functions (FRFs) via the fast Fourier transformation (FFT) operation. Subsequently, a series of FRFs for each of the considered points are then merged and analyzed through a modal parameter estimation module of LMS PolyMAX which enable the modal parameters of natural frequency, damping, and mode shape to be extracted. Figure 3 illustrates the general setup for the EMA testing of the wing model.

3 Results & Discussion

Table 1 tabulates the first six measured natural frequencies. From the results, it can be seen that the sequence of the modes is similar for all the considered wing box models. Note that the natural frequency of the bending modes were found more or less similar for all the considered cases with the percentage of difference between the oriented rib profiles and baseline is less than seven percent. Therefore it can be concluded that the effects of oriented rib profiles is insignificant in the direction of bending modes.

Nevertheless, significant increases in the torsional mode frequencies was visible in the case of best overall configuration when compared with the baseline wing configuration, where increases in natural frequency of more than 35% for the first torsional mode was obtained. This provides an outstanding outcome where the best

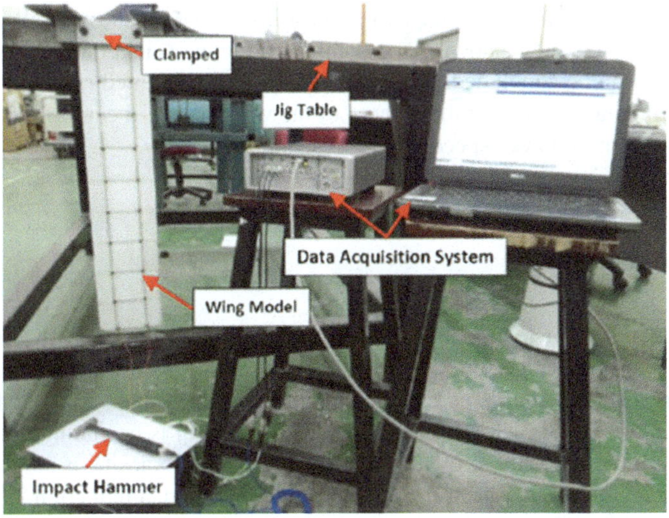

Fig. 3 Experimental modal testing setup

Table 1 Measured natural frequencies

Modes	Baseline (Hz)	Best Individual (Hz)	Best Parallel (Hz)	Best Overall (Hz)
B-1	6.508	6.165 (5.3%)	6.364 (2.2%)	6.149 (5.52%)
T-1	23.079	23.454 (1.6%)	24.419 (5.8%)	31.253 (35.4%)
B-2	40.629	38.562 (5%)	39.638 (2.4%)	38.093 (6.2%)
T-2	78.944	75.269 (4.7%)	81.802 (3.6%)	97.538 (23.6%)
B-3	114.557	108.186 (5.6%)	111.735 (2.5%)	108.390 (5.4%)
T-3	159.416	153.662 (3.6%)	162.861 (2.2%)	177.729 (11.5%)

[a]**B-1** (1st Bending), **B-2** (2nd Bending), **B-3** (3rd Bending), **T-1** (1st Torsion), **T-2** (2nd Torsion), **T-3** (3rd Torsion)
[b]Parenthesis refers to the percentage of difference between the oriented profile and baseline wing configuration

overall oriented rib profile permit in increases of overall torsional rigidity of the wing while maintaining its bending rigidity. Therefore, a much stiffer characteristic in twisting direction can be offered, hence leading a possibility in increases of frequency gap between the flutter modes and subsequently offers in increases of flutter speed. Note that, at higher frequency of torsional modes, the percentage of difference reduces, which demonstrate that the best overall oriented rib profile becoming less impactful at higher modes. In the case of best individual and best parallel oriented rib profiles, both of them showing a very minimal increases in torsional frequencies. Therefore it can be concluded that their corresponding oriented rib profiles do not significantly change the overall stiffness of the wing, hence it can be expected that the improvement in flutter speed will be very minimal if any.

4 Conclusion

This study is set out to investigate the three considered oriented rib profiles via experimental modal analysis. The findings shows that the best overall oriented rib profile offers significant increases in torsional frequencies when compared with the baseline configuration; which demonstrate that the wing is becoming much stiffer in twisting direction. This provides an encouraging outcomes as it may offer into an improvement in flutter performance. On the other hand, a reduction in weight is also possible if the flutter performance of the best overall configuration is to be kept similarly as the baseline configuration.

Acknowledgements The authors acknowledge the financial support from Ministry of Education (MOE) Malaysia for this study through the Fundamental Research Grant Scheme (FRGS/1/2015/TK09/UPM/02/2) with Project Code: 03-01-15-1625FR.

References

1. Argüelles P, Bischoff M, Busquin P (2001) European aeronautics: a vision for 2020. In: Report of the Group of Personalities. Office for Official Publications of the European Communities
2. Darecki M, Edelstenne C, Enders T, Fernandez E, Hartman P, Herteman J, Wörner J (2011) Flightpath 2050. Flightpath 2050 Europe's Vision for Aviation. Publication office of the EU
3. International Civil Aviation Organization (2016) Assembly Resolutions in Force as October
4. Harmin MY, Ahmed AT, Cooper JE, Bron F (2011) Aeroelastic tailoring of metallic wing structures. 52nd AIAA/ASME/ASCE/AHS/ASC Structures. Structural dynamics and materials conference. AIAA, Denver, pp 1–19
5. Jutte CV, Stanford BK, Wieseman CD, Moore JB (2011) Aeroelastic tailoring of the NASA common research model via novel material and structural configurations. In: 52nd Aerospace Sciences Meeting. AIAA, Maryland, pp 1–20
6. Francois G, Cooper JE, Weaver PM (2015) Aeroelastic tailoring using rib/spar orientations: experimental investigation. 56th AIAA/ASME/ASCE/AHS/ASC Structures. Structural dynamics and materials conference. AIAA, Florida, pp 1–22
7. Francois G, Cooper JE, Weaver PM (2016) Impact of the wing sweep angle and rib orientation on wing structural response for un-tapered wings. In: 57th AIAA/ASCE/AHS/ASC structures, structural dynamics, and materials conference. AIAA, San Diego
8. Francois G, Cooper JE, Weaver PM (2017) Aeroelastic tailoring using crenellated skins-modelling and experiment. Adv Aircraft Spacecraft Sci 4(2):93–124
9. Othman MS, Chun OT, Harmin MY, Romli FI (2016) Aeroelastic effects of a simple rectangular wing-box model with varying rib orientations. IOP Conf Series: Mater Sci Eng 152(1)

10. Chan YN, Harmin MY, Rafie ASM (2019) Aeroelastic tailoring via ribs orientation of nasa common research model. IOP Conf Series Mater Sci Eng 270(1)
11. Harmin MY, Othman MS, Romli FI (2018) Parametric study on the flutter characteristics of a simple rectangular wing-box model with varying ribs orientation. Int J Pure Appl Mathe 119(15):3771–3777

Validation of Rotor-Disc Model for Light Autogyro in Steady-State Flight Mode

Shahrul Ahmad and Douglas Thomson

Abstract Most rotorcraft flight dynamic studies employed higher fidelity rotor modelling approach to precisely resemble the dynamic behaviour of the real rotorcraft in simulation. However, high-fidelity calculations always implemented at the expense of other simulation costs, such as the CPU run-time and the machine hardware. This paper is aimed to highlight the used of lower fidelity rotorcraft calculation approach for light autogyros in normal flight mode where extensive calculations are less vital. A lower fidelity helicopter simulation package, named Helicopter Generic Simulation (HGS) is used as the basis where the structural features are modified according to the unique autogyro's kinematics. Due to the common longitudinal stability issues previously experienced by typical light autogyros, validations are made in longitudinal steady-state flight mode against the real flight data of the test autogyro. Simulation results show tolerable steady-state predictions of important parameters, such as the fuselage pitch attitude and the longitudinal shaft tilt. Hence, it is recommended to employ the rotor-disc modelling approach for autogyro's applications that are not subjected to complex rotor blade dynamics, such as in automatic flight control and inverse simulations in normal flight mode.

Keywords Autogyro · Flight dynamics · Rotor-disc model

Nomenclature

a_0	Aerofoil lift-curve slope (1/rad)
a_1	Zero angle of attack lift coefficient
c	Rotor blade chord length (m)
C_l, C_d	Rotor blade lift and drag coefficients

S. Ahmad (✉)
Universiti Kuala Lumpur—MIAT, Selangor, Malaysia
e-mail: shahrul@unikl.edu.my

D. Thomson
University of Glasgow, Glasgow G128QQ, UK

P. Rajendran et al. (eds.), *Proceedings of International Conference of Aerospace and Mechanical Engineering 2019*, Lecture Notes in Mechanical Engineering,
https://doi.org/10.1007/978-981-15-4756-0_42

d, l	Drag and lift force per unit span (N)
$\boldsymbol{f}$	Local force vector per unit span (N)
$\boldsymbol{F}$	External force vector (N)
g	Gravity (m/s^2)
I_R	Rotor moment of inertia (kg m^2)
I_{yy}	Body moment of inertia in pitch (kg m^2)
$\boldsymbol{i}, \boldsymbol{j}, \boldsymbol{k}$	Unit vectors
K_β	Centre-spring rotor stiffness (Nm/rad)
m	Autogyro total mass (kg)
M	Aerodynamic pitch moment (Nm)
$\boldsymbol{M}$	Moment vectors (Nm)
Q	Body axes pitch velocity (rad/s)
Q_R	Rotor torque (Nm)
$\boldsymbol{r}$	Position vector (m)
R	Rotor radius (m)
$\boldsymbol{r}_b$	Local distance from rotor hub (m)
T_R, T_{prop}	Rotor and propeller thrust (N)
U, W	Translational velocity components in body axes (m/s)
$\boldsymbol{v}$	Translational velocity vector (m/s)
v_0, v_{1s}, v_{1c}	Induced velocity components (m/s)
X, Y, Z	External force components in body axes (N)
β	Flapping angle (rad)
χ	Induced airflow wake angle (rad)
δ	Rotor blade profile drag coefficient
Ω	Rotorspeed (rad/s)
ϕ	Rotor blade angle of incidence (rad)
ϕ_s, θ_s	Lateral and longitudinal shaft tilt (rad)
ψ	Rotor blade azimuthal position (rad)
ρ	Air density (kg/m^2)
Θ	Autogyro pitch angle (rad)

1 Introduction

Autogyro is defined as a rotary-wing aircraft (rotorcraft), similar to a helicopter with rotating blades that produces lift that keeps the vehicle airborne [1, 2]. However, one distinct functional feature of an autogyro is the autorotating blades that are aerodynamically powered by relative airflow during flight. The forward thrust is mainly provided by a small engine-driven propeller, either of a 'tractor' or 'pusher' type. This configuration also resembles unique flight dynamic characteristics compared to helicopters and fixed-wing aircraft.

The vehicle has gained its popularity in Europe and the United States in the past 20 years mostly in sports and leisure flying activities. This is due to the advantages this vehicle has over aeroplanes and helicopters; low maintenance cost and stall-proof capability. The popularity, however, slowly diminished with alarming safety records which triggered the Aviation authority to establish thorough investigations on its flight dynamics and stability characteristics [3–5].

In the UK for instance, research was initiated by the Civil Aviation Authority (CAA) since 1992, for more than 15 years that recounted to many published reports as reviewed by Houston and Thomson in Ref. [6]. Due to vigorous flight dynamic conditions to be investigated, a higher fidelity 'Individual Blade' mathematical model was employed in those studies. This comprehensive rotor blade model, however, is often unnecessary concerning the rigid-body behaviour of an autogyro which usually not much affected by the rotor blade dynamics in normal flight conditions.

Therefore, in this research, a lower fidelity autogyro mathematical model has been developed using a 'multiblade' modelling approach (also known as the 'rotor-disc' model), instead of the previous individual blade model. The light autogyro used in the study is the Montgomerie-Parsons research autogyro (shown in Fig. 1), which was formerly used in the previous CAA research programme. Table 1 shows the Montgomerie-Parsons's general configurations obtained from a CAA publication [3].

Fig. 1 Montgomerie-Parsons research autogyro [3]

Table 1 Autogyro general configuration [3]

Parameter	Value
Vehicle mass (m)	355 kg
Moment of inertia	
I_{xx}	72.96 kg m^2
I_{yy}	297.21 kg m^2
I_{zz}	224.25 kg m^2
I_{xz}	0 kg m^2
Rotor parameters	
Aerofoil type (untwisted)	NACA 8-H-12
Blade span (R)	3.81 m
Chord length (c)	0.197 m
Blade mass (m_r)	17.255 kg
Lift curve slope (a_0)	5.75 rad^{-1}

2 Mathematical Model

The light autogyro mathematical model is developed based on the existing rotor-disc type Helicopter Generic Simulation (HGS) model [7], which has been used extensively in helicopter flight dynamics simulations [7–10].

The dynamics of the vehicle is presented in the form of the equations of motion in the longitudinal mode (below), with the body axes originated at the vehicle's centre of gravity (c.g.) as shown in Fig. 2.

$$m(\dot{U} + WQ) = \sum X \sin \Theta \tag{1}$$

$$m(\dot{W} - UQ) = \sum Z + mg \cos \Theta \tag{2}$$

$$I_{yy}\dot{Q} = \sum M \tag{3}$$

$$\dot{\Omega} = -\frac{Q_R}{I_R} \tag{4}$$

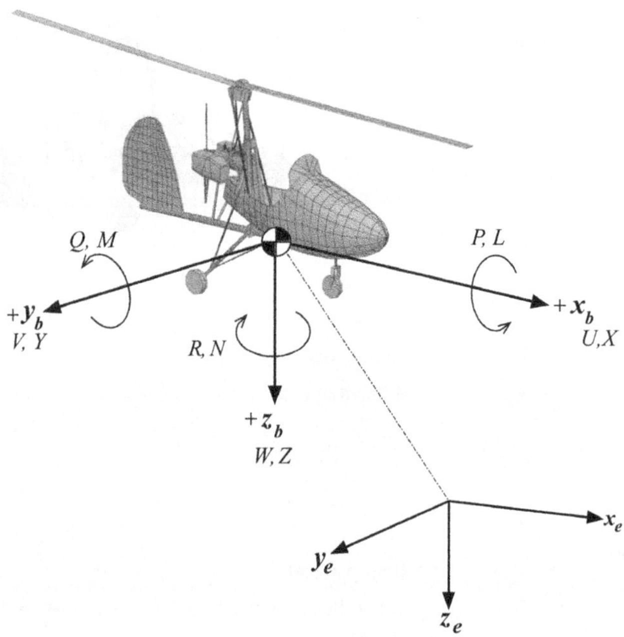

Fig. 2 Rigid-body orientation in *body* axes references

The $\dot{\Omega}$ in Eq. (4) represents the rotorspeed of the autogyro in autorotation which was modified from the helicopter's main rotor governor equation in Ref. [8]. The calculation of the external forces and moment of Eqs. (1)–(4) are obtained by dividing the vehicle into four different subsystems: The *Rotor (R)*, the *Fuselage (fus)*, the *Tailplane (tp)* and the *Propeller (prop)*. The total external forces and moment in longitudinal mode are calculated according to the following (detail calculations are not shown in this paper):

$$\sum X = X_R + X_{fus} + X_{tp} + X_{prop} \tag{5}$$

$$\sum Z = Z_R + Z_{fus} + Z_{tp} + Z_{prop} \tag{6}$$

$$\sum M = M_R + M_{fus} + M_{tp} + M_{prop} \tag{7}$$

2.1 The Rotor Model

The higher fidelity 'individual blade' model captures more detail features of the rotor loads through numerical calculations. In contrast, the lower fidelity 'rotor-disc' approach calculates the aerodynamic loads contributed by the rotor blade in the form of analytical integration over the whole span of the blade. The calculated loads are assumed to be the average for the entire rotor disc, without accounting the nonlinear blade aerodynamics. As such, the lift, drag and Mach number associated with the geometrical point of the blade element are not modelled. In this paper, only distinct features of an autogyro that are adapted from the HGS are highlighted.

Rotor Blade Kinematics. The force of the rotor subsystem is defined as the external force contributed by the rotor blade to the fuselage (c.g.) of the vehicle. This rotor force is also recognised as the local aerodynamic force exerted at the rotor blade in the form of lift and drag due to the relative motion of the blade during flight. These aerodynamic forces are calculated from relative translational velocities and accelerations obtained through kinematic translations from the body axes set.

Rotor Forces and Moments. The local velocity and acceleration at the blade obtained from the above kinematics are defined to be associated with the 'element of mass' at the blade. The blade force is then determined by considering the elemental aerodynamic and inertial forces acting on the 'element of mass' in the *blade* axes, which is written in the unit vector,

$$\boldsymbol{f}_{\mathrm{e}}^{\mathrm{bl}} = -\mathrm{m}_0 a_{\mathrm{x}}^{\mathrm{bl}} \boldsymbol{i}_{\mathrm{bl}} + \left(f_{\mathrm{y}}^{\mathrm{bl}} - \mathrm{m}_0 a_{\mathrm{y}}^{\mathrm{bl}}\right) \boldsymbol{j}_{\mathrm{bl}} + \left(f_{\mathrm{z}}^{\mathrm{bl}} - \mathrm{m}_0 a_{\mathrm{z}}^{\mathrm{bl}}\right) \boldsymbol{k}_{\mathrm{bl}} \tag{8}$$

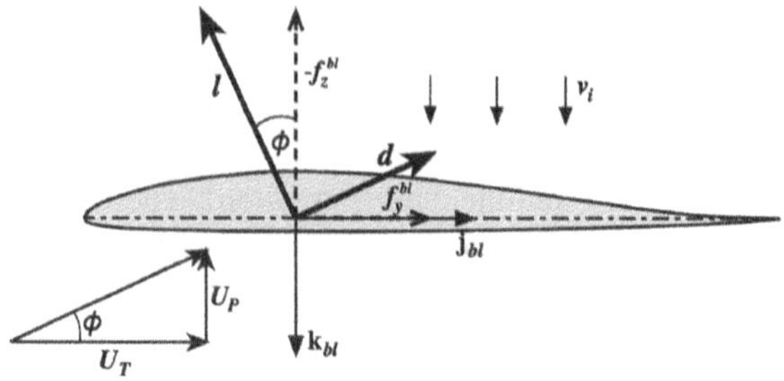

Fig. 3 Normal and tangential velocity components of the blade

The aerodynamic force components in Eq. (8) are defined as the function of *normal* and *tangential* air that flows through the blade, denoted as U_P and U_T in Fig. 3. The two airflows are also related to the local *lift* (l) and *drag* (d) of the blade.

In common understanding, $U_T \gg U_P$, thus the local incidence angle (ϕ) is small. Hence, the aerodynamic force components of the blade in Fig. 3 can be written as the function of lift and drag forces, approximated by:

$$f_{\mathrm{y}}^{\mathrm{bl}} \approx d - l\phi \tag{9}$$

$$f_{\mathrm{z}}^{\mathrm{bl}} \approx -l \tag{10}$$

The normal and tangential velocity airflows are then written by

$$\mathrm{U_P} = v_{\mathrm{z}}^{\mathrm{bl}} - v_i \tag{11}$$

$$\mathrm{U_T} = -v_{\mathrm{y}}^{\mathrm{bl}} \tag{12}$$

where $v_{\mathrm{z}}^{\mathrm{bl}}$ and $v_{\mathrm{y}}^{\mathrm{bl}}$ are velocity components obtained from the blade kinematics. Another velocity component, v_i is the induced airflow that passing through the rotor disc, given in the form of the azimuth angle (ψ) of the blade rotation and written as

$$v_i = v_0 + \frac{r_{\mathrm{b}}}{R}(v_{1\mathrm{s}} \sin\psi + v_{1\mathrm{c}} \cos\psi) \tag{13}$$

The rotor-disc inflow model employs the fundamental Glauert's inflow model which is still applicable for a typical autogyro operating in the lower speed region with relatively low rotor rpm [11]. In this inflow model, the airflow is assumed to be incompressible and steady. Adapting the inflow model from the HGS to an autogyro involved several changes where the most important one is the wake angle (χ_w) created by the induced airflow through the rotor disc. Figure 4a, b illustrates the different between helicopter and autogyro induced airflow model. A conventional helicopter normally flies with the induced air passes through the forward tilt rotor disc. Thus, the wake angle of a conventional helicopter is defined as $\chi_w < 90°$. Unlike helicopters,

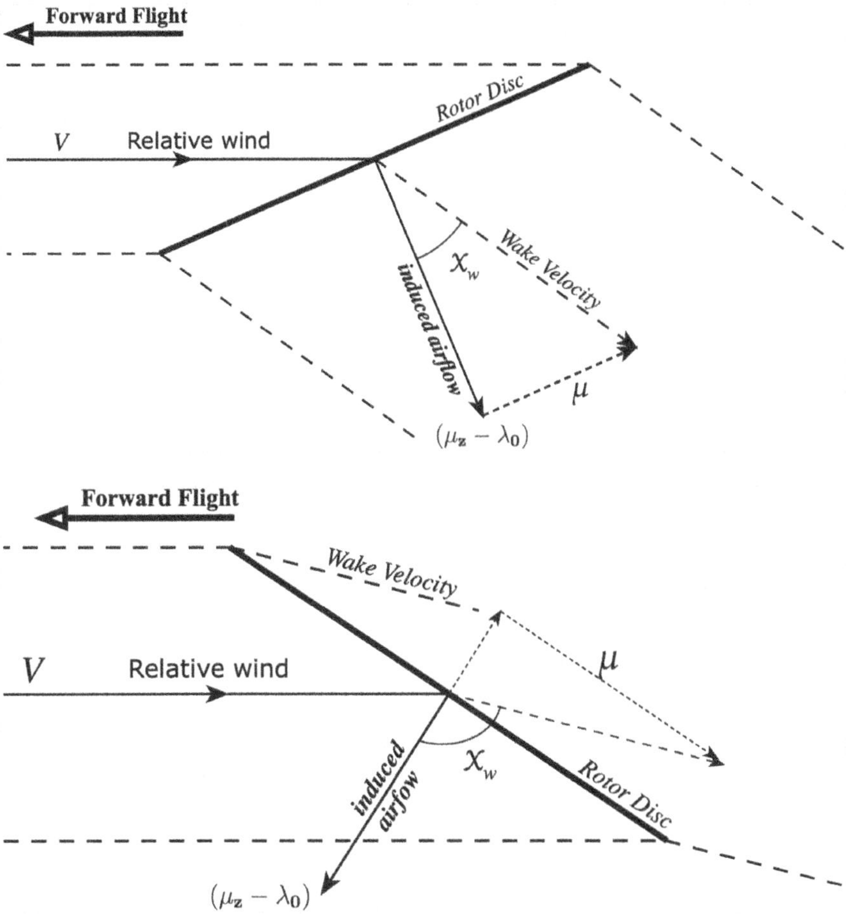

Fig. 4 Changes made on the induced airflow model

the wake angle of a typical autogyro is defined as $\chi_w > 90°$, which also known as the windmill state where the rotor disc is positively tilted (aft tilt) in forward flight.

The aerofoil of a rotor-disc model is expressed in a 2-dimensional thin aerofoil theory where the lift and drag coefficients are given in average values (C_l and C_d) for the entire blade, and written by

$$l = \frac{1}{2}\rho c\left(U_P^2 + U_T^2\right) \sim C_l \tag{14}$$

$$d = \frac{1}{2}\,\rho c\,U_T^2\,C_d \tag{15}$$

Furthermore, typical autogyros like the Montgomerie-Parsons used the untwisted blade ($\theta_{bl} = 0$) as mentioned in Table 1 with the lift-drag coefficients defined according to Schaefer and Smith [12]. Hence, the lift and drag coefficient is given by:

$$C_l = a_0\alpha_{\mathrm{bl}} + a_1 \tag{16}$$

while C_d is defined as the profile drag with a constant mean value δ. Hence, the total aerodynamic forces in the blade axes set is given by

$$\boldsymbol{F}_{\mathrm{A}}^{\mathrm{bl}} = \begin{bmatrix} X_{\mathrm{A}}^{\mathrm{bl}} \\ Y_{\mathrm{A}}^{\mathrm{bl}} \\ Z_{\mathrm{A}}^{\mathrm{bl}} \end{bmatrix} = \frac{1}{2}\rho c \begin{bmatrix} 0 \\ \int_0^{\mathrm{R}} \left(\delta U_{\mathrm{T}}^2 - a_1 U_{\mathrm{P}} U_{\mathrm{T}} - a_0 U_{\mathrm{P}}^2\right) dr_{\mathrm{b}} \\ -\int_0^{\mathrm{R}} (a_0 U_{\mathrm{P}} U_{\mathrm{T}} + a_1 U_{\mathrm{T}}^2) dr_{\mathrm{b}} \end{bmatrix} \tag{17}$$

Finally, the external forces contribution of the rotor to the vehicle's c.g. can be determined through reverse transformations from the blade axes in Eq. (17) to the body axes of the vehicle, that yields the rotor force contributions in the longitudinal mode as,

$$\boldsymbol{F}_{\mathrm{cg}}^{\mathrm{b}} = X_R^b \boldsymbol{i}_{\mathrm{b}} + Z_R^b \boldsymbol{k}_{\mathrm{b}} \tag{18}$$

The rotor *moments* acting on the rotor hub is defined according to the same 'element of mass' at a distance $\boldsymbol{r}_{e/bl}$ from the root of the blade. Hence, considering the $\boldsymbol{f}_{\mathrm{e}}^{\mathrm{bl}}$ in Eq. (8) the rotor hub moment is given by

$$\boldsymbol{M}_{\mathrm{hub}}^{\mathrm{bl}} = \int_0^{\mathrm{R}} \boldsymbol{r}_{e/bl} \times \boldsymbol{f}_{\mathrm{e}}^{\mathrm{bl}} dr_{\mathrm{b}} = \begin{bmatrix} 0 \\ K_\beta \beta \\ Q_{\mathrm{hub}}^{\mathrm{bl}} \end{bmatrix} \tag{19}$$

Similar to the rotor forces, the rotor moment contribution on the autogyro's c.g. is determined by the same reverse transformations from the blade to the body axes set of the vehicle, given in longitudinal form by

$$\boldsymbol{M}_{\mathrm{cg}}^{\mathrm{b}} = M_{\mathrm{R}} \boldsymbol{j}_{\mathrm{b}} \tag{20}$$

Blade Flapping. Flapping is described as the vertical movement of the blade with the axis of oscillation centred at the rotor hub. According to Eq. (19), the blade flapping angle β is limit by the hub stiffness K_β of the blade. This is only true for a conventional helicopter with an articulated rotor blade as featured in the HGS model. For a typical light autogyro, the blade is of a hingeless type where the two blades are rigidly connected and fixed to a teetering bolt. This so-called teetering blade produces a 'see-saw' effect where flapping at one side of the blade causes the opposite movement on the other side of the blade. Therefore, the $K_\beta \beta$ in the equation can be neglected, thus the flapping moment in eq. (19) becomes

$$\int_0^R (f_z^{bl} - m_0 a_z^{bl}) r_b dr_b = 0 \tag{21}$$

3 Rotor-Disc Model Validation

The new autogyro model is validated with the real flight data obtained from ref. [3]. The simulation results of the Montgomerie-Parsons rotor-disc model are compared with the test flight data of the same vehicle in trim conditions. Additionally, the trim results of an Individual Blade model of the same autogyro obtained from ref. [13] are also presented for comparison. Overall trim results demonstrate good agreement between the two models and the flight data, in terms of the trend.

The pitch attitude (Θ) and the longitudinal shaft tilt (θ_s) angles are seen reduced with the forward velocities. This behaviour is known, due to a few possible interrelated factors; the fuselage aerodynamics, the disc angle of attack, and the propeller-rotor thrust relative to the c.g. (T_{prop}–T_R). At lower flight regime, where the fuselage aerodynamics are less influential, the rotor-disc is aft tilted to compensate the weight of the vehicle, causing a larger fuselage pitch angle in maintaining the flight equilibrium. This behaviour, in nature, is predictable for a typical autogyro due to the induced airflow through the rotor disc in a 'windmill' state (refer back to Fig. 4).

Another plausible explanation is the propeller thrust line relative to the c.g. as illustrated in Fig. 5. This configuration is unique, only for a pusher-type autogyro like the Montgomerie, which also plays a major role in the stability and controllability of this vehicle. The propeller thrust line (T_{prop}) of the Montgomerie-Parsons autogyro is located about 1.5 inches above the vehicle's c.g., which contributes to a nose-down moment by default. At lower flight regime, the rotor disc thrust, $T_R(u, w)$ produces more positive pitch moment (aft tilt) to compensate the bigger nose-down moment created by the propeller thrust in maintaining an equilibrium flight. In fact, studies about the relationship between these two pitch moment contributors have been done and reported in various publications [14–17], as cited by Houston and Thomson in Ref. [6].

Moreover, from the trim results in Fig. 6, the good agreement in pitch attitude between the rotor-disc and the individual model is predictable. It is known that the higher fidelity individual blade model captures the higher frequency rotor dynamics. However, this feature is less applicable for an autogyro as the fuselage dynamics is much slower and the rotor blade coupling effect is less compared to helicopters. Thus, higher fidelity simulation in this context is unnecessary, as the complex blade dynamic calculations contribute less to rigid-body dynamics calculations in steady-state conditions.

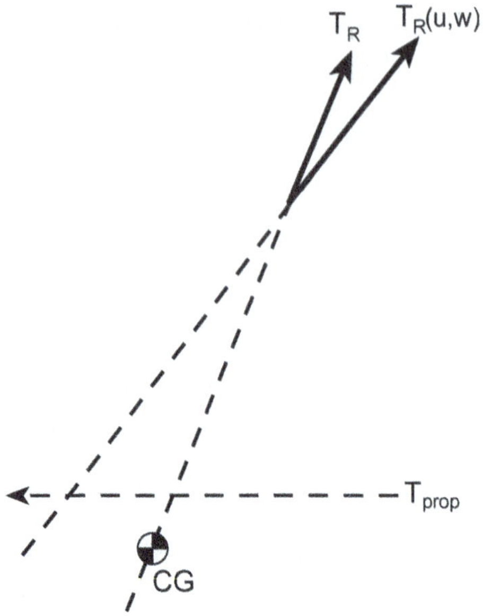

Fig. 5 Rotor and Propeller thrust line relative to the c.g

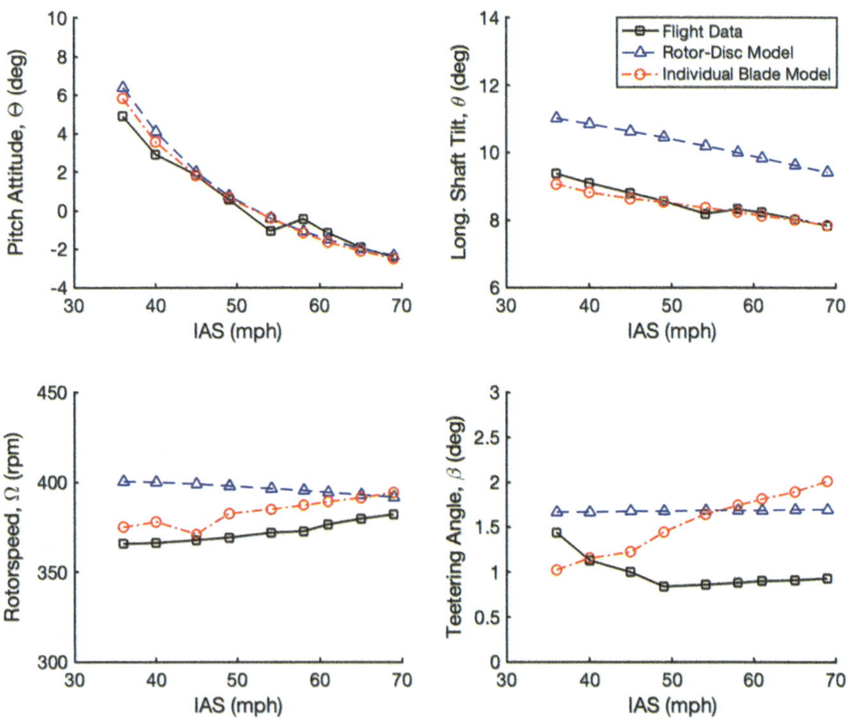

Fig. 6 Trim comparison with real flight data for validation

The rotorspeed result of the rotor-disc model shows disagreement with both; the flight data and the individual blade model. The result manifests an approximate 10% deviation in lower speed region but improves in higher speed region. In the rotor-disc model, rotor forces and moments are estimated as an average of the entire rotor-disc through analytical integrations. It is expected that there is a certain degree of uncertainties with the rotor-disc model, where detail aerodynamic parameters such as the lift-drag and Mach numbers at a different angle of attacks are not modelled. Therefore, more uniform patterns of almost constant rotorspeed are observed for the entire speed range. In contrast, the rotorspeed result of the individual blade model shows better agreement with the flight data, which is expected for the higher fidelity model simulations.

4 Conclusion

The rotor-disc simulation package that was extensively used in helicopter simulations has been successfully implemented in a light autogyro mathematical model. The kinematics of the rotor model is modified to ensemble more complex autogyro translations. The inflow model that represents an autorotating rotor disc of an autogyro is also implemented successfully in the new autogyro model. For the flapping dynamics model, the unique teetering effect of an autogyro rotor blade has also been included, which entirely different from the original HGS flapping dynamics. Finally, the validation of the new rotor-disc model is found adequate versus the flight data. However, the usage of the rotor-disc model is observed to be limited to simulations in normal flight conditions due to the uncertainties that are not captured in the rotor-disc model. As such, any simulation which involved extreme and vigorous flight manoeuvre, the higher fidelity individual blade model must be used. Therefore, for studies that focused on the dynamic behaviour of the rigid body of the vehicle, the rotor-disc model is good enough to be employed. In reality, the simulation run-time and the hardware complexity are significantly reduced with the rotor-disc model in place.

References

1. Leishman JG (2006) Principles of helicopter aerodynamics. Cambridge University Press
2. Anon (2013) CAP 643: British civil airworthiness requirements section T—light gyroplanes. 5
3. Anon (2010) The aerodynamics of gyroplanes
4. Anon (2002) CAP 735 aviation safety review 1992–2001
5. Anon (2008) CAP 780 aviation safety review 2008
6. Houston S, Thomson D (2017) On the modelling of gyroplane flight dynamics. Prog Aerosp Sci 88:43–58. https://doi.org/10.1016/j.paerosci.2016.11.001
7. Thomson DG (1992) Development of a generic helicopter model for application to inverse simulation. Glasgow

8. Padfield GD (2008) Helicopter flight dynamics (Google eBook). Wiley
9. Ferguson K, Thomson D (2015) Maneuverability assessment of a compound helicopter configuration. J Am Helicopter Soc 61:1–15. https://doi.org/10.4050/jahs.61.012008
10. Ferguson K, Thomson D (2015) Flight dynamics investigation of compound helicopter configurations. J Aircr 52:156–167. https://doi.org/10.2514/1.C032657
11. Glauert H (1926) A general theory of the autogyro
12. Schaefer RF, Smith HA (1998) Aerodynamic characteristics of the NACA 8-H-12 airfoil section at six reynolds numbers. Langley, Langley
13. Bradley R, Padfield GD, Murray-Smith DJ, Thomson DG (1990) Validation of helicopter mathematical models. Trans Inst Meas Control 12:186–196. https://doi.org/10.1177/014233129001200405
14. Houston SS (1994) Validation of a nonlinear individual blade rotorcraft flight dynamics model using a perturbation method. Aeronaut J 98:260–266
15. Houston SS (1996) Longitudinal stability of gyroplanes. Aeronaut J 100:1–6
16. Houston S, Thomson D (1995) A study of gyroplane flight dynamics. In: 21st European Rotorcraft Forum. ERF1995, St. Petersburg, Russia
17. Spathopoulos VM (2004) Validation of a rotorcraft mathematical model in autorotation by use of gyroplane flight tests. Aeronaut J 108:51–58

Adsorption Characteristic of a Two Layer Hollow Cylindrical Silica Bed

Pravinth Balthazar, Mohd Azmi Ismail, Lee Chern khai, Muhammad Iftishah Ramdan, Nurul Musfirah Mazlan, and Hussin Bin Mamat

Abstract Desiccant air conditioning systems are using the concepts of desiccant dehumidification and evaporative cooling. Apart from desiccant material, the flow characteristics and inlet conditions also play vital role in improving the dehumidification performance of a desiccant bed. This paper communicates the experimental adsorption characteristics of two layer hollow cylinder silica bed that is used in desiccant-cooling systems for warm and very humid climates, during adsorption processes. Air velocity varied between 1 and 5 m/s and experimental investigation concluded that moisture adsorption ability is increased respect to air velocity for hollow desiccant arrangement. 3.7 m/s air velocity illustrated 2.2 times higher dehumidification rate than 1 m/s.

Keywords Silica bed · Adsorption · Hollow cylinder · Air velocity

1 Introduction

The dehumidification processes of air has diverse application in industrial arena. Recently, alternative to the conventional air-conditioning system have gained attention worldwide, known as Hybrid air conditioning systems (combination of

P. Balthazar · M. A. Ismail (✉) · L. C. khai · M. I. Ramdan
School of Mechanical Engineering, Universiti Sains Malaysia, 14300 Nibong Tebal, Pulau Pinang, Malaysia
e-mail: azmi_meche@usm.my

P. Balthazar
Department of Mechanical Engineering, South Eastern University of Sri Lanka, 32360 Oluvil, Eastern Province, Sri Lanka

N. M. Mazlan · H. B. Mamat
School of Aerospace Engineering, Universiti Sains Malaysia, 14300 Nibong Tebal, Pulau Pinang, Malaysia

P. Rajendran et al. (eds.), *Proceedings of International Conference of Aerospace and Mechanical Engineering 2019*, Lecture Notes in Mechanical Engineering,
https://doi.org/10.1007/978-981-15-4756-0_43

desiccant dehumidification and air conditioning system). Silica gel, activated alumina and molecular sieve are few solid desiccants that are generally used in the purification and dehumidification process. Desiccant material can be regenerate by using lesser grade regeneration heat source such as solar thermal energy [1–4] or waste heat from a cogeneration of other sources [5]. Silica gel can be most suitable example which makes it useful for use with solar energy [6] and a research study on dehumidification illustrated that silica gel transfers about 30% more water per unit dry mass than activated alumina [7].

Kim et al. [8] conducted a study named low compact pressure drop desiccant bed for solar air conditioning applications by using silica gel under laminar flow and demonstrated the advantages of using laminar packing. In 2002, Hamed [9] used burned clay to study the adsorption characteristics of vertical packed porous bed on transient flow. His investigation results showed that the potential ratio exponentially decreases with the increase in dimensionless time. After that, Hamed et al. [10] extended their study on the transient adsorption characteristics of solid desiccant in a vertical fluidized bed. Later, in 2008 Kabeel [11] expanded Hamed et al. [9, 10] researches in adsorption operations of multilayer desiccant packed bed for dehumidification. The experimental study illustrated that silica bed divided into equally eight layers. Depending on the inlet conditions; comparison between the first layer to the eighth layer varied about 200–400% adsorbed water and the adsorption and desorption rates decreased from the first layer to the last layer. Awad et al. [12] have performed an analysis on radial flow using silica gel. Their theoretical and experimental study used hollow cylindrical bed and concluded that 7.2 diameter ratio hollow cylindrical bed showed the highest efficient operation compared to the other four test samples.

The present experimental investigations' objective is to evaluate the effect of the two layer hollow cylindrical packed bed behavior for adsorption process under axial flow. The experimental tests were performed at various operating conditions including air velocity, relative humidity and temperature, in order to investigate the performance of hollow cylindrical desiccant bed.

2 Experimental Apparatus

HVAC experimental rig that available in School of Mechanical Engineering, Universiti Sains Malaysia has been modified in order to investigate desiccant dehumidification performance as illustrated in Fig. 1.

Hollow Cylindrical desiccant bed was used to study the dehumidification ability of the silica gel as solid desiccant adsorbent. The silica gel have average grain diameter between 3 and 4 mm. Two layer desiccant hollow cylindrical bed was fabricated to study the ability of the adsorption of the silica gel under axial flow conditions. The total surface area of the silica gel bed for the systems is 213.63 cm^2. The length of the dehumidifier system is 30 cm as shown in Fig. 2. Diameter of experimental pipe is 10.5 cm and out of two layers, one of the layer had been arranged as 2 cm thickness

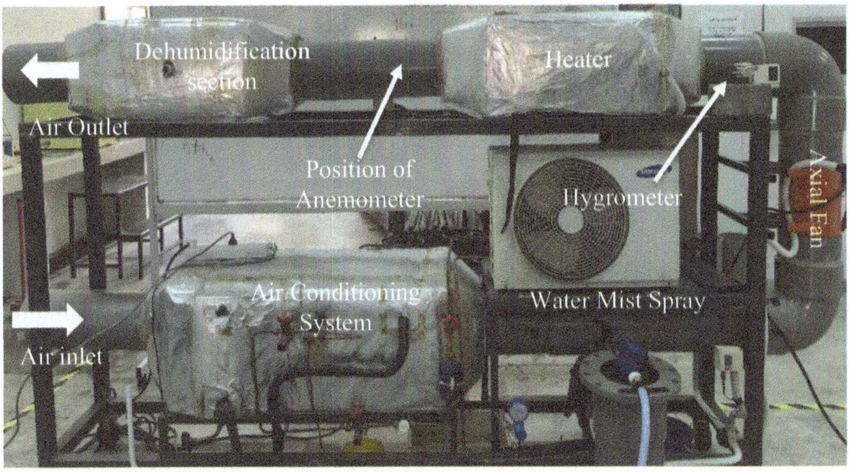

Fig. 1 Experimental rig setup

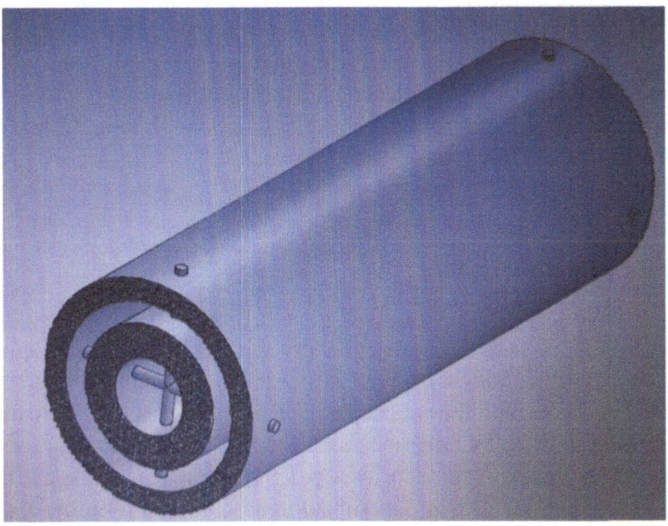

Fig. 2 Two layer desiccant bed

(10.5 and 8.5 cm) from the outer perimeter. Inner layer diameter is 6.5 and 3.5 cm and total weight of dry Silica gel used in experiment is 5 kg.

The experimental setup has four different compartments. Air conditioning compartment were built to conduct future experiments under cool climate condition. Water mist spray was included into the experimental unit to control inlet relative humidity (RH) with respect to air velocity in order to reduce the fluctuation of RH respect to atmospheric condition, and to conduct the experiments in reasonably similar inlet conditions. Heating compartment arrangement was developed for regeneration analysis of the system. Finally, dehumidification compartment was developed for adsorption analysis of the desiccant system which will be investigated in the present studies. The flow meter and hygrometer were used to measure the air velocity, and air relative humidity and it's temperature, respectively. The desiccant bed was installed inside the test section. The dry bulb and wet bulb thermocouples used to measure temperature downstream of desiccant or test section. The reading of RH, dry and wet bulb temperatures at both air inlet and outlet taken for every 10 s interval using data loggers. Humidified air was supplied throughout the dehumidifier system for around two hours. Then, at the exit, four dry bulb and wet bulb thermocouples' average temperature were taken to determine humidity ratio. The difference between inlet and exit was the adsorbed water vapour in that moment. Using that trend average adsorbed capacity, total water vapour adsorbed can be calculated.

Later on, the desiccant bed was dried by heated air (temperature at 90 °C) flow pass through it around two hours before the adsorption process begin for next air velocity. This precaution taken to ensure that any moisture contained in the silica gel inside the test bed were removed and maintained as bench mark for throughout all the air velocities between 1 and 5 m/s. All adsorption experiment was carried out during morning between 8:30 am and 12:00 pm in order to reduce the uncertainty in atmospheric conditions. Average inlet condition (dry bulb and RH) for 1 m/s, 1.5 m/s, 3.7 m/s and 5 m/s air velocity was respectively 28.8 °C & 92.4%, 27.4 °C & 90.9%, 27.8 °C & 88.7% and 27.5 °C & 86.8%.

3 Results and Discussion

Figure 3 illustrates the experimental data of dry and wet bulb thermocouples reading placed at the exit of the hollow cylinder desiccant bed for 5 m/s air velocity. There were four dry and wet bulb thermocouples positioned 90° apart and readings were taken. The graph showing that, thermocouples took around 15 min to produce stable reading even though data logger set to average temperature readings each 10 s interval and plot it every minute. Similar pattern and trend observed for other air velocities as well throughout the experiment.

Figure 3 describes the outlet dry bulb temperature always decreasing throughout the time. This phenomenon happen due to any desiccant material adsorb the moisture, then it releases the thermal energy to the atmosphere. Desiccant material have a lot of porous to fill at the start of the experiment so higher adsorption rate at present; will release more energy, however adsorption rate decreased throughout the time made dry bulb temperature to drop.

Inlet and outlet humidity for 1.5 m/s air velocity is demonstrated in the Fig. 4. The unit g/kg in y axis can be illustrated as amount of moisture in one kilo gram of dry air. Inlet air humidity is plotted by measuring inlet dry bulb and relative humidity of air. Atmospheric condition has changed throughout the experiment forced to

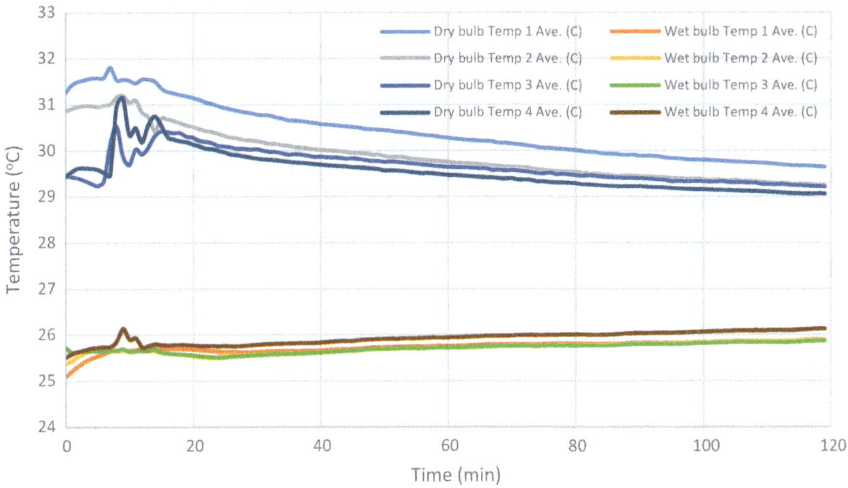

Fig. 3 Experimental data for 5 m/s air velocity

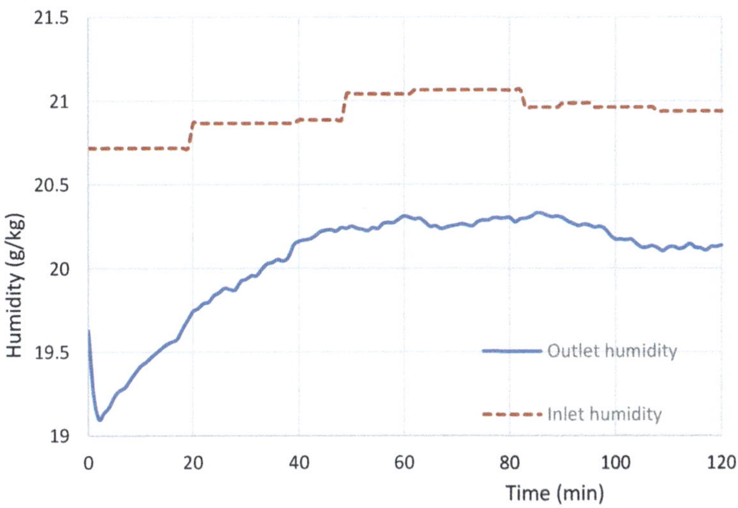

Fig. 4 Inlet and Outlet humidity for air velocity 1.5 m/s

fluctuate inlet humidity even after humidified by water mist sprayer. This can be spotted in the graph after 90 min of the experiment. Desiccant material saturated can be identified if the outlet humidity is almost equal to inlet humidity. However, from 0 to 120 min, the two layers silica gel desiccant hollow cylindrical bed isn't saturated. The total 4.5% humidity increase obtained comparing at the start and end of the outlet air. However, between 80 and 90 min, the time interval region showed 5.4% higher exit air humidity increase.

Figure 5a, b shows the amount of adsorbed humidity against time for 1 m/s and 3.7 m/s air velocity, respectively. Initially, the high temperature of air leaving the desiccant two layer hollow cylindrical bed were observed for both cases. Obviously, this is vital at the beginning of the adsorption process as silica gel is dry and, therefore, holds higher adsorption ability. Similar adsorption characteristics are observed in the both figures. However, for case 1 m/s air velocity, the adsorbed humidity is not reached the stable value. It showed 68.7% higher at the starting comparing to end of 120 min. Similarly, air velocity show 96% higher adsorption per flowing one kilogram of air.

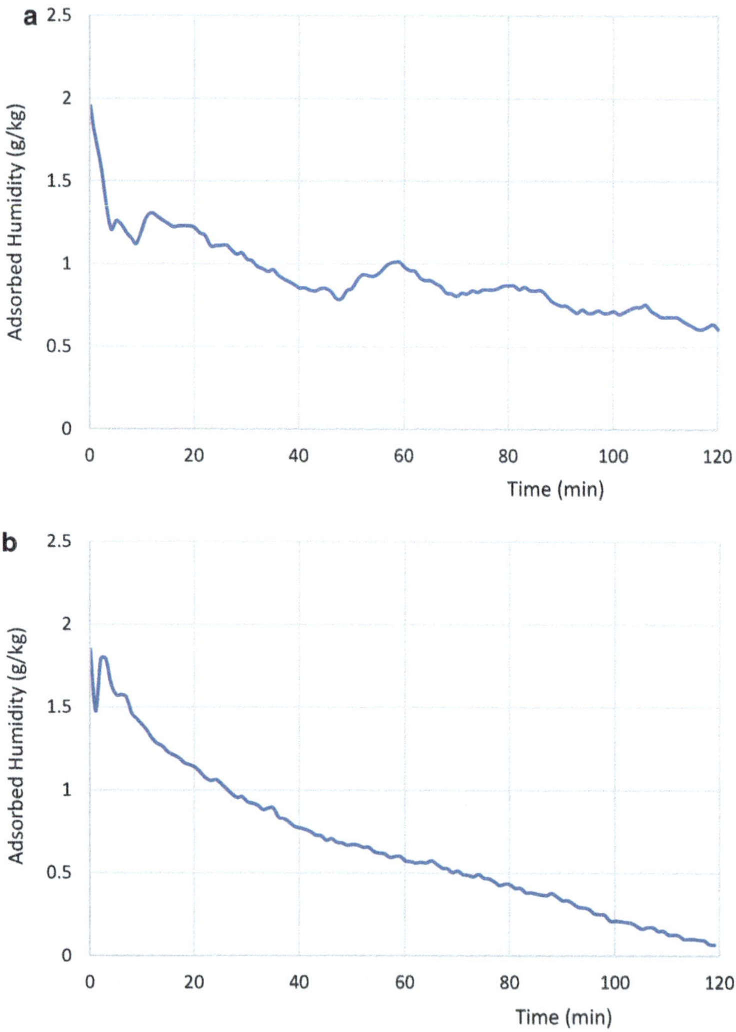

Fig. 5 **a** Adsorbed humidity for air velocity 1 m/s. **b** Adsorbed humidity for air velocity 3.7 m/s

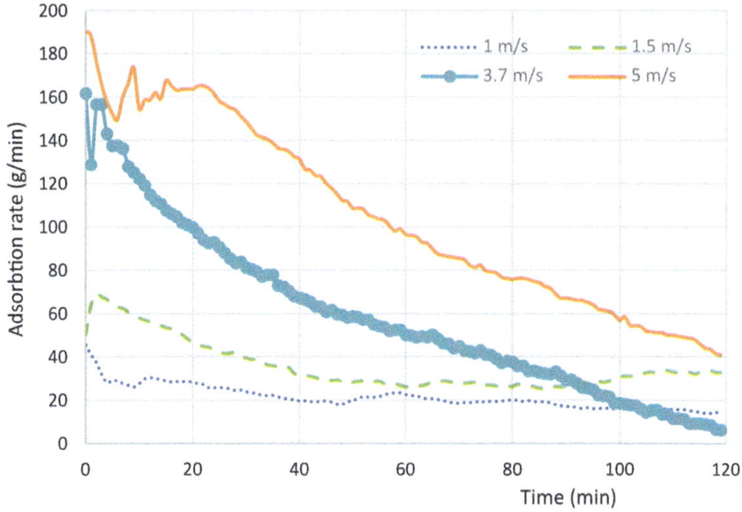

Fig. 6 Comparison of total adsorption rate respect to air velocities

Figure 6 illustrates the influence of the air velocity on the adsorption rate at two layer of the hollow cylindrical bed for nearly 120 min time period. Total weight of desiccant is 5 kg and figure shows the total adsorption of the desiccant bed. Figure 6 displays the adsorption rate increases respect to the increase of velocity throughout the experiment. Also, it shows that the amount of water absorbed at the begining of the experiment decreased as the air velocity decrease. Air velocities of 3.7 m/s, 1.5 m/s and 1 m/s show 15.2, 63.5 and 76% reduction of moisture adsorption, respectively at 5 m/s air velocity.

After 60 min this adsorption rate is about 23 g/min for 1 m/s air velocity, while it is around 50 g/min and 97 g/min for 3.7 m/s and 5 m/s air velocity, respectively. With the increase of time, the difference of the rate of adsorption of layers further decreases for 1.5 and 3.7 m/s. However, air velocity 1 m/s maintained at the same percentage.

4 Conclusions

The present research investigation conclude that;

- Inlet humidity ratio and exit air temperature decreased and exit humidity increased throughout the time.
- Adsorbed humidity is decreasing with the time for all air velocities.
- Air velocity 1 m/s showed 76% lower adsorption rate than 5 m/s.
- Water vapour adsorption rate proportional to air velocity.

Acknowledgments Grant code 304/PMEKANIK/6315150 by Universiti Sains Malaysia, and grant code UGC/VC/DRIC/PG2017 (II)/SEUSL/01 by South Eastern University of Sri Lanka were used to fund this research.

References

1. Angrisani G, Capozzoli A, Minichiello F, Roselli C, Sasso M (2011) Desiccant wheel regenerated by thermal energy from a microcogenerator: experimental assessment of the performances. Appl Energy 88(4):1354–1365
2. Calise F, d'Accadia MD, Roselli C, Sasso M, Tariello F (2014) Desiccant-based AHU interacting with a CPVT collector: Simulation of energy and environmental performance. Sol Energy 103:574–594
3. Wang X, Cai W, Lu J, Sun Y, Ding X (2013) A hybrid dehumidifier model for real-time performance monitoring, control and optimization in liquid desiccant dehumidification system. Appl Energy 111:449–455
4. Qi R, Lu L, Yang H, Qin F (2013) Investigation on wetted area and film thickness for falling film liquid desiccant regeneration system. Appl Energy 112:93–101
5. Angrisani G, Roselli C, Sasso M, Tariello F (2014) Dynamic performance assessment of a micro-tri generation system with a desiccant-based air handling unit in Southern Italy climatic conditions. Energy Convers Manage 80:188–201
6. Development of Solar Desiccant Humidifier (1978) Research Manufacturing Company of California, Technical progress report, No. 87-14957-1
7. Dupont M, Celestine B, Nguyen PH, Merigoux J, Brandon B (1994) Desiccant solar air conditioning in tropical climates: I-dynamic experimental and numerical studies of silica gel and activated alumina. Sol Energy 52(6):509–517
8. Kim S, Biswas P, Mills AF (1985) A compact low-pressure drop desiccant bed for solar air conditioning applications-2: bench scale tests. J SolEnergy Eng 107(2):120–127
9. Hamed AM (2002) Theoretical and experimental study on the transient adsorption characteristics of a vertical packed porous bed. Renew Energy 27(4):525–541
10. Hamed AM, Walaa R, El-Emam SH (2005) Study of the transient adsorption/desorption characteristics of solid desiccant particles in fluidized bed. In: The third Minia international conference for advanced trends in engineering, Minia University, Egypt
11. Kabeel AE (2009) Adsorption–desorption operations of multilayer desiccant packed bed for dehumidification applications. Renew Energy 34(1):255–265
12. Awad MM, Hamed AM, Bekheit MM (2008) Theoretical and experimental investigation on the radial flow desiccant dehumidification bed. Appl Therm Eng 28(1):75–85

The Effect of Flow Control on Wing-In-Ground Craft Hull-Fuselage for Improved Aerodynamics Performance

Irahasira Said, Mohd Rosdzimin Abdul Rahman, Azam Che Idris, Fadhilah Mohd Sakri, and Mohd Rashdan Saad

Abstract Wing-In-Ground (WIG) crafts has become one of the latest technologies in the marine vehicle sector. Improvements on WIG craft to increase aerodynamic performance has been done for decades. However, WIG crafts must be able to overcome significant hydrodynamic drag take-off from water surface and aerodynamic drag during flight. The current design of the hull-fuselage can cause flow separation and increase in drag hence flow control devices can be used to improve flow separation phenomenon on the hull-fuselage of the WIG craft. The objective of this study is to investigate the effect of flow control on the aerodynamics performance of WIG craft through wind tunnel experiments. From the results, C_D values were reduced with the presence of flow control on WIG hull-fuselage. Different configurations on height, angle, and spacing of the flow control device helped in obtaining minimum C_D. The flow control device was proven to reduce CD for up to 25% from the baseline model. This indicates that flow control device is helpful in improving aerodynamics performance of WIG craft.

Keywords Flow control · Wing-In-Ground craft · Wind tunnel · 3D-Printing

1 Introduction

WIG craft is the latest concept of designing fast moving marine craft, which has huge potential in numerous areas such as rescue operations, tourism, transportation, cargo, and military functions which are the alternate solution to gain higher speed [1]. WIG craft is the innovation of craft which can fly near ground using the positive ground effect. Qu et al. [2] in their study has stated that a WIG craft has

I. Said · M. R. Abdul Rahman · A. Che Idris · M. R. Saad (✉)
Department of Mechanical Engineering, Faculty of Engineering, Universiti Pertahanan Nasional Malaysia, Kem Perdana Sungai Besi, 57000 Kuala Lumpur, Malaysia
e-mail: rashdan@upnm.edu.my

F. Mohd Sakri
Aerospace Section, Universiti Kuala Lumpur, Malaysian Institute of Aviation Technology, Dengkil, Selangor 43800, Malaysia

P. Rajendran et al. (eds.), *Proceedings of International Conference of Aerospace and Mechanical Engineering 2019*, Lecture Notes in Mechanical Engineering,
https://doi.org/10.1007/978-981-15-4756-0_44

higher speed, lower drag, and lower fuel consumption, and its cruising speed is less affected by the sea states compared to a ship. However, hull-fuselage of WIG can cause so much drag during flight, most especially on its backward facing step (BFS). BFS creates a shear layer over the step recirculation regions near to the wall and a reattachment area downstream [3]. The sharp geometry at the step edge causes flow separation at this location, whereas at a more distance downstream, the flow reattaches, and a new boundary layer creates. Recently, Hudy and Naguib [4] investigated this "wake mode" by analyzing temporally the growth of large-scale coherent structures at approximately half the reattachment length where maximum reverse flow occurred. This is consistent with the theoretical study by Huerre and Monkewitz [5] and the numerical simulation by Wee et al. [6] showing that the absolute instability takes place in the middle of the recirculation region and disturbances spread the impact on the whole shear layer leading to self-sustained oscillations of the shear layer. Reduction in fuselage drag is possible and has been demonstrated by the application of active flow control as proven by Martin et al. [7] and Ben et al. [8]. Hazen [9] in his study found the technique that has been produced to control the boundary layer, either to maximizing the lift or diminishing the drag and separation delay are grouped under the general heading of flow control. One of the solutions to reduce drag during flight is by using flow control. Saad et al. [10] has conducted studies on micro-ramps for hypersonic flow control and the results show reduction in the upstream interaction length by delaying the pressure rise, hence suppressing the flow separation. Passive flow control techniques were also employed in numerous investigations [11–14] to assist in drag reduction. Godard and Stanislas [15] in their study state that passive flow control is less expensive to use widely and has high potential in drag reduction studies [16, 17]. Hence in this investigation, passive flow control device will be employed on the hull-fuselage of WIG, specific at the Backward Facing Step (BFS) area, and its effect towards drag reduction will be studied.

2 Methodology

In this study, SOLIDWORKS is the software used to produce CAD drawings of WIG hull-fuselage as shown in Figs. 1 and 2. The length of the fuselage is 0.33 m, width is 0.07 m and height is 0.09 m. Different configurations of the flow control device need to be tested in this study. Therefore, to minimize fabrication cost, a slot had been drawn and extrude cut from the fuselage. This part was then installed with MVG of different configurations.

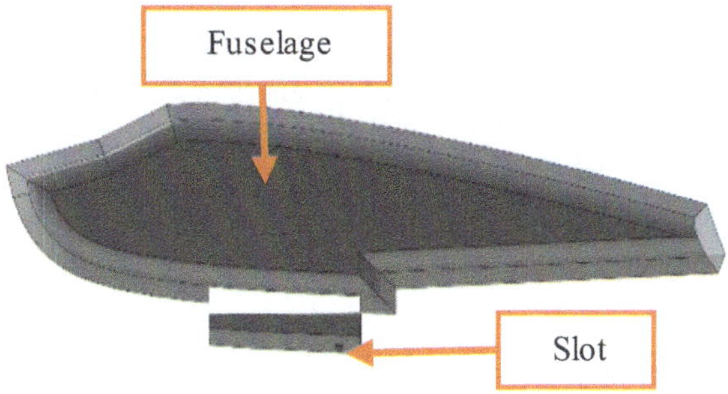

Fig. 1 Isometric assembly view of WIG Craft hull-fuselage with the slot

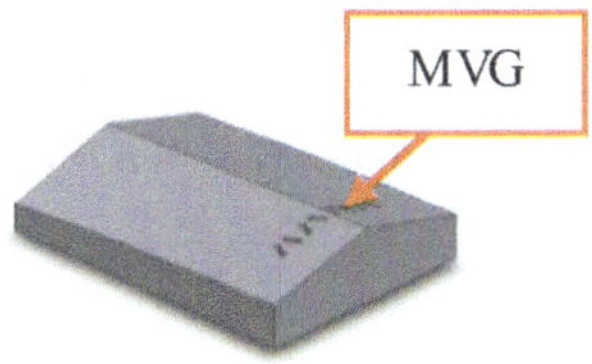

Fig. 2 Isometric view of the slot

2.1 Hull-Fuselage Fabrication

3D Printing (PLA filament material) was used to fabricate a small-scale 3D model of WIG hull-fuselage as shown in Figs. 3 and 4. During the printing process, supports were produced to stabilize the model printing process. After removal of the supports, the surface of the model becomes uneven. The roughness of the surface of the model needed to be cleaned to avoid any surface roughness effects to the flow during the

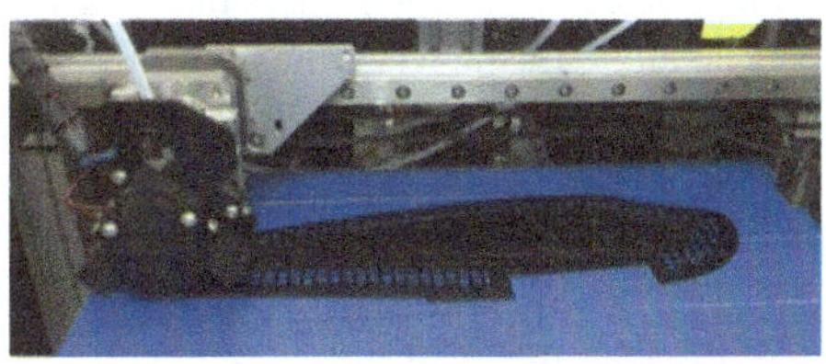

Fig. 3 Printing process of hull-fuselage

Fig. 4 Printing process of slots

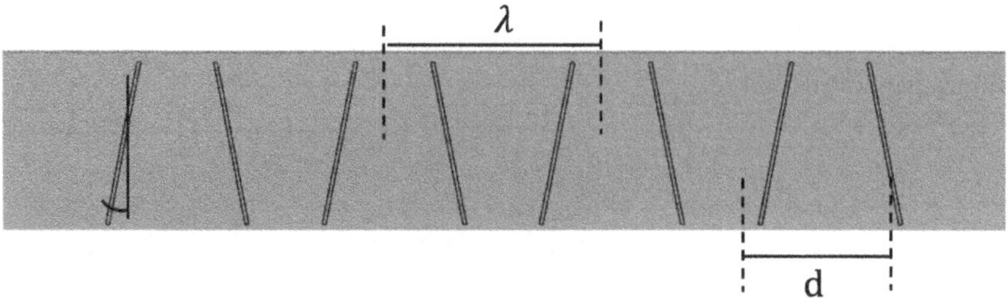

Fig. 5 The arrangement of counter-rotating micro vanes

wind tunnel test. The disturbance in the flow will cause turbulence and will affect the drag force of the WIG Craft hull-fuselage 3D model [18]. The sanding process were done to make sure that the surface of the model was smooth using different grades of sandpapers.

2.2 *Micro Vortex Generator (MVG)*

The type of MVG used in this study is the vane type. The dimensions of the MVG used in this study are based on the previous study on flow separation control with vortex generators by Gardarin et al. [19]. Combination of different heights, angles and spacing of the MVG were tested in this study. In this study, 0.4δ, 0.6δ and 0.8δ of heights were chosen for testing purpose as these heights can obtain the least drag coefficient due to the sub-boundary layer height. δ is the boundary layer thickness of the model at the interaction location and was estimated from CFD studies. The arrangement of the vortex generators is shown in Fig. 5.

The angle of the MVG has a significant effect on the results of drag coefficient. To obtain the best angle of MVG for WIG craft, three different angles were chosen to be installed in the fuselage and tested. The three different angles were 10°, 16° and 23° based on a study conducted by Lin [20] on low-profile vortex generators to control boundary-layer separation. Small spacing between the pairs were chosen to give the best result of drag coefficient. The values of spacing were 3.1δ, 3.6δ and 4.1δ.

2.3 *Wind Tunnel*

Subsonic wind tunnel Longwin LW-9300R located in Wind Tunnel Laboratory, Faculty of Engineering UPNM was used to run the experiments. The range of the operational speed of the open-loop suction type wind tunnel is from 0 to 105 m/s. The dimensions of the test section of the subsonic wind tunnel are 0.3 m wide, 0.3 m

Fig. 6 Subsonic wind tunnel

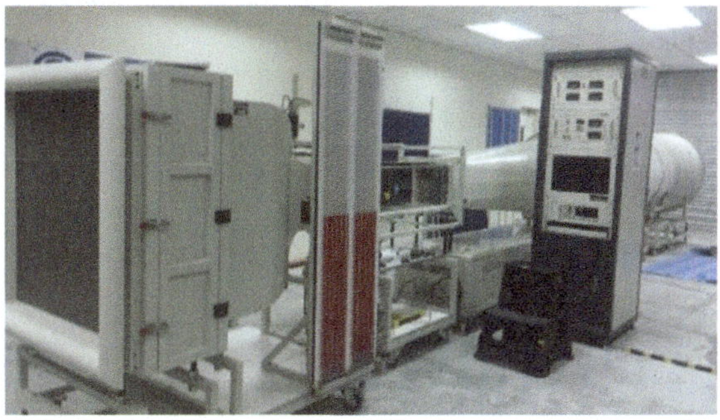

Fig. 7 Control system and test section of wind tunnel

height and 1.0 m in length. The wind tunnel is equipped with 3-component force balance to measure the drag force, F_D and the lift force, F_L. The experiments were conducted in different speeds $u = 5$, 10, 15, 20, 25, and 30 m/s. Figures 6 and 7. Show the wind tunnel system utilised in this work.

3 Results and Discussion

3.1 Ground Effect with MVG

Table 1 shows the data of the drag coefficient, C_D for the model with and without the ground plate. Figure 8 shows the graph of drag coefficient versus speed for baseline experiments (no flow control). It shows that the drag coefficient for the experiment with ground effect is relatively higher compared to the non-presence of ground effect. Even though it is established that the present of the ground effect is good for lift, however the same case does not apply towards drag. In order to reduce drag, flow control is essential. Table 2 then shows the results from an experiment with ground effect for baseline (no flow control) and with MVG. While Fig. 9 is the graph plotted based on data in Table 2. C_D values of the hull-fuselage with MVG were found to be

Table 1 Data of drag coefficient, C_D or ground test

Speed (m/s)	Baseline	
	Without ground	With ground
5	0.1694	0.1694
10	0.1412	0.1970
15	0.1443	0.1883
20	0.1483	0.1871
25	0.1468	0.1853
30	0.1475	0.1914

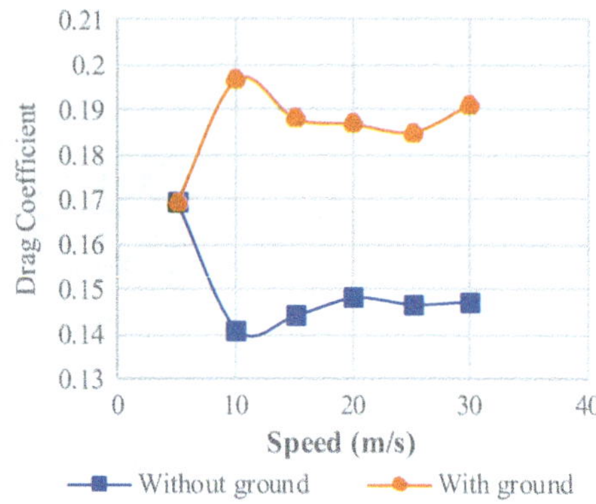

Fig. 8 C_D versus speed (ground test)

Table 2 Data of drag coefficient, C_D for ground test

Speed (m/s)	With ground	
	Baseline	MVG
5	0.1694	0.1694
10	0.1970	0.1694
15	0.1883	0.1632
20	0.1871	0.1659
25	0.1853	0.1672
30	0.1914	0.1679

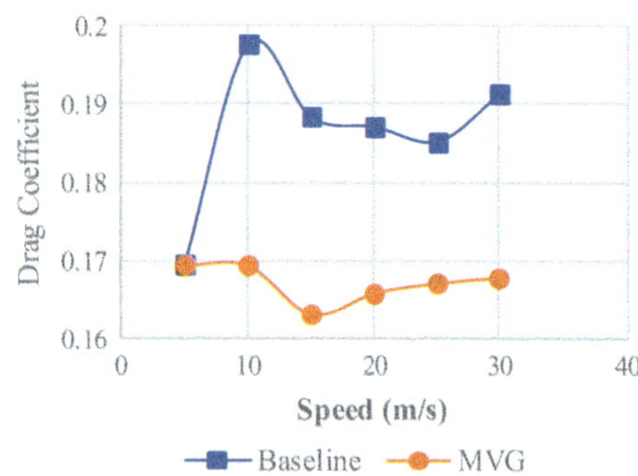

Fig. 9 C_D versus speed (flow control test)

significantly lower than the baseline condition. This shows that the presence of the MVG were able to reduce the C_D of the hull-fuselage in the case of ground effect.

3.2 *MVG Configuration*

From Fig. 10, it shows the trend of C_D for each MVG height is almost similar as the baseline. The experiment deduced that 0.8δ is the best height for MVG application on WIG hull-fuselage as it shows the minimum C_D. The best height of MVG was then used in the next stage of experiments in order to determine the best angle for MVG. Table 3 shows percentages difference between the C_D of the baseline and 0.8δ. From the table, the highest percentage recorded was 7.14% at 10 m/s.

Figure 11 represent the data of the drag coefficient versus velocity for different angle configurations. Both 10° and 23° angles showed improvements in drag coefficient, although 16° angle has the minimum drag coefficient. The graph in Fig. 11 indicates that 16° angle is the best angle that can be applied on WIG hull-fuselage. The best angle and best height of MVG were then used in the final stage of this study.

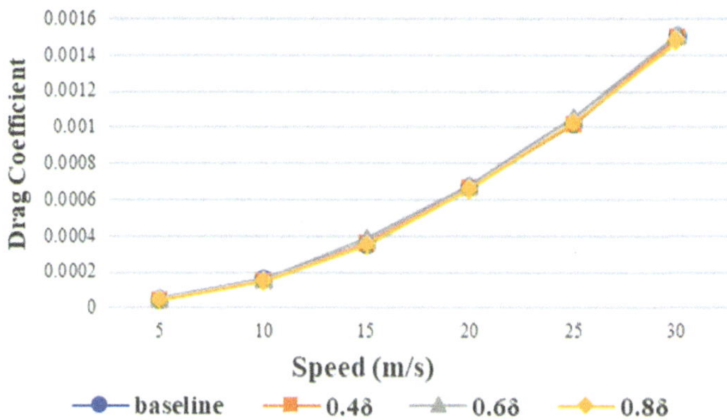

Fig. 10 CD versus speed (height)

Table 3 Percentage different of C_D based on height configuration of MVG

Velocity (m/s)	Drag coefficient, C_D		Percentage different (%)
	Baseline	0.8δ	
5	0.00046	0.00040	0.00
10	0.00016	0.00015	7.14
15	0.00035	0.00035	0.00
20	0.00067	0.00066	1.75
25	0.00102	0.00103	−1.15
30	0.00150	0.00148	1.56

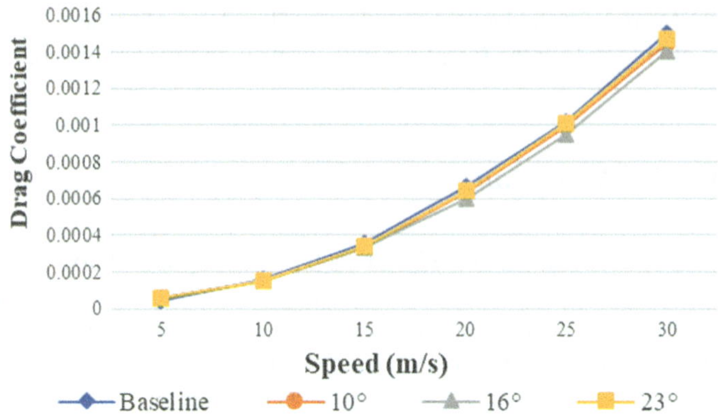

Fig. 11 C_D versus speed (angle)

Table 4 shows percentages of difference between baseline and the best angle. From the table, the highest percentage difference is 10.53% at 20 m/s.

Figure 12 shows the graph of drag coefficient versus velocity for different spacing configurations. The graph shows that drag coefficient increase as the speed increases. Spacing of 3.06δ, 3.6δ and 4.14δ shows the decrease in drag coefficient, with 3.06δ

Table 4 Percentage different of C_D based on angle configuration of MVG

Velocity (m/s)	Drag coefficient, C_D		Percentage different (%)
	Baseline	16°	
5	0.00046	0.00059	−25.00
10	0.00016	0.00015	7.14
15	0.00035	0.00032	6.66
20	0.00067	0.00059	10.53
25	0.00102	0.00098	6.90
30	0.00150	0.0014	7.03

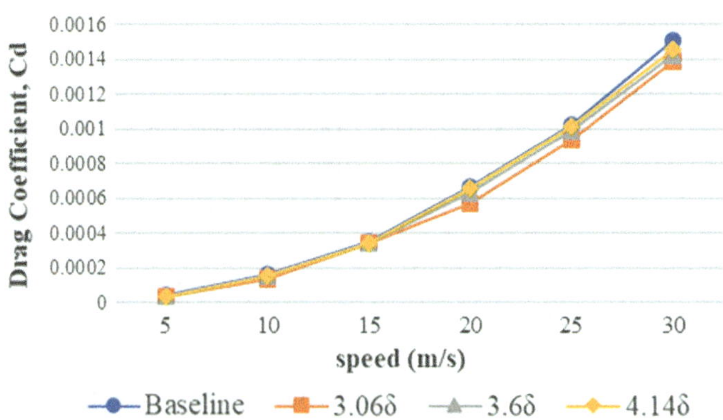

Fig. 12 C_D versus speed (spacing)

Table 5 Percentage different of C_D based on spacing configuration of MVG

Velocity (m/s)	Drag coefficient, C_D		Percentage different (%)
	Baseline	3.06δ	
5	0.00046	0.00035	25.00
10	0.00016	0.00014	14.29
15	0.00035	0.00034	3.33
20	0.00067	0.00058	14.04
25	0.00102	0.00094	8.05
30	0.00150	0.00139	7.81

producing the minimum drag coefficient. The graph indicates that spacing 3.06δ is the best spacing that can be applied on WIG hull-fuselage. Table 5 shows the percentage of difference of baseline and best spacing. From Table 5, the highest percentage of difference is 25% at 5 m/s.

4 Conclusion

The installation of flow control (in this case the MVG) on WIG craft hull-fuselage has shown its effectiveness in reducing drag coefficients. Experiments were conducted on different heights, spacing and angles of the MVG. Hull-fuselage with MVG of 0.8δ height, 16° angle and 3.06δ spacing exhibit the best performance with minimum recorded drag coefficient.

Acknowledgements The authors would like to thank Mr. Muhammad Hafiz Bin Mahmood and the technical staffs of the Wind Tunnel Laboratory, Department of Mechanical Engineering, Faculty of Engineering, Universiti Pertahanan Nasional Malaysia (UPNM) for the technical assistance in using the equipment.

References

1. Tofa MM, Maimun A, Ahmed YM, Jamei S, Priyanto AR (2004) Experimental investigation of a wing-in-ground effect craft. Sci World J 2014
2. Qu Q, Lu Z, Liu P, Agarwal RK (2014) Numerical study of aerodynamics of a wing-in-ground-effect craft. J Aircraft 51(3):913–924
3. M X, Geisler R, Agocs J, Schroder A (2015) Investigation of coherent structures generated by acoustic tube in turbulent flow separation control. Exp Fluids 56(46)
4. Hudy LM, Naguib A (2007) Stochastic estimation of a separated-flow field using wall-pressure-array measurements. Phys Fluids 19:024103
5. Huerre P, Monkewitz PA (1985) Absolute and convective instabilities in free shear layers. J Fluid Mech 159:151–168

6. Wee D, Yi T, Annaswamy A, Ghoniem AF (2004) Self-sustained oscillation and vortex shedding in backward-facing step flow: simulation and linear instability analysis. Phys Fluids 16:3361–3373
7. Martin PB, Tung C, Hassan AA, Cerchie D, Roth J (2005) Active flow control measurements and CFD on a transport helicopter fuselage. In: Annual forum proceedings-American helicopter society, vol 61, no 1, American Helicopter Society, Inc.
8. Ben-Hamou E, Arad E, Seifert A (2007) Generic transport aft-body drag reduction using active flow control. Flow Turbul Combust 78(365)
9. Hazen DC (1967) Boundary layer control. J Fluid Mech 29:200–208
10. Saad MR, Zare-Behtash H, Che-Idris A, Konstatinos K (2012) Micro-ramps for hypersonic flow control. Micromachines 3(2):364–378
11. Sidhu BS, Saad MR, Ahmad KZK, Idris AC (2016) Riblets for airfoil drag reduction in subsonic flow. ARPN J Eng Appl Sci 11(12):7694–7698
12. Baljit SS, Saad MR, Nasib AZ, Sani A, Rahman MRA, Idris AC (2017) Suction and blowing flow control on airfoil for drag reduction in subsonic flow. J Phys Conf Ser 914(1):012009
13. Aliashak MZM, Saad MR, Latif AT, Idris AC, Rahman MRA (2018) Microvane in controlling noise in open cavity flow. IOP Conf Ser Mater Sci Eng 370(1):012016
14. Jumahadi MT, Saad MR, Idris AC, Sujipto S, Rahman MRA (2018) The potential of hybrid micro-vortex generators to control flow separation of NACA 4415 airfoil in subsonic flow. AIP Conf Proc 1930(1):020030
15. Godard G, Stanislas M (2006) Control of a decelerating boundary layer. Part 1 : optimization of passive vortex generators, vol 10, pp 181–191
16. Joslin RD, Miller DN (2009) Brief history of flow control. Brief history of flow control, pp 403–441
17. Vecchia PD, Nicolosi F (2014) Aerodynamic guidelines in the design and optimization of new regional turboprop aircraft. Aerosp Sci Technol 38:88–104
18. Kim BN, Chung MK (1995) Experimental study of roughness effects on the separated flow over a backward-facing step. AIAA J 33:159–161
19. Gardarin B, Jacquin L, Geffroy P (2008) Flow separation control with vortex generators, pp 1–17
20. Lin JC (2002) Review of research on low-profile vortex generators to control boundary-layer separation. Program Aerosp Sci 38:389–420

Review of Hot Air Anti-icing System Inside Bias Acoustic Liner

Lee Chern Khai, Mohd Azmi Ismail, Yu Kok Hwa, Khairil Faizi bin Mustafa, Nurul Musfirah Mazlan, and Pravinth Balthazar

Abstract Ice accumulation and noise dissipation in aircraft nacelle lip–skin are the two main issues faced by the aviation industry nowadays. In order to prevent ice formation and reduce engine noise at the same time, hot air anti-icing system has been incorporated into bias acoustic liner. This paper discusses briefly the introduction on aircraft crashes that happen due to icing issues, literature review on the types of hot air anti-icing systems and acoustic liners used today and its application in computational fluid dynamics as well as a summary on the advantages of bias acoustic liner in nacelle lip–skin.

Keywords Nacelle lip–skin · Hot air anti-icing system · Bias acoustic liner

1 Introduction

Ice accumulation on the aircraft critical surfaces; including wing, nacelle lip, blades and Pitot tube, can be hazardous. Between 2000 and 2008, at least eight aircraft crashes were reported because of icing problems [1]. On 17 January 2008, one aircraft crashed happened at London Heathrow airport that has been reported by The Guardian news [2]. Latest aircraft accident happened due to icing is reported by NSTB in 2014, where the report claims that the small air plane accident in Idaho happened due to ice build-up in the engine [3]. Therefore, the hot air anti-icing system is necessary for the aircraft be protected from ice accumulation. This type of anti-icing is the most reliable and effective anti-icing system. Piccolo Tube Anti-Icing (PTAI) and Swirl Anti-Icing (SAI) are the examples of anti-icing tools using hot bleed air anti-icing system. PTAI is the most popular and effective anti-icing tool, however, it has low average heat transfer efficiency, produces a lot of hotspots along targeted surface and potentially create run-back ice accretion on the downstream area. SAI

L. C. Khai · M. A. Ismail (✉) · Y. K. Hwa · K. F. Mustafa · P. Balthazar
School of Mechanical Engineering, Universiti Sains Malaysia, Penang, Malaysia
e-mail: azmi_mechi@usm.my

N. M. Mazlan
School of Aerospace Engineering, Universiti Sains Malaysia, Penang, Malaysia

P. Rajendran et al. (eds.), *Proceedings of International Conference of Aerospace and Mechanical Engineering 2019*, Lecture Notes in Mechanical Engineering,
https://doi.org/10.1007/978-981-15-4756-0_45

is better in heat distribution along targeted area and average heat transfer efficiency than PTAI, even it less effective than PTAI. Now, Federal Aviation Administration (FAA) necessitates that all aircraft manufacturer to reduce noise level of their aircraft. Some researchers (Ives et al. [4], Jeffrey et al. [5] and Breer et al. [6]) suggested that noise abatement tools, including Acoustic Liner (AL) and Bias Acoustic Liner (BAL), are installed in the nacelle lip–skin in nacelle D-Chamber section. However, ice protection system including hot bleed air anti-icing is installed in this area too. Therefore, this research proposes to investigate a flow of hot air inside nacelle lip–skin. Then, a new knowledge about flow and heat behavior inside nacelle lip–skin emerges from this study. Subsequently, a new theory on how to improve this system will be discussed in this research. Finally, a new empirical correlation will be created for this case according to experimental and simulation results.

2 Literature Review

During taxi, takeoff and initial climb, climb, hold, descent and initial approach and final approach and landing, aircraft accidents could occur [7]. The main reason is ice accumulation on aircraft nacelle, tail and wing leading edge. This is due to formation of super-cooled liquid which leads to aircraft icing. Ice accumulation on the wing, tail and nacelle can be dangerous to the aircraft. Icing on the aircraft tail reduces the stability and control of the aircraft, especially when the aircraft is climbing, cruising and landing at low Mach number [8]. Furthermore, the aircraft will lose control because of the ice accumulating on the surface. The icing on the wing skin will cause lift force to decline, increasing drag and altering moment characteristics. Aircraft fuel consumption will therefore increase and aviation operating costs will escalate. Furthermore, airflow through the fan can be obstructed by ice formation on nacelle lip–skin inner surface, leading to loss in performance and malfunction. The worst case could happen when the broken ice particles are absorbed and ingested into the engine which would damage the engine [7].

In the worst-case scenario, some plane crash incidents occur because of icing. For instance, air crash investigators identified that icing was a factor in the McDonnell Douglas MD-81 twin jet transport aircraft crash. The ice build-up on the wing leading edge in front of the nacelle intake broke during the takeoff and was ingested by engine. Consequently, the fan blades and compressor failed causing the aircraft to go down after four minutes during take-off [4]. Another air crash was reported on February 2005 by the National Transport Safety Board. A Cessna Citation 560 crashed in the U.S. during descent when rime ice was formed on wing leading edge. The aircraft came to a stall and rolled towards the left before crashing [7]. As such, Federal Aviation Agency (FAA) necessitate that all aircraft manufacturers have to demonstrate that the aircraft is able to operate safely in icing condition. Currently, the anti-icing system (AI) and the de-icing system (DI) are the two most common ice protection systems employed in commercial aviation [7]. The AI system prevents ice accumulation at all times. The system includes chemical fluid anti-icing (FAI) and hot air-icing systems. A special liquid, glycol, which has a low freezing temperature, is used in fluid anti-icing (FAI) system. A mixture of ethylene/propylene/diethylene glycol additives and de-ionized fluids/water is pumped through porous panels [9]. A typical concentration of glycol in an FAI system varies from 50 to 80%. Liquid is then impregnated with impacting water droplets resulting in a lower freezing temperature of Supercooled Water Droplets (SWD) than the local/ambient temperature [3].

2.1 Air Anti-icing System in Small Aircrafts

The hot air anti-icing system, particularly the piccolo tube anti-icing (PTAI) is one of the most popular thermal anti-icing systems in modern commercial aviation [10]. Piccolo tube can be found in PTAI system inside wing leading edge and near to the surface consisting of a series of in line and staggered holes. The mechanism in the system operates by supplying high pressure and temperature air from engine compressor to wing and nacelle leading edge through nozzles at high velocity jet. As a result, ice formation on the surface can be prevented by heat conduction from inner surface to outer surface. Although this system requires a lot of energy to supply hot air to the nacelle leading edge, it is the most effective and reliable ice protection device in avoiding ice accumulation on the surface. The advantage of hot air anti-icing system is that it can also prevent very thin ice formation on the surface. This high-level performance is not observed for any other ice protection systems [11]. Thermal performance of PTAT can be controlled by varying the hot air velocity, jet spacing, hole to surface distance and angle of impingement etc. [5].

Unfortunately, PTAT system has several drawbacks including: 1. the heat transfer rate decreases when the distance is greater than optimum length 2. Intense heat transfer rate on the hotspot 3. The non-uniform temperature distribution occurs on the nacelle lip–skin. 4. Run back ice accretion on the downstream surface area. 5. Less efficient compared to the other [12]. The complexity of the PTAI system also causes high manufacturing costs and a weight penalty [13]. The swirl anti-icing (SAI) is another ice protection tool that used hot air anti-icing system. Not many researchers study the performance of SAI in preventing ice accumulation on the nacelle lip–skin. SAI has the potential to provide uniform temperature distribution and prevent runback ice accretion better than PTAI. Besides, SAI is also more simply constructed compared to other hot air anti-icing systems. SAI system has fewer components, simple plumbing, light and inexpensive as compared to PTAI.

2.2 Basics of Noise Abatement System/Acoustic Liner

Nowadays, acoustic liner is one of the excellent noise abatement tools. The main function of the acoustic liner (AL) is to reduce noise produced by the turbo jet engine. Honeycomb structure of the acoustic liner provides the condition with higher impedance to optimize the high dissipation of noise. In commercial aircraft, AL is planted at the nacelle cowl zone. The AL has been developed both to reduce noise and allow cooling of the combustion chamber The AL consists of three major components: solid back-face sheet, honeycomb and perforated face sheet (Fig. 1). The perforated face sheet is used as an exterior surface.

One example of noise abatement tool is bias acoustic liner. Bias acoustic liner is located in nacelle D-chamber with higher heat transfer rate compared to acoustic liner [14]. As shown in Fig. 2 [4, 6], the acoustic liner consists of a multi-layer structure of porous plates that allow hot air to flow through the liner. A perforated back face sheet has been used instead of a solid back face sheet. A perforated intermediate layer has been added between the inner portion honeycomb and outer portion honeycomb.

The figure shows that the face sheet has a much higher percentage of porosity compared to back face sheet. The perforated back face has been applied in the system due to excess bleed hot air from the conduit to honeycomb. In addition, the intermediate perforated layer can add strength and rigidity to the forward liner, which is important near the lip–skin to protect the lip–skin from foreign object damage. Furthermore, the application of an inner portion honeycomb and outer portion, instead of one deep honeycomb, reduces the many risks during honeycomb fabrication and

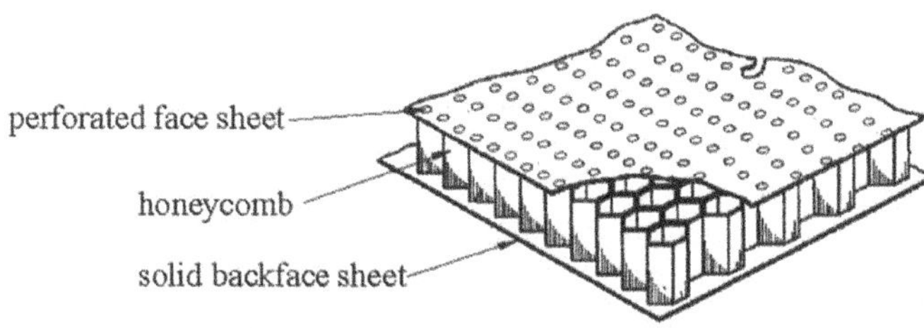

Fig. 1 The commercial acoustic liner system

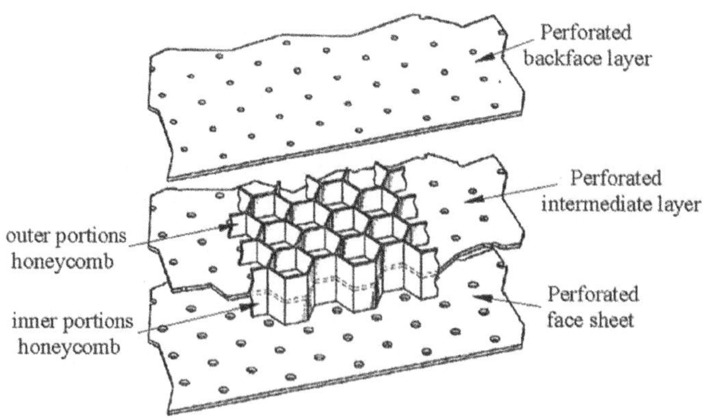

Fig. 2 The bias acoustic liner system [4–6]

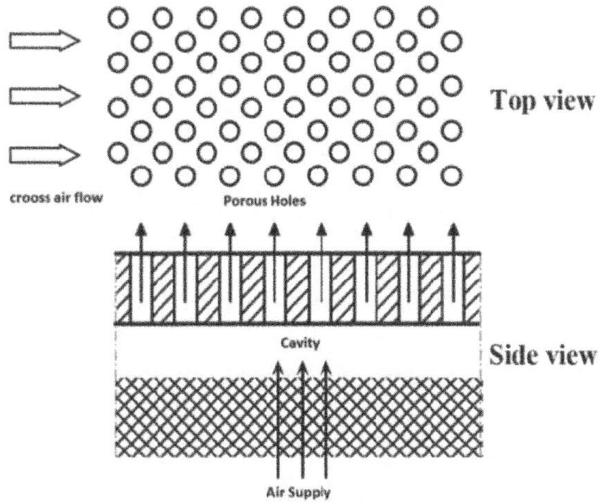

Fig. 3 The bias flow diagram

forming. Figure 3 shows the bias flow diagram for bias acoustic liner. Bias acoustic liner is able to reduce drag force, shear stress and velocity gradient by forming slip condition on nacelle lip. This is important to increase the efficiency of aircraft to save fuel cost [14].

2.3 CFD Application in Anti-icing Study

Nowadays, the commercial CFD packages are widely used in AI studies. Many researchers use CFD due to the complexity of anti-icing experiments and flight tests. Moreover, the cost of experimental study is extremely high, especially for cases involving wet conditions.

The CFD has been popular among researchers since the early 1990s. At that time, researchers employed CFD to optimize the performance of hot air anti-icing systems, especially PTAI. Table 1 summarizes the development of CFD application in anti-icing study.

3 Conclusion

Ice protection systems that are employed in commercial aviation today are anti-icing system and de-icing system. The hot air anti-icing system consists of PTAI system and SAI system. Bias flow is achieved with the use of bias acoustic liner with higher rate of heat transfer than acoustic liner.

Table 1 CFD application in anti-icing study

Year	Researchers	Research title
Early 1990	Khalil et al.	Analyse performance of hot air anti-icing as engine inlet ice protection
	Morency et al.	Analyse heat transfer phenomenon on airfoil surface
	Mattos and Olivera	Investigate conjugate heat transfer on wing slat
1995	Hua and Liu	Predict temperature distribution along wing leading edge
	Planquart et al.	Map heat transfer coefficients in multi impinging jets anti-icing system
	Papadakis and Wong	Study effect of piccolo tube configuration on temperature distribution of wing leading edge
	Saeed and Al-Ghani	Effect of different nozzle arrays on thermal performance of PTAI
2007	Wang et al.	Study PTAI performance on wing slat in wet conditions
	Elangovan and Hung	Predict temperature distribution on wing leading edge in both wet and dry conditions
2014	Campos et al.	Predict effect of shear and bias flow on noise abatement performance
	Legendre and Campos	Study capability of acoustic liner with bias flow to absorb sound
2016	Ismail et al.	Study characteristics that affect temperature distribution on nacelle lip
	Hayashi and Yamamoto	Simulate ice and growth and shedding on fan blades of aircraft
	Asante et al.	Simulate pitot tube de-icing
	Rose and Hamilton	Develop new coating method to overcome icing on aircraft surface
Late 2016	Qummare et al.	Study effect of bias flow on lift and drag forces of aircraft
2017	Thomas and Cassoni	State of the art of aircraft anti-icing and de-icing technique and modelling
	Zhoe et al.	Temperature and runback ice prediction method for three dimensional hot air anti-icing system
	Wang et al.	Passive anti-icing and active de-icing films
	Bu et al.	Numerical simulation of airfoil electro-thermal anti-icing system

Acknowledgements The authors acknowledge funding from the Ministry of Higher Education Malaysia's Fundamental Research Grant Scheme (203.PMEKANIK.6071411) and University Sains Malaysia's Fellowship Scheme.

References

1. https://www.chiltontimesjournal.com/tag/global-aircraft-anti-icing-system-market-2018 and National Transportation Safety Board Homepage. http://www.ntsb.gov/aviation. Last accessed 21 Apr 2019
2. The Guardian Homepage. https://www.theguardian.com/uk/2008/jan/17/worldtheairlineindustry. Last accessed 15 Mar 2019
3. Mercury news, NTSB Homepage. https://www.mercurynews.com/2014/09/03/ntsb-plane-icing-causedidaho-crash-that-killed-san-jose-pilot-relatives/. Last accessed 08 Mar 2019
4. Ives AO (2009) Perforated Honeycomb Acoustic Liner Heat Transfer, PhD thesis. Queen University Belfast
5. Brown JM (1999) Investigation of heat transfer between a staggered array of hot air jets and a jet engine nacelle lipskin surface, PhD thesis. Queens University Belfast
6. Breer MD, Olsen RF, Kunze RK, Riedel BL Jr (2002) Method and apparatus for aircraft inlet ice protection. US Patent 6371411B1
7. Ismail MA, Abdullah MZ (2015) Applying computational fluid dynamic to predict the thermal performance of the nacelle anti-icing system in real flight scenarios. Indian J Sci Technol 8(30)
8. Ronaudo RJ, Batterson JG, Reehors AL, Bonds TH, O'Mara TM (1991) Effect of tail ice on longitudinal aerodynamic derivatives. J Aircraft 28(3):193–199
9. Ross JF, Connolly JT (1993) Contrasting requirements for type II de/anti-icing fluids. J Aircraft 30(1):10–13
10. Wright WB (2004) An evaluation of jet impingement heat transfer correlation for piccolo tube application. In: 42nd AIAA aerospace science meeting and exhibit
11. Raghunathan S, Benard E, Watterson JK, Cooper RK, Curran R, Price M, Yao H, Devine R, Crawford B, Riordan D, Linton A, Richardson J, Tweedle J (2006) Key aerodynamic technologies for aircraft engine nacelles. Aeronaut J 110(17):265–288
12. Herman R (1987) Swirl anti-icing system. US Pattern 4688745
13. Vest MS (2001) Hot air injection for swirling rotational anti-icing system. US Pattern 6267328 B1
14. Qummare A, Ismail MA, Mazlan NM, Bashir M (2016) Numerical comparison of drag coefficient between nacelle lip-skin with and without bias acoustic liner. Int Rev Mechan Eng (I.RE.M.E.) 10(6)

Physical and Mechanical Characterization of Kenaf Fiber Filament Wound Composite Produced Using Vacuum-Bagging and Heat-Shrink Tube Method

Sharifah Fathin Adlina Syed Abdullah, Nurul Zuhairah Mahmud Zuhudi, Khairul Dahri Mohd Aris, Mohd Nazrul Roslan, and Mohamad Dali Isa

Abstract Most natural fiber reinforcements in the industry are based on random discontinuous fibers. This study aims to demonstrate the potential of continuous natural fiber-reinforced composites in structural applications. Kenaf-cotton yarn composites have been manufactured by dry filament-winding process with winding angle that varies from 50 to 80° and wet lay-up assisted by two methods, i.e. vacuum-bagging and heat-shrink tube. A hoop tensile strength test was conducted and winding angle of 80° was found to give the highest tensile strength of approximately 49.3 MPa. Heat-shrink tube method improved composites performance by 36%.

Keywords Kenaf · Natural fiber composites · Filament wound · Tubular · Filament winding

1 Introduction

The growth in bio-composites is moving at a promising pace, with researchers actively working on optimizing natural fiber composites for industrial applications. Kenaf fibers in particular is known to have good attributes which ranges from being a highly resilient plant with short gestation period to having high stiffness and extremely low density [1]. Researchers have also worked on various alkali treatment on kenaf fibers to improve the strength of individual kenaf fibers as well as fiber-matrix adhesion [2].

To exploit the high fiber strength of natural fibers, its composite systems has to be based on continuous fibers [3]. This can be achieved by spinning staple natural fibers into yarns. An aligned technical composite can then be obtained via filament winding. In filament-winding process, continuous yarns is wrapped around a rotating

S. F. A. S. Abdullah (✉) · N. Z. M. Zuhudi · K. D. M. Aris · M. D. Isa
Universiti Kuala Lumpur, Malaysian Institute of Aviation Technology, Sepang, Malaysia
e-mail: sfathinadlina@unikl.edu.my

M. N. Roslan
Faculty of Engineering Technology, Universiti Tun Hussein Onn, Parit Raja, Malaysia

P. Rajendran et al. (eds.), *Proceedings of International Conference of Aerospace and Mechanical Engineering 2019*, Lecture Notes in Mechanical Engineering,
https://doi.org/10.1007/978-981-15-4756-0_46

mandrel, creating a helical winding pattern with the angle of the yarn with respect to the mandrel axis called the wind angle. By adjusting the carriage feed rate and the mandrel's rotational speed, any wind angle between near 0° (i.e., longitudinal winding) to near 90° (i.e., hoop winding) were obtained. A wet lay-up process then produces axisymmetric hollow parts.

Very few have studied the performance of kenaf filament wound composites due to the unavailability of kenaf roving or filaments. Mokhtar et al. studied the strength of kenaf filament wound tubes under axial compression load and observed the effect of winding angle and type of resin used [4]. Angle 45° generated the higher compressive strength followed by other winding angle in the ascending order. Another study by Misri et al. focused on the torsional behavior of filament wound kenaf fiber reinforced unsaturated polyester composite hollow shafts, with a specific focus on the maximum torsion capacity for different winding angle and aluminium reinforcement [5]. It was found that torsion capacity is greatly affected by winding angle.

Dry filament winding and wet lay-up method can be assisted by vacuum bagging, heat shrink tube method [6], or vacuum infusion [4]. These methods can affect fiber volume fraction, void content, porosity, and consequently the mechanical properties of the fabricated composites. Vacuum assisted method provides maximum pressure during curing and minimum resin content, whereas heat shrink method provides a more natural method which allows removal of resin access via a rotational motion and pressure applied by heat-shrink tube.

The aim of the present work is to characterize the strength of kenaf filament wound composite tubes in terms of its winding angle and to study the options for manufacturing process of the composites. Heat-shrink tube method is gaining grounds in tubular composites manufacturing, however no studies of natural fiber filament wound via heat-shrink method is reported. A comparison between heat shrink and conventional vacuum bagging method was done to obtain the best manufacturing method for kenaf filament wound composites.

2 Materials and Methods

2.1 Production of Kenaf-Base Yarn

Kenaf fiber with water retting extraction grade was obtained from Lembaga Kenaf dan Tembakau Negara (Pahang, Malaysia) at the harvesting age of 4 months. The fibers were immersed into Sodium Hydroxyde (NaOH) solution of 6% at the temperature of 90 °C for 2 h, and dried in an oven at 60 °C for a period of 24 h to remove the excess moisture. Kenaf was cut into 4 cm and blended with cotton obtained from China at the ratio of kenaf:cotton 40:60 at opening machine where the fibers were blown together at high speed. The blended fibers were carded, then drawn into sliver of 4 g/m. The sliver was then slightly twisted and drafted at roving machine into sliver of 500 tex. Finally it was spinned in Z-twist into a single ply of 40 tex, and

Fig. 1 Spinning process of kenaf-cotton yarn

doubled in S-twist into a double ply yarn of 80 tex. The fiber treatment, blending, drawing, roving, spinning and doubling processes were as shown in Fig. 1.

2.2 Fabrication of Filament Wound Composite Tube

Cylindrical composite tubes were produced by a combination of dry filament winding from X-Winder filament-winding machine at Aero Composite Cluster Laboratory at Universiti Kuala Lumpur—Malaysian Institute of Aviation Technology, and hand lay-up method. Yarns were wound around a circular stainless steel mandrel of diameter 3.5 cm. The traversing speed of the carriage and the winding speed of the mandrel are controlled to create the desired winding angle patterns. The winding process was set at fast mode with carriage moving at the speed of 32.15 mm/s and rolling drum at 45 rev/min. Yarn orientations were varied at angle of 50°, 60°, 70° and 80°. Upon completion of one layer of winding process, AlphaEpoxy supplied by Aircraft spruce with hardener weight ratio of 2:1 was firmly applied on the kenaf preform while mandrel was rotating on the filament winder. The process is shown in Fig. 2.

Two separate methods were used on the wet preform. The first method was vacuum bagging where the preform was removed from the rotating winding machine and cured in a vacuum bag for 12 h. In the second method the heat shrink tube was placed through the wet preform and heated with a heat gun at 125 °C to remove resin access. Resin was seen dripping from the ends of the tube during the shrinking process. The mandrel was rotated continuously for 12 h at room temperature. This is to eliminate the forming of resin ridge at the bottom of the mandrel due to gravity. This method was adapted from Hamada et al. [6, 7].

Physical characterization of the composite were taken from data of length of the yarn used during the filament winding process, weight of the composite after it was cured, and thickness of the composite. Observation of void and cross-section of the tube composite were taken by using Olympus optical microscope.

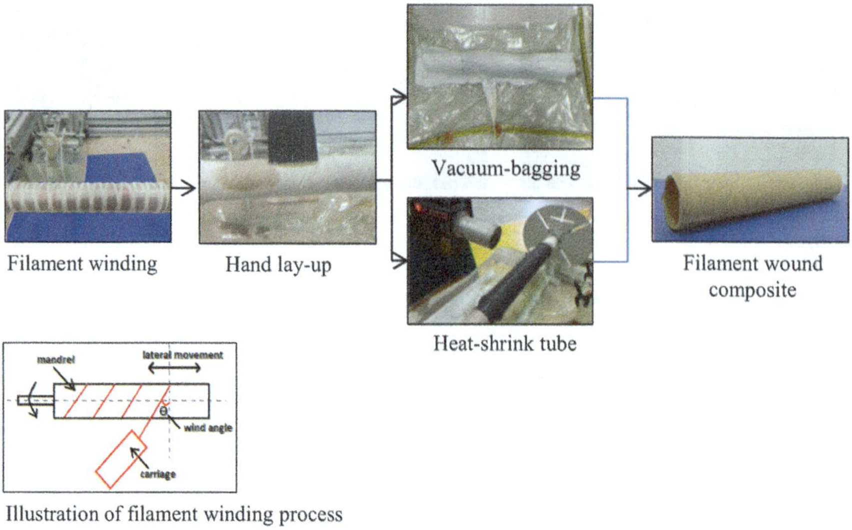

Fig. 2 Filament winding and composite lay-up

2.3 *Split Disk Tensile Test*

The tube was cut to 2.3 cm length ring for hoop tensile strength in accordance with ASTM D 2290-00 Apparent Hoop Tensile Strength of Plastic or Reinforced Plastic Pipe by Split Disk Method [8]. The split disk tests were performed for two composite tube replicates of filament wound composite at angle 50°, 60°, 70° and 80°. Five specimens were prepared from each composite. The inner diameter of the tube is 3.5 cm and the reduced area has to be a minimum of 1.397 cm. The radial apparent tensile strength σ_a (MPa) of the specimens were determined from the maximum force divide by the area of both of the reduced area. Figure 3 showed the prepared specimens for split disk tensile lest.

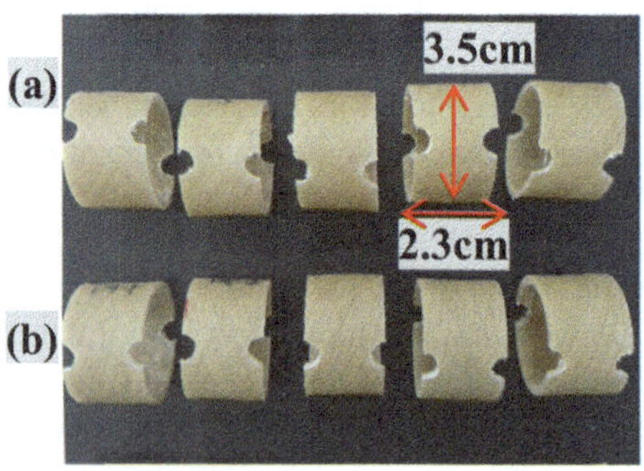

Fig. 3 Specimens from (**a**) vacuum bagging (**b**) heat shrink tube method

3 Results and Discussion

Filament wound composite tube at various winding angle were fabricated using two different processes. Physical and mechanical properties of the fabricated filament composite tube were compared and evaluated.

3.1 *Physical Characterization of Filament Wound Composite*

Physical characterization was conducted to study the effect of manufacturing method (vacuum bagging denoted as VB; heat-shrink tube denoted as HS) and winding angle on the composites strength. The physical characterization which consists of thickness of composites, weight of composites and fiber weight fraction is displayed in Table 1, and the cross section of the tube composite is in Fig. 4.

Table 1 Physical characterization of composites

Winding angle	Yarn length (m)	Thickness (mm)		Mass (g)		Fiber weight fraction	
		VB method	HS method	VB method	HS method	VB method	HS m
50°	53.8	2.87 (9.79)	3.09 (5.27)	44.88 (3.20)	40.75 (2.9)	0.063	0.070
60°	45.9	2.75 (8.52)	3.05 (4.39)	35.50 (3.68)	45.50 (3.25)	0.066	0.057
70°	37.5	2.67 (9.16)	3.18 (3.89)	34.17 (2.39)	46.88 (2.56)	0.065	0.039
80°	28.2	2.48 (8.21)	3.10 (4.27)	31.75 (4.10)	49.00 (3.12)	0.063	0.029

(): cv in the unit of %

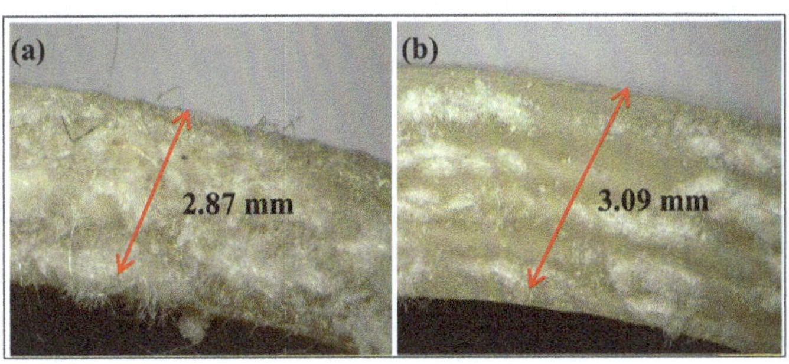

Fig. 4 Magnified composites manufactured by (**a**) vacuum bagging (**b**) heat shrink tube method

Both method successfully produced filament wound composites of different physical character. Vacuum assisted curing produced lighter composites as compared to heat shrink method as more resin access is removed by the applied pressure. As the winding angle increase, it took less yarns during the winding, thus produced thinner and lighter composites. The mass and thickness of composites manufactured via with heat shrink tube however was found to increase at higher winding angle. This is most likely due to the denser preform at higher angle winding angle creating areas which absorb more resin. Winding angle of 80° shows a significantly higher weight due to the noticeably thicker kenaf/cotton preform after winding is completed. As expected vacuum-bagging also produced higher fiber weight fraction specifically at angle above 50°. However as the fiber fraction was not optimized, there can still an improvement on the fiber fraction value. It is also interesting to note that vacuum bagging method produced composites of higher variance, ranging from 8.21 to 9.79 as opposed to only 3.89–5.37 coefficient of variance in thickness of heat-shrink tube method [9]. This is due to the high suction during vacuum-bagging which produced wavy composites surface following the surface of the preform. This contributes to the inconsistency of the composites thickness.

The surface of the composites was observed from the optical microscope at magnification of 20x as in Fig. 5. The images presented formation of voids on the surface of VB cured specimen as compared to a smooth surface of HS composites, although bubbles in HS tube may still indicate porosity. Voids in filament wound composites are common due to interbundle voids throughout the preform. Although appropriate application of vacuum can assist in reduction of voids, voids may still occur due to many reasons. Small spherical voids are often seen in between the fiber tows whereas large elongated voids are present in areas between the tow boundaries [10]. The voids are significantly seen on VB composites although it was not measured and characterized.

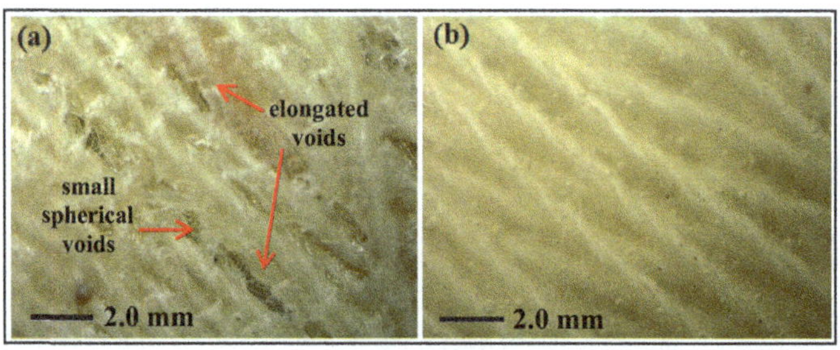

Fig. 5 Surface of composites cured via (**a**) vacuum-bagging (**b**) heat shrink tube method

3.2 *Mechanical Characterization of Filament Wound Composite*

Apparent tensile strength measured the mechanical performance of the composites as displayed in Fig. 6 and Table 2. The overall result showed better tensile performance at higher angle. VB method obtained strength of 32.47 MPa at 80°, which is an increase of 41% from the lowest angle tested, which is angle 50°. HS method obtained strength of 49.3 MPa at 80°, which is an increase of 85% from winding angle of 50°. Comparing both curing processes, it is observed that the composite tube produced using heat shrink method withstand higher tensile strength compared to vacuum bagging method. Heat shrink tube method enhanced the strength of the composites as much as 36%.

It is known that mechanical properties of filament wound or tubular composite is dependent on its winding angle as its failure depends on the direction of load. Higher winding angle produce higher tensile strength as the direction of load

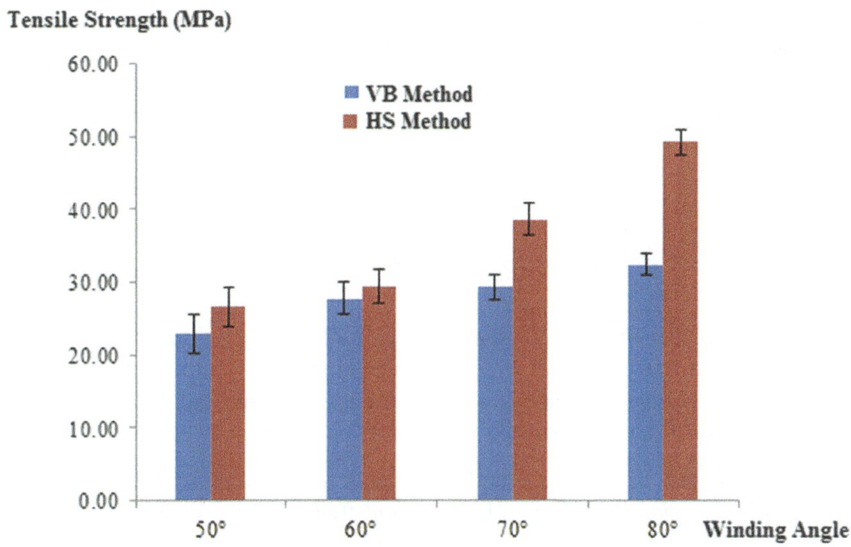

Fig. 6 Tensile strength of kenaf/cotton filament wound composites

Table 2 Tensile strength of kenaf/cotton filament wound composites

Angle	Yarn length	Tensile strength (MPa)	
		VB method	HS method
50°	53.8	22.97	26.60
60°	45.9	27.82	29.33
70°	37.5	29.31	38.57
80°	28.2	32.47	49.30

approaches direction of fiber. Meanwhile, higher winding angle decreases bending and compressive strength [4, 11].

The difference between the two manufacturing method is thickness profile of its composite and the obvious surface voids on VB composites. VB composites were more optimized in terms of its fiber fraction; however the waviness profile of filament wound composites produced resin-starved regions. These areas have insufficient resins to transfer load to the reinforcements and failed earlier. This is more significant on low density yarns as was used in this study. In addition, voids have detrimental effects over the mechanical behaviour of composites since they generally decrease overall static and fatigue strength [12]. Initiation of microcracks in the matrix may be due to high stress concentration at the ends of the void [12].

HS method provide an excellent alternative to conventional vacuum bag as it produced consistent thickness and a void-free surface. The rotational motion during post-curing avoid formation of rich and starved-resin region and the shrinking of the tube provides a method to remove access resin.

The performance of kenaf/cotton yarn as filament wound composites was compared to previous work. Sapiai et al. has fabricated kenaf filament wound composite from ready-made kenaf yarn and obtained composites tensile strength of 49.48 MPa [13], which is comparable to this work. Kenaf-based yarn was successfully spinned for filament wound composites and can be an alternative to the 100% kenaf yarn obtained in the market. Kenaf-base yarn can also be customized to meet the application of filament wound composite.

4 Conclusion and Recommendations

Kenaf yarns of low linear mass density were successfully produced for filament wound composites. Composites were fabricated at various winding angle from 50° to 80° using two separate methods which is vacuum-bagging and heat-shrink tube. From this study, it can be concluded that:

- Heat-shrink tube is a better method for filament wound composites as it produced a more consistent thickness profile and a void-free surface. It improves the tensile strength of the composites as much as 36% as compared to the conventional vacuum bagging method. It is however recommended that the optimized fiber volume fraction is studied.
- A higher winding angle gives better tensile strength as the direction of load approached direction of fibers. Other mechanical tests have to be conducted to obtain the optimized winding angle suitable for its application.

References

1. Ramesh M (2016) Kenaf (Hibiscus Cannabinus L.) fibre based bio-materials: a review on processing and properties. Prog Mater Sci 78–79, 1–92
2. Abdullah SAS, Zuhudi NZM, Anuar NIS, Isa MD (2018) Mechanical and thermal characterization of alkali treated kenaf fibers. IOP Conf Ser Mater Sci Eng 370(1)
3. Muda MKH, Mustapha F, Aris KDM, Sultan MTH (2014) Fabrication technique for bio-composite patch repair on laminated structures of CFRP plate. Appl Mech Mater 564:366–371
4. Mokhtar I, Yahya MY, Saman A, Kader A, Hassan SA (2014) Experimental analysis of kenaf filament wound tubes under axial compression load. Appl Mech Mater 660:778–782
5. Misri S, Sapuan SM, Leman Z, Ishak MR (2015) Torsional behaviour of filament wound kenaf yarn fibre reinforced unsaturated polyester composite hollow shafts. Mater Des 65:953–960
6. Okano M, Sugimoto K, Nakai A, Hamada H (2004) Energy absorption properties of braided composite tubes. In: Proceedings of the fourth Asian–Australasian conference on composite materials (ACCM 4). Woodhead Publishing Limited, pp 466–471
7. Roslan MN, Yahya MY, Ahmad Z, Hani A, Rashid A, Wang W (2018) Energy absorption capacity of basalt sandwich composite cylinder subjected to axial compression loadings. Mater Sci Forum 917:7–11
8. ASTM D2290 (2003) Standard test method for apparent hoop tensile strength of plastic or reinforced plastic pipe. ASTM International, West Conshohocken, PA, pp 1–5
9. Haris MY, Laila D, Zhahir A, Mustapha F, Mohd Aris KD (2012) A comparative study of an aircraft radome closed mold through vacuum infusion technique. Adv Mater Res 576:690–694
10. Stringer LG (1989) Optimization of the wet lay-up/vacuum bag process for the fabrication of carbon fibre epoxy composites with high fibre fraction and low void content. Composites 20(5):441–452
11. Yıldırım H, Pıhtılı H (2016) Investigation of the mechanic behaviours in hollow composite shafts having different fiber reinforcement and orientation angles. Mater Sci Eng Technol 7:646–656
12. Mehdikhani M, Gorbatikh L, Verpoest I, Lomov SV (2018) Voids in fiber-reinforced polymer composites: a review on their formation, characteristics, and effects on mechanical performance. J Composite Mater 53(12):1579–1669
13. Sapiai N, Jumahat A, Mahmud J (2015) Flexural and tensile properties of kenaf/glass fibers hybrid composites. Jurnal Teknologi 76(3):115–120

Flow Visualization Study Using Dye Mixtures on a Hydrokinetic Turbine in a Water Tunnel

Teo Chen Lung, Mohd Badrul Salleh, and Noorfazreena M. Kamaruddin

Abstract A simple gravity-feed dye injection system is developed in this study and used as a flow visualization technique for dye mixtures consisting of dye with water (DW) and dye with water and milk (DWM) with combination ratios of 1:9, 3:7 and 5:5 for DW; and 1:5:4, 3:3:4 and 5:1:4 for DWM. All dye mixtures are investigated under dynamic flow regimes at 0.085, 0.119 and 0.80 m/s water flow speed, respectively. The practicality of the mixtures is analysed in terms of dye clarity, rate of dispersion, and dye flow path to determine the best dye mixture for each flow regime. A hydrokinetic turbine model is placed in the test section of the water tunnel to study the flow structures across the model. At 0.085 m/s, the DW with 1:9 ratio is the best dye mixture provided the turbine is placed not more than 6 cm from the injector. At 0.119 m/s, DWM with a 3:3:4 ratio is the best solution, because milk prolongs the dye diffusion thus retaining the dye traces in the water, whereas at 0.80 m/s, the effect of milk in reducing the dye dispersion is more significant. A clear dye flow pattern can be observed when the dye concentration is increased; thus, DWM with the ratio of 5:1:4 is found to be the best dye mixture for that flow speed. A reliable flow visualization study can be achieved if a suitable dye mixture is used for a specific flow regime.

Keywords Hydrokinetic turbine · Flow visualization · Dye · Water tunnel

1 Introduction

Hydropower is a renewable energy resource with 3402 billion kWh electricity generation in 2010, it is expected to increase about 83% from 2010 to the year 2040 [1]. While stream-based hydrokinetic energy research is not as evolved as its wave energy counterpart, there has been growing interest in this technology over the past few years. The river current hydrokinetic turbine (hydrokinetic turbine) is a potential

T. C. Lung · M. B. Salleh · N. M. Kamaruddin (✉)
School of Aerospace Engineering, Universiti Sains Malaysia, 14300 Nibong Tebal, Penang, Malaysia
e-mail: fazreena@usm.my

P. Rajendran et al. (eds.), *Proceedings of International Conference of Aerospace and Mechanical Engineering 2019*, Lecture Notes in Mechanical Engineering,
https://doi.org/10.1007/978-981-15-4756-0_47

device to provide sustainable electricity in rural areas [2]. For low cost and low load applications, hydrokinetic turbine could be sufficient to meet the power demand. The hydrokinetic turbine directly extracts the kinetic energy of water current and converts it to electrical power. There are various types of hydrokinetic turbine, and each has its own efficiency, some of which are higher than others [3, 4]. The efficiency and reliability of a turbine is highly dependent on how flow dynamics behave across the turbine [5, 6]. Therefore, it is crucial to study and understand the flow structures and dynamics across the turbine so that the efficiency of the turbine can be improved and optimized. This can be achieved by conducting flow visualization study either through numerical or experimental approach.

Flow visualization can provide a quick, qualitative assessment of the flow field, guiding initial concepts and the design of more detailed experiments. Unlike many velocity probes that provide information at only one point or along one line, flow visualization has the advantage of describing the entire flow field [7]. In experimental flow visualization study, there are several techniques that can be used to visualize flow, such as particle tracers, surface flow visualization, and optical techniques [8]. In particle tracer techniques, particles such as dye, smoke or microspheres can be added to a flow medium to trace fluid motion. This technique is simple and low cost, but can provide high quality of pictorial representation [9]. A particle tracer using dye injection is suitable to visualize water flow whereas smoke wire can be used when the flow medium is air. Several studies examining these techniques can be found in [10–12].

It is worthwhile to mention that care must be taken when using these techniques to minimize disruption of the tracer flow path and clarity inside the fluid medium, especially when using dye injection techniques. For instance, the injection velocity should match the water flow velocity, and the density of the dye should be comparable with the fluid medium [9]. One parameter that influences the feasibility of dye injection technique is the dye mixture; thus, it is crucial to find suitable dye mixtures with respect to water flow properties such as flow speed, density and turbulence intensity. It is rather challenging to visualize the flow structures across a hydrokinetic turbine, since the turbine model is in a dynamic state. The rotation of the turbine model highly influences the flow dynamics inside a water tunnel.

Therefore, the present study aims to find the best dye mixtures for flow visualization studies across a hydrokinetic turbine. Several dye mixtures with various combination ratios are proposed and tested under various flow regimes inside a water tunnel. The practicality of the dye mixtures to visualize the flow structures across the turbine is analyzed in terms of dye clarity, rates of dispersion and dye flow path patterns.

2 Methodology

2.1 Dye Injection Method

The flow visualization technique considered in the present study is the dye injection method. This method is selected because it is low cost, can be easily set up on the water tunnel and it can provide high quality pictorial representations [9]. As shown in Fig. 1, the setup for dye injection method consists of dye tanks made of a plastic bottle, flow valves, dye tubes of 4 mm in diameter, and dye injectors using a butterfly needle with 0.5 mm diameter. An L-wand is used to hold and support the dye injectors in the test section. In this setup, the dye tanks are located high above the dye injectors to ensure a continuous flow of the dye into the test section through a gravity-feed system. There are two dye injectors positioned 7 cm apart to accommodate different dye colors during the testing. In order to increase the dye contrast inside the test section, a white background is used.

2.2 Dye Mixtures

Since the present study aims to find the best dye mixture to visualize the flow dynamic across hydrokinetic turbine, two dye mixtures i.e. dye with water (DW) and dye with water and milk (DWM), have been considered and tested. Table 1 lists the dye mixtures together with their respective combination ratios. For every experiment, a total of 40 ml dye mixture is prepared and the ratio of combination is adjusted based on the total volume.

The type of the dye used is Star Brand food coloring with blue and green colors to provide better contrast during flow visualization experiment [8]. It is worth noting that water is added into the dye to dilute the dye solution and to prevent dark concentrated

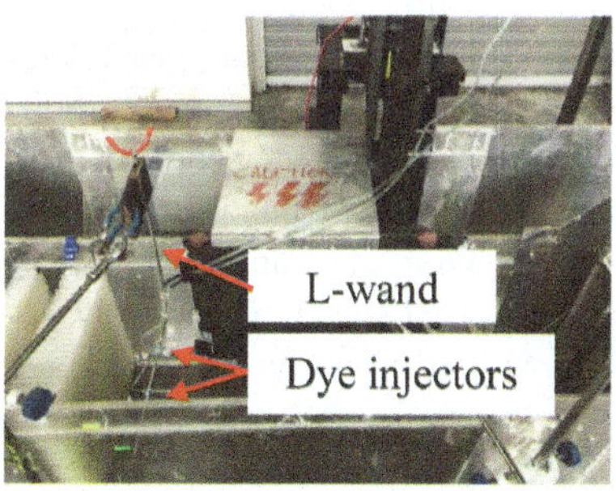

(a) Dye injectors are positioned inside the test section

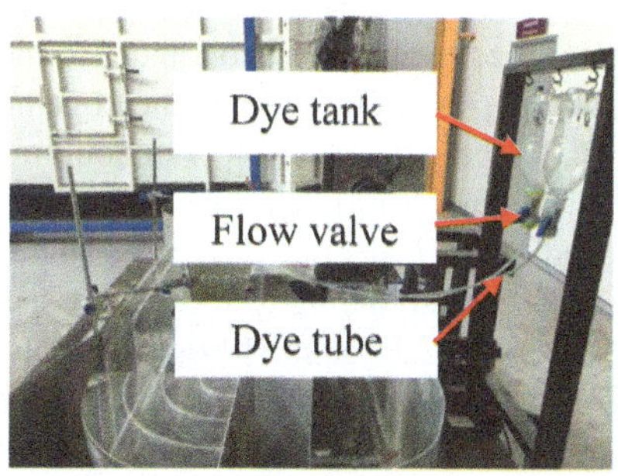

(b) Dye tanks are hang high above the dye injectors

Fig. 1 Dye injection method setup on the water tunnel

Table 1 Dye mixtures and their respective combination ratios

Dye mixtures	Elements	Combination ratios
DW	Dye with water 	Dye to water ratio 3:7 1:9 5:5
DWM	Dye with water and milk  	Dye to water to milk ratio 3:3:4 1:5:4 5:1:4

spots of dye in the flow visualization experiment whereas milk is added into the dye solution to stabilized the dye streak lines so that the lines do not dissolve quickly once the dye mixture is ejected into the water in the test section. It is presumed that the fattiness of the milk retards the diffusion of the dye solution into the main bulk of water, allowing the dye flow pattern to be retained in water before dissolving [8].

2.3 Water Tunnel

The water tunnel was initially designed by Tong [13] in his final year project and modified by Viknaraj [14] and consists of primary components including the inlet flow guide, exit flow guide, test section, X crosspiece, flow conditioning system, and water propeller with motor, and also the support jig to hold the water propeller with motor in place. The water tunnel together with its individual components is shown in Fig. 2, where the water flow direction is counter-clockwise as viewed from the top.

The water flow rate inside the water tunnel is controlled by adjusting the rotational speed of the propeller. The motor has a speed rate knob that can be adjusted from 0 to 5, corresponding to the lowest speed to the highest speed, respectively. A BK Precision DC Regulated Power Supply 1688 A is used to power the motor at 5 V. In order to measure the water flow speed, Nixon Streamflow Velocity Meter is used. The device acquires the flow speed in terms of revolution per minute (RPM). Using a provided calibration chart, the water flow speed in meter per second (m/s) is obtained. In the experiment, the turbulence intensity of water flow inside the test section is about 6.77%.

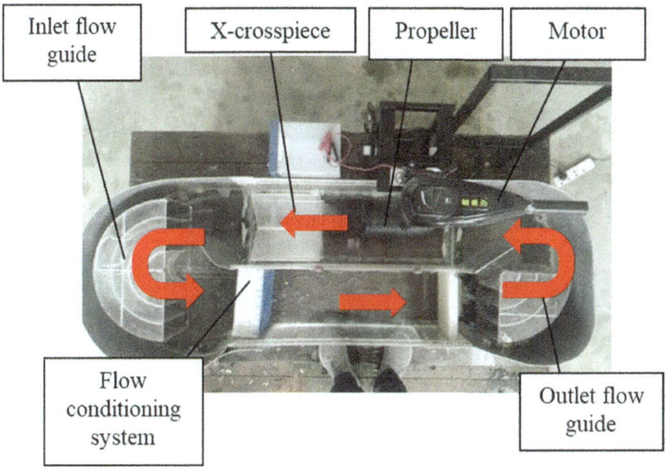

Fig. 2 Top view of water tunnel. Red arrows indicate the water flow direction

2.4 Hydrokinetic Turbine Model

A 3-bladed Savonius hydrokinetic turbine is used as the water tunnel model. It has a simple design which consists of a shaft with the blades attached to the bottom section of the shaft, as shown in Fig. 3. The height of the turbine blade is 8.5 cm and its diameter is 14 cm giving the aspect ratio of 0.61. In cases in which the turbine must be fixed, the shaft is clamped directly using a retort stand. Whenever the turbine is required to rotate freely by the motion of the flow, a bearing is attached to the shaft of the turbine and placed upright in the test section. This allows the incoming stream to come into contact with the blades, forcing the turbine to rotate.

Fig. 3 The 3-bladed Savonius hydrokinetic turbine model

Table 2 Case studies and corresponding flow regimes

Case study	Flow regime	Presence of turbine in test section
A	Dynamic flow (0.085 m/s)	No turbine
B		Turbine fixed at a location without rotating
C	Dynamic flow (0.119 m/s)	No turbine
D		Turbine fixed at a location without rotating
E	Dynamic flow (0.800 m/s)	Turbine is set to rotate freely

2.5 *Experimental Procedure*

In the present study, two dye mixtures, each with different combination ratios of the mixture elements, are tested. These mixtures are tested under the various case studies and flow regimes as listed in Table 2. The dynamic flow regime consists of low speed flow (case study A and B) and high speed flow (case study C and D) corresponds to 46 RPM (0.085 m/s) and 176 RPM (0.119 m/s), respectively. For dynamic flow regime of case study E, the water propeller speed rate is set to 4 with a 10 V power supply to achieve 864 RPM (0.800 m/s) water flow speed. This flow speed allows the turbine to rotate freely. Note that for case studies with a turbine, inside the turbine is placed the test section 6 cm downstream of the dye injectors.

All experimental data are taken and recorded using a phone camera that is built with dual camera 12-megapixel system, f/1.8 aperture, optical image stabilization (OIS) and video stabilization. The images captured undergo analysis in terms of clarity of the flow pattern, rate of dispersion, dye flow direction, as well as the applicability of the dye mixture on respective flow regimes.

3 Results and Discussion

3.1 *Case Study A and B*

The results for dynamic flow regime at 0.085 m/s water flow speed with no turbine are shown in Fig. 4. Both DW mixture with 3:7 and 1:9 ratios show a longer straight line pattern of about 6 cm. The flowing water carries the dye stream forward following the flow direction, which reduces the downward flow pattern of the dye, per Fig. 4a, b. The DW mixture with 1:9 ratio provides better clarity compared to that of 3:7 ratio, since there is no dark concentrated spot as the dye concentration is low. This facilitates flow visualization around the turbine. On the other hand, both DWM

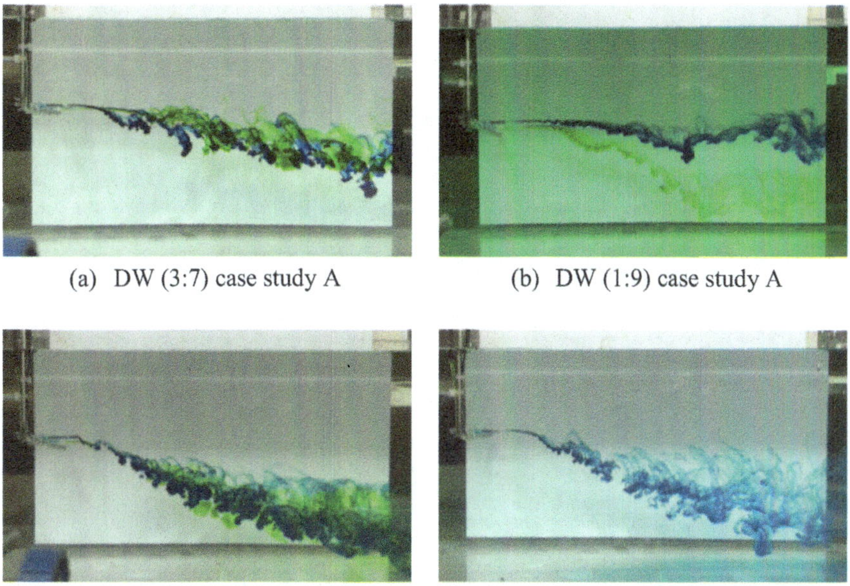

(a) DW (3:7) case study A
(b) DW (1:9) case study A
(c) DWM (3:3:4) case study A
(d) DWM (1:5:4) case study A

Fig. 4 Dye with water (DW) and dye with water and milk (DWM) mixtures under dynamic flow regime at 0.085 m/s water flow speed with no turbine (side view). Water flow direction is from left to right

mixtures exhibit a curvy flow pattern after being carried forward by the water flow at a short distance downstream of the injectors, per Fig. 4c, d. This is because the addition of milk increases the density of the mixture. Therefore, the mixture is not suitable for flow visualization study under this flow regime.

Visualization properties of the DW and DWM mixtures under dynamic flow regime at 0.085 m/s water flow rate with a fixed turbine are shown in Fig. 5. The DW mixture with 1:9 ratio provides better clarity and a less concentrated dye stream. The dye stream is able to reach the turbine placed at 6 cm downstream of the injectors. However, it is not applicable when the turbine is placed downstream further than 6 cm because the dye stream is no longer flow in a straight line along with the water flow direction as discussed in case study A. For DWM mixtures, the presence of milk, which is denser than water, causes the mixtures to flow underneath the turbine. Therefore, the mixtures with milk are not suitable for this type of flow regime.

3.2 *Case Study C and D*

For case study C, it is found that all dye flow patterns have a straight path line. However, the dye clarity of DW mixtures reduces as the dye dissolves rapidly, per

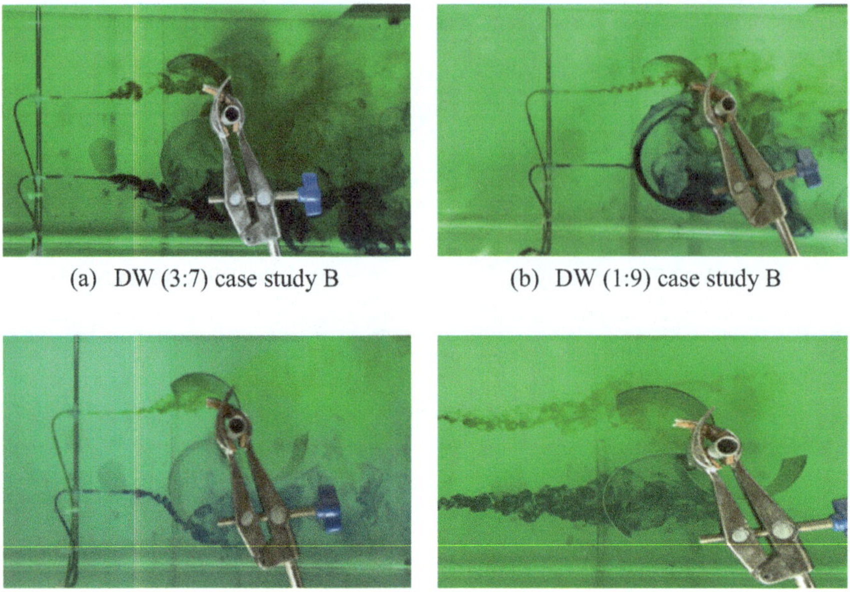

(a) DW (3:7) case study B

(b) DW (1:9) case study B

(c) DWM (3:3:4) case study B

(d) DWM (1:5:4) case study B

Fig. 5 DW and DWM dye mixtures under dynamic flow regime at 0.085 m/s water flow speed with a fixed turbine (top view). Water flow direction is from left to right

Fig. 6a, b. The DW mixture with 1:9 combination ratio has poor clarity because the rate of dispersion is higher than the injection rate and is thus not suitable for this flow speed. For the DWM mixtures, the high water flow speed causes the dye stream to flow in a straight path line, despite having higher density due to the addition of milk. Hence, flow visualization experiment using a solution added with milk can be carried out at higher water flow speed. The milk retards the dye dispersion hence the dye stream path can be observed even at outlet of the test section, per Fig. 6c, d.

Because the DW mixtures dissolves rapidly at 0.119 m/s of water flow speed, the flow pattern cannot be observed clearly across the fixed turbine in case study D. Based on Fig. 7a, b, there is no trace of dye downstream of the turbine. The dye starts to disperse upon reaching the turbine blade. For DWM mixtures, the dye streams are able to retain longer where traces of dye can be observed downstream of the turbine, especially the 1:5:4 DWM mixture even though it is faint, see Fig. 7c, d. Therefore, only the DWM mixture is suitable for a flow visualization study across a turbine under high water flow speed.

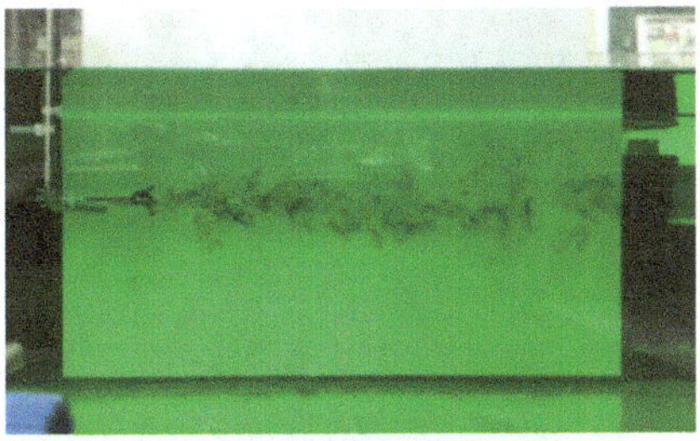

(a) DW (3:7) case study C

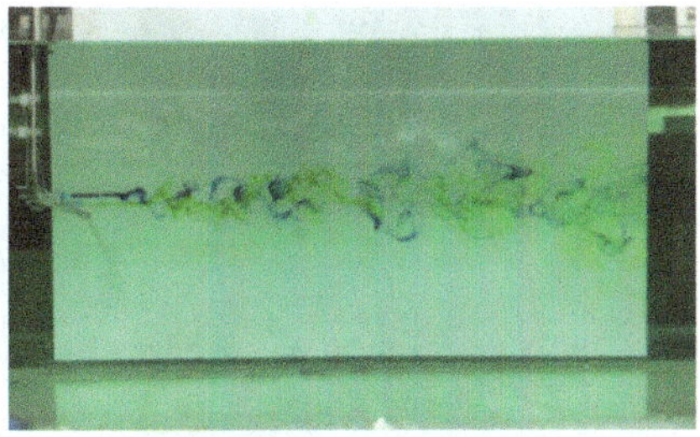

(b) DW (1:9) case study C

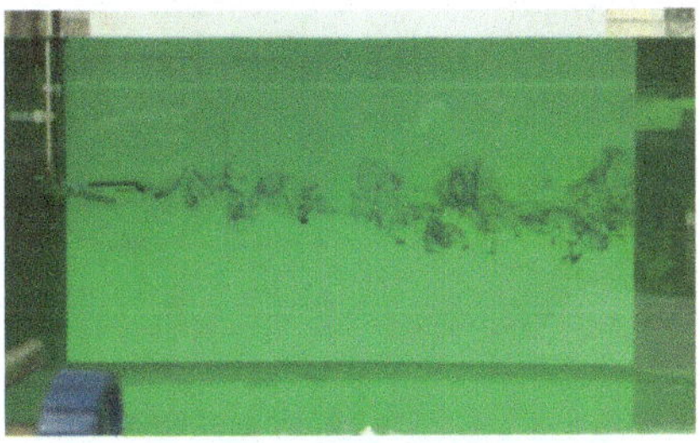

(c) DWM (3:3:4) case study C

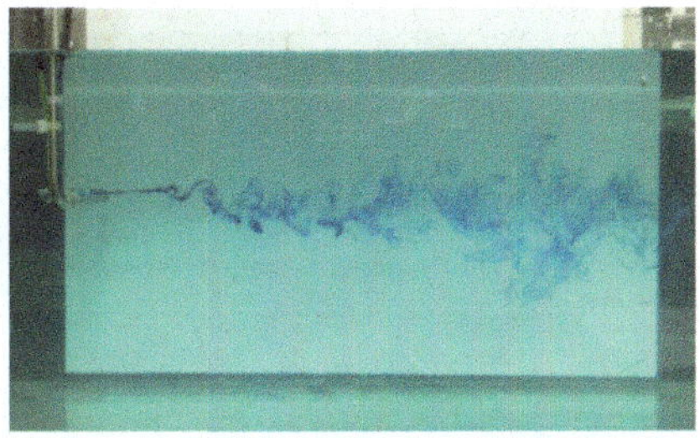

(d) DWM (1:5:4) case study C

Fig. 6 DW mixture under dynamic flow regime at 0.119 m/s water flow speed with no turbine (side view). Water flow direction is from left to right

3.3 Case Study E

As the water flow speed increases up to 0.800 m/s, dye diffuses quickly into the main bulk of water as the increase of kinetic energy of the dye molecules increases the dispersion rate. Therefore, the concentration of the dye for this case is increased for a clearer and observable flow field, as shown in Figs. 8 and 9 for DW with 5:5 ratio and DWM with 5:1:4 ratio, respectively. It is found that the 5:1:4 DWM mixture has a clearer dye flow pattern as compared to that of the 5:5 DW mixture where the dye can be clearly observed downstream of the turbine, as shown in Fig. 9. The addition of milk reduces the rate of dispersion as discussed in previous case studies. Therefore, high dye concentration with milk mixture is practical for a flow visualization study across the turbine, especially to visualize the wake formation downstream of a turbine.

4 Conclusion

In the present study, two types of dye mixtures have been tested in the flow visualization experiment i.e. dye with water (DW) and dye with water and milk (DWM). Each mixture has different combination ratios and the mixtures are being investigated

(a) DW (3:7) case study D

(b) DW (1:9) case study D

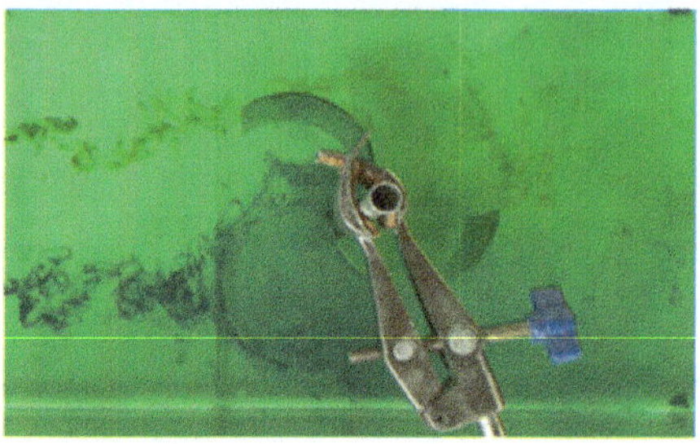

(c) DWM (3:3:4) case study D

(d) DWM (1:5:4) case study D

Fig. 7 DW and DWM dye mixtures under dynamic flow regime at 0.119 m/s water flow speed with a fixed turbine (top view). Water flow direction is from left to right

Fig. 8 Consecutive images showing the 5:5 DW dye flow pattern across the rotating turbine at 0.800 m/s water flow speed (side view). Water flow direction is from left to right

Fig. 9 Consecutive images showing the 5:1:4 DWM dye flow pattern across the rotating turbine at 0.800 m/s water flow speed (side view). Water flow direction is from left to right

under different flow regimes corresponding to different flow speed and the state of the turbine model. It is found that the 1:9 DW mixture gives the best clarity in terms of the flow pattern across the turbine under low flow regime i.e. 0.085 m/s water flow speed provided the turbine is place not more than 6 cm downstream of the dye injector. The DWM with 3:3:4 ratio is the best dye mixture to be used under high flow regime i.e. 0.119 m/s. The addition of milk reduces the rate of dispersion of fast moving dye molecules. Although milk increases the density of the dye mixture, the dye stream is maintained in a straight path line due to high water flow speed. For the case where the turbine is rotating, the best solution is the DWM with 5:1:4 ratio. The ratio of the dye has to be high such that the dye flow pattern is visible and can be retained longer inside the test section. Conclusively, all types of mixtures behave differently in different flow regime. Therefore, it is important to determine the type of flow regime in the water tunnel so that a practical and suitable dye solution can be used for flow visualization study across the hydrokinetic turbine model.

Acknowledgements The authors would like to acknowledge the financial support from the Ministry of Education (MoE) Malaysia under the Fundamental Research Grant Scheme (FRGS): (203.PAERO.6071433).

References

1. Roupp A, Ruprecht A, Riedelbauch S, Arnaud G, Hamad I (2014) Development of a hydro kinetic river turbine with simulation and operational measurement results in comparison. IOP Conf Series Earth Environ Sci 22:1–10
2. Kumar D, Sarkar S (2016) A review on the technology, performance, design optimization, reliability, techno-economics and environmental impacts of hydrokinetic energy conversion systems. Renew Sustain Energy Rev 58:796–813
3. Vermaak HJ, Kusakana K, Koko SP (2014) Status of micro-hydrokinetic river technology in rural applications: a review of literature. Renew Sustain Energy Rev 29:625–633
4. Khan MJ, Bhuyan G, Iqbal MT, Quaicoe JE (2009) Hydrokinetic energy conversion systems and assessment of horizontal and vertical-axis turbines for river and tidal applications: a technology status review. Appl Energy 86(10):1823–1835
5. Kumar A, Saini RP (2017) Performance analysis of a Savonius hydrokinetic turbine having twisted blades. Renew Energy 108:502–522
6. Boudreau M, Dumas G (2017) Comparison of the wake recovery of the axial-flow and cross-flow turbine concepts. J Wind Eng Ind Aerodyn 165:137–152
7. Cobleigh BR, Frate JD (1994) Water tunnel flow visualization study of a 4.4% scale X-31 forebody. In: NASA technical memorandum 104276, pp 1–40
8. Smits AJ, Lim TT (2012) Flow visualization techniques and examples, 2nd edn. Imperial College Press, London
9. Merzkirch W (2012) Flow visualization. Academic Press, Germany. https://www.elsevier.com/books/flow-visualization/merzkirch/978-0-08-050658-6
10. Nakajima M, Iio S, Ikeda T (2008) Performance of Savonius rotor for environmentally friendly hydraulic turbine. J Fluid Sci Technol 3(2):420–429
11. Kalyankar H, Choudhary D, Melwanki R, Chaudhari D, Jethwa S (2015) Design and analysis of low speed water tunnel for flow visualization of bluff body. In: 2nd international conference on advances in mechanical engineering and its interdisciplinary areas (ICAMEI). India

12. Fujisawa N (1992) On the torque mechanism of Savonius rotors. J Wind Eng Ind Aerodyn 40:277–292
13. Tong PE (2018) Design and development of water tunnel for flow visualization study on micro-hydrokinetic turbine. Universiti Sains Malaysia, Final Year Project
14. Viknaraj S (2019) Water tunnel turbulence intensity investigation for hydrokinetic turbine application. Universiti Sains Malaysia, Final Year Project

Backward Bent Duct Buoy (BBDB) of Wave Energy Converter: An Overview of BBDB Shapes

N. I. Ismail, M. J. Aiman, M. R. A. Rahman, and M. R. Saad

Abstract The energy extracted from ocean waves is one of the promising types of renewable energy to meet the energy demands. There are several projects conducted worldwide on various wave energy designs and one of them is floating oscillating water column (FOWC) type. The most investigated FOWC device currently is the backward bent duct buoy (BBDB) concept. However, the full potential of FOWC, is still not fully investigated and understood. Therefore, this paper presents an overview on most of the investigations previously conducted on FOWC. The optimised design parameter of FOWC in terms of main body corner shapes and buoyancy module shapes will be presented and suggested.

Keywords Renewable energy · Wave energy · FOWC · BBDB

1 Introduction

Many researchers are looking for clean energy, most available, and unlimited resources to meet economic vitality need. The largest source of renewable energy is the ocean energy, since over 70% of the surface of our planet is covered with ocean. Ocean energy comprises surface wave energy, offshore wind, tidal and marine current energy and ocean thermal energy conversion [1].

Ocean waves, which originates from the wind currents passing over the open surface of the water, is a promising basis of renewable energy. Moreover, ocean wave energy is the one of the best bases to produce power since the production is up to 90% of the time, compared to 25% for solar and wind devices [2, 3]. Since the early research conducted by Falcao [4] in 1799, the attraction in wave energy conversion system and technology increases and a variety of wave energy conversion technologies [4–9] have emerged endlessly. The different types of wave energy converter

N. I. Ismail · M. J. Aiman · M. R. A. Rahman · M. R. Saad (✉)
Department of Mechanical Engineering, Faculty of Engineering, Universiti Pertahanan Nasional Malaysia, Kem Perdana Sg. Besi, 57000 Kuala Lumpur, Malaysia
e-mail: rashdan@upnm.edu.my

P. Rajendran et al. (eds.), *Proceedings of International Conference of Aerospace and Mechanical Engineering 2019*, Lecture Notes in Mechanical Engineering,
https://doi.org/10.1007/978-981-15-4756-0_48

(WEC) that have been developed are point absorbers [10], attenuators [11], oscillating water columns [6], overtopping [12], and others [13–18] which can also be categorised into two types which are fixed and floating device.

Generally, fixed devices are constructed on the sea floor and its construction process is influenced strictly by the wave, wind, water depth, and tide compared to floating devices whereas no need construction on the sea floor. Most floating devices (Figs. 1 and 2) use the motion between the seawater and a platform to convert the wave energy. The floating device can be divided into three categories (Fig. 3) which are overtopping devices [19] that assisted by the motion between seawater and the platform, oscillating water columns (OWC) that use the relative motion between the platform and the air, and lastly the oscillating bodies [20] which use the motions between several parts of the device and the platform (Fig. 4).

The floating OWC device uses air as the mediator to convert wave energy. It is considered one of the simplest devices since it only comprises of one simple body [21]. Backward Bend Duct Buoy (BBDB) type first proposed by Masuda [22] in 1986, is currently the most popular and most researched OWC floating device. A BBDB converter consists of air chamber, air turbine, L-shaped duct, buoyancy chamber,

Fig. 1 Pelamis wave power [31]

Fig. 2 OPT buoy [32]

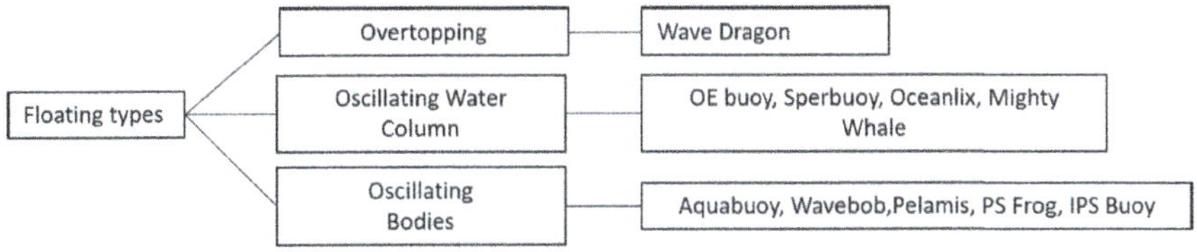

Fig. 3 Classification of floating devices

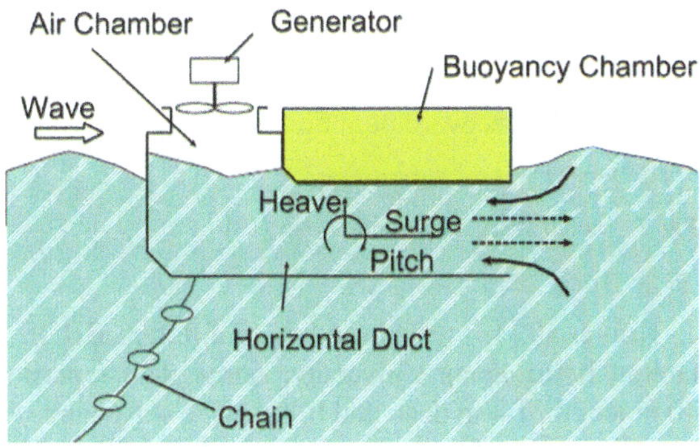

Fig. 4 Schematic diagram for floating BBDB [30]

and electrical generator, as shown in Fig. 2 for the OPT buoy. BBDB technology is expected to accomplish high conversion efficiency than others floating device due to the use of mechanical systems [23, 24].

Since BBDB exhibits high potential in generating ocean energy, increasing number of researchers from Japan, China, India, Korea and Ireland had carried out experimental studies in 2D and 3D wave tanks and also field tests in real sea conditions [23]. However, the real potential of the BBDB type of WEC has not been studied completely yet and the interactions between the design dimensions are still not conclusive. Hence, this article presents an overview on the optimum designs of BBDB from previous studies, focusing on the shape of the main body and the buoy shapes. It is hoped that the optimised dimensions of BBDB will be more understandable and its contribution towards the power generation will be clearer.

2 BBDB Corner Shapes

Many researchers have reported that modifications of the BBDB main body corners can provide better hydrodynamic performance and the ability to produce the highest efficiency [25]. The performance between the round and sharp corners has been previously studied as shown in Fig. 5 [13]. Comparisons between these two shapes were done via simulations and 2-Dimensional tank (2D tank) experiments.

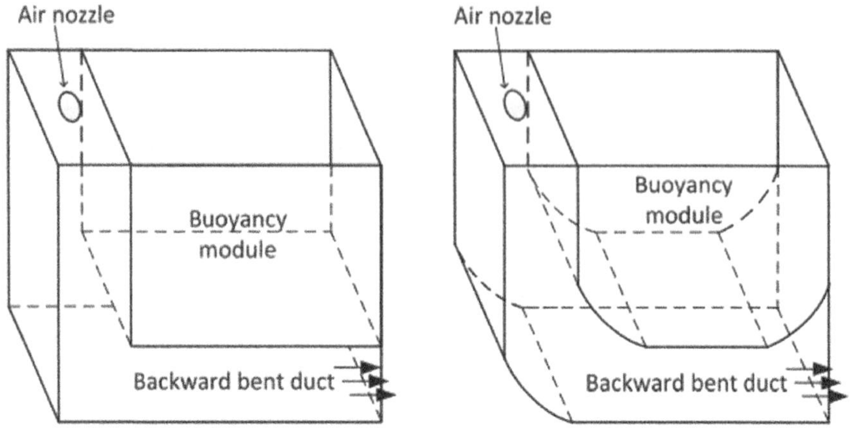

Fig. 5 Model sharp and round-corner [13]

Kim et al. [13] tested models of different BBDB corner shapes in the 2D wave tank at Kwan-Dong University, Korea. The experiments were conducting using a 40 m length, 1.4 m deep and 0.7 m wide wave tank. An active flap-type wave maker was installed in front of the wave tank to generate propagating waves in the water depth of 40 cm. The specifications of the models are listed in Table 1.

Figure 6 shows the comparison of power output for each model with different BBDB corner shapes. The result showed that the power output for BBDB sharp corner shape is lower compared to round corner. For the BBDB sharp corner shape, the maximum power generated occurs near $\lambda/L \sim 4$, and $\lambda/L \sim 8.5$. Thus, it is observed that BBDB round corner shape has more accessible wave power than the sharp corner, as there is less energy loss due to motion and body shape. It is proven that BBDB round corner is better than a sharp corner due to its ability to produce the best power output, while sharp corner will subsidize more energy losses due to viscous effects [13, 25].

Table 1 Specifications of two different corner BBDB models [13]

	Sharp corner BBDB (m)	Round corner BBDB (m)
Length (L)	0.35	0.35
Breadth (B)	0.48	0.48
Body height	0.24	0.24
Draft	0.16	0.16

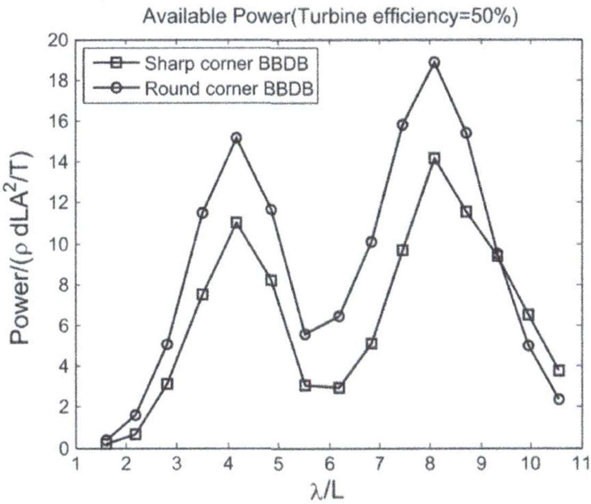

Fig. 6 Comparison power for the sharp and round-corner BBDBs [13]

3 BBDB Buoy Shapes

Sheng [26] carried out the experiment by using 2D wave tank with 1.2 m wide, 56 m long, 1.2 m deep and the water depth is 0.92 m to test different BBDB buoy shape models such as non-buoyancy module with cylindrical buoyancy module (Model A), triangle buoyancy module (Model B), and wider triangle buoy (Model C) as shown in Fig. 7. The regular waves with wave height of 0.1 m are used in this experiment to obtain the Capture Width Ratio (CWR) value. The CWR value is the parameter measured to determine the best BBDB buoy shapes. The shape that exhibits high CWR value was considered as the best shape. The CWR value results for all three models is shown in Fig. 8.

Based on the graph in Fig. 8, it shows that the highest CWR value among the models is Model C with 121% of CWR in the 2D tank. For Model A and B, the CWR value obtained were 80.8% and 82.5%, respectively [26]. As for 3D tank, it is proven that, wider buoyancy module can contribute in producing highest value of CWR compared to another module, based on the results in Fig. 9.

In a similar research carried out separately by Martins et al. [27], a model of 3.345 times the size of Model C resulted in the highest CWR of up to 130.4% when being

Fig. 7 The models for Model A, Model B, and Model C [26]

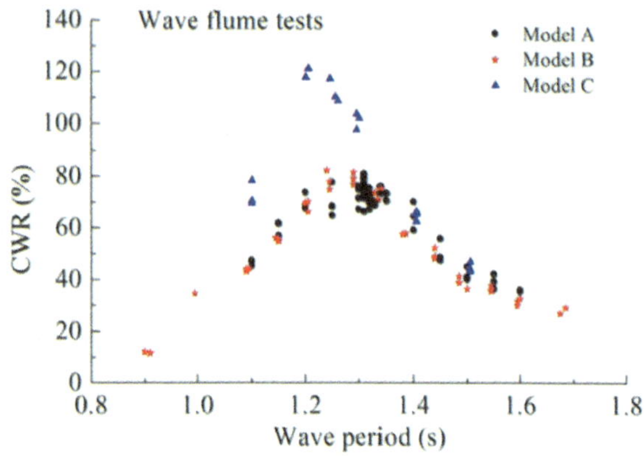

Fig. 8 The CWR value for Model A, Model B, and Model C [26]

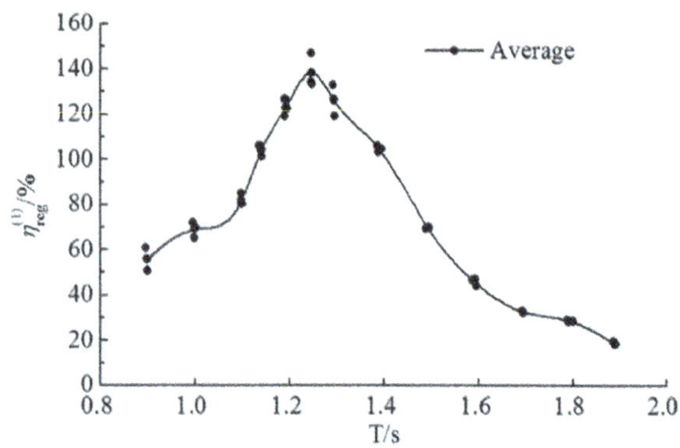

Fig. 9 CWR of the model in the 3D tank [27]

tested in the regular 3D tank, 42% of efficiency from wave-to-wire and 32.2% from pneumatic power.

On the other hand, the extended version of Model C was investigated and discussed by Falcao [4] by conducting 2D and 3D regular wave tank experiments. The wave flume which is a regular 2D wave tank of 56 m long, 1.2 m in width and 0.9 m in water depth is located at Guangzhou Institute of Energy Conversion, Chinese Academy of Sciences (GIEC, CAS). The 3D wave basin is located at School of Civil Engineering and Transportation in South China University of Technology with features of 40 m long, 30 m in width and 0.75 m in depth.

The experiment of extended version of Model C resulted in 119.10% of CWR value in 2D tank, while 146.8% for 3D tank as shown in Fig. 9 [27]. However, the result of CWR for the 2D tank increases to 130.78% when the model was moored at the intersection front as discussed by Chen et al. [14]. Liang et al. [28] and Abbas [29] strengthened the finding by also stating that different buoyancy module shape will definitely affect the power output.

In 2010, Toyota et al. [30] carried out experimental studies on the difference buoyancy module, which is rectangular with rear semi-cylindrical (Type C_1) and

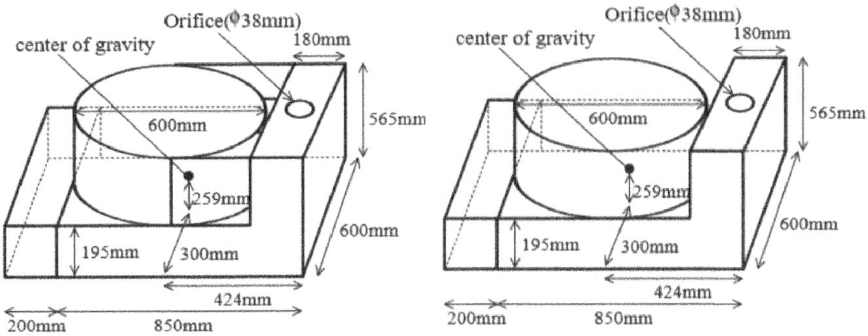

Fig. 10 The models with different buoyancy module, type C_1 (left) and type C_3 (right) [30]

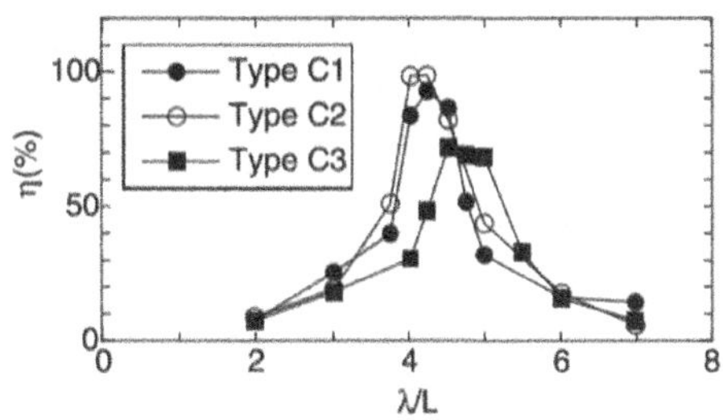

Fig. 11 The efficiency t of different buoyancy module [30]

cylindrical buoyancy (Type C_3) module as shown in Fig. 10. The outcome of the experiments as demonstrated in Fig. 11, showed that the highest efficiency value was achieved by C_1 with 95% when λ/L equals 4.0 while for Type C_3, the efficiency value obtained was 60% [30]. It is proven that, rectangular with rear semi-cylindrical would perform better than all cylindrical buoyancy module and producing greater efficiency value.

4 Conclusion

BBDB is a device based on the concept of floating oscillating water column. The best shape of BBDB is crucial and there is a need to consider its shape before accomplishment of experiment in order to obtain highest efficiency value and highest power output. Dimensions such as the corner shape and buoy shapes are vital parameters to consider in designing BBDB device. For corner shape, the round corner can be concluded as the best shape based on the performance compared to the sharp corner models, since it yields the highest efficiency compared to sharp corner. The different shapes of the buoyancy module placed on the BBDB also affects the power output. The semi-cylindrical buoyancy module is the best shape for BBDB buoy in terms of the power output produced compared to the all cylindrical module.

References

1. Aderinto T, Li H (2018) Ocean wave energy converters: status and challenges. Energies 11(5):1–26
2. Ocean homepage. https://www.renewableenergyworld.com/oceanenergy/tech.html. Last accessed 16 May 2019
3. Crespo AJC, Altomare C, Domínguez JM, González-Cao J, Gómez-Gesteira M (2017) Towards simulating floating offshore oscillating water column converters with smoothed particle hydrodynamics. Coastal Eng 11–26
4. Falcao FDO (2010) Wave energy utilization : a review of the technologies. Renew Sustain Energy Rev 899–918
5. Lindroth S, Leijon M (2011) Offshore wave power measurements—a review. Renew Sustain Energy Rev 15(9):4274–4285
6. Heath TV (2012) A review of oscillating water columns. Philos Trans Royal Soc 235–245
7. Astariz S, Iglesias G (2015) The economics of wave energy: a review. Renew Sustain Energy Rev 397–408
8. Falcão AFO, Henriques JCC (2016) Oscillating-water-column wave energy converters and air turbines: a review. Renew Energy
9. Zhang D, Li W, Lin Y (2009) Wave energy in China: current status and perspectives. Renew Energy 34(10):2089–2092
10. Do HT, Dang TD, Ahn KK (2018) A multi-point-absorber wave-energy converter for the stabilization of output power. Ocean Eng 337–349
11. ITTC-Recommended Guidelines (2014) Wave energy converter model test experiments. In: International towing tank conference, pp 1–13
12. Wang L, Kolios A, Cui L, Sheng Q (2018) Flexible multibody dynamics modelling of point-absorber wave energy converters. Renew Energy 790–801
13. Kim S, Koo W, Kim M (2015) Nonlinear time-domain NWT simulations for two types of a backward bent duct buoy (BBDB) compared with 2D wave-tank experiments. Ocean Eng 584–593
14. Chen T, Wu B, Li M (2017) Flume experiment study on capture width ratio of a new backward bent duct buoy with a pentagon buoyancy cabin. Ocean Eng 12–17
15. Pietra L, Tello M, Bhattacharjee J, Soares CG (2012) Review and classification of wave energy converters, pp 1–10
16. Viviano A, Naty S, Foti E, Bruce T, Allsop W, Vicinanza D (2016) Large-scale experiments on the behaviour of a generalised oscillating water column under random waves. Renew Energy 875–887
17. De Chowdhury S, Nader J-R, Sanchez AM, Fleming A, Winship B, Illesinghe S, Toffoli A, Babanin A, Penesis I, Manasseh R (2016) A review of hydrodynamic investigations into arrays of ocean wave energy converters
18. Drew B, Plummer AR, Sahinkaya MN (2009) A review of wave energy converter technology. Proc Inst Mech Eng Part A J Power Energy 887–902
19. Pelamis wave power. http://www.emec.org.uk/about-us/wave-clients/pelamis-wave-power/. Last accessed 16 May 2019
20. OPT. https://www.oceanpowertechnologies.com/. Last accessed 16 May 2019
21. Wu B, Chen T, Jiang J, Li G, Zhang Y, Ye Y (2018) Economic assessment of wave power boat based on the performance of 'Mighty Whale' and BBDB. Renew Sustain Energy Rev 946–953
22. Masuda Y (1986) Experience in pneumatic wave energy conversion in Japan. In: Specialty conference on utilization of ocean waves-wave energy conversion, pp 1–3
23. Nagata S, Toyota K, Imai Y, Setoguchi T, Mamun MAH, Nakagawa H (2011) Numerical analysis on primary conversion efficiency of floating OWC-type wave energy converter, pp 578–585
24. Imai Y, Toyota K, Nagata S (2011) An experimental study on generating efficiency of a wave energy converter "Backward Bent Duct Buoy". In: EWTEC, pp 1–8

25. Wu B, Li M, Wu R, Chen T (2018) BBDB wave energy conversion technology and perspective in China. Ocean Eng 281–291
26. Sheng W (2019) Motion and performance of BBDB OWC wave energy converters : I, hydrodynamics. Renew Energy 106–120
27. Martins JC, Goulart MM, Gomes MDN, Souza JA, Rocha LAO, Isoldi LA, dos Santos ED (2018) Geometric evaluation of the main operational principle of an overtopping wave energy converter by means of constructal design. Renew Energy 727–741
28. Liang NJX, Wang W, Du B (1997) Experimental research on backward bent duct wave power generation buoy model. Marine Eng 17(4):55–63
29. Abbas MS, Abbas MZ, Pasha RA, Suleman Z, Butt Z (2015) Design and fabrication of wave energy power plant using oscillating water column technique. Power Gener Syst Renew Energy Technol (PGSRET) 1–4
30. Toyota K, Nagata S, Imai Y, Oda J, Setoguchi T (2010) Primary energy conversion characteristics of a floating OWC 'Backward Bent Duct Buoy'. In: Proceedings of the twentieth international offshore and polar engineering conference, pp 850–855
31. Pelamis Wave Power. https://www.bbc.com/news/uk-scotland-scotland-business-30151276. Last accessed 24 May 2019
32. OPT Wave Power. https://oceanpowertechnologies.gcs-web.com/static-files. Last accessed 24 May 2019

Turbulence Intensity in Water Flow for Hydrokinetic Turbine Application

Viknaraj Subramaniam and Noorfazreena M. Kamaruddin

Abstract A water tunnel is used to investigate a hydrodynamic behavior of an immersed body in the fluid dynamics study. The operational principle of water tunnel is identical to a wind tunnel but with a different working fluid and higher flow-pumping capacity. Tests such as flow visualization in wind tunnel are more challenging, as turbulent flows in wind dissipate quickly whilst water tunnel is more suitable for such purpose due to its higher fluid viscosity. However, the turbulence in the water tunnel test section must be low in order to produce a valid result. The present work focuses to improve the previous water tunnel performance to mimic a river flow that has been developed for micro hydrokinetic application particularly to reduce the turbulence intensity and improve its flow uniformity. To achieve this objective, several primary components of the water tunnel were modified, fabricated and assembled. Quantitative measurement on the velocity profiles and uniformity of the water flow at the test section were analyzed with small increments of 2 cm between the measured points. This was performed by collecting the data of the velocity at every point in the cross-section with increment of 1 cm. The results obtained were then used to plot a uniformity graph and analyze the turbulence intensity in the water tunnel test section. The turbulence intensity in the water tunnel test section after modification produced a minimum result of 6.8% in comparison to the previous version around 9.7%. Thus, the turbulence intensity in the water tunnel was reduced by 29.9%

Keywords Hydrokinetic turbine · Turbulence intensity · Water tunnel

V. Subramaniam · N. M. Kamaruddin (✉)
School of Aerospace Engineering, Universiti Sains Malaysia, 14300 Nibong Tebal, Penang, Malaysia
e-mail: fazreena@usm.my

P. Rajendran et al. (eds.), *Proceedings of International Conference of Aerospace and Mechanical Engineering 2019*, Lecture Notes in Mechanical Engineering,
https://doi.org/10.1007/978-981-15-4756-0_49

1 Introduction

A water tunnel is an experimental facility used for the testing of the hydrodynamic behavior of submerged bodies in flowing water [1]. It is very similar to a wind tunnel but the fluid inside the tunnel is restricted to water. Water tunnel is important for investigating the phenomena of the boundary layer, such as the vortex induced, flow separation, turbulence and flow visualization [2].

The aim of this project was to improve on an already existing water tunnel in terms of its turbulence intensity for a Micro-hydrokinetic turbine. The existing water tunnel was designed and fabricated by Tong [3]. Micro-hydrokinetic turbine is a new technology that extracts kinetic energy from the flowing water current of the river to generate electricity, instead of the potential energy of falling water from the dam and reservoir required by micro hydropower generation [4]. Hence, the water tunnel designed and developed where the micro-hydrokinetic turbine model will be placed should be capable to mimic river flow conditions in Malaysia [5].

To investigate the flow characteristic in the water tunnel, we can characterize performance based on turbulence intensity. Turbulence intensity is a scale characterizing turbulence expressed as a percentage. Turbulence intensity should be low in this case, because high turbulence will cause the water flow to shift from laminar to turbulent that will trigger an unfavorable transition which eventually affect the accuracy of measurements. Idealized air flow with no fluctuations have a Turbulence Intensity of 0%, which is impossible to be achieved in reality. Thus, with minimum turbulence intensity in the water tunnel, the fluctuations can also be minimized [6].

To improve the performance of a previously developed water tunnel, the current study aims to reduce turbulence intensity by characterizing the velocity profile and the flow uniformity at the test section to clearly visualize and analyze the flow patterns.

2 Methodology

The current study aims to improve the performance of a previous developed water tunnel that capable to mimic controllable river flow conditions in Malaysia. The finalized design of the water tunnel consists of few primary components as shown in Table 1, which are the frame, the contraction zone, diffuser, and honeycomb inlet flow guide, exit flow guide, test section, X crosspiece, flow conditioning system, water propeller with motor and also the support jig to hold the water propeller with motor. The modifications performed to the flow conditioning system are also discussed.

Table 1 Design specification of the previous water tunnel

Water tunnel [3]	Design specifications
Overall dimension	1.5 m × 0.6 m × 0.25 m
Test section dimension	0.85 m × 0.25 m × 0.25 m
Contraction zone	X
Diffuser	X
Flow conditioning system	Tube bundles (0.06 m diameter)
Motor and propeller	SUNELEXE Motor Boat Propeller 24 lbs
Final turbulence intensity	9.77%

2.1 Previous Developed Water Tunnel

The CAD design of the existing model (pre-modification) is shown in Fig. 1. This design did not incorporate the usage of a contraction zone and a diffuser. These components are vital in suppressing the turbulence intensity of a flow.

2.2 Water Tunnel

The water tunnel in the current study includes all of the components stated above. The overall dimensions of the water tunnel were 1.83 m × 0.71 m × 0.3 m. The test section was 0.85 m × 0.25 m × 0.3 m as shown in Fig. 2.

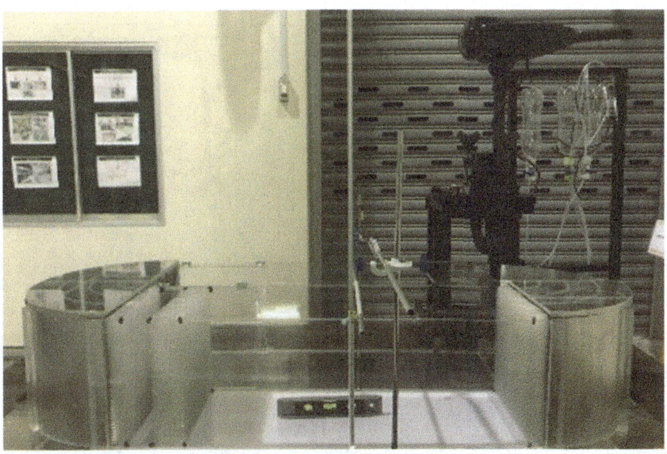

Fig. 1 Water tunnel designed by Eng [3]

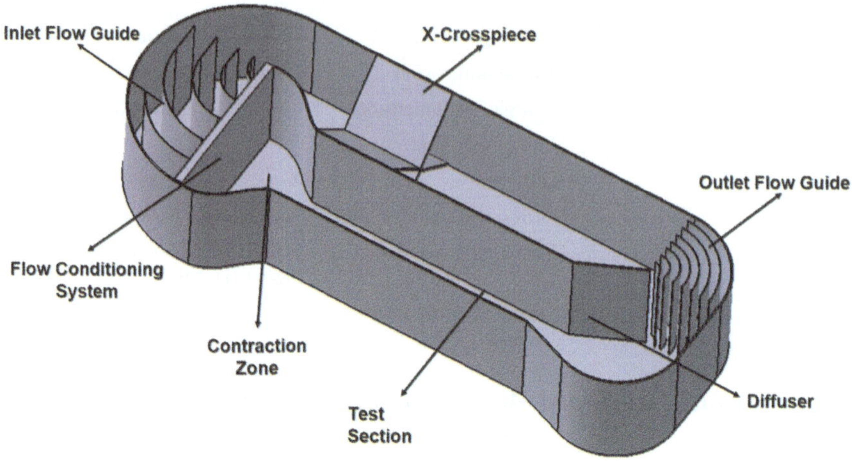

Fig. 2 New water tunnel design in the current study

2.3 *Fabrication of the Water Tunnel*

The fabrication of the water tunnel was carried out in three separate sections, which were later assembled to form the final water tunnel as shown in Fig. 3. The first part was fabrication of the contraction zone; the 2nd part was the mid-section of the water tunnel; and the last part was the diffuser section. A flow conditioning system was also added. The walls of the contraction zone were made using a metal sheet with a thickness of 2 mm.

Fig. 3 Fabricated Modified Water Tunnel

Table 2 Combination of components of water tunnel measured for velocity profile

	Combinations						
Components of water tunnel	1	2	3	4	5	6	7
Water tunnel frame	✓	✓	✓	✓	✓	✓	✓
Contraction zone		✓	✓	✓	✓	✓	✓
Diffuser			✓	✓	✓	✓	✓
Flow conditioning tube bundles 1 and 2				✓	✓	✓	✓
X Crosspiece					✓	✓	✓
Inlet flow guide						✓	✓
Exit flow guide							✓

2.4 *Experimental Procedure*

In this study, the turbulence intensity was measured with various combinations of the water tunnel components as shown in Table 2. The velocity profile was then analyzed for each combination.

3 Results and Discussion

3.1 *Combination 1: Water Tunnel Frame (Measure Cross-Sectional Area of Test Section)*

The turbulence intensity for this configuration in the test section was 79.2%, which was very high. The high turbulence intensity was high due to the corner flow effect at a 180° angle when the water changes direction drastically without the assistance of flow guide vanes. As there was a centrifugal acceleration from the curve, the velocity of the flow was increased without any flow conditioning. The lack of flow guide vanes, coupled with the centrifugal acceleration and lack of flow straighteners, was the reason for the high turbulence intensity of 79.2%. The mean water velocity was found to be 0.398 m/s at the test section. The water velocity at the outer wall was higher, and slowly decreased towards the inner wall of test section. The range of water velocity was 0.08–0.76 m/s as shown in Fig. 4.

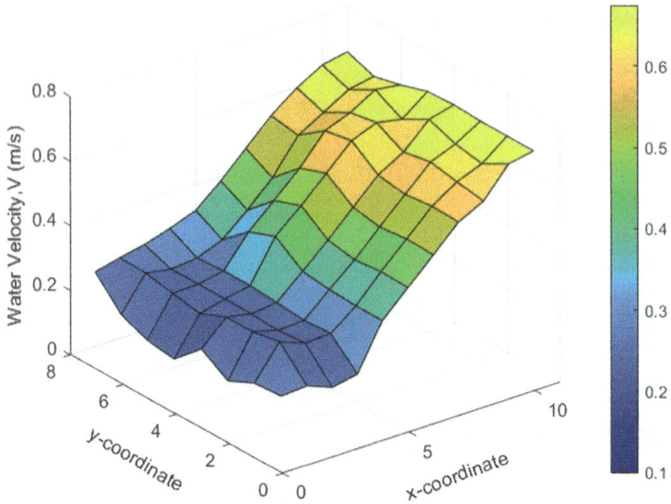

Fig. 4 Velocity profiles of combination 1 of the water tunnel components

3.2 Combination 2: Water Tunnel Frame and Contraction Zone

As the contraction zone was added to the water tunnel, the turbulence intensity of the velocity profile was reduced to 49.4%. Although the turbulence intensity was still high, there had been a significant reduction of from the previous configuration of 79.2%. This reduction of turbulence intensity was due to the addition of the contraction zone. The contraction zone functioned by causing a sudden reduction in cross-sectional area, which in turn increased the flow speed while keeping the turbulence low [7, 8]. The mean water velocity at test section was 0.489 m/s, with a range of 0.23–0.64 m/s as shown in Fig. 5.

3.3 Combination 3: Water Tunnel Frame with Contraction Zone and Diffuser

As the diffuser was added to the water tunnel, the turbulence intensity in the test section was further reduced to 17.9%. As the turbulence intensity was reduced, the mean water velocity at test section increased to 0.51 m/s. The addition of diffuser to the water tunnel was the reason behind the reduction of turbulence intensity. This was because the diffuser minimizes the pressure loss in the system, which stops the

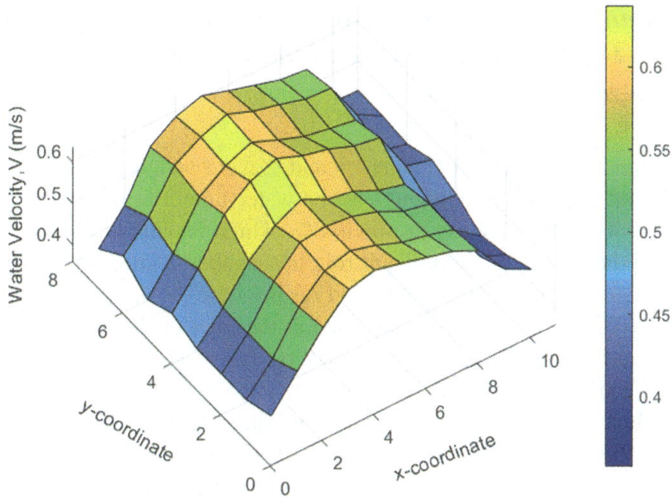

Fig. 5 Velocity profiles of combination 2 of the water tunnel components

water from flowing back towards the test section [8]. A diffuser also was developed to slow down the flow and regain the pressure of the flow [9]. This provides a smooth transition of water flow from contraction zone to test section, then the diffuser. The mean water velocity at the test section ranged from 0.36 to 0.57 m/s as shown in Fig. 6.

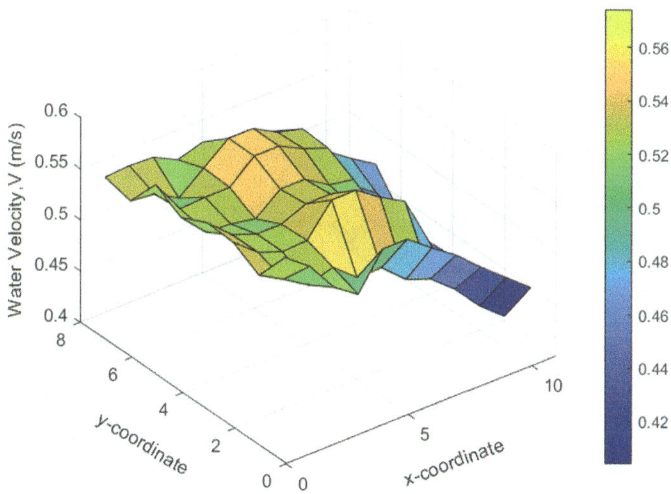

Fig. 6 Velocity profiles of combination 3 of the water tunnel components

3.4 Combination 4: Water Tunnel Frame with Contraction Zone, Diffuser and Flow Conditioning System (Honeycomb 1 and 2)

In this combination, flow straighteners 1 and 2 were added to the configuration. The addition of the flow straighteners to the water tunnel further reduced the turbulence intensity of the velocity profile to 8%, which is quite low. This shows that the flow straighteners had a significant role in turbulence intensity reduction. The flow straighteners eliminated the swirl and restored the distorted flow due to bends and curves [10]. The flow straightener used at the inlet of the test section was the 4 mm tube bundle, which was made using hollow tubes. The flow straighteners used at the outlet of the test section were a readily available 6 mm diameter tube bundle. With the turbulence intensity of 8%, the 4 mm diameter tube bundle was efficient in reducing the turbulence intensity. A similar test was performed by replacing the 4 mm diameter tube bundles with 6 mm diameter tube bundles in the same setup, where the turbulence intensity obtained was 11.3%. Comparatively, it can be concluded that the 4 mm diameter tube bundle is superior to the 6 mm diameter tube bundle in suppressing the swirls and reducing the turbulence intensity as shown by Seo [10]. Although the flow straighteners reduced the turbulence intensity, it is still considered as an obstruction in the water tunnel, thus slightly reducing the mean water velocity at the test section to 0.47 m/s as shown in Fig. 7. The range of water velocity was from 0.43 to 0.52 m/s.

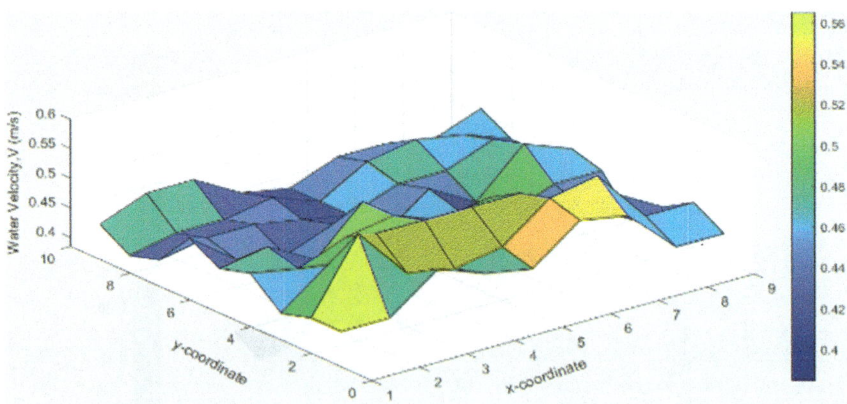

Fig. 7 Velocity profiles of combination 4 of the water tunnel components

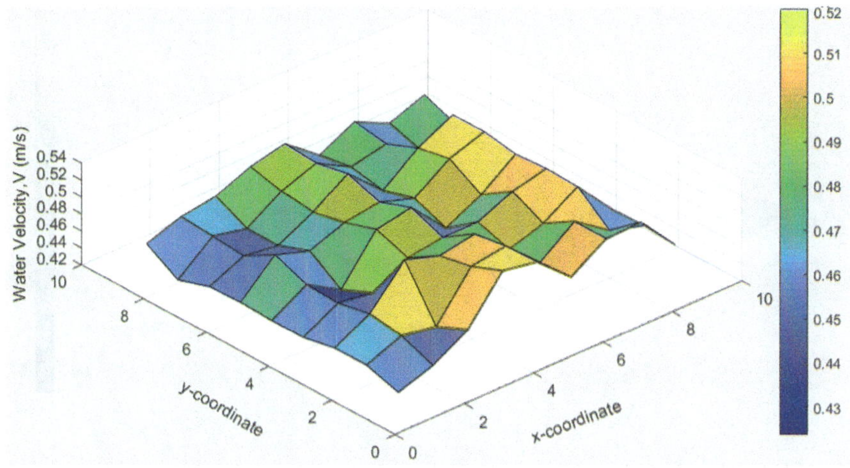

Fig. 8 Velocity profiles of combination 5 of the water tunnel components

3.5 Combination 5: Water Tunnel Frame with Contraction Zone, Diffuser Flow Conditioning System (Honeycomb 1 and 2), and X Crosspiece

With the placement of the X Crosspiece added to the water tunnel, the turbulence intensity in the test section was reduced to 7.7%. However, the X crosspiece also slightly reduced the mean water velocity at the test section to 0.46 m/s, with a range of 0.38–0.57 m/s as shown in Fig. 8. The X crosspiece functions by separating and aligning the water flow into four separate compartments.

3.6 Combination 6: Water Tunnel Frame with Contraction Zone, Diffuser Flow Conditioning System (Honeycomb 1 and 2), X Crosspiece and Inlet Flow Guide

As the inlet flow guide was placed in the water tunnel, the turbulence intensity pattern showed a slight increase to 11.5%. This increase in turbulence intensity may have been caused by the limited time given for the component to settle into the flow. The reason for using an inlet flow guide is to reduce the cornering effect as the water flow is guided by 4 vanes into the contraction zone. The mean water velocity was 0.42 m/s, with a range of 0.32–0.52 m/s as shown in Fig. 9.

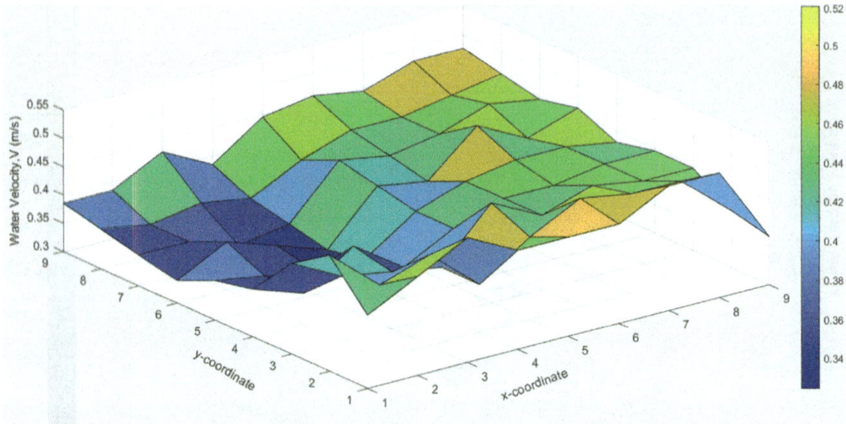

Fig. 9 Velocity profiles of combination 6 of the water tunnel components

3.7 Combination 7: Water Tunnel Frame with Contraction Zone, Diffuser Flow Conditioning System (Honeycomb 1 and 2), X Crosspiece, Inlet Flow Guide and Exit Flow Guide

This is the final configuration of the water tunnel. With the addition of Exit Flow Guide into the water tunnel, the turbulence intensity observed was the lowest of all configurations at 6.8%. The mean water velocity in the test section was 0.42 m/s and ranged from 0.34 to 0.54 m/s. The turbulence intensity reduction can be explained by the addition of the exit flow guide, as this component works in coherent with the diffuser, aiding in straightening the water flow stream at the test section and reducing the turbulence effect of the water flow stream entering the acceleration zone. Besides, the corner flow effect at the diffuser section was eliminated by the addition of the exit flow guide which prevented the backflow of water into the test section. Thus, the lowest turbulence intensity achieved was 6.8%, with a mean water velocity of 0.42 m/s in this final configuration of the water tunnel as shown in Fig. 10.

4 Conclusion

The best configuration of the modified water tunnel yielded a finalized turbulence intensity of 6.8% with a mean water velocity of 0.42 m/s. This configuration consists of contraction zone, inlet flow guide, diffuser, exit flow guide, flow conditioning system (honeycomb 1–4 mm diameter; honeycomb 2–6 mm diameter). Turbulence intensity is considered low if it is less than 10%; thus, it can be concluded that the turbulence intensity achieved in the current study is superior to that of previous

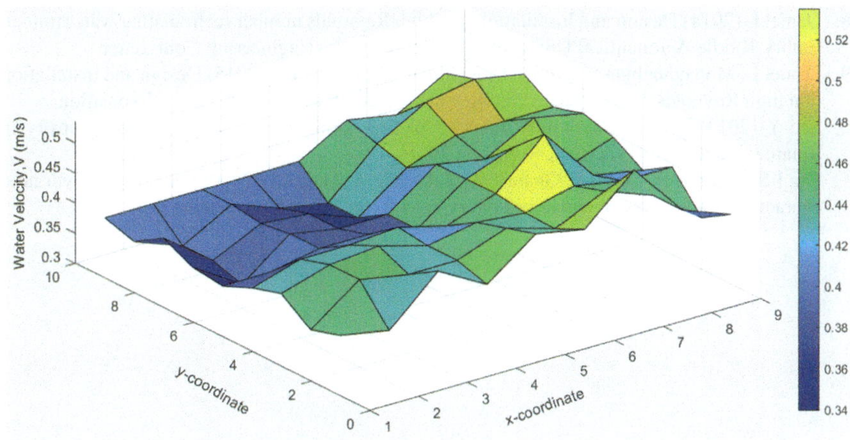

Fig. 10 Velocity profiles of combination 7 of the water tunnel components

versions. The turbulence intensity before the modifications and the old configuration was 9.7%. There was a reduction in the turbulence intensity of 29.9%. The maximum water velocity in this water tunnel was 0.88 m/s, whereas river flow velocity condition in Malaysia are around 0.1–0.6 m/s [11]. This indicates that this water tunnel can be used to mimic river flow conditions in Malaysia for the hydrokinetic turbine applications.

Acknowledgements The authors would like to acknowledge the financial support from the Ministry of Education (MoE) Malaysia under the Fundamental Research Grant Scheme (FRGS): (203.PAERO.6071433).

References

1. Kalyankar H, Melwanki R, Choudhary D, Jethwa S, Chaudhari D (2015) Design and analysis of low speed water tunnel for flow visualization of bluff body. In: International conference on advances in mechanical engineering and its interdisciplinary areas, pp 1–9
2. Dehghan Manshadi M (2011) The importance of turbulence reduction in assessment of wind tunnel flow quality. In: Wind tunnels and experimental fluid dynamics research
3. Eng TP (2018) Design and development of water tunnel for flow visualization study on micro-hydrokinetice turbine. Final Year Project, Universiti Sains Malaysia
4. Hoq MT, Nu A, Islam MN, IbneaSina, Syfullah MK, Rahman R (2011) Micro hydro power: promising solution for off-grid renewable energy source. Int J Sci Eng Res 2(12)
5. Rahman M, Faizal H, Saat A, Abd Wahid M (2017) Renewable resource based decentralized power system to a remote village in Malaysia: optimization and techno-economic evaluation. J Adv Res Fluid Mech Thermal Sci, 1–17
6. Dehghan Manshadi M (2011) The importance of turbulence reduction in assessment of wind tunnel flow quality. Wind tunnels and experimental fluid dynamics research
7. Weiselsberger C (1925) On improvement of air flow in wind tunnels. J Soc Mech Eng Jpn (98):1–98

8. Daniel L (2014) Design and installation of a high Reynolds number recirculating water tunnel. Embry-Riddle Aeronautical University, OSPE Oklahoma Engineering Conference
9. Daniel L, Mohagheghian S, Dunlap D, Ruhlmann E, Elbing BR (2015) Design and installation of a high Reynolds number recirculating water tunnel. Int Mech Eng Congr Exposition
10. Seo Y (2013) Effect of hydraulic diameter of flow straighteners on turbulence intensity in square wind tunnel. HVA C&R Research, 19
11. Hin LS, Bessaih N, Ling LP, Ghani AA, Zakaria NA, Seng MY (2008) A study of hydraulic characteristics for flow in equatorial rivers. Int J River Basin Management 6(3):213–223

Statistical Overview of CubeSat Mission

N. M. Suhadis

Abstract The number of CubeSats launched since the first group of six CubeSats was launched in March 2003 has increased. Several reliable databases that collect CubeSat information show that 1035 CubeSats have been launched as of 31 December, 2018. The author has taken the opportunity to use the information published in these databases and presents a statistical overview of CubeSat missions in this paper. This overview will hopefully be beneficial to the readers.

Keywords CubeSat mission · CubeSat launcher · CubeSat deployer

1 Introduction

The term CubeSat refers to a standard cube-shaped satellite that is popularly known as 1U. It has 10 cm × 10 cm × 10 cm dimensions and a mass of no greater than 1.33 kg. This CubeSat design standard was an output of CubeSat development programme in 1999, which was a collaboration work between Bob Twiggs of Stanford University and Jordi Puig-Suari, a professor at California Polytechnic State University [1, 2]. CubeSat not only offers affordable access to space but also can be used for various missions, such as education, remote sensing, science and defence. These advantages have attracted the attention of many universities, organisations and group of enthusiasts from around the world to develop their own CubeSat. In 2003, the first group of six CubeSats each developed by universities was successfully launched and deployed using Russian launcher called Rockot [3]. By the end of 2018, more than one thousand CubeSats developed by various organisations from 59 countries have been launched using 18 types of different launchers.

N. M. Suhadis (✉)
School of Aerospace Engineering, Universiti Sains Malaysia,
14300 Nibong Tebal, Penang, Malaysia
e-mail: nurulasikin@usm.my

P. Rajendran et al. (eds.), *Proceedings of International Conference of Aerospace and Mechanical Engineering 2019*, Lecture Notes in Mechanical Engineering,
https://doi.org/10.1007/978-981-15-4756-0_50

2 Statistical Data Overview

The data presented in this paper cover information of CubeSats that were launched from 2003 until 31 December, 2018. These data have been taken from two online sources that compile CubeSats detail information, namely, Nanosat Database [4] and Saint Louis University's CubeSat Database [5]. In the author's estimation, these online sources have the most complete information regarding few CubeSat categories that interests the author, such as number, size, mission, developer, launcher, deployer and status. The statistical overview of these CubeSat missions where their current global situation regarding these categories can be identified should be provided [6–8]. Therefore, the statistical overview of these CubeSat missions is presented in Sects. 2.1 to 2.4 and is based on the categories and their trends.

2.1 Number of Missions

Countries that have launched CubeSats, along with their respective number of CubeSats launched and the years they were launched, are tabulated in Table 1. These data show that USA, Japan, Denmark and Canada are amongst the countries that successfully launched the first six CubeSats. As of 31 December, 2018, 1035 CubeSats have been launched by 59 countries. Amongst these countries, the USA leads with 746 CubeSats and contributes to 72% out of the total launch.

The number of CubeSats launched per year per their launch status is plotted in Fig. 1. An exponential increase is observed since the first six CubeSats were launched. However, the number of launching dropped in 2007, 2015, 2016 and 2018 due to launcher failure the year before, which probably delayed the scheduled launches.

2.2 Mission Configuration

Multiple 1U CubeSats have been combined to form larger satellites for carrying complex mission. Thus, CubeSat mission can be 1U, 2U, 3U and so forth in size and weigh less than 1.33 kg per U. Figure 2 shows that most missions at the beginning was launched using 1U CubeSat platform. However, starting in 2013, most missions were launched using 3U CubeSat platform due to the increasing demand on fulfilling the diversification of the missions together with the advancement of the CubeSat technologies.

As shown in Fig. 3, 3U CubeSat is the mostly used configuration with 646 launches, followed by 1U, 2U, 6U and 1.5U with 190, 76, 63 and 50 launches, respectively. The largest configuration is 12U with only a single launch. The data collected show that 3U CubeSat platform has been mostly used for earth observation

Table 1 Number of CubeSats launched per year by country from 2003 to 2018

Country	Launching Year																Total
	2003	2004	2005	2006	2007	2008	2009	2010	2011	2012	2013	2014	2015	2016	2017	2018	
USA	1			12	6	2	4	11	9	12	51	109	102	57	223	147	746
China													5	7	7	23	42
Japan	2		1	2		1		3		3		4	1	7	2	7	33
Germany			1			1	2				5			1	3	5	18
Russia												2	2		7	7	18
South Korea				1							3				3	10	17
Denmark	2					1					2	2	2	1		3	13
UK											2	1	1	1	5	2	12
Italy										2		1	1	2	2	2	10
Australia															4	5	9
Spain										1	2			1	1	2	7
Canada	1					1								1	1	2	6
Netherland						1					2				1	2	6
France										1					3	1	5
India								1	1					2		1	5
Singapore											1	2	2				5
Turkey							1				1				2	1	5
Brazil												1	2			1	4
Finland															2	2	4
Israel												1			3		4

(continued)

Table 1 (continued)

Country	Launching Year																Total
	2003	2004	2005	2006	2007	2008	2009	2010	2011	2012	2013	2014	2015	2016	2017	2018	
South Africa											1				2	1	4
Belgium												2		1			3
Ecuador											2				1		3
Kazakhstan															1	2	3
Lithuania												2			1		3
Norway			1	1							1						3
Peru											1	2					3
Swirtzeland							1	1								1	3
Argentina											2						2
Colombia					1											1	2
Greece															2		2
Malaysia																2	2
Poland										1						1	2
Sweden															2		2
Taiwan												1			1		2
UAE															1	1	2
Ukraine												1			1		2
Vietnam										1	1						2
Austria															1		1

(continued)

Table 1 (continued)

Country	Launching Year																Total
	2003	2004	2005	2006	2007	2008	2009	2010	2011	2012	2013	2014	2015	2016	2017	2018	
Bangladesh															1		1
Belarus																1	1
Bhutan																1	1
Bulgaria																1	1
Chile															1		1
Costa Rica																1	1
Czech															1		1
Estonia											1						1
Ghana															1		1
Hungary										1							1
Jordan																1	1
Kenya																1	1
Mongolia															1		1
Nigeria															1		1
Pakistan											1						1
Phillipines																1	1
Romania										1							1
Slovakia															1		1
Thailand																1	1
Uruguay												1					1
Total	6	0	3	16	7	7	8	16	10	22	78	132	118	81	289	240	1035

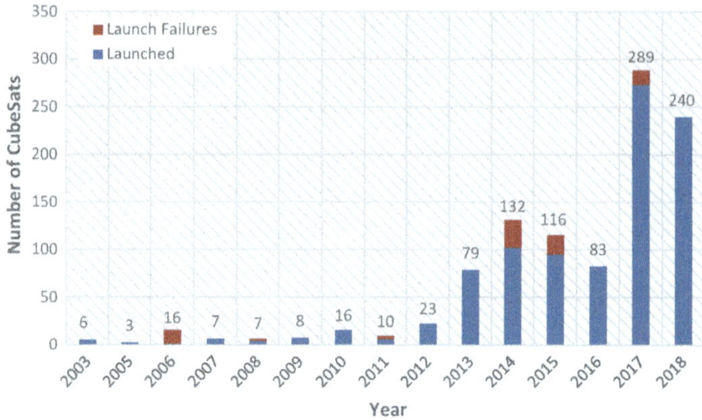

Fig. 1 Number of CubeSats launched per year

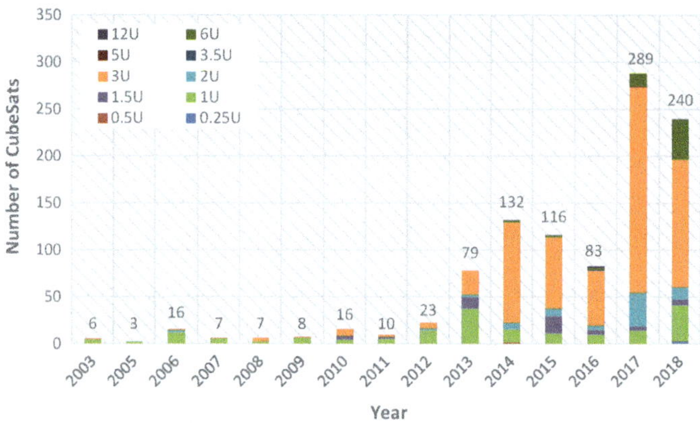

Fig. 2 Number of CubeSats launched per year by type

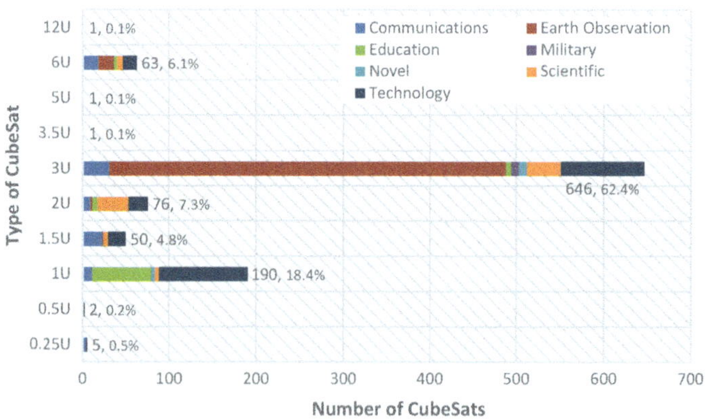

Fig. 3 Number of CubeSats launched per type and per mission

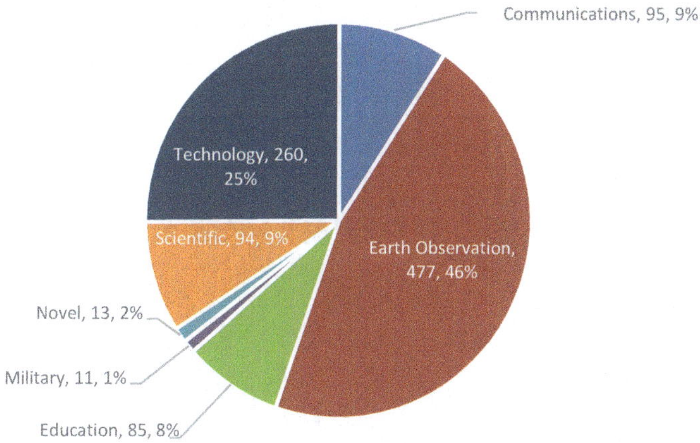

Fig. 4 Distribution of CubeSats by mission

missions, whilst 1U CubeSat platform has been mostly utilised for education missions. The distribution of CubeSats according to mission is shown in detail in Fig. 4. The fraction shows that CubeSat platform has been used for various missions, such as earth observation (46%), technology (25%), communications (9%), scientific (9%), education (8%), novel (2%) and military (1%).

2.3 *Developer*

The number of CubeSats launched per year by developer is plotted in Fig. 5. The trend shows that most missions launched at the beginning were dominated by university. However, in 2014, companies started to become interested in launching CubeSats.

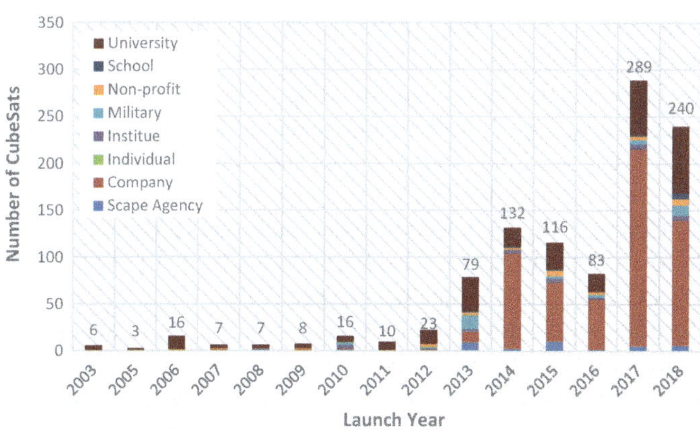

Fig. 5 Number of CubeSats launched per year by developer

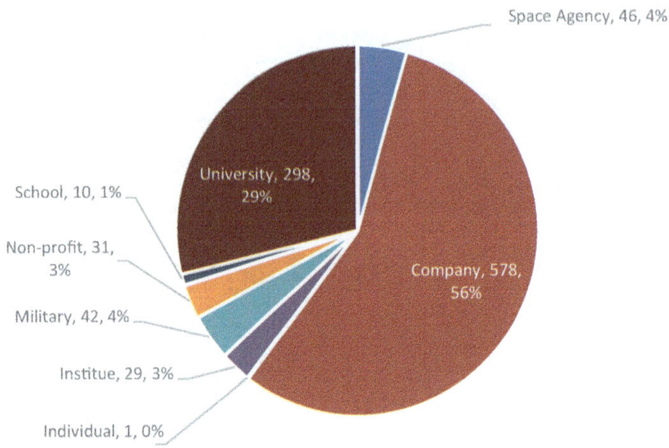

Fig. 6 Distribution of CubeSats by developer

Since then, most missions launched were dominated by companies. The distribution of CubeSats by developer is shown in detail in Fig. 6. Apart from companies and university, space agency, institute, non-profit organisation, military, school and individual also have been taking part in launching CubeSats. However, the percentage is very low at only 15%. Companies lead the fraction by 56%, followed closely by university with 29%.

2.4 *Launcher*

CubeSats have been launched into the orbit either through launch services that allow CubeSats to hitch a ride or through cargo that is sent on a resupply mission to the space station and released into space from there [9]. Either way, CubeSats will be attached to an ejection system that is called deployer, which will deploy the CubeSat to the orbit.

The main launchers that have been providing CubeSat launch services should be identified. As shown in Fig. 7, Antares dominated the CubeSat launching services in 2014. In 2017, PSLV and Sojuz started to join and were followed by Falcon the next year. Overall, Fig. 8 shows that PSLV leads and has been chosen as a launcher by 222 (22.2%) CubeSats. It is followed by other important launchers, namely, Atlas V, Antares, Sojuz and Falcon, which have been chosen by 141 (13.6%), 136 (13.1%), 133 (12.9%) and 118 (11.4%) CubeSats, respectively.

Figure 9 shows the distribution of CubeSat by deployer in detail. As observed, Quad-pack and NRCSD play an important role and have been chosen by 327 (31.6%) and 235 (22.7%) CubeSats, respectively. Quadpack can accommodate CubeSats onboard Dnepr, PSLV, Soyuz and Falcon 9 launchers where it can house up to a maximum volume of 12U. In the meantime, NRCSD is used to deploy satellite from the ISS and can house to a maximum volume of 6U.

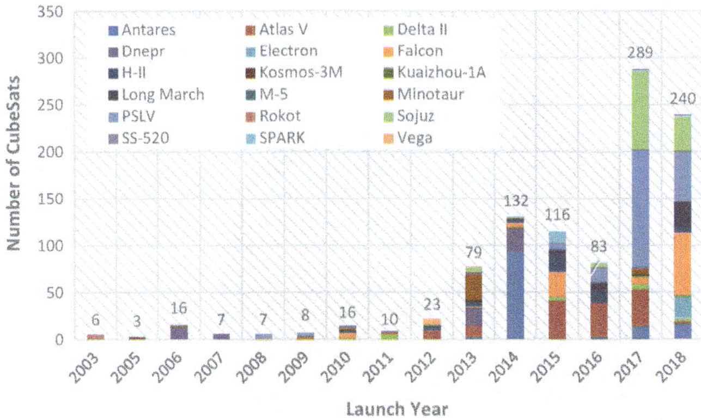

Fig. 7 Distribution of CubeSats by launcher

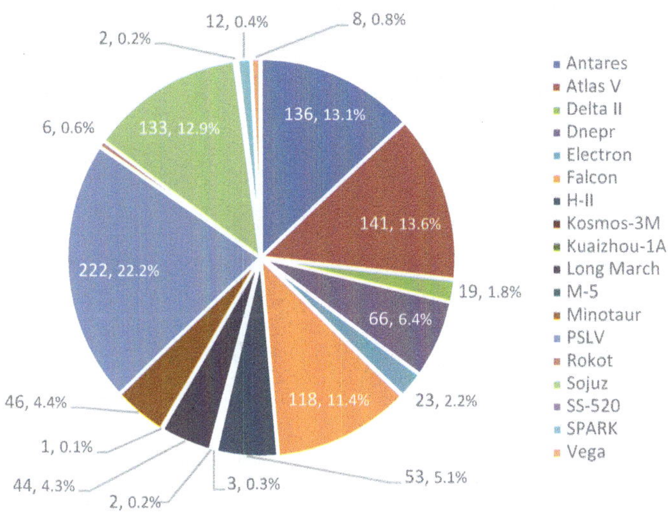

Fig. 8 Distribution of CubeSats by launcher

2.5 *Mission Status*

CubeSat mission status per type for all CubeSats launched until 31 December, 2018 is shown in Fig. 10. A mission indicated as 'fail' means it fails due to launch failure. The data show that the success rate for 0.25U, 0.5U, 3.5U, 5U and 6U CubeSats are 100%. The lowest success rate is 84% for 1.5U CubeSats.

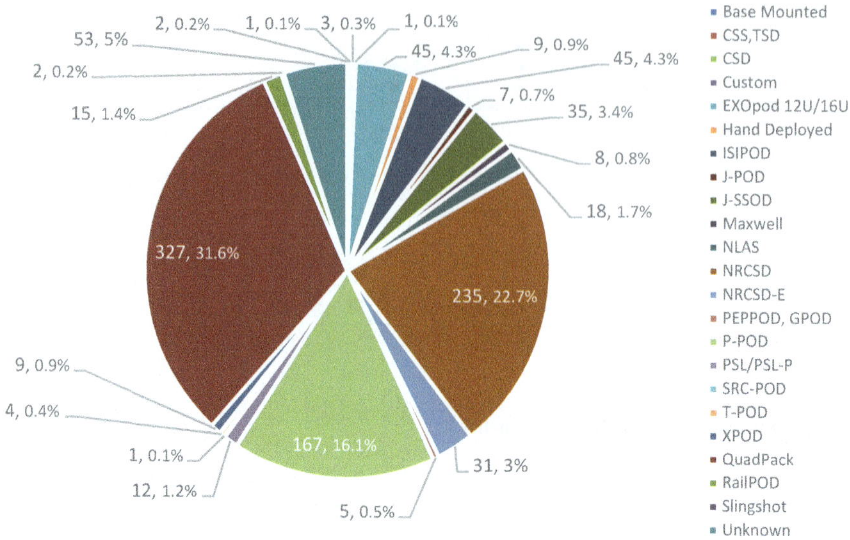

Fig. 9 Distribution of CubeSats by deployer

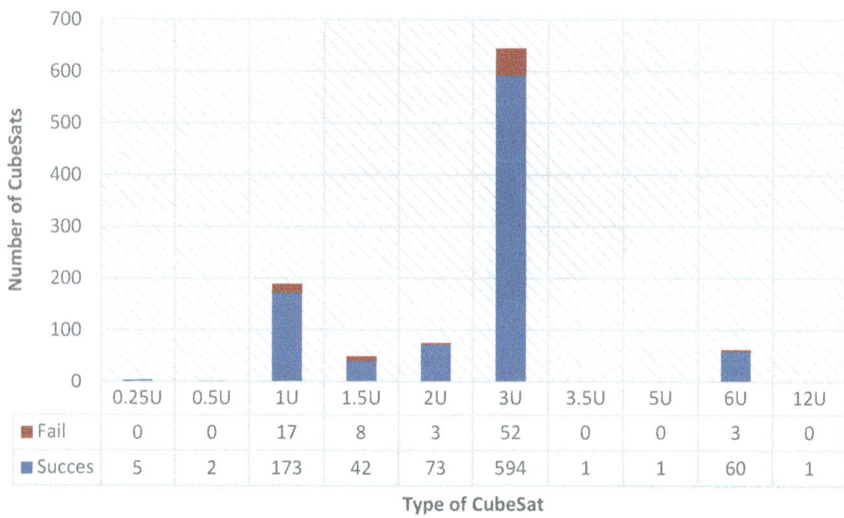

Fig. 10 CubeSats mission status per type

3 Conclusion

Statistical overview of CubeSat number, size, mission, developer, launcher, deployer and status for all CubeSats launched until 31 December, 2018 is presented. The trends are also identified. A total of 1035 CubeSats have been launched, and only 83

CubeSats fail due to launcher failure. This success rate shows that CubeSat platform is an interesting option for future missions.

Acknowledgements The work presented in this paper has been supported by the USM Research University Grant (1001/PAERO/8014108).

References

1. Twiggs R (2008) Origin of CubeSat. In Helvajian, H, Janson, S (eds) Small satellite: past, present and future, pp 151–173
2. Heidt H, Puig-Suari J, Moore A, Nakasuka S, Twiggs R (2000) CubeSat: a new generation of picosatellite for education and industry low-cost space experimentation. In: 14th annual/USU conference on small satellites, Logan Utah, 21–24 Aug 2000
3. Chin A, Coelho R, Nugent R, Munakata R, Puig-Suari J (2008) CubeSat: the pico-satellite standard for research and education. In: AIAA SPACE 2008 conference & exposition, San Diego, California, 9–11 Sept 2008
4. Kulu E (2019) Nanosats database. https://www.nanosats.eu/. Last accessed 19 Sept 2019
5. Saint Louis University, CubeSat Database, https://sites.google.com/a/slu.edu/swartwout/home/cubesat-database/. Last accessed 19 Sept 2019
6. Swartwout M (2011) Attack of the CubeSats: a statistical look. In: 25th annual/USU conference on small satellites, London
7. Langer M, Bouwmeester J (2016) Reliability of CubeSats—statistical data, developers' beliefs and the way forward. In: 30th annual/USU conference on small satellites, Logan Utah, 6–11 Aug 2016
8. Swartwout M (2018) Reliving 24 years in the next 12 minutes: a statistical and personal history of university-class satellites. In: 32nd annual/USU conference on small satellites, Logan Utah, 4–9 Aug 2018
9. Chin J, Coelho R, Foley J, Johnstone A, Nugent R, Pignatelli D, Pignatelli S, Powell N, Puig-Suari J (2013) Cubesat 101: basic concepts and processes for first-time CubeSat developers, San Luis Obispo
10. Swartwout M (2013) The first one hundred CubeSats: a statistical look. J Small Satell 2(2):213–233
11. Villela T, Costa CA, Brandão AM, Bueno FT, Leonardi R (2019) Towards the thousandth CubeSat: a statistical overview. Int J Aerosp Eng

A Physical Characterisation for the Preliminary Fabrication of Natural Woven Fabric Prepreg and Their Composites

Fadzil Adly Ishak, Wan Muhammad Izzat Wan Zaludin, Nurul Zuhairah Mahmud Zuhudi, and Khairul Dahri Mohd Aris

Abstract In order to find alternative composite material that possesses the ability for a great revival of structural application, cheap and environmentally friendly, natural woven fabric prepreg have earned its place as a replacement to traditional synthetic fabric prepreg. However, the limitation is the applicability of techniques of fabrication, which are compatible with the different properties of natural woven fabrics. The investigation on the individual material of natural woven fabric and resin are also essential to recognise the factor that influences the fabrication process of prepreg product. A preliminary experiment conducted on three different types of natural woven fabric, i.e., flax, bamboo, and jute in the form of a single-layer prepreg product which combined with Acrodur polyester resin. The resin content and their dispersion, prepreg, fibre and resin area weight are the parameters evaluated. The produced natural woven fabric prepregs have retained the resin content more than 50% (55.7–63.6%) for jute and flax, but only 47.6% resin content for bamboo material.

Keywords Natural woven fabric · Prepreg · Hot bonder curing

1 Introduction

Nowadays, the widely used renewal resources such as natural fibres have gained advantages due to its strength, multifunctional and recyclability. Natural fibre, especially plant fibre is now more preferred and viable as the reinforcement composites based on studies done by many researchers [1, 2]. Plant fibres function as reinforcing agents of the composite, which benefit the eco-system that shows a positive impact on the environment and society due to its biodegradability and safer handling behavior, respectively. Plant fibres can be considered as reinforced composites having cellulose embedded in the lignin matrix, and combination with other components such as hemicelluloses, pectin, and wax [3]. Kenaf, bamboo, sisal, jute, flax, ramie, sisal,

F. A. Ishak (✉) · W. M. Izzat Wan Zaludin · N. Z. Mahmud Zuhudi · K. D. Mohd Aris
Universiti Kuala Lumpur, Malaysian Institute of Aviation Technology, Shah Alam, Malaysia
e-mail: fadziladly@unikl.edu

P. Rajendran et al. (eds.), *Proceedings of International Conference of Aerospace and Mechanical Engineering 2019*, Lecture Notes in Mechanical Engineering,
https://doi.org/10.1007/978-981-15-4756-0_51

and hemp are the example of plant fibre that found much suitable for replacement of synthetic fibres such as fibreglass which are comparatively heavy, high cost, and not environmental friendly [2]. Hojo et al. [4] has studied tensile property especially on low cycle fatigue behaviour of three natural fibres mat (bamboo, jute and kenaf) reinforced unsaturated polyester composites. The natural fibre mats reinforced composite have shown a substantial tensile property as well as the potential to replace the petroleum-based composite. In general, the best mechanical properties of composites are obtained when the fibre alignment parallel to the direction of the applied load. This alignment of natural fibre, however is quite challenging to achieve for short fibre when compared to long fibre. The natural fibre can be carded and placed manually in a sheet before matrix impregnation [5].

Alternatively, processes such as braiding, knitting and weaving are common to fibre textile processing, can also be employed on natural fibre. Woven fabric is now popular as the reinforcement to a polymer which provides balanced properties in the fabric plane compared to non-woven or unidirectional fibre. Through the study done by Hani et al. [6], the combination of resin system with textile fibre, yarn or fabric system had shown that yarn types and fabric structures have significant effects on the performance of impact and flexural properties. Woven structure parameters such as yarn weight, fabric count and weave pattern also influence the strength and stiffness of fabric composites [7]. Natural fibres may also apply the weave geometries pattern such plain, twill and satin weave that usually utilised for synthetic fibre, either in the form of a straight filament or twisted as to hold the fibre together.

The selection of composite manufacturing techniques is also vital to have a good quality product and meet the desired requirement and characteristics. Prepreg process has revealed its readiness of curing that highly control ratio of combinations of fibre-resin material. The prepreg process is an advantage to production to speed up the process of fabrication of composite products where it is more efficient and easy to handle [8]. Prepreg product can be in the form of unidirectional, woven or stitched cloth fibre in which embedded by whether thermoset or thermoplastic resin that partially cured (semi-finished product). Aircraft composite manufacturers often used prepreg to produce composites for high and uniform volume fraction and freedom from voids and porosity. The prepreg material is design for an elevated thermal cure through a specific curing cycle of high temperatures. The curing reaction depends on the resin system applied, e.g., typically cured at 180 °C with positive or negative pressure for 2 or 3 h [9]. Most of the natural fibres prepreg composites studies frequently focusing on hot press curing or oven, but not on other alternative sources such hot bonder system.

In this study, the research focuses on the fabrication process of impregnation of woven fabric natural fibre and thermoset resin to form a partially cured prepreg material by using the laminating machine. There are a few types of reinforced material have been investigated, such as hemp and kenaf, to form a prepreg product which combined with Acrodur polyester resin [10–12]. The factors that influence the quality of prepreg, especially material geometrical/architectures of fabric and machine parameters such as conveyor line speed, hot tables temperature, and laminate contact pressure have also been monitored and controlled. The criteria of the prepregs and composites for quality evaluation such as resin content and dispersion, prepreg, fibre and resin area weight were selected and determined. The final composites' physical characteristics are also being evaluated, especially on the void appearance on both surfaces (mould and bagged side) of the prepreg composite. The natural woven fabric prepreg can be used in various area of the composite application without requiring high technology equipment or machinery such as autoclave curing. The use of oven and vacuum bagging process is sufficient to assist in generating the semi-structural composite component. The hot bonder application can be utilised for any 'in-situ' repair on any natural fibre composite parts that makes the product cheaper and potentially marketed as "easy to use" products.

2 Materials and Methods

2.1 Material

Three types of woven fabric such as jute, flax, and bamboo were used during the natural fabric prepreg manufacturing. The weave pattern, thickness, warp and weft yarn diameter, weight of fabric sample and crimp percentage of woven fabric were measured and tabled as shown in Table 1. The Acrodur resin, DS3530 from BASF was used as a matrix resin. It's an aqueous acrylic resin-based having a viscosity of 300–1,500 mPa s at 23 °C and miscibility with water. Under the condition of the pre-cure state, the resin exhibits thermoplastic properties and confers the prepreg material excellent storage stability compared to other thermoset resins. It's a formaldehyde-free binder and is adapted to wood and plant fibres such as flax, sisal or jute. Acrodur enables to overcome adhesion problems with natural fibres and reduces the processing costs by avoiding surface treatments. Curing in the presence of moisture activates bonding by an esterification reaction between the acid group of Acrodur and hydroxyl groups of the natural fibre constituents [12].

Table 1 Properties of natural woven fabric

Natural fabric	Bamboo	Flax	Jute
Weave pattern	2/1 TWILL	2/2 TWILL	1/1 PLAIN
Thickness (cm)	0.022	0.028	0.071
Weight (g)	5	4	6
Warp diameter (average), mm	0.321	0.582	1.067
Weft diameter (average), mm	0.306	0.388	1.298
Crimp percentage (%)	4.9	4.7	4.8

2.2 Fabrication of Woven Natural Fabric Prepregs

The prepregs were fabricated using hand layup and followed by the laminating machine process. All dry fibres, bamboo, jute and flax were cut into 8 in × 4 in per ply. The mass of each ply was recorded to identify the difference of mass before and after the manufacturing process is done. Generally, the coating of the pre-pregging process was conducted either through the process of the hot-melt method or the dipping solvent method [9]. However, since the experiment is performed for lab scaled sampling, the application of Acrodur resin by manual layup (wet layup) seems appropriate to simulate the process as a dipping method with resin completely penetrated (woven fibre wetted thoroughly). The wetted woven fabric was then be wrapped with backing film (polyethene protector) individually (Fig. 1b) and went through the laminating machine by controlling the speed conveyor, pressure and temperature. During the prepreg laminating process, the force applied could squeeze the access resin out in order to control the dispersion ratio of resin to the woven fabric of prepreg. It is vital to control the pressure amount to avoid any massive access resin to flow out to the machine conveyor. The control and adjustment of pressure, temperature and conveyor speed were done on the laminating control panel. The conveyor speed and pressure were fixed throughout the prepreg fabrication, with 0.33 m/min and 5 N/m^2, respectively. Three sets of temperatures were set differently to customise the Degree of Cure (DOC) of prepreg. The temperature of 60, 75, and 90 °C were selected for the experiments.

The prepreg fabrications were grouping into seven samples; namely, Sample 1 (B1) as a sample of bamboo prepreg, Sample 2, 3 and 4 (F2, F3 and F4 respectively) are samples of flax prepreg and Sample 5, 6 and 7 (J5, J6 and J7 respectively) are samples of jute prepreg. The laminating machine was set to the temperature of 90 °C

(a)

(b)

Fig. 1 Lamination machine system used for prepreg fabrication, **a** complete set of laminating machine system **b** wetted prepreg with polyethene protector on machine conveyor for all three types of natural fibre fabric (B1, F2 and J5) as the first trial run. The second run (F3 and J6) and the third run (F4 and J7) of prepreg were only focusing on the flax and jute woven fibres. The temperature of 75 and 90 °C was set for second and third run prepreg fabrications respectively. The prepregs were kept in the freezer (0 °C) before the commencement of the final layup and curing process. All the natural woven fabric prepregs produced are weighted and determined individually. The resin content of prepreg is determined in weight percentage relative to fibre areal weight and provides the total weight of prepreg when added to the fibre areal weight [9].

2.3 Composite Fabrication and Curing

The fabrication of the woven natural fabric composites was carried out using the vacuum bagging process for the curing. The preparation of the prepreg materials was carried out prior to the vacuum bagging process in which the arrangement of the ply stacking sequences was stacked together with six plies of woven fabric prepreg (in the earlier stage) on a flat metal plate (mould). The woven fabric prepreg plies were taken out from the freezer for thawing for approximately about 10–15 min. All protective release films (separators) to be removed from the prepreg before lay-up. The mould was cleaned using acetone. The release agent was applied to prevent prepreg resin from bonding to mould surface, and the cured composites are easily de-moulded. Seven samples of prepreg composite were laminated into the group as described before (B1, F2, F3, F4, J5, J6 and J7). The mass of each ply was recorded to identify the difference of mass before and after the manufacturing process is done. A debulking process was not performed throughout the hand layup process due to the thin layer of plies. The specific curing recipe was applied using the hot bonder curing process by using of Heatcon HCS9200B, hot bonder system. The elevated temperature was set to a rate of 3 °C/min until reaching 150 °C. Curing time was set for 3 h, and cooling at a rate of 3 °C/min. The vacuum pressure was recorded and maintained a minimum of 22 in hg throughout the curing process under vacuum bagging.

2.4 Physical Characterisation

The characterisation of the natural woven fabric prepreg and their composites were conducted using the USB Digital Microscope Dyno Lite Edge model AM4815ZT to analyse the surface structure of the materials and the composites. The natural woven fabric prepreg and their composites were inspected for the resin dispersion throughout the woven fabric. The surface panels, mould and bagged surfaces, were divided into 9 zonal areas, and each zone area was subdivided into 10 × 10 as shown in Fig. 2. This procedure permitted visualisation of each sample, and it's revealing

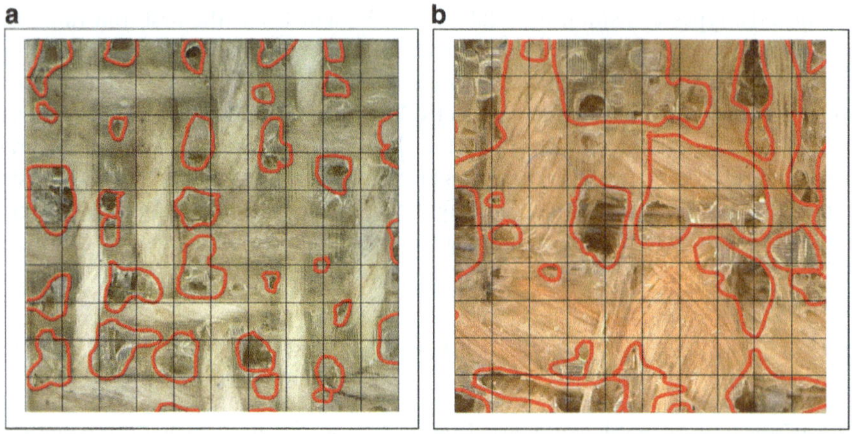

Fig. 2 Example of sub-division of every zone. **a** 10 × 10 quadrat of Flax zone (left), **b** 10 × 10 quadrat of Jute zone (right)

the distribution of voids that allowing the void estimation measurement by quadrat sampling method.

3 Results and Discussion

3.1 Physical Characteristics of Natural Woven Fabric

It can be observed that the woven bamboo has a very fine fabric surface topology compared to Jute and Flax Fabric (Fig. 2). Bamboo fabric depicted a very tight structure of fabric with a small diameter of yarn. The summary of the characteristic of bamboo, jute and flax woven fabrics are shown in Table 1. Azrin Hani et al. [6] has demonstrated that the tightness and crimp effect of the fabric structure is influenced by the inter-yarn porosity, fibre yarn thickness and the weft wavelength where a tight woven structure can be achieved by a small yarn diameter, lower wavelength with the lower inter-yarn porosity.

3.2 Physical Characteristic of Natural Woven Fabric Prepreg

The produced prepregs, either woven of bamboo, Jute and flax are a semi-finished product consisting of viscous matrix and continuous reinforcing fibres. The prepreg was still in wet condition but considerably tacky. Dispersion of resin throughout the natural woven fabric prepregs was also observed under the magnification of USB

Digital Microscope Dyno lite. One can found from the global view of the sample that resin was scattered unevenly on the single fabric surface area, with few spotted areas visibly resin-rich on prepreg surface. Void appearances and starved-resin regions were observed due to the irregular pattern, inconsistent and wide distance of warp and weft yarn alignment. The smooth and clean surface can only be observed on the bamboo prepreg. No void formation can be detected or recognised by using Dyno Lite on the bamboo prepreg surfaces. Refer to Fig. 3.

Khalfallah et al. [12] obtained the glass transition temperature of 67 °C for flax tape reinforced Acrodur Biocomposites. The temperature of 100 °C was set in the experiment to transform the prepreg to thermoplastic film to drive water off from the Acrodur system. This becomes the reference of temperature setting for prepreg fabrication. The temperature setting of 60, 75, and 90 °C were selected for the laminating of prepreg. The prepreg materials required for a warm-up that beyond the dew point temperature to avoid condensation of air humidity on the unprotected material surface [13]. Based on the first trial run finding, since the fabric tightness of flax and jute are very low, and the fabric porosity was visibly witnessed. The second and the third run of prepreg fabrication were only focusing on the woven flax (F3 and F4) and woven jute (J6 and J7), respectively.

Table 2 summarised the properties of prepreg produced, namely Prepreg Areal Weight (PAW), Fibre Areal Weight (FAW), Prepreg Resin Contents and Resin Areal Weight (RAW). The prepreg areal weight measurement is presented in the average value for a single ply of the woven prepreg. The prepreg resin content was determined, which specifies the percentage of the resin by weight of the woven fabric prepreg, and results are within the range of 47.6–63.6%. For the synthetic fibre type, The objective of the new generation prepreg is to minimise the bleeding excess resin during curing. This target can be achieved by controlling the net resin content of prepreg, which the resin should be pre-impregnated to almost the same resin content as the final cure part [14, 15].

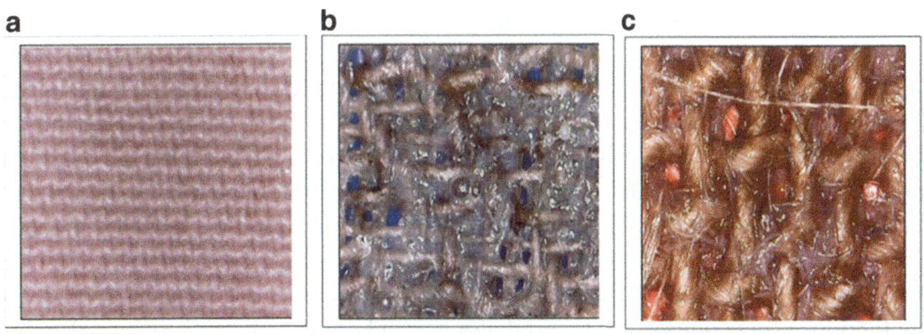

Fig. 3 Natural woven fabric prepreg under magnification of Dyno Lite, **a** Bamboo Prepreg (left), **b** Flax Prepreg (centre) and **c** Jute Prepreg (right)

Table 2 Physical properties of natural woven fabric prepreg (average value)

Properties	Bamboo	Flax			Jute		
	S1	S2	S3	S4	S5	S6	S7
Prepreg areal weight (g/cm^2)	0.052	0.051	0.048	0.051	0.081	0.071	0.077
Prepreg resin content (%)	47.6	61.8	60.00	61.8	63.6	55.7	62.0
Fibre areal weight (g/cm^2)	0.027	0.019	0.019	0.019	0.029	0.031	0.029
Resin areal weight (g/cm^2)	0.025	0.031	0.029	0.031	0.052	0.040	0.048

3.3 Physical Characteristic of Natural Woven Fabric Composites

The result of void measurement are presented in Table 3. The measurements were carried out only on the bagged side of the composite panel. Jute and flax prepreg composites had more void surface contents due to the coarse fibres yarn and a bigger gap between the adjacent warp and weft yarn. Dry composite surfaces were observed on the mould side of the composite panel compared to the bagged side of composite. The dry surface happened may due to the prolonged exposure of the composite panel to a hot metal plate (mould) that highly condenses the water content in reinforced fabric and resin itself. The Acrodur system initially had about 50% water content, thus the effect of elevated rate temperature that evaporates the solution (water) and detriment the curing process of the composite. In this case, the dry surface might lead to a poor surface finish, and therefore it is unacceptable as of finished part. Besides, some vapour is released and may create blisters and voids in the final composite laminates during high curing temperatures [12].

Figures 4 and 5 showed the appearance of the void on composites panel surfaces (mould side) and cross-section of composites panel respectively.

The curing process of the resin matrix would be the most critical manufacturing step of the prepreg process, where the polymerisation reaction takes place during the consolidation of composite plies under pressure (positive or negative). Temperature and pressure are the two main parameters that have to take into consideration with the suitable curing cycle selection. The matrix provides the cured composite with toughness, damage tolerance, impact and abrasion resistance. Void or laminate porosity is one of the highlight issues that are concerned by industries processing thermosetting composites material along with honeycomb crush and exothermic reaction issues [13]. Medina et al. [10] have reported that using prepregs of natural fibres/Acrodur containing 20% weight of moisture leads to a composite with bubbly face due to trapped water vapour. The origin of void formation is primarily due to trapped volatiles, either entrapped during the lay-up process or more commonly apart of the prepreg systems such as absorbed water from the pre-pregging process.

Table 3 Void percentage measurement on composites

Composite Prepreg	Percentage of void (%)										
Types	Samples	Zone 1	Zone 2	Zone 3	Zone 4	Zone 5	Zone 6	Zone 7	Zone 8	Zone 9	Average
Bamboo	B1	0	0	0	0	0	0	0	0	0	0
Flax	F2	22	26	29	36	39	33	31	30	35	31.2
	F3	27	37	28	31	44	46	44	40	35	36.9
	F4	23	31	30	32	44	31	31	33	24	31.0
Jute	J5	42	53	59	49	54	53	50	46	50	50.7
	J6	60	42	56	65	57	50	51	50	71	55.8
	J7	34	39	26	62	41	62	72	38	50	47.1

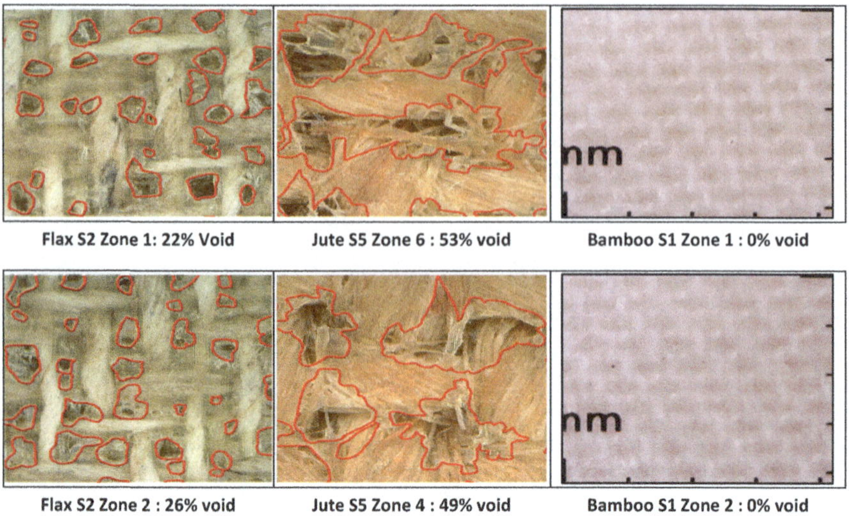

Fig. 4 A sampling of void distribution measurement of the zone for flax S2 (left), jute S6 (centre) and bamboo S1 (right) on the bagged side of composites

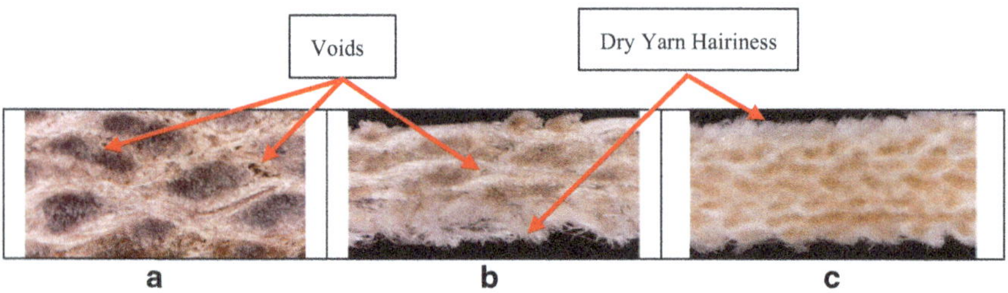

Fig. 5 Cross-section of natural woven fabric composites panel (left) **a** Jute composite (left), **b** flax (centre) and c) bamboo composite (right)

4 Conclusions

It can be concluded that the preliminary experiments of manufacturing the natural woven fabric prep-preg and their composites were successfully conducted using the bamboo, jute and flax fabric. The woven prepreg made from the jute and flax has shown the high porosity (gap) compared to bamboo. It is also observed that the resin dispersion was not distributed evenly due to the irregular pattern, inconsistent and wide distance of warp and weft yarn alignment. The experiment to be further conducted to optimise the manufacturing process with the control of the materials and process parameters. The experiment also showed that there is a potential of the impregnation process of the natural woven fabric to be used as potential reinforcement material of the semi-finished products such as prepreg, thus can reduce the dependency of the synthetic based reinforcements. Furthermore, the physical and mechanical characterisation needs to be conducted in the future to ensure the quality of the product and enhance the potential of the prepreg made from natural fabric based.

References

1. Alavudeen A, Thiruchitrambalam M, Venkateshwaran N, Athijayamani A (2011) Review of natural fiber reinforced woven composite. Rev Adv Mater Sci 27:146–150
2. Cicala G, Cristaldi G, Recca G, Latteri A (2010) Composites based on natural fibre fabrics. Woven Fabric Engineering, Polona Dobnik Dubrovski, IntechOpen
3. Hojo T, Zhilan XU, Yang Y, Hamada H (2014) Tensile properties of bamboo, jute and kenaf mat-reinforced composite. Energy Proc 56:72–79
4. Pickering KL, Efendy MGA, Le TM (2016) A review of recent developments in natural fibre composites and their mechanical performance. Compos Part A Appl Sci Manuf 83:98–112
5. Hani ARA, Ahmad R, Mariatti M (2013) Influence of laminated textile structures on mechanical performance of NF-epoxy composites. Int J Mater Textile Eng 1131–1137 (2013)
6. Nicolais L, Canonico P (2012) Woven composite structures. Wiley Encyclopedia, Compos
7. Shaghaghi S, Beheshty MH, Rahimi H (2011) Preparation and rheological characterization of phenolic/glass prepregs. Iran Polym J 20(12):969–977
8. Lengsfeld H et al (2015) Prepreg technology. Carl Hanser Verlag GmbH & Co, KG
9. Medina L, Schledjewski R, Schlarb AK (2008) Process related mechanical properties of press molded natural fiber reinforced polymers. Special issue 12th European conference on composite materials ECCM 2006 69(9):1404–1411
10. Salim MS, Ariawan D, Ahmad Rasyid MF, Ahmad Thirmizir MZ, Mat Taib R, Mohd. Ishak ZA (2019) Effect of fibre surface treatment on interfacial and mechanical properties of non-woven kenaf fibre reinforced acrylic based polyester composites. Polym Compos 40:E214–E226
11. Khalfallah M et al (2014) Innovative flax tapes reinforced Acrodur biocomposites: a new alternative for automotive applications. Mater Des 64:116–126
12. Pansart S (2013) Prepreg processing of advanced fibre-reinforced polymer (FRP) composites. Woodhead Publishing Limited (2013)
13. Wolff-Fabris F, Lengsfeld H, Krämer J (2015) Prepregs and their precursors. Compos Technol 11–25
14. Campbell FC (2003) Manufacturing processes for advanced composites. Elsevier Science (2003)
15. Seferis JC, Hillermeier RW, Buehler FU (2000) Prepregging and autoclaving of thermoset composites. Compr Compos Mater 2:701–736

Vibration Signal Separation Technique by Using the Combination of Adaptive Filter and Hilbert Transform for Defect Machine Component in Rotating Machinery

K. A. Wahid, M. N. F. Saniman, M. A. Khairul, and I. A. Azid

Abstract This paper focused on enhancement defect signal by separating the unwanted signal by using combination of Adaptive Filter and Hilbert Transform. This combination is performed via simulation as well in experimental work. A growth index is introduced to serve as indicator to verify the fault detection capability. The result of this combination is compared with previous approach which are combination between wavelet packet and Hilbert Transform. The results showed that the unwanted signal is successfully separated, and sign of failures is greatly enhanced.

Keywords Adaptive filter · Hilbert transform · Noise separation · Machine diagnosis

1 Introduction

The critical component such as bearing, gear and time-belting in manufacturing equipment could result the serious damage to the machine system if the defects occur during operation. Hence, an early fault detection is essential. The common method for fault detection is by examining the changing of their time waveform signal characteristics such as vibration and sound signal [1]. However, the change of signal characteristics is hard to be detected and evaluated since the defect machine signal usually will be disrupted by unwanted signal that coming from another machine component and environment. Regarding to this problem, these unwanted signal needs to be separated from the machine defect signal so that the early sign of failure easily can be detected and evaluated.

K. A. Wahid (✉) · M. N. F. Saniman · M. A. Khairul
Mechanical Engineering Section, Universiti Kuala Lumpur (UniKL), Malaysia France Institute (MFI), 43650 Bandar Baru Bangi, Selangor, Malaysia
e-mail: Khairulanuarabdwahid@unikl.edu.my

I. A. Azid
Mechanical Engineering, Universiti Kuala Lumpur (UniKL), Malaysia Spanish Institute (MSI), Kulim Hi-Tech Park, 09000 Kulim, Kedah, Malaysia

P. Rajendran et al. (eds.), *Proceedings of International Conference of Aerospace and Mechanical Engineering 2019*, Lecture Notes in Mechanical Engineering,
https://doi.org/10.1007/978-981-15-4756-0_52

There are several available methods that can be used to separate the unwanted signal. One of the well-known methods is the basic filter such as lowpass filter, high pass filter, bandpass filter, and band stop filter [2]. These basic filter methods are applied to separate the signal based on frequency. However, the basic filter has several serious drawbacks where range of frequency component that need to be separated from the signal need to be determine first. Otherwise, the useful signal component from deterioration signal will be removed unintentionally due to the overlapping bandwidth with unwanted signal.

In order to overcome this limitation, Blind Source Separation (BSS) technique has been proposed. BSS algorithm collects all the signals from each source and then used them as reference signal that needs to be separated [3]. This observation process can be carried out by attaching the sensor at each signal source However, attaching sensor at each source is difficult and computationally expensive for complex machine.

Aside BSS, the Wavelet Transform (WT) also show the promising approach to separate the unwanted signal. [4] employed Wavelet Transform to suppress the unwanted signal which in turn able enhance the sensitivity of mechanical defect. [5] enhance the SNR of low-speed rolling element bearing, [6] clean the noise signal in diesel cylindrical vibration and [7] clean the heavy noise signal that masked the gears and roller bearing signal. These widely application is regarding to the advantages of Wavelet Transform where it can separate the signals based on multi-scale both in time and frequency domain. However, wavelet transform also suffer several limitations such as interference terms, border distortion and energy leakage. These limitations will generate a lot of undesired small spikes all over the frequency scales and made the separation results confusing and difficult to be interpreted.

Due to wavelet transform limitations, a Hilbert Huang Transform (HHT) has been proposed. HHT is an extension study from the principle of empirical mode decomposition and Hilbert Transform [8]. When HHT is adopted, the intrinsic mode function (IMF) is applied to determine all the instantaneous frequencies from the complex signal. Compared with wavelet transform, the computational time is less, thus believed suitable to separate the complex noisy signal. However, HHT also suffer from several shortcomings such as misinterpretation at low frequency, and unable to detect low amplitude of signal.

To solve this problem, a Wavelet packet transform is used to separate the signal component into a set of narrow band signals. Hence, the frequency components that contain low energy are easier to be identified at different narrow bands. The HHT then will be applied to separate these narrow band signals. [9] extend this study by combining HHT with discrete wavelet transform. In this study, multi-level signal decomposition is applied in order to separate the signal into different frequency band. However, multi-level decomposition has several limitations where it can be iterated indefinitely [10]. Thus, the optimum of number level of decomposition is unknown for specific. Moreover, it is quite challenge for wavelet to detect the fault frequency from wide band frequency components since more level decomposition will be needed if the specific frequency band is intended to inspect. This becomes harder since the specific band of noise signal usually is unknown or difficult to determine in a real complex machine system.

To overcome this problem, an adaptive filter is proposed in this paper to combine with Hilbert transform as a new processing tool for unwanted signal separation technique. The faulty bearing will be used to verify the proposed method. By using this combination, the noise separation is greatly reduced from the bearing defect signal and the amplitude at specific fault frequency will be greatly enhanced.

2 Methodology

Adaptive filter can overcome other noise separation technique since the capability of this technique to identify the noise signal component that embedded in machine defect signal efficiently without need determine any cut-off frequency setting or range of signal bandwidth. This regarding to principle operation of adaptive filter where reference signals that high correlate with noise component is used. Based on this reference signal, adaptive filter will automatically adjust the filter coefficients until optimal. Once optimal, the noise component that embedded in machine defect signal will remove from the machine defect signal. Least mean square algorithm (LMS) and Recursive Least algorithm (RLS) is commonly used to dictate how the filter parameters are adjusted. However, based on simplicity and less in time computational, the LMS is the most preferred. A brief explanation on adaptive filter flow operation is described based on Fig. 1.

Figure 1 shows the flow operation of adaptive filter model. The adaptive filter needs two input signals which are the primary input signal and the reference input signal. The primary input signal $u(n)$ is a mixed signal which contains machine defect signal $s(n)$, and noise signal $v(n)$ as shown in Eq. (1) while the reference input signal $v_o(n)$ is the signal that high correlate with the noise $v(n)$ is shown in Eq. (2). Reference signal can be obtained directly from the noise source at the machine body.

$$u(n) = s(n) + v(n) \tag{1}$$

$$v_o(n) \sim v(n) \tag{2}$$

2.1 *Adaptive Filter Algorithm*

As shown in Fig. 1, the reference signal is drive in into adaptive filter. With the initial set of filter coefficient, the adaptive filter will adjust the reference signal statistical characteristic and force to close as possible with the noise signal component in primary input. This filtered reference signal is defined as $y(n)$ and described as following.

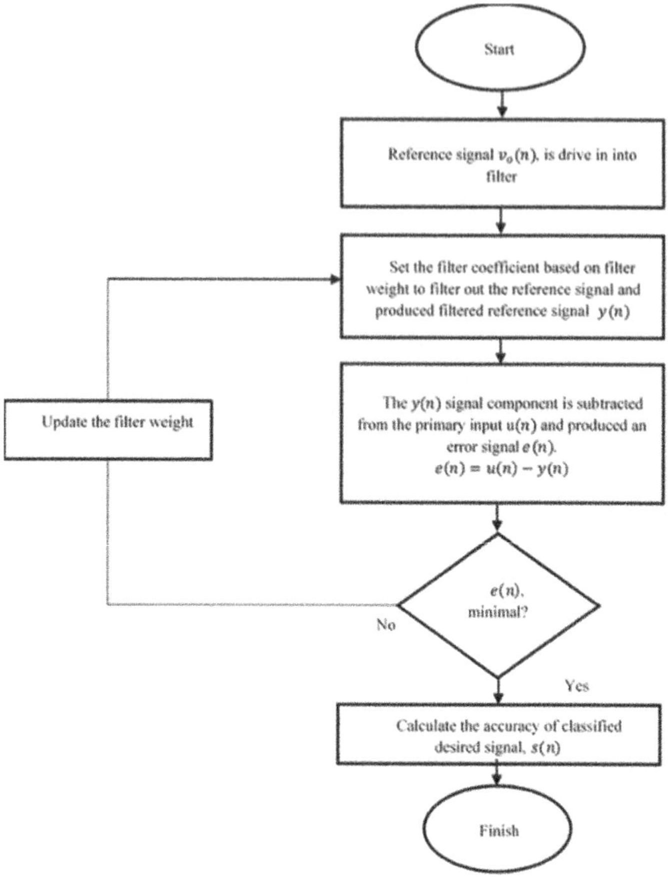

Fig. 1 Adaptive filter algorithm

$$y(n) \approx v(n) \tag{3}$$

The filtered reference signal for adaptive filter can be obtained by using Eq. (4) respectively.

$$y(n) = \sum\nolimits_{i=0}^{M-1} \vec{w_i}(n)u(n-i) \tag{4}$$

where M is filter length, $\vec{w_i}(n)$ is the filter coefficient, and $u(n-i)$ is primary input with previous adjustment step.

This filtered reference signal $y(n)$ need to close as possible with noise signal component in primary input so the noise component in primary input can be eliminated through subtraction process as following equation.

$$e(n) = u(n) - y(n) \tag{5}$$

$$e(n) = [s(n) + v(n)] - y(n) \tag{6}$$

The error signal $e(n)$, is produced in this subtraction process. This error signal $e(n)$ then can be used to indicate how well the noise is eliminate from the primary input signal by using statistical performance function known as mean square error which can be calculated as following.

$$J_{\min} = E\left[|e(n)|^2\right] \tag{7}$$

$$\varepsilon = \frac{J_{\min}}{\sigma_{ref}^2} \tag{8}$$

where J_{min} is a mean square, σ_{ref}^2 is the variance of the reference signal, and ε is the scale of optimal filter. The target of this statistical performance function is to minimize the mean square of error signal until minimal. The mean square needs to be minimal since it will affect the scale of optimal filter design and indicate how close the signal y (n) with v (n). The scale of optimal filter is known in between the range $0 \le \varepsilon \le 1$. If ε is zero, the adaptive filter with the optimum filter coefficient has perfectly eliminate noise from the primary input signal whereas on the other hand if ε is 1, the adaptive filter does not eliminate the noise at all. Therefore, if ε is not minimal, that's mean the y (n) signal is not close enough with v (n) thus need a new setting of filter coefficient to filter out the reference signal.

Regarding to this matter, the filter coefficient needs to re-adjust by updating the filter weight based on following equation.

$$\vec{w}(n+1) = \vec{w}(n) + \mu \cdot e(n) \cdot \vec{\mu}(n) \tag{9}$$

where μ is the step size of the adaptive filter, $\vec{w}(n)$ is the filter coefficient vector, $e(n)$ is an error signal, and $\vec{\mu}(n)$ the filter input vector.

The filter coefficient will keeps adjusted until the optimal scale of filter design, ε is minimal to zero.

2.2 Hilbert Transform

Hilbert transform can be defined as follows:

$$H[x(t)] = \frac{1}{\pi}\frac{x(\tau)}{t-\tau}d\tau \tag{10}$$

where t and τ are time and translation parameters. Hilbert has known not affecting by the non-stationary characteristic of modulating signals which usually generate from the defect component such in real application. Therefore, in order to identify the fault signatures, the signal needs to demodulate which can be done by adopting following equation.

$$B(t) = x(t) + iH[x(t)] = b(t)e^{\emptyset(t)} \tag{11}$$

where $b(t) = \sqrt{x^2(t) + H^2[x(t)]}$, $A = \pi r^2$, $\emptyset(t) = arctan\frac{H[x(t)]}{x(t)}$, and $i = \sqrt{-1}$. $b(t)$ is the envelope of $B(t)$.

In this research, it was found that the amplitude of machine defect signal spectrum gradually can be increased accordingly to the ith time envelope process, where $i = 1, 2, 3 \ldots n$. Thus, it very useful to increase the amplitude sign of failure so the evaluation of machine component can be done easier.

The collected signal $s(t)$ which known has mixed with other signals component which defined as noise need to be cleaned first by using adaptive filter. Afterward, the cleaned $s(t)$ signal needs to pre-process in order to reduce the residual of random variables by using following equation.

$$x(t) = \frac{y(t) - \bar{y}}{\sigma} \tag{12}$$

where $x(t)$ is the pre-processed signal, $\bar{y}$ is the mean value of $s(t)$ and σ is the standard deviation of cleaned $s(t)$.

The amplitude of signal characteristic and harmonic of faulty bearing usually will become dominant over other frequency components. However, the observed signal may present a modulation phenomenon caused by the resonance vibration. As a result, the fault signature cannot be seen. Thus, modulated signal can be demodulated by using Eq. (11) and thus the fault related signatures can be exhibited. The flow chart of the proposed method in this paper is shown in Fig. 2 which includes the following steps:

Step 1. Collect the machine defect signal by attaching sensor at the target component and the second sensor used to collect the signal from part of machine that produce the noise. Perform the noise separation by using adaptive filter to obtain the cleaned signal $s(t)$.

Step 2. Relieve the remaining random variables at cleaned noise signal $s(t)$ by using Eq. (12) to obtain $x(t)$.

Step 3. Perform the envelope process where Hilbert Transform is adopted to obtain the analytic signal of $x(t)$. The envelope process is necessary to repeat ith time since it capable to increase the amplitude at fault frequency accordingly. In this research, the envelope process repeats fourth time.

Step 4. Perform the spectrum analysis of enveloped signal.

Step 5. Fault feature extraction is carried out.

In order to verify the proposed method, the results in this research is compared with method proposed by work performed by [11] where his applied continuous

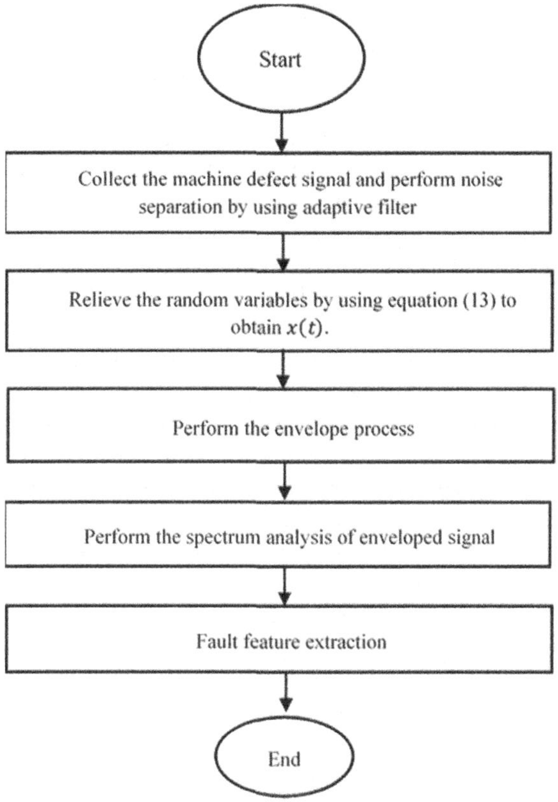

Fig. 2 The flow chart of proposed method

wavelet transform, and Hilbert transform to enhance the bearing fault signature. In his work, the signal is decomposed into four stages and an envelope process is repeated twice. The growth index is proposed in this paper in order to compare the efficiency of proposed method. The growth index can be constructed as follow:

Step 1. Choose the highest amplitude located at fault characteristic frequency and its harmonic. The highest amplitude is denoted as $A_{k\times f}$ where k is the harmonic order and f is fault characteristic frequency.

Step 2. $Index(k) = \frac{A_{k\times f}}{M}$ where M represents the length of envelope spectrum.

3 Result and Discussion

3.1 Simulation Result

The impulse function generate from the MATLAB is adopted and used as synthetic signal that denotes as bearing fault signal. To make it more realistic, the impulse signal is mixed with random signal that represents as noise. In order to show the efficiency of proposed method, two level of noise has been generated.

First level noise is created at low noise density and mixed with impulse signal. As a result, the impulse signal is mixed with the noise signal but the sharp transient still can be seen. On the other hand, second level noise generated with high density noise thus the sharp transient is totally buried under noise. The original impulse signal, impulse signal mixed with low density noise and high-density noise are plotted in Fig. 3a–c respectively. Proposed method and traditional method are applied to separate those noises and both results are compared.

The spectrum of envelope analysis and the growth index obtained for both mixed signals is shown in Figs. 4a, b and 5a, b respectively for impulse signal mixed with low noise density and high noise density. Figures 4a and 5a showed the amplitude of fault signature for every harmonic seem higher than the amplitude obtained from traditional method. This indicate that the envelope process repeated fourth time greatly increase the amplitude of fault signal. Moreover, by close look the plot in Figs. 4a and 5b, we can notice that the noise floor of proposed method is clearly reduced. This proof that, the adaptive filter is capable to reduce noise effectively rather than wavelet transform. The calculated growth index at Fig. 4b and 5b clearly showed that the proposed method can increase the amplitude of fault bearing signal significantly than traditional method. A nice curve of growth index according to the harmonic order for Fig. 4b indicates that both method able to reduce the noise at low density and subsequently each peak of transient impulse signal is recovered properly. However, for the impulse signal that mixed with high noise density, it shows that the growth indexes for traditional method is unstable and fluctuate as shown in Fig. 5b. This indicate that the traditional method unable to recover the peak of impulse signal and their harmonic order in high density noise. Meanwhile, growth index for proposed method is successfully recovered the impulse signal and their harmonic in high density noise.

3.2 *Experimental Result*

In this experiment, a set of real bearing data was obtained through industry bearing running on a test rig as shown in Fig. 6. Bearing at B2 is used in this study where the artificially defect is created at the ball bearing. By created defect at ball bearing, an impact will be generated during machine rotation. In separate study, the signal at B1 seem greatly influence the signal at B2 rather than other noise source such as noise source at B3 and M. This probably, the location of B1 closes to weighted pulley. It deems the shaft connected from pulley make the shaft connected at B1 is greatly increasing the pressure between ball bearing element at B1 and shaft surface. Thus, the signal generate at B1 is dominate and easily transfer to bearing at B2. Because of this reason, the signal at B1 is used as reference signal to reduce the noise that mixed with signal from bearing B2.

Figure 7a, b show the collected signal from faulty bearing at B2 and reference sig-

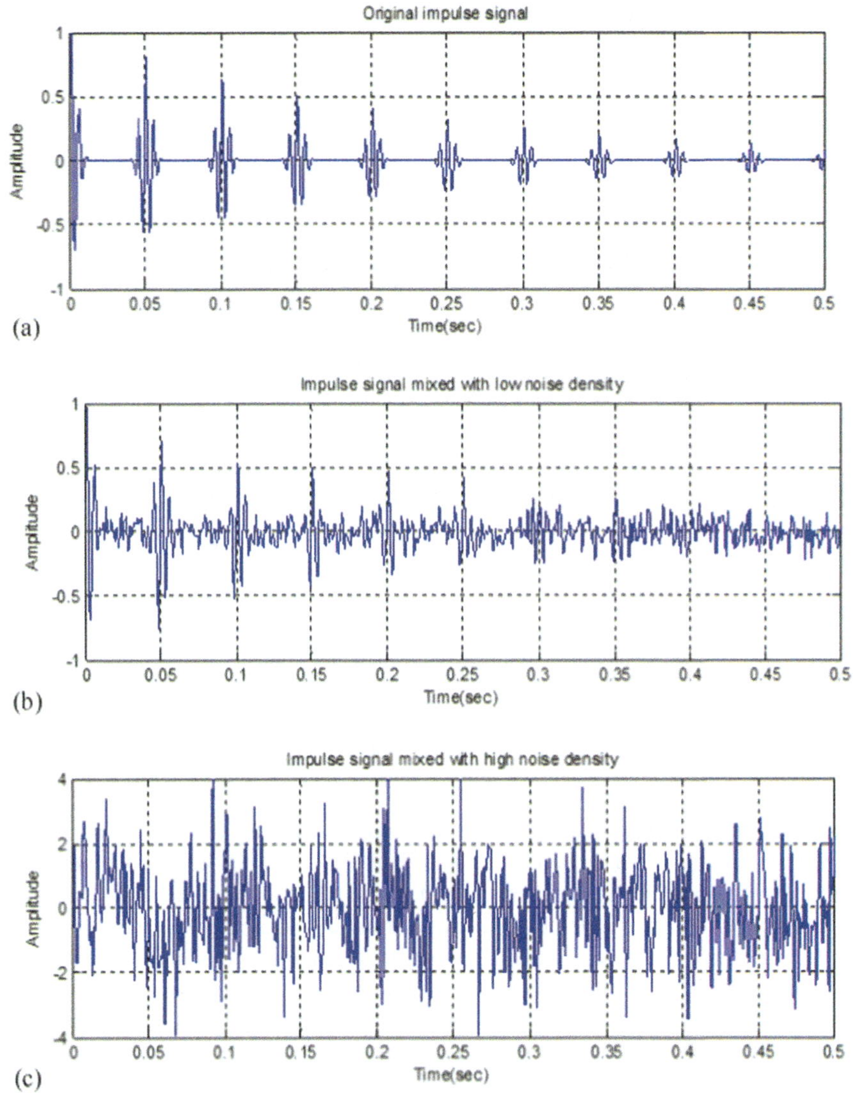

Fig. 3 Simulation of fault bearing signal. **a** A signal containing impulse signal. **b** Impulse signal mixed with low density noise and **c** Impulse signal mixed with high density noise

nal at B1 respectively. As shown in Fig. 7a, it clearly shows that the signal contained many sharp impulse signals that indicate the bearing element suffer from the defect at ball bearing. The signal in 7(a) also clearly can be seen modulated by the resonances vibration thus make the truth level of amplitude signal hard to be determined precisely. Figure 7a also shows that the bearing signal B2 is polluted by the noise signal. Figure 7b shows the reference signal obtain from the B1. By closely look

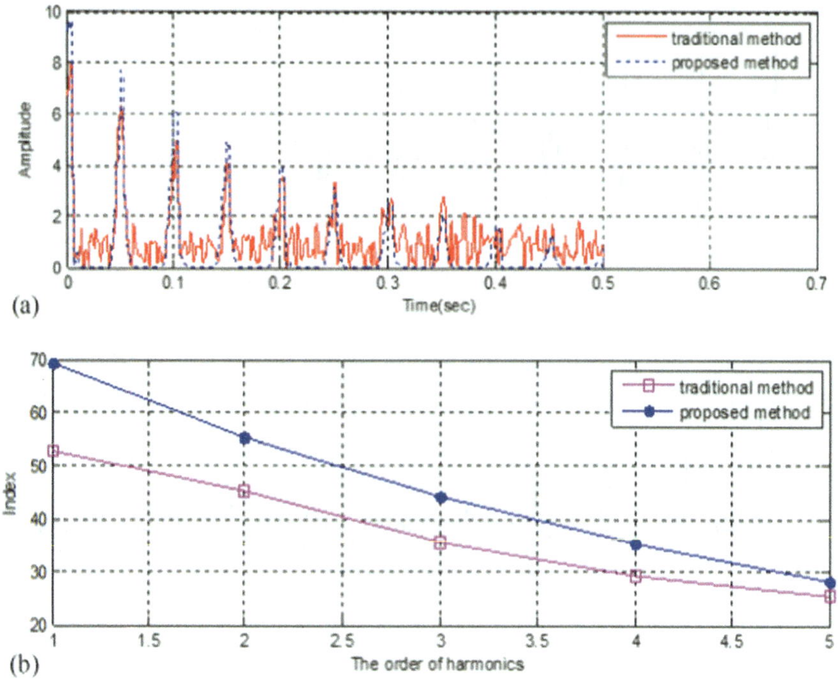

Fig. 4 An analysis for impulse signal mixed with low noise density. **a** Spectrum of envelope analysis, **b** The growth index for fault bearing detection

at the amplitude level of B1 signal; we can roughly indicate that B1 signal is high correlate with noise that embedded in B2 signal. Proposed and traditional method is applied to demodulate the signal, reduce the noise and enhance the amplitude signal at fault characteristics frequency and result from both is compared.

The spectrum of envelope and the growth index obtained from fault bearing signal is shown in Fig. 8a, b. As expected, the amplitude signal at bearing fault frequency and their harmonics obtained by the proposed method is greatly enhanced compared with traditional method. This indicates that the proposed method is more efficient to enhance the amplitude of defect signal. The harmonic order of fault bearing also clearly can be seen and measured even though the original signal of B2 is seriously contaminated by noise and distorted by modulation phenomena. Moreover, by closely look at Fig. 8a, it obviously shown that the proposed method also able to enhance the amplitude signal at fault frequency without increase the noise floor. This in turn make the signal-noise-ratio of the proposed method is greatly increased. The growth index plotted in Fig. 8b shows that the proposed method is larger than traditional method. This indicate that the proposed method able to enhance the amplitude of fault signal significantly. Hence it is very useful when came the demand to identify early signal of failure that usually generate a very low impulse signal. A nice curve of

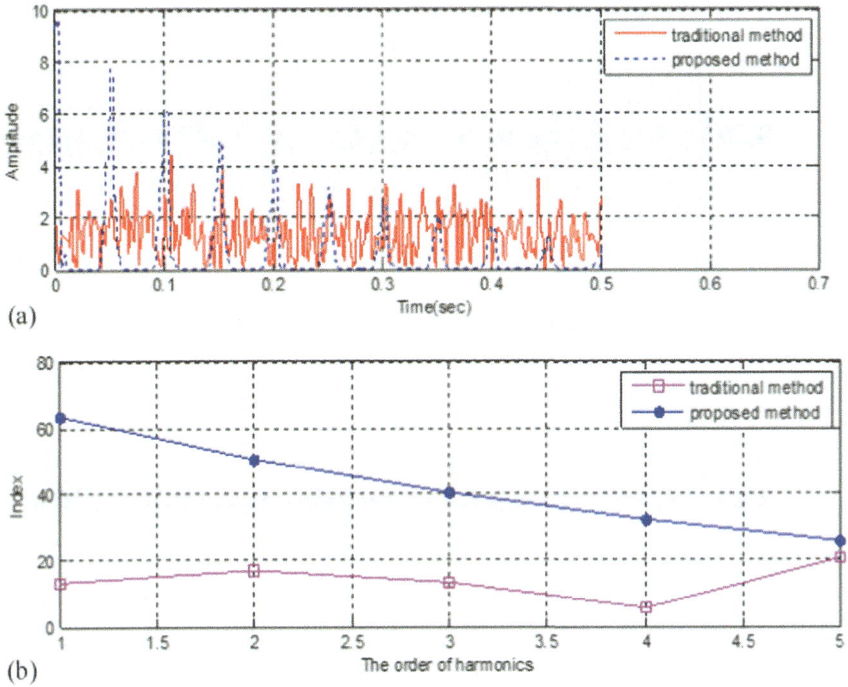

Fig. 5 An analysis for impulse signal mixed with high noise density. **a** Spectrum envelope analysis, **b** The growth index for fault bearing detection

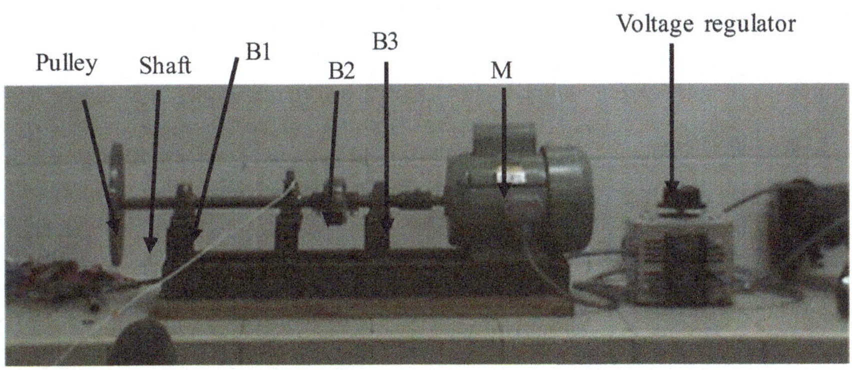

Fig. 6 Fault bearing in test rig

growth index accordingly with bearing harmonic order also show that both method able to recover the impulse signal of fault signal at bearing properly.

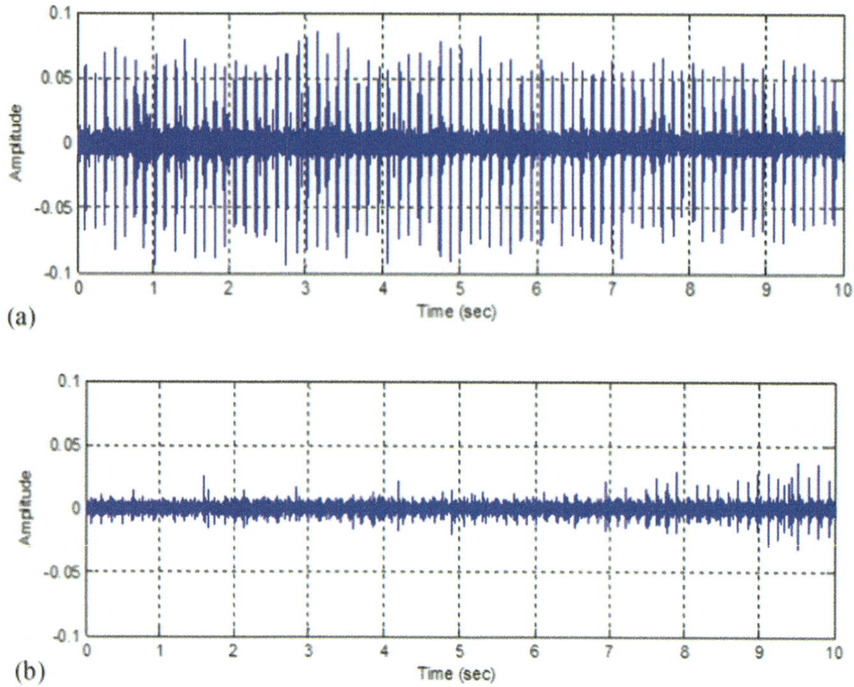

Fig. 7 The collected signal from test rig. **a** Faulty bearing signal at B2, **b** Reference signal from B1

4 Conclusion

Fault detection at bearing component is one of the core researches in machine diagnosis. The work conducted, proposing the method of the bearing fault identification based on combination between adaptive filter and Hilbert Transform. This proposed method is comparing with previous method where they combine the wavelet transform with Hilbert Transform. The result shows that the proposed method has great potential from real applications in complex machine system. This regarding the capability of proposed method that able to enhance the amplitude of fault signal significantly. In the other hand, the noise embedded in fault signal is greatly reduce thus make the signal-noise-ratio of proposed method has a good increment. Regarding to these advantages, the growth index shows that the combination of Adaptive Filter with Hilbert Transform is more efficient.

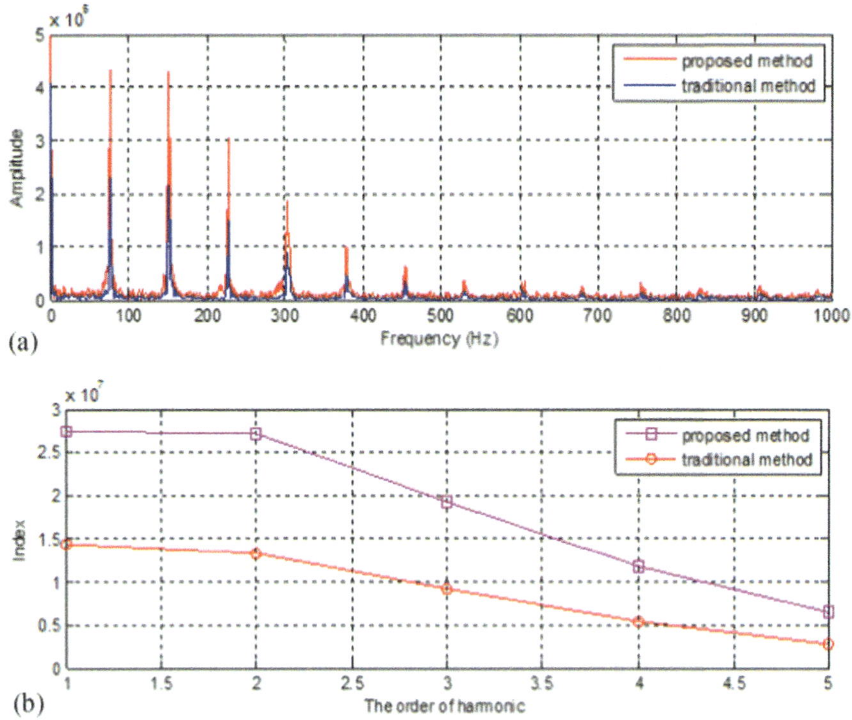

Fig. 8 Fault bearing detection. **a** The analysis of fault bearing, **b** Fault growth index

References

1. Manghai A, Jegadeeshwaran TM, Sugumaran V (2017) Brake fault diagnosis through machine learning approach-A review. SDHM 12(1):43–67 (2017)
2. Pactitis SA (2007) Active filter theory and design, 1st edn. Taylor & Francis Group, Boca Raton
3. Mulugeta AH, Dykas B (2015) Blind source separation for vibration based diagnostics of rotorcraft bearings. J Vibr Control 22
4. Hu Y, Li F, Li H, Liu C (2017) An enhanced empirical wavelet transform for noisy and non-stationary signal processing. Digital Signal Process 60:220–229 (2017)
5. Nguyen P, Kang M, Kim JM, Ahn BH, Ha JM, Choi BK (2015) Robust condition monitoring of rolling element bearing using de-noising and envelope analysis with signal decomposition techniques. Expert Syst Appl 42(22):9024–9032
6. Figlus F, Gnap J, Skrucany T, Sarkan B, Stoklosa J (2017) The use of denoising and analysis of the acoustic signal entropy in diagnosing engine valve clearance. Entropy 18(7):253
7. Wang Y, Xu G, Liang L, Jiang K (2015) Detection of weak transient signals based on wavelet packet transform and manifold learning for rolling element bearing fault diagnosis. Mech Syst Signal Process 54–55:259–276
8. Elbouchikhi E, Choqueuse V, Amirat Y, Benbouzid MEH, Turri S (2017) An efficient Hilbert-Huang Transform based bearing faults detection in induction machines. IEEE Trans Energy Conversion 32(2):401–413
9. Das B, Pal S, Bag S (2016) A combined wavelet packet and Hilbert Huang transform for defect detection and modelling of weld strength in friction stir welding process. J Manuf Process 22:260–268

10. Korneev VG, Langer U (2017) Domain decomposition and preconditioning. In: Encyclopedia of Computational Mechanics, 2nd edn. Wiley Online Library
11. Wang D, Miao Q, Fan X, Huang H (2009) Rolling element bearing fault detection using an improved combination of Hilbert and wavelet transforms. J Mech Sci Technol 23:3292–3301

Experimental Assessment of Nonlinear Modal Behaviour of Flexible Beam

A. R. Bahari, M. A. Yunus, M. N. Abdul Rani, W. I. I. Wan Iskandar Mirza, and M. Z. Nuawi

Abstract Nonlinearity is a frequent occurrence in engineering structures. The nonlinearity detection in the operating mechanical systems is important because the nonlinearity can change the dynamic behaviour. The aim of this work is to experimentally measure the nonlinearity of a flexible beam structure due to the different applied dynamic load configurations. In this study, the modal parameters were obtained from the modal testing. The excitation approach by adopted the burst random excitation through an electromagnetic shaker and the vibration response was measured using accelerometers. The frequency response function (FRF) at different level of dynamic excitation was utilized to exhibits the nonlinear behaviour. The elastic vibration characteristics which are bending and torsional modes were compared to identify the nonlinear. As a conclusion, the results showed that the nonlinear property of the flexible beam is due to the geometric nonlinearity.

Keywords Nonlinear · Structural dynamics · Experimental modal analysis · Frequency response function

1 Introduction

Designing and operating mechanical systems and engineering structures requires accurate structural modelling, computing and testing. It is a compulsory requirement to perform the dynamic analysis in order to accurately predict the dynamic response of the systems and structures [1]. In the dynamic mechanical and environmental loading conditions with large displacement amplitudes regime, the structure exhibits nonlinear phenomenon [2].

A. R. Bahari · M. A. Yunus (✉) · M. N. Abdul Rani · W. I. I. Wan Iskandar Mirza
Structural Dynamics Analysis & Validation, Faculty of Mechanical Engineering, Universiti Teknologi MARA (UiTM), 40450, Shah Alam, Selangor, Malaysia
e-mail: mayunus@uitm.edu.my

M. Z. Nuawi
Department of Mechanical and Materials Engineering, Faculty of Engineering and Built Environment, Universiti Kebangsaan Malaysia, 43600 UKM, Bangi, Selangor, Malaysia

P. Rajendran et al. (eds.), *Proceedings of International Conference of Aerospace and Mechanical Engineering 2019*, Lecture Notes in Mechanical Engineering,
https://doi.org/10.1007/978-981-15-4756-0_53

In the aerospace applications, various physical mechanisms contribute to the nonlinearities of the aircraft. The structural geometrical nonlinearities is significant from the steady-state or large motion of the wing [3] which depends on the kinematic of the body. Aerodynamic geometrical nonlinearities are due to the dependence of the pressure at a given point on the disturbances throughout the surface [4].

The improvement on the lift-to-drag ratio due to the reduction of aerodynamic induced drag is obtained by increasing the wing aspect-ratio [5]. However, this will result on the structural design issues where susceptible higher structural flexibility or deflections and higher stress level create at the wing root at the operating conditions [6]. This can subsequently cause the effect on the dynamic behaviour (modal properties) and consequently the aero-elastic behaviour. For instance, very high aspect ratio wings with both structural and aerodynamic nonlinearities, delta wings with geometrical plate nonlinearities, nonlinear structural damping, airfoils with stiffness nonlinearities and aerodynamic flows with large shock motions and flow separation [7].

In modal analysis, the equation of motion for the dynamic response of a linear model to describe the structural motion is a second-order differential equation [8, 9].

$$M\ddot{x}(t) + C\dot{x}(t) + Kx(t) = F(t) \tag{1}$$

where M, C and K are the mass, damping and stiffness matrices. $\ddot{x}(t)$, $\dot{x}(t)$ and $x(t)$ are the acceleration, velocity and displacement vectors, respectively. $F(t)$ is an external force vector. For a nonlinear structure, the equation of motion [10] consists of a nonlinear parameter.

$$M\ddot{x}(t) + C\dot{x}(t) + Kx(t) + f_{NL}(\dot{x}(t), x(t)) = F(t) \tag{2}$$

where f_{NL} is a function of nonlinear. For an undamped hardening or softening nonlinear restoring force spring with cubic nonlinearities, the equation of motion is known as Duffing's equation [11].

$$M\ddot{x}(t) + Kx(t) + K_{NL}x^3(t) = 0 \tag{3}$$

In this work, a frequency response function from the dynamic behaviour based on the modal analysis is used to evaluate the nonlinear behaviour of a beam structure. The proposed experimental method was implemented in this study. A quantitative comparison on the experimental modal analysis is given, followed by a discussion of the nonlinear behaviour.

2 Methodology

Experimental modal analysis was conducted to determine the modal parameters of the beam which are the natural frequency and damping ratio [12]. The experimental setup for the modal testing is shown in Fig. 1. The set up consists of a beam under the test was fixed on the test rig by using bolted plate at one side of the beam to obtain fixed-free boundary condition. Spectral testing by burst random excitation was applied on the longitudinal axis of the mild steel beam by an electromagnetic shaker through a stinger at fixed point B. This fixed point was chosen in order to excite both the bending and torsional modes. The measurement of the applied excitation by using force transducer (PCB Piezotronics 208C03).

The vibration response was measured using two uniaxial shear accelerometers (Dytran Model 3032A). One accelerometer (location point B) was used for measuring the bending and torsional vibration response at the driving point of the excitation. Another mounted accelerometer (location point A) to measure the vibration response at the center of the beam. Those accelerometers have a low weight characteristic for avoiding the mass loading issue. Each accelerometer weighs 2 g (which amounted to 0.73% of the total mass of the beam), therefore no mass distribution loading affect to the beam. The equipment packages used to extract the modal parameters is the LMS system data acquisition. The measured data were processed using LMS Test.Lab software. Three dynamic excitation voltages were considered from 0.1 up to 2.0 V. The bandwidth frequency is 150 Hz to investigate the first five elastic modes of the

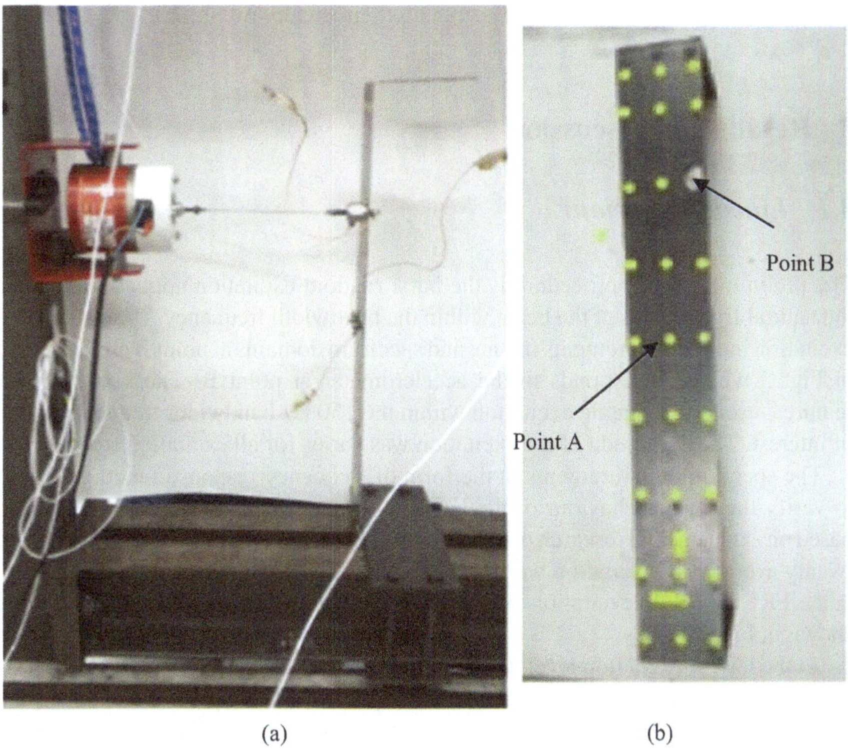

Fig. 1 Photograph of the modal testing **a** experimental set up **b** flexible beam

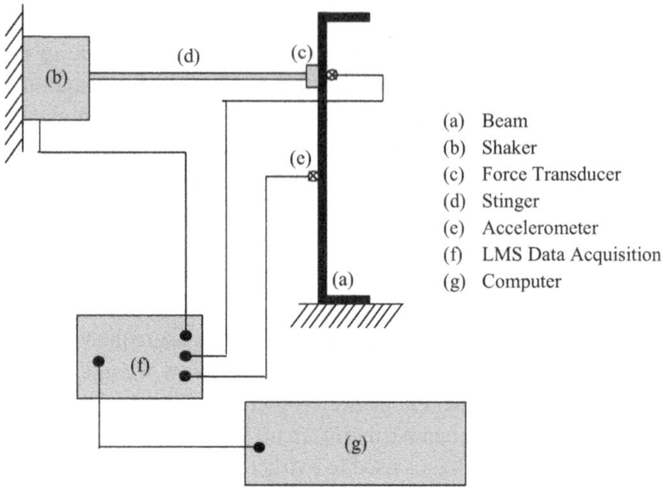

Fig. 2 Schematic diagram of the modal testing

beam. Time samples of 512 were recorded for each level with 0.3 Hz resolution. The assembly of the modal testing is illustrated schematically in Fig. 2.

3 Results and Discussion

3.1 Linear Behaviour

The preliminary work procedure is the burst random excitation applied to identify the natural frequencies of the beam within the bandwidth frequency. Typical voltage excitation input measurements in time and spectrum domains at point B are presented in Fig. 3 which corresponds to the accelerometer at point B. The voltage input features resemble dynamic excitation within the 150 Hz bandwidth frequency range of interest. The amplitude of the excitation was varies for all excitation frequencies.

The spectrum measurements in the form of frequency response function is used to verify the linear behaviour of the beam. In this case, the detection was defined based on experimental research observations. The measured vibration responses from twenty averages of excitation were utilised to improve the quality of FRF. The inertance FRF of 0.1 V excitation was measured using accelerometer as graphically shown in Fig. 4.

At the lowest excitation level (0.1 V), the structure behaves linearly. FRF analysis extracts the structures linear with lightly damped natural frequencies and damping ratios. The identified natural frequencies correspond to the vibration elastic modes of the beam are listed in Table 1 under fixed-free boundary condition. From Fig. 4, it can be observed that location point A at the center of the beam shows less significant

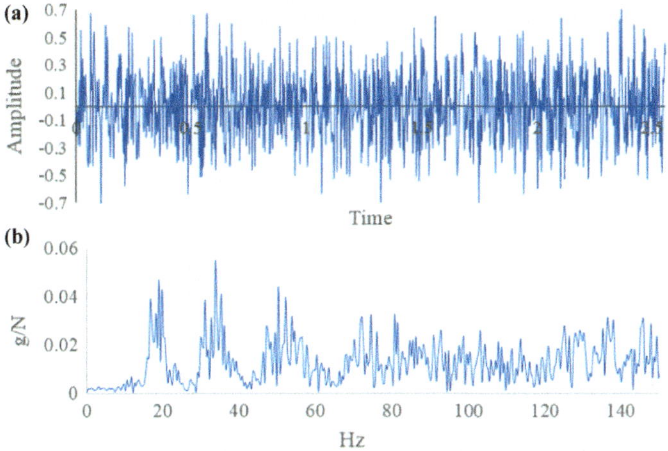

Fig. 3 Typical burst random excitation voltage used in the modal testing at point B **a** time domain **b** spectrum domain

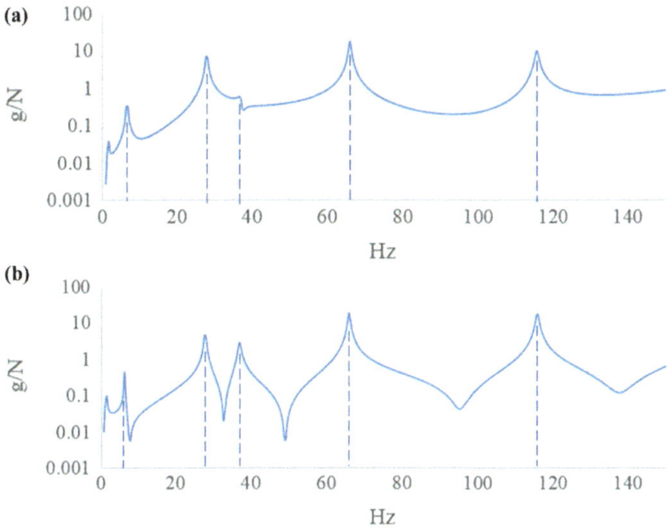

Fig. 4 FRF measured on beam at 0.1 V excitation **a** point A **b** point B

peak at mode 3 which is torsional mode, while location point B at the side of the beam shows high visibility peak of the torsional mode.

Table 1 Natural frequencies and damping ratios of the linear beam

Vibration mode	Natural frequency (Hz) Point A	Damping ratio (%)	Natural frequency (Hz) Point B	Damping ratio (%)
1	6.44	2.63	6.43	1.06
2	27.96	0.91	27.94	0.69
3	37.09	1.18	37.08	1.02
4	66.04	0.42	66.011	0.41
5	115.71	0.40	115.55	0.32

3.2 *Nonlinear Behaviour*

In order to detect the nonlinear behaviour of the beam, the previous test was repeated with higher excitation level (1.0 and 2.0 V). Figure 5 compares the resultant inertance FRF for points A and B for all excitation voltage levels. The FRF plot shows the energy-dependencies of the nonlinearities induced in the beam. Note that point A at the center of the beam while point B is driving point of the excitation. By analysing this particular mapping FRF for all modes, it can be observed that the nonlinear structural parameter is identified at mode 1, mode 3 and mode 5.

The peaks skewed and shifted downward in natural frequency to the lower value. The peaks become shorter and wider due to the increasing in damping parameter. The measured natural frequencies at mode 2 and mode 4 were found to remain constant for all excitation levels, showing the linear part of the structural parameter. Determination

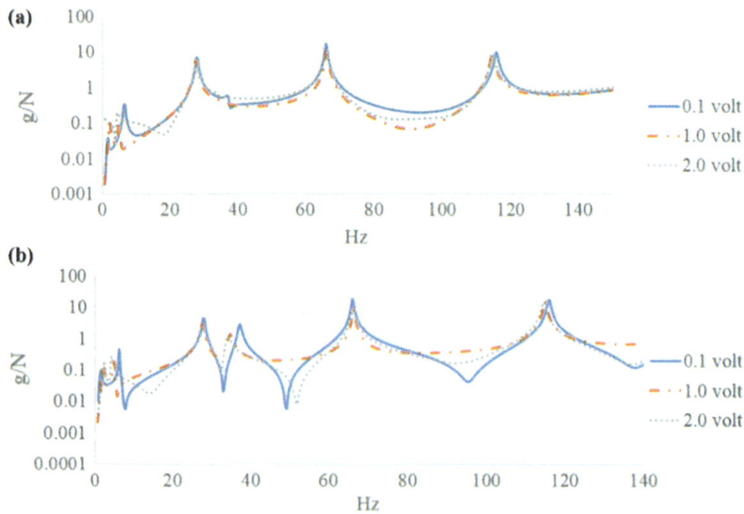

Fig. 5 Comparison of identified nonlinearity from the measured inertance FRF **a** point A **b** point B

Table 2 Natural frequencies of point A and B for three different excitation voltages

Vibration mode	Natural frequency (Hz) Point A			Natural frequency (Hz) Point B		
	0.1 V	1.0 V	2.0 V	0.1 V	1.0 V	2.0 V
1	6.44	5.22	4.34	6.30	5.21	4.33
2	27.96	27.66	27.63	27.94	27.67	27.62
3	37.09	34.76	34.06	37.08	34.76	34.05
4	66.04	66.44	66.23	66.011	66.44	66.42
5	115.71	114.80	114.50	115.55	114.81	114.50

of the characteristic of the modes is described in Table 2 which compares the natural frequency values for each excitation voltage level, in order to identify the degree of nonlinearity.

Specifically, at first bending mode 1, the natural frequencies decrease from around 6 Hz to 4 Hz as the excitation voltage increase. This reflects that mode 1 has a softening characteristic with strong nonlinearity. At second bending mode 2, the linear natural frequencies remain almost constant around 27 Hz even the increasing excitation voltage. The first torsional mode 3 shows that as the excitation voltage level increases, the drop of natural frequencies value from 37 Hz to 34 Hz has been detected. This identification indicates that mode 3 has also a softening characteristic in the dynamic response with strong nonlinearity. The third bending mode 4 shows unchanged natural frequencies. And the fourth bending mode 5, it is shown that there is a reduction of the natural frequencies value from 115 Hz to 114 Hz. This describes the softening nonlinearity. However, this mode 5 has a weak characteristic based on the small decrement in the natural frequency. Overall detections found a decreasing stiffness of this beam structure on two bending modes (mode 1 and mode 5) and one torsional mode (mode 3). Based on Eq. 3, the decreasing of the natural frequencies values for these three modes were correspond to the negative value of the nonlinear stiffness K_{NL} in the nonlinear equation of motion. It is identified that the geometrically nonlinear characteristic of the flexible beam is obtained from the skewed peak of the frequency response function.

4 Conclusions

Nonlinear dynamics detection from the modal parameters of the flexible mild steel beam structure has been successfully conducted in this present study. Burst random spectral test has been utilized as an excitation approach to demonstrate the nonlinear characteristics. The inertance FRF results indicate that some natural frequencies of the beam are significantly changes with the level of dynamic excitation. Based on the FRF analysis, it has been found that both bending (mode 1 and 5) and torsional modes (mode 3) skewed to show the nonlinearity. Finally, the observability of the geometric nonlinearity of the flexible beam via softening characteristics has been assessed.

Acknowledgements The authors acknowledge the financial support of the Fundamental Research Grant Scheme 600-IRMI/FRGS 5/3 (465/2019).

References

1. Rezaiefar A, Galal K (2018) Free vibration of thin rectangular steel plates with geometrically-nonlinear load-displacement behavior. Thin-Walled Struct 129:381–390
2. Chatterjee A (2010) Identification and parameter estimation of a bilinear oscillator using volterra series with harmonic probing. Int J Non-Linear Mech 45:12–20
3. Xie C, Liu Y, Yang C, Cooper JE (2016) Geometrically nonlinear aeroelastic stability analysis and wind tunnel test validation of a very flexible wing. Shock Vibr 1–17
4. Patil MJ, Hodges DH (2004) On the importance of aerodynamic and structural geometrical nonlinearities in aeroelastic behavior of high-aspect-ratio wings. J Fluids Struct 19:905–915
5. Mian HH, Wang G, Ye Z-Y (2014) Numerical investigation of structural geometric nonlinearity effect in high-aspect-ratio wing using CFD/CSD coupled approach. J Fluids Struct 49:186–201 (2014)
6. Afonso F, Vale J, Oliveira E, Lau F, Suleman A (2017) A review on non-linear aeroelasticity of high aspect-ratio wings. Prog Aerosp Sci 89:40–57
7. Dowell E, Edwards J, Strganac T (2003) Nonlinear aeroelasticity. J Aircraft 40(5):857–874
8. Sulaiman MSA, Yunus MA, Bahari AR, Abdul Rani MN (2017) Identification of damage based on frequency response function (FRF) data. MATEC Web Conf 90(01025)
9. Hassin NM, Yunus MA, Abdul Rani MN, Sulaiman MSA, Wan Iskandar Mirza WII, Mat Isa AA (2016) Experimental and finite element analysis of riveted joints structure of a simplified model of aluminium crash box. J Eng Appl Sci 11(7):1451–1455 (2016)
10. Worden K, Tomlinson GR (2001) Nonlinearity in structural dynamics: detection, identification and modelling. IOP Publishing Ltd (2001)
11. Cveticanin L (2001) Vibrations of a coupled two-degree-of-freedom system. J Sound Vib 247(2):279–292
12. Omar R, Abdul Rani MN, Yunus MA, Mat Isa AA, Wan Iskandar Mirza WII, Mohd Zin MS, Roslan L (2018) Investigation of mesh size effect on dynamic behaviour of an assembled structure with bolted joints using finite element method. Int J Automotive Mech Eng 15(3):5695–5708 (2018)

The manufacturer's authorised representative in the EU is Springer Nature Customer Service Centre GmbH, Europaplatz 3, 69115 Heidelberg, Germany. If you have any concerns regarding our products, please contact ProductSafety@springernature.com

Printed and bound by CPI Group (UK) Ltd, Croydon, CR0 4YY
30/09/2026
02225509-0016